Vitaly V. Slezov

Kinetics of First-order Phase Transitions

Related Titles

Ostrikov, K.

Plasma Nanoscience

Basic Concepts and Applications of Deterministic Nanofabrication

2008

Hardcover

ISBN: 978-3-527-40740-8

Skripov, V. P., Faizullin, M. Z.

Crystal-Liquid-Gas Phase Transitions and Thermodynamic Similarity

2006

Hardcover

ISBN: 978-3-527-40576-3

Kelsall, R., Hamley, I. W., Geoghegan, M. (eds.)

Nanoscale Science and Technology

2005

Hardcover

ISBN: 978-0-470-85086-2

Schmelzer, J. W. P.

Nucleation Theory and Applications

2005

Hardcover

ISBN: 978-3-527-40469-8

Champion, Y., Fecht, H.-J. (eds.)

Nano-Architectured and Nanostructured Materials

Fabrication, Control and Properties

2004

Hardcover

ISBN: 978-3-527-31008-1

Vitaly V. Slezov

Kinetics of First-order Phase Transitions

WILEY-VCH Verlag GmbH & Co. KGaA

The Author

Prof. Vitaly V. Slezov
National Science Center, Kharkov
Institute of Physics and Technics
Kharkov, Ukraine
itp@kipt.kharkov.ua

Consultant Editor

Jürn W. P. Schmelzer
Physics Department
University of Rostock
Rostock, Germany
juern-w.schmelzer@uni-rostock.de

Cover Picture

Phase separation in a strontium-borate glassforming melt in the stage of coarsening
Copyright Dr. Irina G. Polyakova

Library of Congress Card No.: applied for

British Library Cataloguing-in-Publication Data
A catalogue record for this book is available from the British Library.

Bibliographic information published by the Deutsche Nationalbibliothek
The Deutsche Nationalbibliothek lists this publication in the Deutsche Nationalbibliografie; detailed bibliographic data are available on the Internet at http://dnb.d-nb.de.

Typesetting Uwe Krieg, Berlin
Printing betz-druck GmbH, Darmstadt
Binding Litges & Dopf GmbH, Heppenheim

Printed in the Federal Republic of Germany
Printed on acid-free paper

ISBN: 978-3-527-40775-0

Contents

Kinetics of First-order Phase Transitions. Vitaly V. Slezov

ISBN: 978-3-527-40775-0

Foreword

Phase transitions of first-order are phenomena, widely occurring in nature. Among them are: evaporation and condensation, melting and solidification, sublimation and condensation into a solid phase, some structural transitions in the solid state, transitions connected with the decomposition into different phases in multicomponent liquid and solid systems, etc.

The classical explanation of the questions why and when phase transitions of first-order take place was based on thermodynamic concepts, which has been developed already more than hundred years ago. In the first half of the 20th century, huge efforts have been undertaken to determine not only why and when the phase transition takes place, but how it proceeds. To answer this question not only thermodynamics but also kinetic theories had to be developed and applied. An example was the classical theory of nucleation of the evolving phase which goes back to the 30th of the last century and is due to Becker and Döring, Kaischew and Stranski, Frenkel and Zeldovich and others.

First-order phase transformations in a system starting from a metastable initial state proceed via the new-phase nucleation mechanism. The kinetics of such phase transformation can be usually divided into three stages. Let us consider a system supersaturated with certain species inducing a diffusive mass transfer (e.g. by the atoms of a dissolved material in the process of precipitation of other phases from a supersaturated solid or liquid solution; or by vacancies and interstitial atoms in the growth of pores and dislocation loops, or by the atoms of a gas in the growth of gas-filled bubbles etc.). The first stages of decomposition, when the supersaturation, for example, with point defects is large enough, is characterized by intensive generation of viable nucleation centers larger than some critical size. At this stage, the amount of material in the nucleation centers is small, compared with that in the solution, and the supersaturation is essentially constant.

The second transient, or intermediate, stage of the decomposition process begins when the amount of material in the new phase becomes comparable with the initial quantity thus resulting in a decrease of the supersaturation. At this stage, the number of precipitates is practically constant and the volume of the new phase increases mainly through the independent growth of the precipitates.

Finally at the third, late stage of the phase transition, when the already formed aggregates of the newly evolving phase become large enough to allow to essentially decrease the supersaturation, surface tension and the conservation laws for atom species or point defects begin to play a crucial role in the phase transformation, thus resulting in a specific mechanism of the kinetics of new phase growth. This stage of the phase transformation was originally discovered in the analysis of decomposition of metastable solutions by Ostwald in 1900. This late stage of diffusive decomposition of dispersed systems is characterized by an increase in

Kinetics of First-order Phase Transitions. Vitaly V. Slezov

ISBN: 978-3-527-40775-0

the mean size of new phase macroscopic centers, as a result of diffusive mass transfer from the smaller- to the larger-sized centers, the larger-sized centers "devouring" the smaller ones. From a thermodynamic point of view, this behavior is due to a decrease of the free energy of the system as a consequence of a reduction of the interfacial area and the surface energy contributions to the thermodynamic functions. Stochastic generation of new stable nucleation centers at this stage is highly improbable since they must be macroscopic in size. A considerable "diffusive" interaction between grown-up centers of the new phase appears, since each particular center "feels" the self-consistent diffusion field of the entire ensemble of point- and macrospecies of the new phase.

This phenomenon is commonly denoted as "Ostwald ripening" or, more frequently, as "coarsening", or sometimes as "coalescence", though the latter term is, in fact, inadequate. Although the late stage of the phase transition (or decomposition of the originally existing phase), determined by the diffusive interaction between new phase centers, has been analyzed by many authors, an incomplete set of equations has usually been solved, giving size distribution functions which did not obey the law of conservation of point defects. The detailed kinetics of a dispersed system cannot be revealed within such a reduced theoretical framework. The author, together with I.M. Lifshitz, had the opportunity to work out the theory of this late stage in the 50th of the last century giving a first correct solution of these highly non-linear problems.

The book presents the complete description of all three stages of first-order phase transitions, thus allowing one to model the whole course of the first-order phase transition kinetics. Special attention is given to transient stages in nucleation characterized by the establishment of steady-state conditions of nucleation and the determination of the time required for its approach and period of existence of the different stages of the nucleation-growth process.

Phase transformation processes may also proceed through the process of spinodal decomposition of an initially unstable phase. To this end the system should be quickly driven into the totally unstable state. The last chapter of the book deals with the kinetics of the spinodal decomposition. It is interesting that also in this case the whole process can be subdivided into three stages, in some way analogous to the transition in metastable system. Moreover, it is shown that both nucleation-growth and spinodal decomposition processes can be described in a unique way in terms of a generalized cluster model accounting appropriately for both size and composition (or density) changes of the clusters of the newly evolving phase in the course of their evolution to the respective macrophases.

The theoretical results obtained are illustrated in the book by experimental evidences. First of all it concerns the processes of phase decomposition in multicomponent systems, including isotope mixtures of solid helium.

In the course of the work on different aspects of the kinetics of phase formation, I had the pleasure to work together with a number of colleagues. To all of them I would like to express here my sincere thanks. In particular, it is a pleasure to thank the Scientific Editor of this book, Dr. Jürn W. P. Schmelzer, for his advices and gracious assistance in so many ways in the preparation of the present book for publication.

Kharkov, December 2008 *Vitaly V. Slezov*

Preface

The present monograph is written by an outstanding worldwide highly recognized specialist in the field of the theory of first-order phase transitions, Prof. Vitaly V. Slezov. Vitaly Slezov studied physics at the Leningrad Polytechnical Institute, Russia. Since 1954, his scientific activities were permanently connected with the Institute of Physics and Technology in Kharkov now in the Ukraine, first as a PhD student and since 1973 as the head of a department at the institute. He was awarded twice the State Price of the Ukraine for Research (in 1978 and 1993) and elected as a Corresponding Member of the National Academy of Sciences of the Ukraine in 1995.

Vitaly V. Slezov started his scientific career under the guidance of Ilya M. Lifshitz. In the course of the work on his PhD thesis, he developed in cooperation with I. M. Lifshitz the theory of the late stage of coarsening of solid solutions (Ostwald ripening), nowadays, well-known as Lifshitz–Slezov (LS) or Lifshitz–Slezov–Wagner (LSW) theory. The analysis of the properties of multicomponent solid solutions and the kinetics of phase transformation processes in them he continued then in the following decades covering the whole course of these processes from the initial stages of nucleation via independent growth and, finally, coarsening.

Vitaly V. Slezov performed original research also in a variety of other fields of theoretical physics like superconductivity, solid state physics, the behavior of solids under irradiation. However, the analysis of the properties of solid solutions was always in the center of his interests. Hereby he combined theoretical work with experimental applications. The results of his work covering a period of five decades and related developments are summarized in the present monograph.

It is a special pleasure to note that, since 1988, the editor of the present book had the opportunity and pleasure to perform a variety of common analyses in close cooperation with Vitaly V. Slezov and coworkers of his research group. Here, in particular, the kinetics of coarsening under the influence of elastic stresses, the analysis of the first stages of first-order phase transitions, and several problems of the theory of spinodal decomposition have to be mentioned. In the analysis of nucleation–growth processes, widely the classical approach to the description of these processes was utilized treating the aggregates of the newly evolving phase to a large extent as small pieces of the respective macroscopic phase. However, the alternative approach to phase formation denoted as spinodal decomposition solves another problem, it shows how the bulk properties of the clusters evolve in time. Comparing the different methods, Vitaly V. Slezov always posed the question whether it is possible to formulate a theory allowing one a description of the kinetics of phase formation incorporating both cluster growth and dissolution and the change of the bulk properties of the clusters in dependence on time and size of the clusters. Such a theoretical approach could be formulated by the editor of the present monograph in cooperation with coworkers of Vitaly V. Slezov and is included in the present monograph as well.

The results of the long-standing, highly original investigations of Vitaly V. Slezov and his group have been presented by the author and his coworkers and discussed in almost every of the Research workshops *Nucleation Theory and Applications*, which take place in Dubna (near Moscow) at the Bogoliubov Laboratory of Theoretical Physics of the Joint Institute for Nuclear Research and are organized by the editor of the present book each year since 1997 till now. The present book gives, for the first time, the opportunity to the interested reader to get a comprehensive overview in English on the investigations performed by the author and his coworkers on the theoretical description of first-order phase transitions and applications of the theory to experiment. I am sure that the present monograph will be of outstanding interest for all colleagues dealing with the analysis of these intriguing phenomena.

Finally, I would like to express my sincere gratitude to Drs. Alexander S. Abyzov and Leonid N. Davydov for their assistance and the huge work they performed in the preparation of the manuscript for publication.

Rostock (Germany), Dubna (Russia), Kharkov (Ukraine)

December 2008 *Jürn W. P. Schmelzer*

1 Introduction

The present book is devoted to the theoretical description of the kinetics of first-order phase transitions covering the whole course of the processes of nucleation and cluster growth. The outline of the theory is supplemented by a variety of applications.

The book is organized as follows: In Chapter 2, the basic equations are summarized allowing the description of both nucleation and growth. In particular, a new method of determination of the emission coefficients of single particles from clusters of arbitrary sizes – being an essential ingredient of the kinetic equations describing nucleation and growth processes – is developed. This method does not require the application of so-called equilibrium or constraint equilibrium distributions and the principle of detailed balancing to nonequilibrium states. It is applicable generally to any kind of phase transformation processes (condensation of gases, segregation processes in solid and liquid solutions, bubble formation in liquids, pore formation in solids, crystallization in melts, etc.) both for one-component and multicomponent systems. As it turns out from the analysis, the final equations obtained are widely similar to those employed traditionally in classical nucleation theory, where these relations are utilized without a sufficient theoretical foundation. This way, the approach outlined here gives a theoretical foundation of some basic assumption of the classical approach in the description of nucleation–growth processes. Based on the method outlined, the kinetic equations describing nucleation–growth processes are formulated and some further consequences are discussed. It is shown, in particular, that, under quite general conditions, the set of kinetic equations describing nucleation–growth processes in multicomponent systems can be reduced to relations for the description of these processes in one-component systems. However, the thermodynamic and kinetic parameters in the resulting set of kinetic equations depend on the kinetic and thermodynamic parameters of all of the components involved in the process. The respective expressions are derived and outlined in the contribution as well.

Based on analytical solutions of the Frenkel–Zeldovich equation, in Chapter 3 a description of the whole course and of basic characteristics of the first stages of nucleation–growth processes in first-order phase transitions – the stage of establishment of steady-state conditions in certain ranges of cluster sizes, the stage of steady-state nucleation and simultaneous growth, the stage of dominating independent growth of the supercritical clusters – is developed. Analytical expressions are derived for the time evolution of the cluster-size distribution function and the flux in cluster-size space both for kinetic- (ballistic) and diffusion-limited growth modes. In addition, detailed analytical expressions are given for the duration of the different stages and further basic characteristics of this process.

In particular, estimates for the number of clusters, N, formed in nucleation-growth processes in first-order phase transformations, and their average size, $\langle R \rangle$, at the end of the stage

Kinetics of First-order Phase Transitions. Vitaly V. Slezov

ISBN: 978-3-527-40775-0

of independent growth of the supercritical clusters are derived. The results are extended to the description of nucleation and growth in multicomponent systems when aggregates with a given stoichiometric composition are formed first for the simplest case of ideal solutions. The generalization to nonideal solutions is discussed in Chapter 8. As shown, particular thermodynamic properties of the system under consideration have to be taken into account only in evaluating the final expressions of the theory in application to particular experimental conditions. This way, the possibility is opened for a straightforward application of the theory to the interpretation of experimental results.

As an additional application, the theory gives the possibility of testing alternative approaches in the determination of the work of critical cluster formation based on measurements of the number of clusters formed in nucleation–growth processes and their average sizes at the end of the stage of independent growth. Moreover, the number of clusters and their average sizes determined in this way supply us at the same time with the initial conditions for the process of coarsening, outlined in detail in Chapter 4. Here the basic results of the Lifshitz–Slezov theory are outlined in a comprehensive way.

Possible shapes of cluster-size distributions evolving in the course of nucleation–growth processes are analyzed in Chapter 5 giving simultaneously numerical illustrations of the analytical results obtained in Chapters 3 and 4. In the analysis (i) a critical discussion of the applicability of statistical cluster models for an interpretation of experimental results is given; (ii) criteria for the evolution of thermodynamically stabilized monodisperse distributions are formulated; (iii) kinetic equations for the description of the nucleation–growth process are developed avoiding the application of the so–called equilibrium distribution of classical nucleation theory and the principle of detailed balancing (cf. Chapter 2); (iv) numerical solutions of the kinetic equations describing the nucleation–growth process under different conditions are presented. Possible extensions and further applications are discussed.

Chapter 6 is devoted to the theoretical description of coarsening under the influence of elastic stresses. In the present analysis, stresses are considered which are due to cluster–matrix interactions. The theory of Schmelzer and Gutzow is briefly summarized allowing one to obtain expressions for the time dependences of the average cluster size and the number of clusters in coarsening under the influence of such types of stresses. This review is followed first by the outline of the theoretical description of the coarsening process of an ensemble of clusters in a system of nondeformable pores of equal size. The evolution of the cluster-size distribution and related quantities like average cluster size, critical cluster size and the number of clusters is analyzed in detail. It is shown that via an intermediate bimodal distribution, a stable monodisperse distribution of clusters is established. Possible fields of application of the outlined theory are segregation processes in porous materials like vycor glasses and zeolites, spatially inhomogeneous materials or highly viscous melts or polymers. The theory is extended then to the case of so-called weak pores and coarsening in pores with some given pore-size distribution.

In Chapter 7, the kinetics of phase transformation processes in solid or liquid solutions is investigated for two cases when coarsening is affected by different factors from outside of the system. As a first such phenomenon we consider the case that monomeric building units are added homogeneously to the system with a constant rate Φ_a. The analysis is carried out both for diffusion and kinetically limited growth modes. Characteristic quantities like the average ($\langle R \rangle$) and the critical cluster sizes (R_c) are discussed as well as the time evolution of the cluster-size distribution function. It is shown that for diffusion-limited growth the number of clusters, N, evolving in the system is a linear function of the rate of addition of monomers, while the average cluster radius grows, in the asymptotic stages, as $\langle R \rangle \propto t^{1/3}$. For kinetic-limited growth we find $\langle R \rangle \propto t^{5/12}$ and $N \propto t^{-1/4}$. Latter results imply that in the asymptotic stage of kinetically limited growth the quantity $\langle R \rangle^{3/5} N$ approaches a constant value. The results can be used to generate cluster-size distributions with definite properties by varying the rate of input fluxes of monomeric building units of the segregating phase. An analytic description of such processes is performed for the case of void ripening in the presence of bulk vacancy sources. The evolution of the void ensemble is considered for the case that vacancy sources are homogeneously distributed in the system. The asymptotic properties of the size distribution function of voids are investigated in the case of damped, constant, and increasing regimes. The applicability limits of the obtained formulas are determined.

As a second phenomenon, cluster growth and coarsening under the influence of radiation is analyzed in detail. The growth of second phase precipitates from the supersaturated solid solution under irradiation is investigated taking into account a new mechanism of precipitate dissolution. This mechanism is of a purely diffusion origin, i.e., it is based on diffusion out-fluxes of point defects produced by irradiation within the precipitates into the host matrix, provided that the interface boundary is transparent for the point defects. The point defect production rate within a precipitate is proportional to its volume while the total diffusion influx of substitutional impurity atoms is proportional to its radius meaning that there exists a maximum size at which the precipitate growth rate equals the rate of its radiation-induced dissolution. This size is shown to be a stable one implying that under irradiation a stationary state can be achieved far away from the thermodynamic equilibrium.

In Chapter 8, the kinetics of phase transformation processes in multicomponent real solutions is analyzed in detail extending the analysis given in Chapter 3. A kinetic theory of nucleation and growth of a newly evolving phase with a given stoichiometric composition in a multicomponent solid solution is developed. It is assumed naturally that the new phase grows as a result of individual atom incorporation into the new phase domain in a stoichiometric ratio. As is shown, for the case of phase formation in a multicomponent system the basic kinetic equations, describing the nucleation–growth process, can be reduced formally to the respective expression derived for nucleation–growth processes in one-component systems. However, the effective diffusion coefficients and the effective supersaturation are expressed as nontrivial combinations of the thermodynamic and kinetic parameters of the different components involved in the phase formation process. In the determination of these properties, the theory is not restricted in its applicability to perfect solutions but extended to phase formation in real mixtures. Thus, the theory may be applied directly toward the interpretation of experimental data. In particular, the influence of solute–solute interactions on segregation processes in multi-component solid solutions leading to the formation of a new phase with a given stoichiometric composition is investigated. Expressions for the nucleation and growth rates are derived. Estimates are developed for the time required to establish a steady-state nucleation rate in the system and the time interval for which such a steady state can be sustained. Based on these results a method for an experimental determination of the parameters describing the interaction of the solute components is anticipated. Moreover, the kinetic equation describing the evolution of the cluster-size distribution function is generalized to account for stochastic effects both due to fluctuations in the growth rate and due to possible spatial correlations of the evolving clusters in the matrix. The possible influence of such stochastic effects (thermal noise and random coalescence) on coarsening described by such additional terms is discussed briefly.

In Chapter 9, the theory is applied to the case of formation of bubbles and their further growth. In particular, bubble nucleation and growth in low-viscosity liquids supersaturated with gas is theoretically studied. It is shown that, in a certain parameter range, the bubble size is adjusted to the amount of gas in a bubble, and the state of the bubble can be described using one variable: the bubble size or the number of gas atoms in a bubble. Expressions for the nucleation time, bubble-size distribution function, flux of nuclei in bubble-size space, and the maximum number of bubbles formed in the system are determined. After the nucleation period an intermediate stage of the process starts when the number of bubbles per unit volume remains virtually constant, whereas the amount of the gas dissolved per unit volume of the solution significantly decreases, almost attaining the equilibrium value. At the late stage (coalescence), small subcritical bubbles disappear due to gas transfer to large supercritical bubbles; as a result, the number of bubbles in the liquid decreases. For all these stages, the kinetics of evolution of the bubble-size distribution function and the amount of gas per unit volume of the solution is determined.

In Chapter 10, the theory is applied to phase formation processes in helium at temperatures near to the absolute zero. It has been emphasized frequently in the literature that helium and its isotopic mixtures hold much promise as model systems for studying phase transitions. One of the problems generally occurring in the experimental analysis of nucleation phenomena consists in the necessity to generate the conditions for homogeneous nucleation. This task is usually realized by establishing high supersaturations in the system so that homogeneous nucleation dominates. However, in dilute liquid ^{3}He–^{4}He mixtures it is impossible to attain a large supersaturation during cooling because of the limited solubility of ^{3}He at $T \to 0$. In the case of solid helium, these problems might be solved, however, provided that high-quality impurity-free samples are available. The solid helium isotope system is also attractive because the separation process occurs on an accessible time scale: slower than that in fluids, but faster than that in conventional solids. This is a consequence of the unique nature of the atomic motion in solid helium where quantum exchange results in a temperature-independent diffusion coefficient, intermediate between that of a solid and a liquid. Chapter 10 presents NMR measurements in a solid ^{3}He–^{4}He mixture as the temperature was lowered in steps in the course of phase segregation, as well as precision measurements of the pressure during phase separation in solid mixtures of ^{4}He in ^{3}He allowing one to obtain characteristic times of the phase decomposition process. The described results evidence that homogeneous nucleation is realized in ^{3}He–^{4}He solid solutions at significant supercoolings and heterogeneous nucleation at the smallest supercoolings. From a comparison with theory, the surface tension at the interface of the phase-separated clusters is determined.

In Chapter 11, first spinodal decomposition processes in adiabatically closed systems are analyzed. The adiabatic closure of the system results in changes of the temperature and a nonlinear feedback of the phase formation process on the state of the ambient phase. As shown, such effect leads to a similar scenario of the decomposition process as compared with nucleation–growth and coarsening processes. As a next step, basic features of spinodal decomposition, on one side, and nucleation, on the other side, and the transition between both mechanisms are analyzed within the framework of a generalized thermodynamic cluster model based on the generalized Gibbs approach. Hereby the clusters, representing the density or composition variations in the system, may change with time both in size and in their intensive state parameters (density and composition, for example). In the first part of the analysis, we consider phase separation processes in dependence on the initial state of the system for the case when changes of the state parameters of the ambient system due to the evolution of the clusters can be neglected as this is the case for cluster formation in an infinite system. As a next step, the effect of changes of the state parameters on cluster evolution is analyzed. Such depletion effects are of importance both for the analysis of phase formation in confined systems and for the understanding of the evolution of ensembles of clusters in large (in the limit infinite) systems. The results of the thermodynamic analysis are employed in both cases to exhibit the effect of thermodynamic constraints on the dynamics of phase separation processes.

2 Basic Equations: Determination of the Coefficients of Emission in Nucleation Theory

2.1 Introduction

Phase transformations play an important role in a variety of processes ranging from nucleation and growth in the atmosphere [98, 135, 193], nucleation and growth in expanding gases [242, 316], bubble formation in liquids [160, 271], and phase formation in solids [51, 110, 124, 128] to phase transitions in nuclear matter [57, 240, 247] and the early universe [112]. Despite a number of modern developments [28, 33, 93, 145, 358], the theoretical interpretation of experimental results on phase transformations is carried out till now widely based on classical nucleation theory, its modifications and extensions (see, e.g., Refs. [71,94,271,337]). According to the classical picture, the phase transformation proceeds via the formation of clusters representing precursors of the newly evolving phase. Hereby it is assumed in an approximation which in most cases is sufficiently accurate that the growth or dissolution of the clusters proceeds via incorporation or emission of single atoms or molecules.

In order to develop a kinetic description of nucleation–growth processes in the framework of the classical approach, one has to know, consequently, the coefficients of aggregation and emission of single particles for clusters of arbitrary sizes. Moreover, one has to construct the so-called work of formation of clusters of arbitrary sizes, i.e., one has to determine the change of the characteristic thermodynamic potential if a cluster is formed in the system. An example for the dependence of the work of cluster formation on cluster size, as it is commonly assumed in classical nucleation theory, is shown in Figure 2.1 for a binary stoichiometric system. However, although the classical theory was formulated in its basic premises already in the 1930s, till now a number of problems both of fundamental character and with respect to possible applications are not solved finally.

In particular, one of the most debated points in nucleation theory is the method of determination of the emission coefficients. These coefficients are commonly specified by deriving so-called equilibrium or constraint equilibrium distributions with respect to cluster sizes and applying the principle of detailed balancing to thermodynamic nonequilibrium states (cf. Refs. [352,353] and references cited therein). Such an approach is, however, highly questionable [240,247,252]. In application to thermodynamic nonequilibrium states such distributions are artificial constructs; they are not realized in nature. Moreover, the principle of detailed balancing holds for equilibrium but not for nonequilibrium states. For this reason, different attempts have been developed to overcome such difficulties. The most straightforward solution of the problem of determination of the emission coefficients would consist, of course, in the application of microscopic statistical–mechanical approaches (cf., e.g., Refs. [215,216]).

Kinetics of First-order Phase Transitions. Vitaly V. Slezov

ISBN: 978-3-527-40775-0

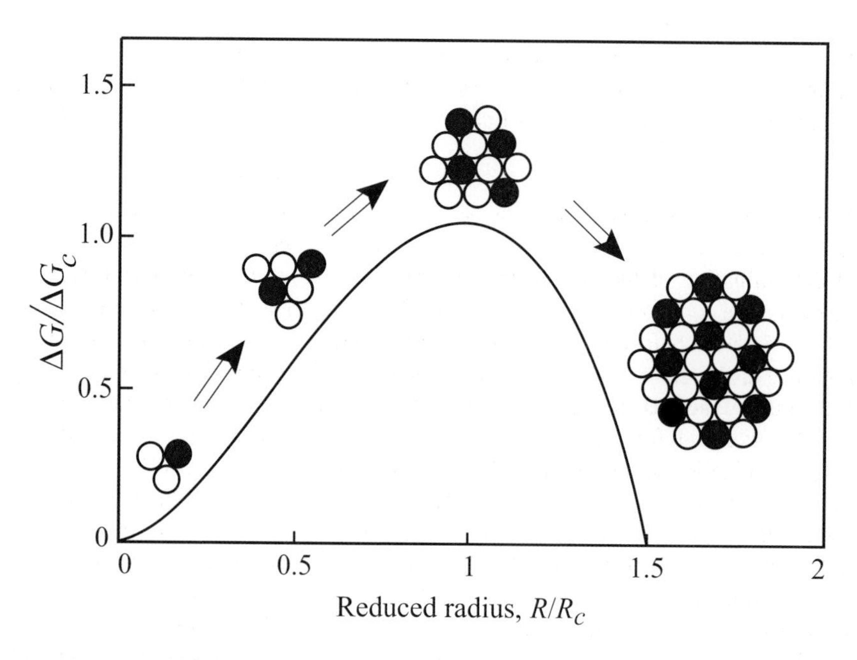

Figure 2.1: Work of formation (or the change of the Gibbs free energy, ΔG) in dependence on cluster size (R is the radius of the cluster) as it is assumed commonly in classical nucleation theory. Nucleation is the process of formation of supercritical clusters with sizes $R > R_c$ capable to a further deterministic growth. In the calculations it is assumed in accordance with the classical approach that bulk and surface properties of the clusters are widely the same as the respective properties in the newly evolving macroscopic phase. The work of cluster formation can then be expressed via Eq. (2.16).

Due to the enormous problems in applying this method to real systems it is however, till now more an interesting possibility rather than a practical tool (cf. also Ref. [352]).

Macroscopic approaches, like the traditional method discussed above, have the advantage that specific properties of the system under consideration enter the description only via the specific expressions for the characteristic thermodynamic functions. These or other thermodynamic characteristics applied can be determined much easily. For the mentioned reasons, it is of use to further develop macroscopic methods for the determination of the emission coefficients retaining their advantages but avoiding highly questionable arguments. One of such approaches was developed by Katz et al. (cf., e.g., Refs. [117–121]) in application to both vapor condensation and nucleation and growth in condensed media. In this approach, the coefficients of emission of single particles are determined first for the state of a saturated system. In a next step, by applying appropriate expressions for the cluster-size distributions evolving in saturated systems and certain additional assumptions (e.g., independence of the emission coefficients on the state of the ambient phase [117–119]), the emission coefficients for the supersaturated systems are determined. In another development, which goes back to Becker and Döring [24], the emission rates are determined by applying the Gibbs–Thomson (or Kelvin's) equation (cf. Refs. [120, 121, 352]).

The problem described was discussed some years ago in detail by Wu [355]. He came to similar conclusions and mentioned: "*The constrained equilibrium hypothesis (CEH) formulated as an extrapolation of fluctuation thermodynamics is ... not valid. A different justification is required ... Since CEH is central to nucleation theory, it is not likely to go away until something better comes along ....*" It is one of the aims of the present chapter to show that these mentioned approaches may be generalized arriving, indeed, at some justification of the methods commonly employed. We show that a mesoscopic statistical–mechanical method for the determination of the coefficients of emission can be developed without relying on such questionable concepts like constraint equilibrium distributions and the application of the principle of detailed balancing to nonequilibrium states. This method does not involve the assumption that the emission rates are independent of the state of the ambient phase. It is applicable quite generally to a variety of phase transformation processes in gases as well as in condensed matter both for one-component and multicomponent systems. It allows a straightforward generalization to nonisothermal phase formation processes as well.

The chapter is organized as follows. After a brief introduction of the basic kinetic equations (Section 2.2), the newly developed method for the determination of the emission coefficients is outlined. It is shown that the method is applicable both to one-component (Section 2.3) and multicomponent (Section 2.4) systems. As an example, the derivation of the respective dependences is given in detail for the case when the external pressure p and temperature T are kept constant. It is shown further that the method can be applied with minor modifications also if the phase formation proceeds at any external conditions other than isothermal or isobaric (Section 2.5). The resulting relations for the determination of the emission coefficients are independent of the boundary conditions applied. Employing these results, the basic sets of kinetic equations describing nucleation–growth processes both in one- and multicomponent systems are completed, the initial conditions for their solution are specified (Section 2.6) and some further consequences are discussed (Section 2.7). For the first time, this approach has been outlined in [293, 304] and elaborated in more detail in [306, 315]. Below in this chapter we present a comprehensive outline of the basic ideas as well as results of the approach developed and an analysis of some further developments.

2.2 Basic Kinetic Equations

We discuss here first processes of formation of a new phase in an ambient phase with high thermal conductivity. In this case, heat sources, connected with transitions of the basic elementary units of the system (atoms, molecules, aggregates of a given stoichiometric composition, etc.) from one phase to another, practically do not change the temperature T of the part of the system, where the aggregate of the new phase evolves. Therefore, with a sufficiently high precision, the nucleation–growth process may be considered as isothermal. Moreover, in addition, constancy of the pressure p is supposed. To be definite, the discussion is carried out in application to condensation processes in gases (in an inert carrier gas). However, the basic assumptions are applicable quite generally. Thus, the results are equally valid for a variety of different phase formation processes as discussed above, e.g., for segregation in solid or liquid solutions, crystallization of melts or bubble formation in liquids.

As generally assumed in nucleation theory, the condensation or evaporation of the clusters (or droplets) occurs by aggregation or emission of single particles, only. Therefore, the kinetic equation will be of the standard form [144, 293, 298, 304, 306], i.e.,

$$\frac{\partial f(n,t)}{\partial t} = w^{(+)}_{n-1,n} f(n-1,t) - w^{(-)}_{n,n-1} f(n,t) + w^{(-)}_{n+1,n} f(n+1,t) - w^{(+)}_{n,n+1} f(n,t) . \tag{2.1}$$

Here $f(n,t)$ is the distribution function of clusters of the new phase, containing n single particles; $w^{(+)}_{n-1,n}$ and $w^{(+)}_{n,n+1}$ are the average number of events that per unit time one particle is absorbed and the number of particles in a cluster is increased from $(n-1)$ to n and n to $(n+1)$, respectively; $w^{(-)}_{n,n-1}$ and $w^{(-)}_{n+1,n}$ are the average number of events for a cluster to release one particle per unit time and to be transferred to the states with $(n-1)$ and n particles, respectively.

By introducing the fluxes J_n, we may rewrite Eq. (2.1) in the form of an ordinary continuity equation in cluster-size space as

$$\frac{\partial f}{\partial t} = J_{n-1} - J_n , \tag{2.2}$$

where the fluxes J_n are determined by

$$J_{n-1} = w^{(+)}_{n-1,n} f(n-1,t) - w^{(-)}_{n,n-1} f(n,t) , \tag{2.3}$$

$$J_n = w^{(+)}_{n,n+1} f(n,t) - w^{(-)}_{n+1,n} f(n+1,t) . \tag{2.4}$$

Once we have derived the kinetic equations in the general form, we have, now, to determine the kinetic coefficients.

2.3 Ratio of the Coefficients of Absorption and Emission of Particles

The condensation or absorption coefficients $w^{(+)}_{n-1,n}$ and $w^{(+)}_{n,n+1}$ are determined by the kinetic mechanism by which the droplets (or, in general, clusters of the new phase) grow. As will be shown later, they can be determined directly from the macroscopic growth rates. Knowing how to obtain the absorption coefficients it is also necessary to possess the methods for the determination of the emission coefficients, $w^{(-)}$. For such purposes, let us note that the particles of the new phase may be divided into two groups: First, for the particles with a number of atoms less than the critical cluster size, $n < n_c$, the ambient phase is undersaturated. Here n_c is the critical cluster size in nucleation. In a macroscopic deterministic description, such clusters shrink and disappear. In other words, the concentration of single particles in the ambient phase, capable to be incorporated into the clusters of the newly evolving phase, is too small to retain a dynamic equilibrium (cf. Figure 2.1).

Second, for droplets with supercritical sizes, $n > n_c$, the ambient phase is supersaturated and the clusters of the new phase grow in a deterministic description. It means that the concentration of single particles has such values that the average number of aggregation processes per unit time interval exceeds the respective value for emission from a given cluster of the new phase. Particles with critical sizes, $n = n_c$, are in (unstable) thermodynamic equilibrium with the ambient phase of given composition (here the vapor phase). Hence, for critical clusters the average numbers of aggregation and emission processes coincide.

The methods for the determination of the coefficients $w^{(-)}$ differ in our approach in dependence on the range of cluster sizes considered, i.e., which class of clusters is considered, super- or subcritical ones. However, before we explain our method, the traditional approach is briefly revisited and discussed. Hereby, in addition, some general thermodynamic relationships are derived employed in the subsequent analysis.

2.3.1 Traditional Approach

In thermodynamic equilibrium states, a time-independent statistical cluster-size distribution is established in the course of time. This distribution is described by Eq. (2.1) with $(\partial f/\partial t) = 0$. This way, in order to find the equilibrium cluster-size distributions we have to find the time-independent solutions of the set of equations (2.1).

In the search for this solution, we may follow a different path as well. The equilibrium distribution of clusters, $f^{(\mathrm{eq})}(n)$, may be determined by general methods of statistical physics as [144]

$$f^{(\mathrm{eq})}(n) = A \exp\left(-\frac{R_{\mathrm{rev}}(n)}{k_B T}\right). \tag{2.5}$$

Here k_B is the Boltzmann constant and T the absolute temperature. $R_{\mathrm{rev}}(n)$ is the work of formation of a cluster consisting of n particles performed in a reversible process [144]. For processes proceeding at constant values of pressure and temperature (as it is assumed here), it equals the change of the Gibbs free energy, ΔG [144]. This quantity can be expressed, generally, as

$$\Delta G(n) = G_d(n) - n\mu_v(p,T). \tag{2.6}$$

Here $G_d(n)$ is the contribution of the cluster (drop) of size n to the value of the thermodynamic potential of the whole system, $\mu_v(p,T)$ is the chemical potential per particle in the ambient bulk (vapor) phase and n is the number of atoms in the cluster. In other words, $\Delta G(n)$ is the difference of the thermodynamic potentials of the system consisting of a cluster and the vapor compared with the homogenous initial vapor state.

In Eq. (2.6), the particular expression for $G_d(n)$ is not specified, so this relation can be quite universally applied. If, for example, the capillarity approximation and certain additional assumptions are employed [94, 248] for the determination of $\Delta G(n)$, Eq. (2.6) takes the form

$$\Delta G(n) = n\left[\mu_d(p,T) - \mu_v(p,T)\right] + 4\pi\sigma\left(\frac{3}{4\pi}\omega_d\right)^{2/3} n^{2/3}. \tag{2.7}$$

Here $\mu_d(p,T)$ is the chemical potential of one single particle in the bulk liquid phase at a pressure p and a temperature T, ω_d is the volume per atom in the newly evolving phase, σ is the surface tension at the (planar) interface between the ambient phase (vapor) and the newly evolving phase. For the particular case as expressed by Eq. (2.7) we have, consequently,

$$G_d(n) = n\mu_d(p,T) + 4\pi\sigma\left(\frac{3}{4\pi}\omega_d\right)^{2/3} n^{2/3}. \tag{2.8}$$

Note, however, that the further derivation is independent of any particular choice of the expression for $G_d(n)$.

Using the time-independent form of the kinetic equations (Eq. (2.1)) we get (by applying the principle of detailed balancing $J_n = 0, n > 1$)

$$\frac{w^{(+)}_{n,n+1}}{w^{(-)}_{n+1,n}} = \frac{f^{(\mathrm{eq})}(n+1)}{f^{(\mathrm{eq})}(n)} \tag{2.9}$$

or, finally, after substituting the expression for the work of cluster formation, as given by Eq. (2.6), into the distribution function $f^{(\mathrm{eq})}(n)$ (Eq. (2.5)) describing heterophase fluctuations,

$$\frac{w^{(+)}_{n,n+1}}{w^{(-)}_{n+1,n}} = \exp\left(-\frac{[\Delta G(n+1) - \Delta G(n)]}{k_B T}\right). \tag{2.10}$$

It is evident that the knowledge of the specific form of the preexponential coefficient A in Eq. (2.5) is, in general, not required. It has to be supposed only that its dependence on the size and the composition of the cluster is weak as compared with the exponential term in Eq. (2.5).

With the general relation (2.6) we may further write

$$-[\Delta G(n+1) - \Delta G(n)] = \mu_v(p,T) - \left.\frac{\partial G_d(n)}{\partial n}\right|_{n=n+1} = \mu_v - \mu_d(n+1). \tag{2.11}$$

Here $\mu_d(n)$ is the chemical potential per atom of the drop of size n including surface energy and other possible size effects. It is defined by

$$\mu_d(n) = \frac{\partial G_d(n)}{\partial n}. \tag{2.12}$$

For the special choice of $G_d(n)$, as expressed by Eq. (2.8), we have, in particular,

$$\mu_d(n) = \frac{\partial G_d(n)}{\partial n} = \mu_d(p,T) + \frac{8\pi\sigma}{3}\left(\frac{3}{4\pi}\omega_d\right)^{2/3} n^{-1/3}. \tag{2.13}$$

Generally, with Eq. (2.11) we may also write

$$\frac{w^{(+)}_{n,n+1}}{w^{(-)}_{n+1,n}} = \exp\left(\frac{\mu_v - \mu_d(n+1)}{k_B T}\right). \tag{2.14}$$

In this way, it has been shown that in the thermodynamic equilibrium state the relation between the coefficients of emission and absorption is given by either Eq. (2.9), (2.10), or (2.14).

In application to the thermodynamic equilibrium state, the function $f^{(\mathrm{eq})}(n)$ has a real physical meaning. It represents the equilibrium distribution of heterophase fluctuations. Moreover, in equilibrium the principle of detailed balancing holds. Therefore, for the considered so far states the method of derivation of the emission coefficients is fully satisfactory. However, even in this region of applicability some uncertainty remains connected with the properties of the preexponential factor A in Eq. (2.5).

However, considering thermodynamically unstable, nonequilibrium states, where nucleation-growth processes may occur, the situation becomes quite different. Distributions of the type as given by Eq. (2.5) can be derived in a correct way only for equilibrium but not for nonequilibrium states. This remark refers both to classical thermodynamic and statistical-thermodynamic approaches. Frenkel [71] derived his well-known distribution based on the assumption of a minimum of the Gibbs free energy (cf. also the more detailed discussion given here below and in Chapter 5). However, latter condition is applicable to equilibrium states, only. Fisher [69] obtained similar slightly modified expressions by applying methods of equilibrium statistical physics. Thus, the extrapolation of these results to thermodynamic nonequilibrium states is, again, not correct. Generally, the application of methods of equilibrium statistical mechanics for a description of fluctuation processes [144] is valid in the thermodynamic equilibrium state, only.

For the determination of the emission rates of single particles from the clusters and the steady-state nucleation rate, in general, a somewhat artificial model introduced originally by L. Szilard (see Figure 2.2) is commonly utilized. It is assumed that, once a cluster reaches an upper limiting size, $g \gg n_c$, it is instantaneously removed from the system. Moreover, according to Szilard's model, simultaneously to the removal of a g-sized cluster, g single particles are added to the system. In this way, the total number of particles is kept constant. Starting with a state consisting of single particles only, after some time interval (denoted commonly as time lag in nucleation) a time-independent steady-state cluster-size distribution is established in the system.

Assuming that (i) clusters of different sizes can be considered as different components in a multicomponent perfect solution (or a mixture of perfect gases for vapor condensation), (ii) the number of particles aggregated in the clusters is small as compared with the total number of solute particles, (iii) conservation of the total number of solute particles is fulfilled, (iv) the change of the cluster size is possible by emission or aggregation of monomers, only, Frenkel [71] obtained an expression for the stationary cluster-size distribution function $f^{(e)}(n)$ as

$$f^{(e)}(n) = f(1) \exp\left\{ -\frac{\Delta G(n)}{k_B T} \right\}, \tag{2.15}$$

widely similar to Eq. (2.5). Indeed, for the thermodynamic boundary conditions chosen, the work of cluster formation is given by $\Delta G(n)$. The factor $f(1)$ is the concentration (number per unit volume) of single particles which will be also denoted as c.

The distribution (2.15) is commonly denoted as equilibrium or constraint equilibrium distribution with respect to cluster sizes [98,135,352,353,355]. Note, however, that this notation is misleading. The time-independent state of the model system is in reality a nonequilibrium steady state. Therefore, the procedure applied in the derivation of Eq. (2.15) lacks any thermodynamic foundation. Moreover, one has to take into account that the distribution refers

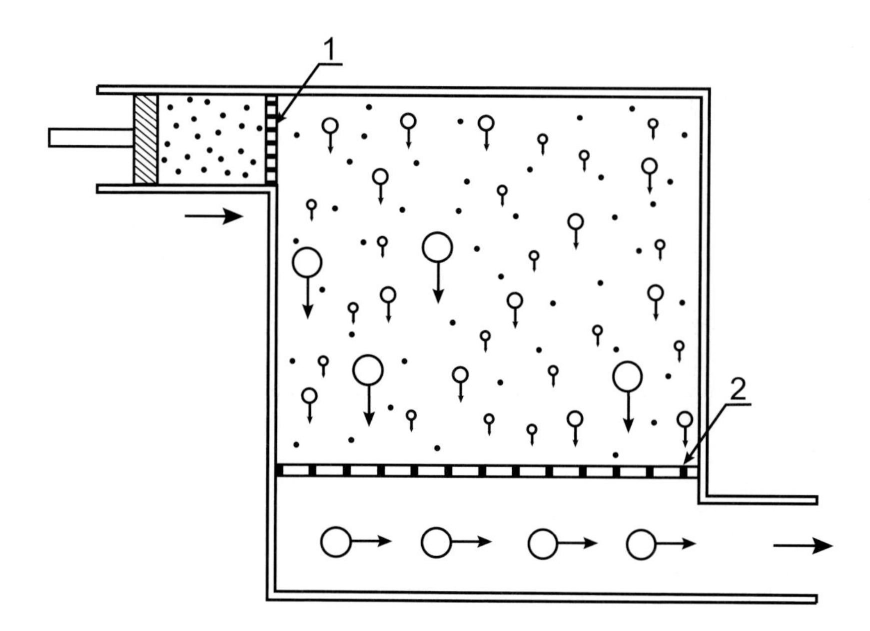

Figure 2.2: Schematic representation of Szilard's model used in classical nucleation theory. Clusters (or droplets) with numbers of monomers $g \gg n_c$ are instantaneously removed from the chamber via a membrane (2) impermeable for clusters of smaller sizes. Simultaneously to the removal of such a cluster, g monomers are added to the chamber (1). In such a way, a constant supersaturation is sustained in the system and a steady state with a constant nucleation rate may be established in the course of time.

to Szilard's artificial model system which is not realized in nature (except for artificial conditions or by assuming some kind of "Szilard's demon" in analogy with Maxwell's demon (cf., e.g., [117–121])). Therefore, the often found identification of the so-called equilibrium distribution with respect to cluster sizes with real distributions evolving in nucleation–growth processes in thermodynamically unstable systems is, in general, incorrect. Even in interpreting such expressions in terms of Szilard's model (cf., e.g., Refs. [240, 247]), the resulting distributions refer not to equilibrium but to stationary nonequilibrium steady states. For such states, the principle of detailed balancing does not hold, in general, as well.

For thermodynamically unstable initial states, ΔG may be written, employing certain approximations, as (e.g., [94], cf. also Eq. (2.7) and Chapter 5)

$$\frac{\Delta G}{\Delta G_{(c)}} = 3\left(\frac{n}{n_c}\right)^{2/3} - 2\left(\frac{n}{n_c}\right). \tag{2.16}$$

The so-called equilibrium distribution function $f^{(e)}(n)$ (cf. Eq. (2.15)) gets in this case the form

$$\left(\frac{f^{(e)}(n)}{f(1)}\right) = \exp\left\{-\frac{\Delta G_{(c)}}{k_B T}\left[3\left(\frac{n}{n_c}\right)^{2/3} - 2\left(\frac{n}{n_c}\right)\right]\right\}. \tag{2.17}$$

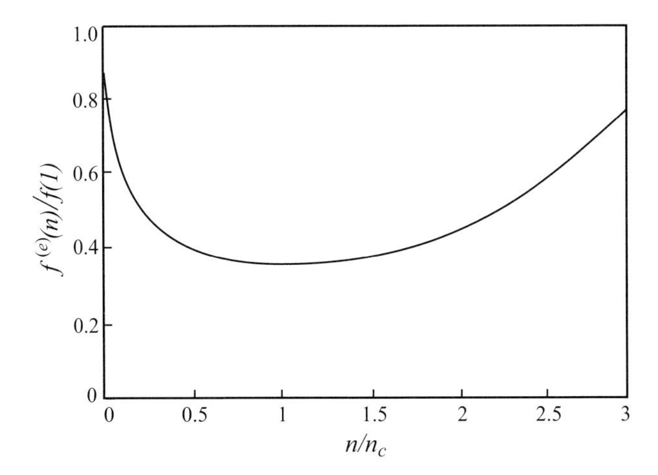

Figure 2.3: *Equilibrium* cluster-size distribution $f^{(e)}(n)/f(1)$ of classical nucleation theory in relative coordinates (n/n_c) for thermodynamically unstable initial states (according to Eq. (2.17)). $(\Delta G_{(c)}/k_B T)$ was chosen equal to 1 for convenience. Fisher's statistical droplet model [69] leads to a widely equivalent dependence except in the immediate vicinity of the critical point.

It is qualitatively presented in Figure 2.3. The function has a minimum for $n = n_c$ and diverges for large values of (n/n_c).

The same conclusions can be drawn with respect to any other similar expressions resulting from different approaches in the determination of the work of cluster formation, ΔG (cf. [98, 135] and Chapter 5). Such distributions may be of use in order to determine the emission coefficients from the expressions for the coefficients of aggregation by applying the principle of detailed balancing to an artificial model state (with all the problems involved in such a procedure (cf. [31,61,69,74,125,144,199,248,293,298,302,304,306,319])). However, the application of these expressions to the description of real cluster-size distributions formed in nucleation–growth processes is, in general, incorrect. To some extent, the results can be considered, however, as a reasonable approximation also for thermodynamic nonequilibrium states, but here only for clusters of subcritical sizes $n < n_c$. Since the system is undersaturated for these aggregates, they are in similar conditions as heterophase fluctuations in thermodynamic equilibrium states. This statement means that the relations between the coefficients of absorption $w^{(+)}_{n,n-1}$ and emission $w^{(-)}_{n,n+1}$ for these aggregates may be the same as in the case of heterophase fluctuations in equilibrium states.

This idea gets additional support by a comparison between the distribution given with Eq. (2.5) and the steady-state cluster-size distribution (cf., e.g., [94, 307, 309]). The latter distribution is the asymptotic solution of the set of kinetic equations provided the concentration of single particles is kept constant by some appropriate mechanism (cf., e.g., Ref. [298]). In the range of cluster sizes $n < n_c$, both distributions coincide with an accuracy to a factor in the range (0.5–1). However, the degree of accuracy of such an extrapolation remains unclear. The situation is getting even more complex for new phase aggregates with $n > n_c$, because in this case we have a fully developed nonequilibrium situation. A reference to heterophase fluctuations is impossible for such clusters and another approach has to be employed for the determination of the emission coefficients $w^{(-)}$.

2.3.2 A New Method of Determination of the Coefficients of Emission

2.3.2.1 General Remarks: Real and Virtual States of the Ambient Phase

After the discussion of the limitations of the traditionally employed approach, we go over, now, to the description of a new general method of determination of the emission coefficients. This method (i) avoids the application of the so-called equilibrium distributions of classical nucleation theory to nonequilibrium states (i.e., to the states of interest in nucleation–growth processes), and (ii) does not employ the principle of detailed balancing, which is valid for thermodynamic equilibrium but not for nonequilibrium states. As mentioned already in the introduction, it is our aim to develop a macroscopic method of determination of the emission coefficients retaining its advantages but avoiding the application of not well founded or even incorrect (for the considered nonequilibrium states) concepts.

One of the basic assumptions, central to our method, is the following. We assume that the clusters of the newly evolving and the ambient phases are both in states of internal thermodynamic equilibrium. Such assumption resembles the concept of a local equilibrium widely employed in the thermodynamics of irreversible processes [96]. It is the basis for an appropriate description of the thermodynamics of heterogeneous systems [85], in general, and thermodynamic (macroscopically based) analyses of cluster formation processes [220, 230], in particular. Moreover, it also gives the foundation to speak about well defined values of the kinetic coefficients. This way, this first assumption is not a serious restriction but the precondition of any mesoscopic approach to the analysis of cluster formation and growth processes. It makes the problem well defined. Remember, however, that the system as a whole is in a nonequilibrium state. The clusters are, in general, not in equilibrium with the surrounding ambient phase.

As a second ingredient in the analysis, we employ the concept of *virtual* and real states of the ambient phase. A virtual state is an idealized model state. It is constructed in the following way. Suppose we have a cluster of given size. Then the question is what should be the state of the ambient phase in order to attain a thermodynamic equilibrium between the cluster considered and the ambient phase. Such *virtual* or possible states of the ambient phase will differ, in general, from the real state of the system. Moreover, for clusters of different sizes, the possible *virtual* states of the ambient phase are different. The coefficients of emission will be determined in our approach by considering the differences between real and virtual states of the system. Hereby it plays no role, as in many other applications of thermodynamics, whether the different *virtual* states may be realized in practice or not. One has to take care only that the models employed do not contradict in their consequences the basic laws of thermodynamics, in particular, or physics, in general. This property is fulfilled by our model.

This concept is a generalization of the application of Kelvin's equation in the determination of the coefficients of emission for vapor condensation [94]. In this method, the same question is asked, i.e., what should be the (virtual) state of the gas phase that a drop of given size is in equilibrium with the vapor. From such considerations, then, conclusions are derived concerning the values of the coefficients of evaporation for drops of given size (aggregation and emission rates have to coincide for the *virtual* state of the vapor). While from the basic idea our approach resembles the mentioned one, it is much more general and applicable to condensation in nonideal one- and multicomponent gas systems as well as to phase formation processes in condensed media both for isothermal and nonisothermal conditions.

2.3.2.2 Clusters of Supercritical Sizes

Going over, now, to a more detailed outline of our method of determination of the emission coefficients, we start with the region of supercritical cluster sizes. In order to proceed with our task we consider, as mentioned, in addition to the real also a *virtual* state of the ambient (vapor) phase. This *virtual* state of the ambient phase is defined in such a way that the chemical potential of the condensing particles is equal to the chemical potential of the building units in the considered aggregate of the new phase. The *virtual* state of the ambient phase depends thus on the size of the cluster (droplet) considered. For clusters of different sizes, different *virtual* states of the ambient phase have to be introduced. Since we are considering, in application to vapor condensation, droplets with sizes $n > n_c$, the chemical potential per particle μ in the real vapor is larger than that in the assumed *virtual* states ($\mu > \widetilde{\mu}$). Here and subsequently, the parameters of the virtual state are specified by a tilde. Therefore, in order to attain a dynamic equilibrium, a certain part of the gas particles has to be fixed in its spatial positions. Consequently, the real and virtual vapor (or, generally, the real and virtual states of the ambient phase) differ by the number of single particles fixed in them.

For the determination of the ratio between the coefficients of aggregation and emission and the specification of the virtual state of the ambient phase, we consider a reference system connected with the chosen cluster. The immobile particles are fixed in their spatial positions with respect to the chosen cluster (drop). Such a choice of the reference system allows us to apply the method also to systems like droplets in gases, where the motion of the clusters affects the aggregation rates. It is required only in the subsequent considerations that the state of the ambient phase in the vicinity of the chosen cluster is at any time in a local thermodynamic equilibrium. This condition implies that the characteristic time scales of changes of the state of the system (including the motion of the cluster) have to be large as compared with the respective times of aggregation or emission of single particles. The assumption of a local equilibrium is inherent in most methods of determination of the coefficients of aggregation. Therefore, we stay here inside the range of commonly accepted quite reasonable approximations. For large drops and most applications of nucleation theory in condensed matter physics, the clusters practically do not move. In these cases, the reference system coincides with the usual laboratory system.

If we describe the vapor as a perfect gas, then the *virtual* vapor state coincides with a state of the gas at a density of mobile particles corresponding to the equilibrium of the vapor with the droplet of the given size (Figure 2.4). In dense vapors, one also has to take into account the interaction of the gas particles. In both cases, the immobile gas particles in the virtual states create the environment for the mobile particles, identical to the environment in the real vapor state. It follows that the condensing particles (in both really existing and *virtual* vapor states) have the same energetic barriers for condensation. More generally, the kinetic conditions for aggregation are the same in both the real and virtual states.

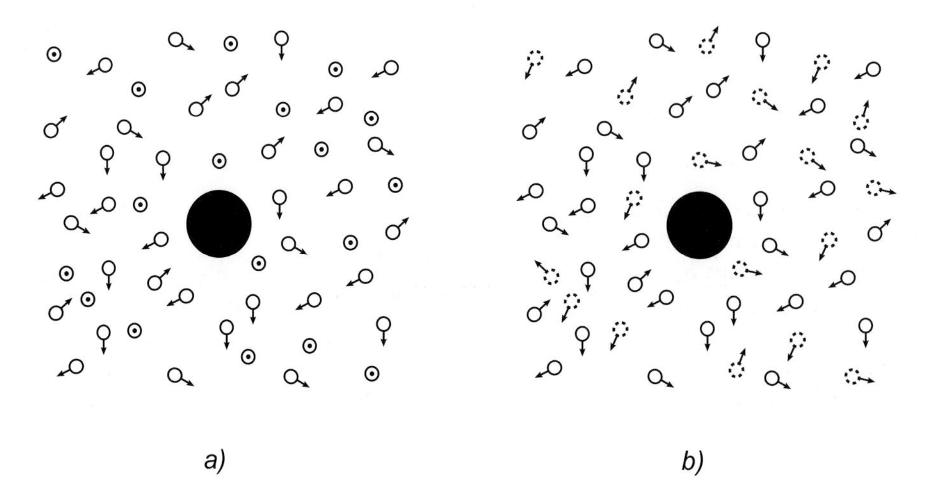

Figure 2.4: Illustration of the model of virtual states (see the text). (a) Supercritical clusters: Particles marked with a dot are made immobile in order to retain a dynamic equilibrium. (b) Subcritical clusters: A number of additional particles (specified by dotted circles) are added in order to retain a dynamic equilibrium.

By definition of the *virtual* vapor state, the gas must contain such a number of mobile ambient phase particles that their chemical potential, taking into account the interaction also with the fixed gas atoms, is equal to the chemical potential of a particle in a drop (or cluster) of size n, i.e., $\widetilde{\mu} = \mu_d(n)$ holds. Under such conditions, the virtual gas (or ambient) phase is in a dynamic equilibrium with a chosen drop (cluster) of the liquid (newly evolving) phase, i.e., the relation $w^{(-)}_{n,n-1} = \widetilde{w}^{(+)}_{n-1,n}$ is fulfilled. Note that this relation is a consequence of the definition of the *virtual* states. It is thus quite different from Eq. (2.9) where rates of aggregation and emission of clusters of different sizes in thermodynamic equilibrium states are compared. Moreover, the virtual vapor state is characterized by equal probabilities per unit time for one selected mobile atom to be added to the liquid drop as compared with a particle at the same state (position and velocity) in the real vapor at the same overall concentration. This property is due to the fact that the environment of the mobile atoms, by definition, is the same as the one in the real vapor phase, or, in other words, in both virtual and real vapors, the kinetic conditions for condensation are the same for the mobile atoms. It follows that the probabilities $w^{(+)}_{n-1,n}$ of aggregation in the real and $\widetilde{w}^{(+)}_{n-1,n}$ in the virtual vapor have the same kinetic preexponential factors. They differ only by the probabilities p_r and $\widetilde{p}_r$ of occurrence of favorable configurations for the realization of these processes with respect to a given mobile particle.

As already mentioned, with sufficiently high precision we may assume that both real and virtual states of the ambient phase (vapor) are each in its internal thermodynamic equilibrium.

Hence, applying the conditions $w^{(-)}_{n,n-1} = \widetilde{w}^{(+)}_{n-1,n}$ and the above given considerations, we get

$$\frac{w^{(+)}_{n-1,n}}{w^{(-)}_{n,n-1}} = \frac{w^{(+)}_{n-1,n}}{\widetilde{w}^{(+)}_{n-1,n}} = \frac{p_r}{\widetilde{p}_r}. \tag{2.18}$$

Here p_r and $\widetilde{p}_r$ are the probabilities of occurrence of a favorable configuration for an incorporation of one mobile particle to the droplet from the real and virtual vapors at a given average (thermodynamic) energy of the corresponding system (cluster or droplet surrounded by a sufficiently large fraction of the ambient phase). Since in the virtual state of the ambient phase some part of the atoms is fixed in their positions, the probabilities of occurrence of favorable configurations obey the condition $\widetilde{p}_r < p_r$.

Furthermore, we employ a basic relationship of statistical thermodynamics [144], i.e.,

$$S = k_B \ln W. \tag{2.19}$$

Here S is the entropy of a given macrostate of a system and W the number of microstates (or thermodynamic probability) referring to the respective macrostate. In application to the considered problem we may write

$$\frac{p_r}{\widetilde{p}_r} = \exp\left[\frac{\left(\Delta S - \Delta\widetilde{S}\right)}{k_B}\right]. \tag{2.20}$$

Here $\Delta\widetilde{S}$ and ΔS are the entropies per mobile particle in the virtual and real vapor (virtual or real states of the ambient phase) for each of the systems considered. The macrostates refer here to such configurations which allow incorporation of a single mobile particle to the cluster.

As mentioned both the virtual and real states of the ambient phase may be considered to be in an internal thermodynamic equilibrium at the respective concentrations of the mobile particles. Moreover, the temperature of both systems is the same. The latter condition implies that the difference of the entropies $\Delta S - \Delta\widetilde{S}$ is equal to the change of the entropy ΔS_n, resulting from the transfer of one mobile particle from the virtual to the real state of the ambient phase. Moreover, the number of configurations allowing incorporation of a mobile particle to a given cluster is larger in the real state of the ambient phase than in the virtual one. Therefore, in the range of cluster sizes $n > n_c$ the relation

$$\Delta S - \Delta\widetilde{S} = \Delta S_n > 0 \tag{2.21}$$

holds. In addition, one can always add to ΔS and $\Delta\widetilde{S}$ the entropies S_0, respectively, $\widetilde{S}_0$ of the rest of the systems (we consider, now and furtheron, only the respective parts of the ambient phase in the real and the virtual states). These contributions do not change in the course of the transfer of a mobile particle between both states. ΔS_n may be treated, therefore, also as the total change of the entropy in such a transfer resulting from the respective changes ΔS and $\Delta\widetilde{S}$ in each system.

At given values of the external thermodynamic parameters, the change of the entropy of a thermodynamic subsystem may be connected with the work R_{rev}, which is the work one has to perform in a reversible process to create the same changes of the system state (the same change of the entropy). Generally, we may write (cf., e.g., Ref. [144])

$$\Delta S = \left(\frac{\Delta S}{\Delta U}\right) R_{\text{rev}} = \frac{1}{T} R_{\text{rev}} . \tag{2.22}$$

Here the thermodynamic relation

$$\frac{\partial S}{\partial U} = \frac{1}{T} \tag{2.23}$$

has been employed, in addition. U in Eqs. (2.22) and (2.23) is the internal energy of the considered system.

For a system at fixed values of the external pressure, p, and temperature, T, the work performed in a reversible process equals the change of the Gibbs free energy Δg in the respective process. Taking into account, moreover, that the temperatures are the same both in the real and the virtual states of the ambient phase, we obtain from Eqs. (2.22) and (2.21)

$$\Delta S_n = \frac{1}{T} R_{\text{rev}} = \frac{\Delta g}{T} . \tag{2.24}$$

Here Δg is the total change of the Gibbs free energy in the transfer of the considered particle from the virtual to the real state.

For one-component systems we may write immediately

$$\Delta g = \mu_v - \widetilde{\mu} . \tag{2.25}$$

Taking into account the relation $\widetilde{\mu} = \mu_d(n)$, we have

$$\Delta S_n = \frac{\mu_v - \mu_d(n)}{T} . \tag{2.26}$$

Here, as earlier, μ_v is the chemical potential of a particle in the real state of the ambient phase, $\mu_d(n)$ is its value in a cluster of size n (including interfacial and other possible finite size effects).

Combining Eqs. (2.18), (2.20), (2.21), (2.24), (2.25), and (2.26), we get for the ratio of absorption and emission coefficients the result

$$\frac{w_{n-1,n}^{(+)}}{w_{n,n-1}^{(-)}} = \exp\left(\frac{\Delta S_n}{k_B}\right) = \exp\left(\frac{\mu_v - \mu_d\,(n)}{k_B T}\right) . \tag{2.27}$$

It follows that the relation between the coefficients of aggregation and emission for supercritical clusters is the same as obtained earlier by applying the traditional method for thermodynamic equilibrium states (cf. Eq. (2.14)), i.e.,

$$\frac{w_{n,n+1}^{(+)}}{w_{n+1,n}^{(-)}} = \exp\left(\frac{\mu_v - \mu_d\,(n+1)}{k_B T}\right) . \tag{2.28}$$

However, in the present derivation no reference was made to the so-called equilibrium distributions and the (nonapplicable in nonequilibrium states) principle of detailed balancing.

Moreover, taking into account the generally valid thermodynamic relationships (Eq. (2.11)), we may write also

$$\frac{w^{(+)}_{n,n+1}}{w^{(-)}_{n+1,n}} = \exp\left(-\frac{[\Delta G(n+1) - \Delta G(n)]}{k_B T}\right). \tag{2.29}$$

This relation is identical in the form to Eq. (2.10). By introducing an auxiliary function $f^{(*)}(n)$ as

$$f^{(*)}(n) = \exp\left(-\frac{\Delta G(n)}{k_B T}\right), \tag{2.30}$$

we may express the relation between the considered coefficients in another equivalent form as the ratio of the functions $f^{(*)}(n+1)$ and $f^{(*)}(n)$, again, i.e. (cf. Eq. (2.9)),

$$\frac{w^{(+)}_{n,n+1}}{w^{(-)}_{n+1,n}} = \frac{f^{(*)}(n+1)}{f^{(*)}(n)}. \tag{2.31}$$

Note that here these functions do not have any physical meaning, however. They have to be considered as auxiliary mathematical quantities.

2.3.2.3 Clusters of Subcritical Sizes

In the first case considered, a certain part of the particles in the ambient phase was fixed in its positions. The kinetic conditions for aggregation for the remaining mobile particles remain thus the same as in the real state of the ambient phase. For clusters of subcritical sizes the concentration of aggregating particles in the ambient phase is too small to reach a dynamic equilibrium. For this reason, in order to construct the virtual state of the ambient phase we have to add, now, mobile particles. The total concentration of mobile particles is defined, again, by the condition $\widetilde{\mu} = \mu_d(n)$. However, to retain, again, the requirement that the kinetic conditions for aggregation have to be the same for any arbitrary particle, the newly added particles do not interact with themselves and with the particles present in the real state of the ambient phase. All further derivations can be carried out then step by step in the same way as explained in detail for the range of cluster sizes $n > n_c$. Therefore, Eqs. (2.28), (2.29), and (2.31) hold equally well for clusters of subcritical sizes.

Summarizing this part, we may conclude: we have developed a method of determination of the coefficients of emission in terms of a macroscopic approach to the description of cluster properties in nucleation–growth processes. The expressions like Eqs. (2.29) and (2.31) are widely used in nucleation theory but so far without any satisfactory theoretical foundation. Such foundation is available now via the method outlined here and resulting into Eq. (2.28). The development of a theoretical foundation of this widely employed approach is one of the basic results of the present chapter. In the further analyses, we will analyze some applications, further generalizations, and consequences of this result.

2.3.3 Applications

Once the kinetic coefficients $w^{(-)}$ have been determined, they may be substituted into the kinetic equations (2.1). For the solution of particular problems, finally, the coefficients $w^{(+)}$ have to be specified. Instead of applying the set of kinetic equations (2.1)–(2.4), we may go over also to a continuous description in the form of a Fokker–Planck equation. It provides a sufficiently accurate description of the initial stages of the condensation process. During the initial stage, for $1 \leq n \leq n_c$, the second derivative of the distribution function $\left(\partial^2 f/\partial n^2\right)$ is important, because it determines the growth of the nucleus from the size $n < n_c$ to a size $n > n_c$. The higher order derivatives are smaller than the second-order one, if $n_c \gg 1$. It means that the main part of the nuclei spectrum lies then in the range $n \gg 1$. The spectrum of droplet sizes in the range $n \sim 1$ may also be well described by this differential equation. This is due to the fact that the character of the solution provides qualitatively correct results, even if we apply the differential equation. Having introduced $f^{(*)}(n) = \exp(-\Delta G(n)/k_B T)$ we get from Eq. (2.3)

$$\begin{aligned} J_{n-1} &= w^{(+)}_{n-1,n} f^{(*)}(n-1) \left[\frac{f(n-1,t)}{f^{(*)}(n-1)} - \frac{f(n,t)}{f^{(*)}(n)}\right] \\ &= J_n - \frac{\partial}{\partial n}\left\{w^{(+)}_{n,n+1} f^{(*)}(n) \left[\frac{f(n,t)}{f^{(*)}(n)} - \frac{f(n-1,t)}{f^{(*)}(n-1)}\right]\right\} \\ &= J_n + \frac{\partial}{\partial n}\left\{w^{(+)}_{n,n+1}\left[\frac{1}{k_B T}\frac{\partial \Delta G(n)}{\partial n} f(n,t) + \frac{\partial f(n,t)}{\partial n}\right]\right\}. \end{aligned} \tag{2.32}$$

Therefore, the difference equation (2.2) acquires the form

$$\frac{\partial f(n,t)}{\partial t} = \frac{\partial}{\partial n}\left\{w^{(+)}_{n,n+1}\left[\frac{1}{k_B T}\frac{\partial \Delta G(n)}{\partial n} f(n,t) + \frac{\partial f(n,t)}{\partial n}\right]\right\}. \tag{2.33}$$

A comparison with the ordinary continuity equation, $(\partial\rho/\partial t) + \mathrm{div}\mathbf{j} = 0$, shows that the deterministic growth rate may be expressed as

$$\frac{\mathrm{d}n}{\mathrm{d}t} = -w^{(+)}_{n,n+1}\left\{\frac{1}{k_B T}\left(\frac{\partial \Delta G(n)}{\partial n}\right)\right\}. \tag{2.34}$$

This equation allows one to determine the coefficients of aggregation $w^{(+)}$ if the macroscopic growth rates $\mathrm{d}n/\mathrm{d}t$ of the clusters are known.

2.4 Generalization to Multicomponent Systems

Assume that both ambient and newly evolving phases may be composed of k different components. The distribution function with respect to cluster sizes is then a function of all numbers n_j of particles of the different components j $(j = 1, 2, \ldots, k)$ in the cluster, i.e., $f = f(n_1, n_2, \ldots, n_k, t)$ holds. Instead of the kinetic equation (2.2) we now have

$$\frac{\partial f(n_1, n_2, \ldots, t)}{\partial t} = -\sum_{j=1}^{k}[J(n_j, t) - J(n_j - 1, t)] \tag{2.35}$$

with

$$J(n_j,t) = w^{(+)}_{n_j;n_j+1} f(n_1,n_2,\ldots,n_j\ldots,n_k,t) - w^{(-)}_{n_j+1;n_j} f(n_1,n_2,\ldots,n_j+1,\ldots,n_k,t). \tag{2.36}$$

As a next step, again, the coefficients of emission have to be specified. This procedure can and will be carried out in the same way as done so far for one-component systems.

2.4.1 Traditional Approach

For thermodynamic equilibrium states in multicomponent systems, the cluster-size distribution may be approximated by the expression for heterophase fluctuations as (cf. again, Ref. [144])

$$f^{(\mathrm{eq})}(n_1,n_2,\ldots,n_k) = A\exp\left(-\frac{\Delta G(n_1,n_2,\ldots,n_k)}{k_BT}\right). \tag{2.37}$$

Here $\Delta G(n_1,n_2,\ldots,n_k)$ is the work of formation of a cluster consisting of the respective number of particles of the different components. Assuming that the preexponential factor A depends only weakly on the composition and size of the clusters, we obtain similar to the one-component case

$$\frac{w^{(+)}_{n_j;n_j+1}}{w^{(-)}_{n_j+1;n_j}} = \exp\left\{-\frac{[\Delta G(n_j+1)-\Delta G(n_j)]}{k_BT}\right\}. \tag{2.38}$$

Here all values of the numbers of particles of the different components in the cluster except the component j are fixed.

As a next step, we introduce the change of the Gibbs free energy, ΔG, of the system when, at constant values of the external pressure, p, and temperature, T, a cluster of a given composition is formed in it. We have

$$\Delta G(n_1,n_2,\ldots,n_k) = G_d(n_1,n_2,\ldots,n_k) - \sum_{j=1}^{k} n_j\mu_{jv}. \tag{2.39}$$

$G_d(n_1,n_2,\ldots,n_k)$ is the contribution of the cluster to the thermodynamic potential including interfacial and other possible correction terms. With Eq. (2.39), we get for example

$$-\big[\Delta G(n_1,n_2,\ldots,n_j+1,\ldots,n_k) - \Delta G(n_1,n_2,\ldots,n_j,\ldots,n_k)\big] = \mu_{jv} - \mu_{jd}(n_1,n_2,\ldots,n_j+1,\ldots,n_k). \tag{2.40}$$

Indeed, we may write

$$\Delta G(n_j + 1) - \Delta G(n_j) = G_d(n_j + 1) - \sum_i n_i\mu_{iv} - \mu_{jv} - \left[G_d(n_j) - \sum_i n_i\mu_{iv}\right]$$
$$= G_d(n_j + 1) - G_d(n_j) - \mu_{jv} \tag{2.41}$$
$$= \left(\frac{\partial G_d(n_j + 1)}{\partial n_j}\right) - \mu_{jv} = \mu_{jd}(n_j + 1) - \mu_{jv}.$$

The dependences (2.40) allow us to reformulate Eqs. (2.38) as

$$\frac{w^{(+)}_{n_j;n_j+1}}{w^{(-)}_{n_j+1;n_j}} = \exp\left\{\frac{[\mu_{jv} - \mu_{jd}(n_1, n_2, \ldots, n_j + 1, \ldots, n_k)]}{k_B T}\right\}. \tag{2.42}$$

Again, for systems in thermodynamic equilibrium states this approach is quite satisfactory provided the additional assumption (weak dependence of the prefactor A on cluster size) is fulfilled. However, the method cannot be applied, as discussed in detail earlier, to thermodynamic nonequilibrium states.

2.4.2 New Approach

The alternative method of determination of the relation between the kinetic coefficients in application to thermodynamically unstable initial states relies, again, on the consideration of the probabilities of formation of favorable configurations for a single particle to be incorporated into the cluster (both in the real and virtual states). The derivations as outlined above can also be applied without any qualitative modification to multicomponent systems. Consequently, as will be shown, qualitatively the same results are obtained as derived in Section 2.3 for phase formation processes in one-component systems. In particular, similar to Eq. (2.27) we arrive at

$$\frac{w^{(+)}_{n_j-1,n_j}}{w^{(-)}_{n_j,n_j-1}} = \exp\left(\frac{\Delta S_n}{k_B}\right) = \exp\left(\frac{\Delta g}{k_B T}\right). \tag{2.43}$$

Remember that Δg is the total change of the Gibbs free energy of both subsystems (real and virtual ones) if a mobile particle is transferred from the virtual to the real state. Taking into account this meaning of Δg, we have

$$\Delta g = g_{\text{final}} - g_{\text{initial}}, \tag{2.44}$$

$$\Delta g = \tilde{g}(n_j - 1) - g(n_j + 1) - [\tilde{g}(n_j) + g(n_j)] \tag{2.45}$$

or

$$\Delta g = -\frac{\partial \tilde{g}}{\partial n_j} + \frac{\partial g}{\partial n_j} = \mu_{jv} - \tilde{\mu}_j. \tag{2.46}$$

Here g or $\widetilde{g}$ are the Gibbs free energies of the ambient phase in the two considered states, real and virtual ones. n_j denotes the number of particles of the component j in the subsystems. Moreover, the well-known relations

$$\mu_j = \frac{\partial g}{\partial n_j}, \qquad \widetilde{\mu}_j = \frac{\partial \widetilde{g}}{\partial n_j} \tag{2.47}$$

are employed. With $\widetilde{\mu}_j = \mu_{jd}$ (consequence of the definition of the *virtual* states), we immediately obtain

$$\frac{w^{(+)}_{n_j;n_j+1}}{w^{(-)}_{n_j+1;n_j}} = \exp\left\{\frac{[\mu_{jv} - \mu_{jd}(n_1, n_2, \ldots, n_j+1, \ldots, n_k)]}{k_B T}\right\} \tag{2.48}$$

and as a consequence

$$\frac{w^{(+)}_{n_j;n_j+1}}{w^{(-)}_{n_j+1;n_j}} = \exp\left\{-\frac{[\Delta G(n_j+1) - \Delta G(n_j)]}{k_B T}\right\}. \tag{2.49}$$

In Eq. (2.48), μ_{jv} is the chemical potential of a particle of the jth component in the ambient vapor phase, while $\mu_{jd}(n_1, n_2, \ldots, n_j+1, \ldots, n_k)$ is its value in a cluster of the specified composition. As in the one-component case, the value of μ_{jd} accounts for surface and other possible finite size effects. In Eq. (2.49), the values of $n_i, i \neq j$, are kept constant, again.

By the introduction of an auxiliary function $f^{(*)}(n_1, n_2, \ldots, n_k)$ via

$$f^{(*)}(n_1, n_2, \ldots, n_k) = \exp\left\{-\frac{\Delta G(n_1, n_2, \ldots, n_k)}{k_B T}\right\}, \tag{2.50}$$

Eqs. (2.38) and (2.49) may be written, generally, in the form

$$\frac{w^{(+)}_{n_j;n_j+1}}{w^{(-)}_{n_j+1;n_j}} = \frac{f^{(*)}(n_j+1)}{f^{(*)}(n_j)}. \tag{2.51}$$

Here $f^{(*)}$ is an auxiliary mathematical function without any physical meaning, again. Thus, the method of derivation of the rates of emission of single particles from clusters of the newly evolving phase can be extended to multicomponent systems.

2.4.3 Applications

Once the coefficients $w^{(-)}$ are specified, they can be substituted into the kinetic equations. Moreover, one can go over, again, to a continuous description in the form of a Fokker–Planck equation. By the same method as demonstrated in Section 2.3, we obtain

$$\frac{\partial f}{\partial t} = \sum_j \frac{\partial}{\partial n_j}\left\{w^{(+)}_{n_j;n_j+1}\left[\frac{1}{k_B T}\frac{\partial \Delta G(n_1, n_2, \ldots, n_k)}{\partial n_j} f(n_1, n_2, \ldots, n_k, t) + \frac{\partial f(n_1, n_2, \ldots, n_k, t)}{\partial n_j}\right]\right\} \tag{2.52}$$

or, equivalently,

$$\frac{\partial f(n_1, n_2, \ldots, n_k, t)}{\partial t} = -\sum_j \frac{\partial J(n_j, t)}{\partial n_j}, \tag{2.53}$$

$$J(n_j, t) = -\left\{ w^{(+)}_{n_j;n_j+1} \left[\frac{1}{k_B T} \frac{\partial \Delta G(n_1, n_2, \ldots, n_k)}{\partial n_j} f(n_1, n_2, \ldots, n_k, t) + \frac{\partial f(n_1, n_2, \ldots, n_k, t)}{\partial n_j} \right] \right\}. \tag{2.54}$$

An inspection of Eqs. (2.53) and (2.54) leads to the conclusion that the macroscopic (deterministic) growth rates are given by

$$v_j = \frac{\mathrm{d}n_j}{\mathrm{d}t} = -w^{(+)}_{n_j;n_j+1} \left[\frac{1}{k_B T} \frac{\partial \Delta G(n_1, n_2, \ldots, n_k)}{\partial n_j} \right], \tag{2.55}$$

while the diffusion coefficients $D_j^{\{n\}}$ in the space of independent variables $\{n_j\}$ are determined by

$$D_j^{\{n\}} = w^{(+)}_{n_j;n_j+1}. \tag{2.56}$$

Note that the theory of Langer [145] results in similar dependences for the description of the evolution of the characteristic parameters of a system undergoing a first-order phase transformation. Differences between his and the theoretical approach employed here occur only in the way the macroscopic parameters of the system are introduced, and the critical cluster size in nucleation and the steady-state nucleation rate are determined.

2.5 Generalization to Arbitrary Boundary Conditions

Summarizing the results obtained so far we come to the following conclusions: it follows from the analysis outlined that the basic relationships between the coefficients of aggregation and emission are given by Eqs. (2.28) (for the one-component case) and (2.48) (for a multicomponent system). By applying thermodynamic identities, these relations can be transformed into Eqs. (2.29) and (2.49). In these expressions, the ratio of the kinetic coefficients is expressed via the differences of the thermodynamic potentials. Furthermore, one may introduce auxiliary functions $f^{(*)}$ (Eqs. (2.30) and (2.50)) in order to express the ratio of the kinetic coefficients as the ratio of these auxiliary functions (Eqs. (2.31) and (2.51)). Latter results are similar in the form to the respective dependences obtained by the traditional approach. However, in our method no reference is made to the so-called equilibrium distributions and the principle of detailed balancing is not applied to nonequilibrium states. As an additional advantage, the problem of determination of the coefficient A in the expressions for the distributions of heterophase fluctuations has not occurred so far at all.

In order to extend the range of applicability of the method, we have to show that similar results can also be obtained for other boundary conditions as constancy of pressure and temperature. In order to proceed in this direction we start with the basic intermediate result of our approach, i.e., with Eq. (2.24) or

$$\Delta S_n = \frac{1}{T} R_{\mathrm{rev}}. \tag{2.57}$$

Assume, now, not external pressure and temperature are fixed but another set of thermodynamic parameters. Then the work R_{rev} one has to perform in a reversible process to create the same change of the entropy ΔS_n is now given by $\Delta\phi$ (and not by Δg). Here ϕ is the appropriate thermodynamic potential for the selected (arbitrary) boundary conditions. Instead of Eq. (2.24) we then obtain

$$\Delta S_n = \frac{1}{T} \Delta\phi. \tag{2.58}$$

Now, proceeding in the same way as in the derivation of Eq. (2.24), we get (cf. Eqs. (2.44)–(2.47))

$$\Delta S_n = \frac{1}{T} \left[\mu_{jv} - \mu_{jd}\right]. \tag{2.59}$$

Here it was taken into account, again, that the relation

$$\mu_j = \frac{\partial \phi}{\partial n_j} \tag{2.60}$$

holds, provided the other appropriate variables except n_j are kept constant. As a result we obtain Eqs. (2.28) or (2.48), again.

Similar to Eq. (2.39), we may express the change of the characteristic thermodynamic function in cluster formation as

$$\Delta\Phi(n_1, n_2, \ldots, n_k) = \Phi_d(n_1, n_2, \ldots, n_k) - \sum_{j=1}^{k} n_j \mu_{jv}. \tag{2.61}$$

Proceeding in the same way as earlier, we arrive at

$$\begin{aligned} &-\left[\Delta\Phi(n_1, n_2, \ldots, n_j + 1, \ldots, n_k) - \Delta\Phi(n_1, n_2, \ldots, n_j, \ldots, n_k)\right] \\ &\qquad = \mu_{jv} - \mu_{jd}(n_1, n_2, \ldots, n_j + 1, \ldots, n_k) \end{aligned} \tag{2.62}$$

with similar consequences.

Finally, in the derivation of the basic equations for the determination of the emission coefficients, it was not utilized that the temperature in the cluster has to be the same as in the surrounding ambient phase. Therefore, the method is equally well applicable to phase formation under nonisothermal conditions. In this case, the values of the chemical potential have to be taken at the respective temperatures of the clusters and the ambient phase. In addition, the basic kinetic equations have to be supplemented by relations describing the heat flow between clusters and ambient phase (cf., e.g., Ref. [306]).

In this way, a regular method of formulation of the kinetic equations for the description of nucleation–growth processes is developed. The method does not depend on the boundary conditions applied, it can be employed both for isothermal and nonisothermal nucleation. Moreover, since the derivation of the relation between the kinetic coefficients does not rely on any specific features of vapor condensation but only on very general thermodynamic arguments, it is equally well applicable generally to the description of any first-order phase formation processes proceeding via nucleation and growth.

2.6 Initial Conditions for the Cluster-Size Distribution Function

The method of determination of the kinetic coefficients, developed here, can be employed without any reference to the so-called constraint equilibrium distributions. Such distributions may enter, nevertheless, the description but in a reduced much less significant way via the determination of the initial conditions for the solutions of the kinetic equations describing nucleation and growth. Indeed, it can be assumed in a variety of applications that the initial cluster-size distribution after the quench into the unstable state corresponds to some extent to the spectrum of heterophase fluctuations existing in the initial equilibrium system before the quench took place. Alternatively, one may suppose that for small cluster sizes the distribution is more or less well expressed by the respective expressions for heterophase fluctuations even in thermodynamically unstable states.

For the determination of the initial conditions, following such an argumentation, only the value of the preexponential factor A in the expressions for heterophase fluctuations [144]

$$f(n_1, n_2, \ldots, n_k) = A \exp\left(-\frac{\Delta G(n_1, n_2, \ldots, n_k)}{k_B T}\right) \tag{2.63}$$

has to be known or, in other words, the limit of the respective distributions for very small cluster sizes $n_j \to 0$. Once the initial cluster-size distribution is determined the further evolution is governed by the kinetic equations themselves. It is also only the value of A, i.e., the limit of the distribution for small sizes of the clusters of the newly evolving phase, which has to be known in order to derive expressions for the steady-state nucleation rate and the steady-state cluster-size distribution (cf., e.g., Ref. [298]). From such a point of view, limiting consistency has to be considered as a fundamental property, i.e., the cluster-size distributions at small cluster sizes have to be expressed accurately, while the shape of the expressions, like those given by Eq. (2.63), for large cluster sizes is of no relevance for nucleation (cf. also [355]).

Moreover, (i) since relations of the type as given by Eq. (2.63) do not reflect real equilibrium distributions which may evolve in thermodynamically unstable states, (ii) they are applied exclusively in order to determine possible initial states for real cluster-size distributions in the considered nonequilibrium states; the fulfillment of the mass action law is no longer considered, to our opinion, as a necessary fundamental condition for the validity of the respective distribution (see, in contrast, Refs. [352, 353]). The coefficients A may or may not obey such property, at part, in dependence on whether these distributions are determined mainly by the initial equilibrium state before the quench took place or by the way the system is transferred into the considered nonequilibrium state. From such a point of view, the fulfillment of the mass action law is not an appropriate starting point for a possible redetermination of the value of the parameter A and the formulation of different specific versions of nucleation theory.

In order to have a guide for the determination of possible initial conditions for the cluster-size distributions, let us summarize, finally, some attempts in the determination of the parameter A for different special cases (cf. also [355]).

(i) For one-component systems, following Frenkel [71], the parameter A may be determined from the limiting condition as (cf. Eq. (2.5))

$$A = c. \tag{2.64}$$

Here c is the volume concentration of single particles in the ambient phase.

(ii) For clusters of arbitrary composition, following Reiss [206], the relation

$$A = c = \sum_{j=1}^{k} c_j \tag{2.65}$$

could be taken as a first approximation. Here c_j denotes the volume concentration of the different components in the ambient phase, which are able to enter the new phase. Equation (2.65) can be founded, similar to Eq. (2.64), via the basic ideas of fluctuation theory as representing the number of particles able to act as centers of condensation. The probability that, indeed, a cluster of some definite but arbitrary composition is formed at a given nucleation site is determined then by the exponential term.

(iii) An extended discussion of different attempts for a proper determination of this coefficient for one-component and binary systems was given by Wilemski and Wyslouzil [352, 353]. They proposed (in application to binary systems) the relation

$$A = c \prod_{j=1}^{k} (x_{j\beta})^{x_{j\alpha}} \tag{2.66}$$

as a better approximation. This expression fulfills the limiting conditions for one-component clusters and, in addition, the mass action law. In Eq. (2.66), $x_{j\beta}$ is the molar fraction of the different components in the ambient phase, while $x_{j\alpha}$ is the molar fraction of the different components in the cluster considered. It is determined as

$$x_{j\alpha} = \frac{n_j}{\sum_{i=1}^{k} n_i}. \tag{2.67}$$

(iv) For clusters of a given stoichiometric composition (cf. also Refs. [293, 304]) the total number of nucleation sites is equal to the number of particles in the system able to enter the new phase. We have for that number

$$c = \sum_{j=1}^{k} c_j . \tag{2.68}$$

A heterophase fluctuation with a given composition can be formed in the ambient phase if a favorable configuration of particles of the different components is developed. Considering the motion of the different particles as independent, the probability of such an event is equal to the product of the molar fractions $x_{i\beta}$ of the different components i in the ambient phase, each of them taken to the power $x_{i\alpha}$. As a result we obtain the following expression for A (cf. Refs. [293, 304]):

$$A = c \prod_{j=1}^{k} x_{j\beta}^{x_{j\alpha}} . \tag{2.69}$$

Most of the above-mentioned considerations concerning the initial state of the cluster-size distribution function can be carried out equally well without any reference to expressions like Eq. (2.63). Therefore, even for the determination of the initial conditions the reference to distributions like that given by Eq. (2.63) may be avoided at all.

2.7 Description of Cluster Ensemble Evolution along a Given Trajectory

2.7.1 Motivation

The set of kinetic equations as outlined above allows us to determine the evolution of the cluster-size distribution function for phase formation in multicomponent systems in a complete way. However, with an increase of the number of components in the system, the computation times increase dramatically. For this reason, a comprehensive description of the whole course of nucleation–growth processes based on the numerical solution of the sets of kinetic equations is possible presently for one-component and binary systems, only (cf., e.g., [251]).

In a number of cases, the problems can be reduced significantly. Indeed, following the classical approach to nucleation–growth processes, one can distinguish the case that the clusters of the newly evolving phase have widely the same composition and structure as the newly evolving macroscopic phase. In these cases, which are illustrated in Figure 2.1, the task to be solved is reduced to a one-dimensional problem. The clusters consist here of units with a given well-defined composition, $\{x_{i\alpha}\}$. This problem has been analyzed in detail for the first time in [293, 304]. However, in general, the composition and state of the clusters will change with cluster size and variations of the state of the ambient phase. For this reason, the classical assumption of independence of the state of the clusters on their sizes can be considered, in general, as a crude approximation, only.

Quite recently a new general approach to the description of nucleation–growth processes has been developed [253–255, 260] (cf. also [257, 258]). This approach allows a theoretically founded determination of the path of evolution of the clusters in the space of cluster parameters or, in other words, a description of the changes of the state parameters of the clusters with their sizes. The respective situation is illustrated in Figure 2.5. The details of this approach are outlined in the papers cited. Here it is only of importance that in this very general case a reduction of the description from the determination of the distribution function $f(n_1, n_2, \ldots, n_k, t)$ to a description in terms of a distribution function $f(n, t)$ is possible as well. The respective transformations are described below.

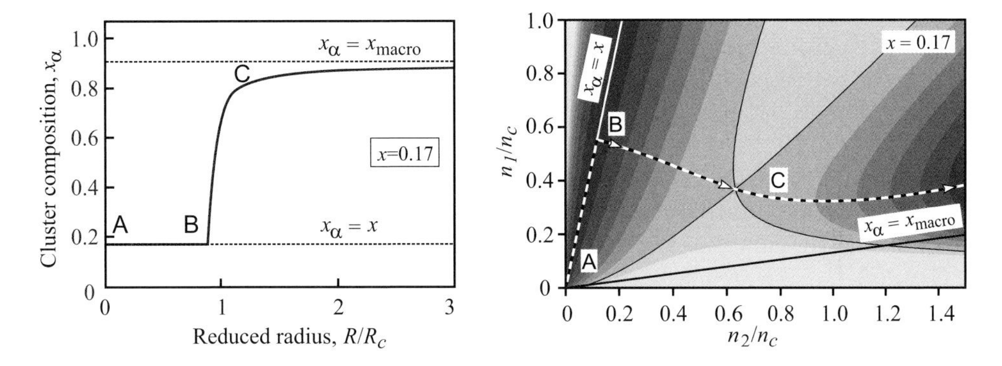

Figure 2.5: Illustration of the evolution of the cluster in the space of cluster parameters for segregation processes in regular solutions [253–255].

2.7.2 Effective Diffusion Coefficients

The variation of the number of particles of the ith component in the cluster is determined via (cf. Eq. (2.55))

$$\frac{dn_i}{dt} = -4\pi R^2 j_i = -w^{(+)}_{n_i;n_i+1} \left[\frac{1}{k_B T} \frac{\partial \Delta G(n_1, n_2, \ldots, n_k)}{\partial n_i} \right] . \tag{2.70}$$

Moreover, the relations

$$n_i = n n_b x_{i\alpha}, \qquad V_\alpha = n \omega_\alpha, \qquad \omega_\alpha = \sum_{i=1}^{k} n_b x_{i\alpha} \omega_{i\alpha} \tag{2.71}$$

hold. Here $\omega_{i\alpha}$ is the volume per particle of the ith component in the cluster, n is the total number of structural units of the newly evolving phase in the cluster of volume V_α, or radius R. n_b denotes the number of particles in a structural unit of the newly evolving macrophase in equilibrium with the ambient phase, and ω_α is the volume of a structural unit of the newly evolving phase in a cluster. For the case where the clusters can be considered as multiples of a basic unit of the newly evolving phase, n_b can be identified with the number of particles in such a unit. If the clusters change their composition with cluster sizes, n_b loses its definite physical meaning and will be determined such as to allow an expression of the basic kinetic equations in the most simple form. In any of the considered cases, it is a constant parameter.

Further, based on Eq. (2.71) (we take the derivative of the identity $n_i = nn_b x_{i\alpha}$ with respect to time) and taking into account that the composition of the clusters depends uniquely on cluster radius, R, or particle number in the cluster, n, we may write

$$\frac{\mathrm{d}n}{\mathrm{d}t} = -4\pi R^2 \left(\frac{j_i}{\nu_{i\alpha}} \right), \qquad \nu_{i\alpha} = n_b \left(x_{i\alpha} + n \frac{\mathrm{d}x_{i\alpha}}{\mathrm{d}n} \right). \tag{2.72}$$

Since the left-hand side of the first of Eqs. (2.72) does not depend on the particular component considered, the right-hand side must have the same value for each of the components. This way, we get

$$\frac{j_i}{\nu_{i\alpha}} = \frac{j_k}{\nu_{k\alpha}} \qquad \text{for} \qquad i = 1, 2, \ldots, k-1. \tag{2.73}$$

As is evident from the derivation, we assume here that the different components may move, in principle, independently. Nevertheless, the motion is coupled by the requirement that the composition of the clusters is a well-defined function of its size.

Equation (2.72) allows us to express the change of the number of ambient phase units, n, via the fluxes of any of the components in the system. Now, as the next step, the terms $(\partial \Delta G / \partial n_i)$ in Eq. (2.70) will be replaced by the change of the characteristic thermodynamic potential $G(n+1) - G(n)$, if the cluster size is increased from n to $(n+1)$. Instead of $G(n)$ and $G(n+1)$, we will use here the differences $\Delta G(n) = G(n) - G_{\mathrm{hom}}$ between the respective heterogeneous and homogeneous initial states. This procedure will be performed here for the case where pressure and temperature are kept constant. As can be checked easily, the derivation can be repeated in a similar form for any other boundary conditions with identical results.

The change of the thermodynamic potential in such a process can be written as

$$G(n+1) - G(n) = \frac{\mathrm{d}\Delta G(n)}{\mathrm{d}n} \Delta n = \sum_{i=1}^{k} \left(\frac{\partial \Delta G(n_1, n_2, \ldots, n_k)}{\partial n_i} \right) \Delta n_i. \tag{2.74}$$

Hereby the values of Δn_i cannot be chosen arbitrarily; they are uniquely determined via Δn as

$$\Delta n_i = \nu_{i\alpha} \Delta n. \tag{2.75}$$

With the condition $\Delta n = 1$ and Eq. (2.75), we arrive at

$$\frac{\mathrm{d}\Delta G(n)}{\mathrm{d}n} = \sum_{i=1}^{k} \left(\frac{\partial \Delta G(n_1, n_2, \ldots, n_k)}{\partial n_i} \right) \nu_{i\alpha}. \tag{2.76}$$

Finally, Eqs. (2.70) yields

$$\frac{4\pi R^2}{w^{(+)}_{n_i;n_i+1}} j_i = \frac{1}{k_B T} \frac{\partial \Delta G(n_1, n_2, \ldots, n_k)}{\partial n_i}. \tag{2.77}$$

Multiplying both sides of Eq. (2.77) with $\nu_{i\alpha}$ and taking the sum over all components, we get with Eqs. (2.72), (2.73), and (2.76)

$$4\pi R^2 \left(\frac{j_i}{\nu_{i\alpha}}\right) \sum_{i=1}^{k} \frac{\nu_{i\alpha}^2}{w_{n_i;n_i+1}^{(+)}} = \frac{1}{k_B T} \frac{\mathrm{d}\Delta G(n)}{\mathrm{d}n} \tag{2.78}$$

and

$$\frac{\mathrm{d}n}{\mathrm{d}t} = -\frac{1}{\sum_{i=1}^{k} \left(\frac{\nu_{i\alpha}^2}{w_{n_i;n_i+1}^{(+)}}\right)} \frac{1}{k_B T} \frac{\mathrm{d}\Delta G(n)}{\mathrm{d}n}. \tag{2.79}$$

A comparison with Eq. (2.34) shows that the growth rates for the clusters can be written, again, in a one-dimensional form as

$$\frac{\mathrm{d}n}{\mathrm{d}t} = -w_{n,n+1}^{(+)} \left\{ \frac{1}{k_B T} \left(\frac{\mathrm{d}\Delta G\,(n)}{\mathrm{d}n} \right) \right\} \tag{2.80}$$

with

$$w_{n,n+1}^{(+)} = \frac{1}{\sum_{i=1}^{k} \left(\frac{\nu_{i\alpha}^2}{w_{n_i;n_i+1}^{(+)}}\right)}. \tag{2.81}$$

Similarly, we can also express the change of the volume of a cluster of the new phase consisting of n ambient phase units. The change of the volume of a cluster of the new phase can generally be written as

$$\frac{\mathrm{d}V_\alpha}{\mathrm{d}t} = -4\pi R^2 \sum_{i=1}^{k} \omega_{i\alpha} j_i. \tag{2.82}$$

With Eq. (2.70), we get

$$\frac{\mathrm{d}V_\alpha}{\mathrm{d}t} = \sum_{i=1}^{k} \omega_{i\alpha} \frac{\mathrm{d}n_i}{\mathrm{d}t}. \tag{2.83}$$

Equation (2.71) yields further

$$\frac{\mathrm{d}n_i}{\mathrm{d}t} = \nu_{i\alpha} \frac{\mathrm{d}n}{\mathrm{d}t} \tag{2.84}$$

resulting in

$$\frac{\mathrm{d}V_\alpha}{\mathrm{d}t} = \frac{\mathrm{d}n}{\mathrm{d}t} \left\{ \sum_{i=1}^{k} \omega_{i\alpha} \nu_{i\alpha} \right\}. \tag{2.85}$$

Following Refs. [293, 304], one can now obtain expressions for the rates of growth of the aggregates of the new phase or the kinetic coefficients $w^{(+)}_{n_i;n_i+1}$ for the different mechanisms of cluster growth of interest. Taking into account that D^*_i/a^2_β is the frequency with which a particle of the ith component hits the interface of a cluster of radius R and $4\pi R^2 a_\beta c^{(s)}_{i\beta}$ is the number of particles of the considered component capable to reach the interface in one step of motion, we obtain

$$w^{(+)}_{n_i;n_i+1} = \frac{D^*_i}{a_\beta}\left(4\pi R^2 c^{(s)}_{i\beta}\right) . \tag{2.86}$$

Here D^*_i are the partial diffusion coefficients of the different components in the ambient solution near the interface while a_β is a characteristic length scale of the ambient phase defined via

$$\omega_\beta = \sum_{i=1}^{k} \omega_{i\beta} x_{i\beta} , \qquad a_\beta = \left(\frac{3\omega_\beta}{4\pi}\right)^{1/3} . \tag{2.87}$$

The parameters $\omega_{i\beta}$ describe the characteristic volumes of different components in the ambient phase. They are connected with the respective size parameters, $a_{i\beta}$, via

$$\omega_{i\beta} = \frac{4\pi}{3} a^3_{i\beta} . \tag{2.88}$$

Assuming steady-state conditions, the volume concentration of particles of the ith component near the interface, $c^{(s)}_{i\beta}$, can be determined by the balance of diffusional fluxes and fluxes from the ambient phase to the cluster. We get after some algebra

$$c^{(s)}_{i\beta} = c_{i\beta}\left\{\frac{1}{1+\left[\left(\frac{D^*_i}{D_i}\right)\left(\frac{R}{a_\beta}\right)\right]}\right\} . \tag{2.89}$$

Here $c_{i\beta}$ is the average volume concentration of the respective component in the ambient phase. D_i are the partial diffusion coefficients of the respective components in the bulk. The diffusion coefficients D^*_i and D_i are connected by the relations $D^*_i = D_i \alpha_i$, where α_i obeys the inequality $\alpha_i \le 1$. With Eq. (2.71) and $a_\alpha = (3\omega_\alpha/4\pi)^{1/3}$ or $R = a_\alpha n^{1/3}$, respectively, we can express the coefficients $w^{(+)}_{n_i;n_i+1}$ in the form (see also Chapter 5 and [247])

$$w^{(+)}_{n_i;n_i+1} = 4\pi D^*_i c_{i\beta} a_\alpha n^{1/3}\left\{\frac{\left(\frac{a_\alpha}{a_\beta}\right) n^{1/3}}{1+\left[\left(\frac{D^*_i}{D_i}\right)\left(\frac{a_\alpha}{a_\beta}\right) n^{1/3}\right]}\right\} . \tag{2.90}$$

Finally, the volume concentration of the ith component in the ambient phase, $c_{i\beta}$, can be expressed as

$$c_{i\beta} = \frac{n_{i\beta}}{V} = c_\beta x_{i\beta} , \qquad c_\beta = \frac{n_\beta}{V} , \qquad x_{i\beta} = \frac{n_{i\beta}}{n_\beta} , \qquad n_\beta = \sum_{j=1}^{k} n_{j\beta} . \tag{2.91}$$

In the general form, the expression for $w^{(+)}_{n,n+1}$ is then given via

$$w^{(+)}_{n,n+1} = \frac{4\pi c_\beta a_\alpha^2 n^{2/3}}{a_\beta} \left\{ \sum_{i=1}^{k} \frac{\nu_{i\alpha}^2 \left[1 + \left(\frac{D_i^*}{D_i} \right) \left(\frac{a_\alpha}{a_\beta} \right) n^{1/3} \right]}{D_i^* x_{i\beta}} \right\}^{-1} . \tag{2.92}$$

Equation (2.92) represents the most general expression for the determination of the quantity $w^{(+)}_{n,n+1}$. It is a rather nontrivial function of the kinetic and thermodynamic parameters of the different components in the multicomponent solution considered.

As already mentioned, in the case of formation of a new phase with a given stoichiometric composition, the total number of particles, n_b, in a basic unit of the new phase is well defined. For the more general case considered here that the composition of the cluster changes with cluster size, such well-defined units do not exist. For this reason, we will set n_b equal to one (cf. Eq. (2.71)). With such definition, n gets the meaning of the total number of particles in a cluster. We then have

$$w^{(+)}_{n,n+1} = \frac{4\pi c_\beta a_\alpha^2 n^{2/3}}{a_\beta \sum_{i=1}^{k} \left\{ \frac{\widetilde{\nu}_{i\alpha}^2 \left[1 + \left(\frac{D_i^*}{D_i} \right) \left(\frac{a_\alpha}{a_\beta} \right) n^{1/3} \right]}{D_i^* x_{i\beta}} \right\}} \tag{2.93}$$

with (cf. Eq. (2.72))

$$\widetilde{\nu}_{i\alpha} = \left(x_{i\alpha} + n \frac{dx_{i\alpha}}{dn} \right) \tag{2.94}$$

or, equivalently,

$$w^{(+)}_{n,n+1} = \frac{4\pi D_{\text{eff}} c_\beta a_\alpha^2 n^{2/3}}{a_\beta} , \tag{2.95}$$

$$\frac{1}{D_{\text{eff}}} = \sum_{i=1}^{k} \left\{ \frac{\widetilde{\nu}_{i\alpha}^2 \left[1 + \left(\frac{D_i^*}{D_i} \right) \left(\frac{a_\alpha}{a_\beta} \right) n^{1/3} \right]}{D_i^* x_{i\beta}} \right\} . \tag{2.96}$$

For the case of kinetically limited growth (if the conditions $1 \gg (D_i^*/D_i) n^{1/3}$ are fulfilled for any of the components), we have

$$w^{(+)}_{n,n+1} = \frac{4\pi D_{\text{eff}}^k c_\beta a_\alpha^2 n^{2/3}}{a_\beta} , \qquad \frac{1}{D_{\text{eff}}^k} = \sum_{i=1}^{k} \frac{\widetilde{\nu}_{i\alpha}^2}{x_{i\beta} \alpha_i D_i} . \tag{2.97}$$

For diffusion-limited growth of the clusters (if the conditions $1 \ll (D_i^*/D_i) n^{1/3}$ are fulfilled for any of the components), we similarly get

$$w^{(+)}_{n,n+1} = 4\pi c_\beta a_\alpha D_{\text{eff}}^d n^{1/3} , \qquad \frac{1}{D_{\text{eff}}^d} = \sum_{i=1}^{k} \frac{\widetilde{\nu}_{i\alpha}^2}{x_{i\beta} D_i} . \tag{2.98}$$

With $\omega_\beta = V/n_\beta = 1/c_\beta$, we can always make the replacement $4\pi c_\beta = 3/a_\beta^3$ in the above equations.

2.7.3 Evolution of the Cluster-Size Distribution Functions

According to theoretical developments discussed, we may write down the following expressions for the determination of the evolution of the cluster-size distribution with time

$$\frac{\partial f(n,t)}{\partial t} = w_{n-1,n}^{(+)} \left\{ f(n-1,t) - f(n,t) \exp\left[\frac{\Delta G(n) - \Delta G(n-1)}{k_B T}\right]\right\} \tag{2.99}$$
$$+ w_{n,n+1}^{(+)} \left\{ -f(n,t) + f(n+1,t) \exp\left[\frac{\Delta G(n+1) - \Delta G(n)}{k_B T}\right]\right\}.$$

For the distribution function $f(n=1,t)$ at $n=1$, we employ the relation

$$f(n=1,t) = c_\beta \prod_{j=1}^{k} x_{j\beta}^{x_{j\alpha}}, \qquad c_\beta = \frac{3}{4\pi a_\beta^3}. \tag{2.100}$$

Here c_β is the total volume concentration of the particles of the different components in the ambient phase, $x_{j\beta}$ the molar fraction of the different components in the ambient phase and, $x_{j\alpha}$ the composition of the cluster phase in the limit $R \to 0$ (or $n \to 1$).

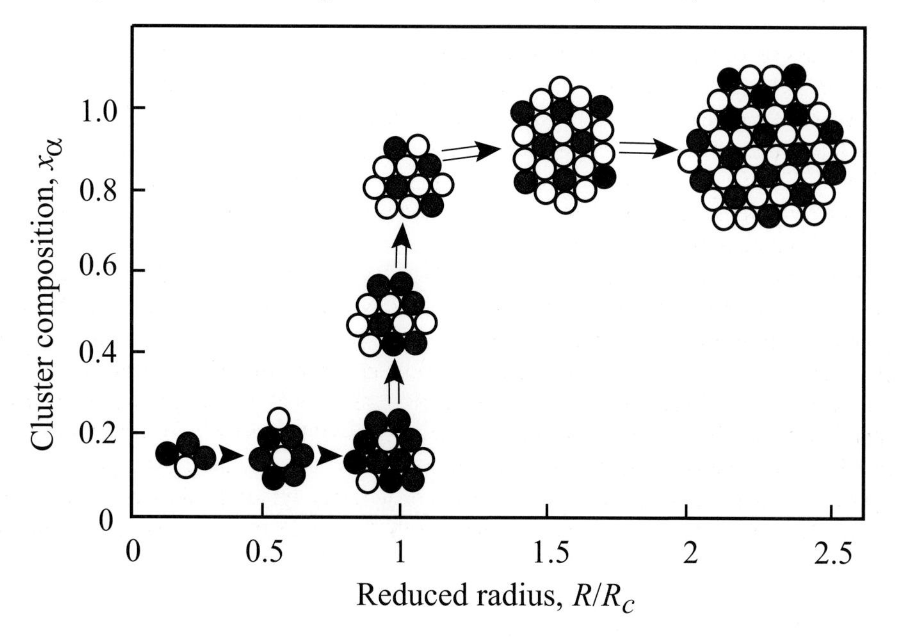

Figure 2.6: Illustration of the cluster evolution if the newly developed approach to nucleation–growth processes is utilized as outlined in detail in Refs. [253–255] (cf. also [257,258]).

The effective values of the coefficients of aggregation, $w^{(+)}_{n,n+1}$, are given by Eqs. (2.95) and (2.96). Assuming, in addition, $D_i = D_i^*$, we get

$$w^{(+)}_{n,n+1} = \frac{4\pi D^{d}_{\mathrm{eff}} c_\beta a_\alpha^2}{a_\beta} \left\{ \frac{n^{2/3}}{1 + \left(\frac{a_\alpha}{a_\beta}\right) n^{1/3}} \right\} . \tag{2.101}$$

This expression is reduced to previously discussed cases under well-defined limiting conditions.

According to the analysis of segregation processes in regular solutions illustrated partly in Figure 2.5 (cf. [253–255, 260]), initially the preferred cluster composition is equal to the composition of the ambient phase. In such cases, however, the cluster cannot be distinguished from the ambient phase. This way, the real cluster evolution starts only when the parameters start to deviate from the respective values they have in the ambient phase. As evident from the figure, such processes start at cluster sizes near the critical one. In the further evolution the composition of the cluster changes then dramatically until values close to the respective macroscopic parameters are reached. This situation is illustrated in Figure 2.6. It deviates dramatically from the classical picture of nucleation and cluster growth shown in Figure 2.1.

2.8 Conclusions

In the discussion of our results we would like to stress two points. First of all, we would like to underline once more that a regular method has been developed allowing one to determine the emission coefficients once the coefficients of aggregation are known. The basic equation (2.48) can be widely applied independent of the application and the boundary conditions considered. The method allows one to eliminate such artificial constructs as constraint equilibrium distributions from the theory as well as the incorrect application of the principle of detailed balancing to nonequilibrium states.

Second, as an intermediate step in our analysis, we introduce an idealized model, as we called it, *virtual* states. According to the definition, *virtual* states are such states for which a dynamic equilibrium between cluster and ambient phase is established. This is the only property employed in addition to the widely accepted assumption of the existence of a local equilibrium in the ambient phase in the vicinity of the clusters of the new phase. In this respect our approach is similar but more general as compared with the application of Kelvin's (or the Gibbs–Thomson) equation. In applying Kelvin's equation the same question is posed as in our analysis, i.e., what should the state of the system be to attain dynamic equilibrium for a cluster of a given size. Starting with our basic equation (2.28) (e.g., in application to one-component systems)

$$\frac{w^{(+)}_{n,n+1}}{w^{(-)}_{n+1,n}} = \exp\left(\frac{\mu_v - \mu_d(n+1)}{k_B T}\right) , \tag{2.102}$$

applying the perfect gas law

$$\mu_v(p,T) = \mu^{(*)} + k_B T \ln\left(\frac{p}{p^{(*)}}\right) , \tag{2.103}$$

and the equilibrium conditions for a drop of given size in the gas

$$\mu_d(n+1) = \mu_v[p_v(n+1),T] = \mu^{(*)} + k_B T \ln\left[\frac{p_v(n+1)}{p^{(*)}}\right] , \tag{2.104}$$

we arrive at

$$w^{(-)}_{n+1,n} = w^{(+)}_{n,n+1}\left[\frac{p_v(n+1)}{p}\right] . \tag{2.105}$$

Here $p^{(*)}$ is the pressure in some reference state and $p_v(n+1)$ is the pressure in the vapor if a drop of size $(n+1)$ exists in dynamic equilibrium in the vapor phase. Similarly condensation in multicomponent perfect gases or segregation in multicomponent perfect solutions may be discussed. Here approaches based on Kelvin's equation and our method lead to equivalent results. Latter one is, however, more straightforward and applicable directly also to phase formation in nonideal vapors, nonideal solutions and beyond.

Finally, having developed a theoretical foundation of the expressions for the coefficients of emission in terms of a mesoscopic approach to nucleation–growth processes by general thermodynamic methods, we have not even touched the problem of determination of the appropriate expressions for the chemical potential of a particle in a cluster of given size or, in other words, the work of formation of a cluster of given size. Such a further step in the analysis is required in order to apply the theory to particular problems. Here a detailed thermodynamic analysis for each specific kind of phase transformation is further necessary. In the present book we cannot and will not make the analysis more explicit but refer to the overviews given in Refs. [3, 253–255, 257, 258]. In the subsequent chapters, the assumptions concerning the thermodynamic properties of the systems analyzed will be specified in each considered case in detail. Here we will start with the simplest classical approach assuming that the clusters of the newly evolving phases have properties widely similar to the properties of the newly evolving ambient phase.

3 Kinetics of Nucleation–Growth Processes: The First Stages

3.1 Introduction

The modern theory of nucleation was worked out in its basic premises first by Stranski, Kaischew (1934) [113], Becker and Döring (1935) [24], Volmer (1939) [337], Frenkel (1946) [71], Zeldovich (1942) [357] and others. With modifications it is till now the most widely applied tool in the interpretation of experimental results on nucleation–growth processes (cf., e.g., [245]).

The mentioned theoretical approach was developed originally in order to determine the steady-state nucleation rate, J, i.e., to estimate the number of supercritical clusters formed per unit time interval in a unit volume of a thermodynamically metastable system. In the further development of the theory, nonsteady-state effects have been intensively studied which are due to the finite time required for the system to reach steady-state conditions (for an overview see, for example, Zettlemoyer [358], Binder and Stauffer [28], Binder et al. [29,32], Gunton et al. [93], Trinkaus and Yoo [329], Shi et al. [265], Shneidman and Weinberg [267,268], Demo and Kozisek [60], Gutzow et al. [94,95]).

There exists, however, also another reason why the determination of the steady-state nucleation rate cannot give a comprehensive information concerning the course of nucleation processes. In most practical applications, a steady state can be established in a system only for a limited period of time. This effect is due to a depletion of the state of the system, i.e., the decrease of the number of particles which can be incorporated into the new phase (see also Tunitskij [330], Wakeshima [344], Binder and Stauffer [28], Rusanov [217], Gunton et al. [93], Schmelzer [220], Schmelzer and Ulbricht [230], Ulbricht et al. [331], Barrett and Clement [20], Grinin [91], Kuni et al. [139, 141]). As has been shown (cf., e.g., Schmelzer et al. [220, 230, 331]), depletion effects affect nucleation quantitatively and determine qualitatively the whole course of first-order phase transformations proceeded by nucleation and growth. As a result of such depletion effects, in particular, only a finite number of clusters develop in the system (cf. Chapter 5 and Schmelzer et al. [164, 247], Slezov et al. [298]). In fact, the situation is similar in phase formation processes proceeded by spinodal decomposition. Here also changes of the state of the system in the course of the transformation determine qualitatively the whole course of this process (cf. [238, 245, 250]).

The determination of the number of clusters formed in a thermodynamically metastable system in dependence on the initial supersaturation and the average size of the clusters observed initially in nucleation experiments is one of the tasks which are beyond the scope of classical nucleation theory and most of its modifications and generalizations. The knowledge

Kinetics of First-order Phase Transitions. Vitaly V. Slezov

ISBN: 978-3-527-40775-0

of these characteristics is, however, of a great technological importance allowing one to vary the dispersity of the newly evolving phase in the ambient phase and in this way the properties of the respective materials. In addition to classical applications like in materials science (see, e.g., Johnson et al. [110], Koiwa et al. [128]) or condensation of gases (Kulmala and Wagner [135], Hale and Kulmala [98], O'Dowd and Wagner [193]), the knowledge of the dependence of the number of clusters formed in nucleation–growth experiments on the initial supersaturation may shed some light also on such problems as the understanding of fragment or cluster-size distributions in molecular or nuclear clustering processes (see, e.g., Müller et al. [186], Schmelzer et al. [240], Smirnov [316]).

The aim of the present chapter consists in the analytical description of the initial stages of first-order phase transitions proceeded by nucleation–growth processes. Analytical expressions for the time evolution of the cluster-size distribution function and the flux in cluster-size space are derived. In addition, estimates of basic characteristics of this process like the number of clusters, N, formed in nucleation–growth experiments and their average size, $\langle R \rangle$, in dependence on the initial supersaturation in the system, the duration of the different stages of the nucleation–growth process are developed. In this way, a quantitative description of all of the basic characteristics of the initial stages of first-order phase transformations is obtained.

The method employed here is a generalization of previous attempts (Slezov et al. [298]). In the present approach, different approximations in the solution of the Frenkel–Zeldovich equation have been employed which are appropriate for the respective ranges of cluster sizes. This method allows us to determine the basic characteristics of the initial stages of the considered process in a relatively simple, but at the same time sufficiently accurate, way. Moreover, the method used here is not restricted in its applicability to phase formation in perfect gases or mixtures. It can be applied quite generally independent of any particular properties of the systems under consideration. Such specific properties enter the description only in the evaluation of the final analytical expressions. This property allows a straightforward application of the results to the interpretation of experimental findings. In addition, the accuracy in the approximations involved could be improved. Moreover, the analysis is carried out simultaneously both for diffusion and kinetically limited growth. Simultaneously, also the generalization to phase formation in multicomponent systems is given for the cases, when the newly evolving phase has a given stoichiometric composition (cf. also [243, 293, 304, 309]). This way, a major part of possible applications to the interpretation of experimental results is covered. As mentioned, the method of analysis has the advantage that relatively simple expressions are obtained for the different characteristics of the first stages of the nucleation–growth process. Nevertheless, extended mathematical derivations are required to get the desired results. For this reason, the basic results are summarized at the end of the chapter in a compact form.

The present chapter is organized as follows: in Section 3.2, the basic equations required for the subsequent derivations are summarized. Section 3.3 deals with nonsteady-state effects in the initial stage of the nucleation–growth process, the determination of the steady-state nucleation rate, the steady-state cluster-size distribution and the way it is established, as well as with the specification of characteristic time scales of the first stages of the transformation. Based on these results, in Section 3.4 expressions for the time dependence of the flux in cluster-size space and cluster-size distribution in the range of cluster sizes larger than the critical one are established. In Section 3.5, the duration of the time interval for steady-state nucleation is determined. The knowledge of this time interval finally gives the possibility of determining the number of clusters formed in the system, the time interval of dominating independent growth, and the average size of the clusters (Section 3.6) after the transformation has reached a certain degree of completion (cf. Binder and Stauffer [28]), i.e., at the end of the stage of dominating independent growth of the clusters. In Section 3.7, a comparison of our results with the Kolmogorov–Avrami-type approaches [13,49,129,191] to the description of the kinetics of phase transformation processes is given. In particular, the relation between the time of steady-state nucleation and the so-called induction periods of the transformation is discussed. In Section 3.8, the results are extended to phase formation in multicomponent systems with a given stoichiometric composition. In Sections 3.9 and 3.10, a summary and discussion of the results is given and some further possible developments of the theoretical approach are anticipated.

3.2 Basic Kinetic Equations

In a continuum approximation, the evolution of an ensemble of clusters formed by nucleation and growth processes can be described by a Fokker–Planck equation of the form (see, e.g., Chapter 2 and [293,298])

$$\frac{\partial f(n,t)}{\partial t} = \frac{\partial}{\partial n}\left\{ w^{(+)}_{n,n+1}\left[\frac{\partial f(n,t)}{\partial n} + \frac{f(n,t)}{k_B T}\frac{\partial \Delta\Phi(n)}{\partial n}\right]\right\}. \tag{3.1}$$

This equation in commonly denoted as Frenkel–Zeldovich equation. Note that, as was shown in Chapter 2, this equation can be obtained without applying the concept of constraint equilibrium distributions and the principle of detailed balancing in the determination of the coefficients of emission from the clusters (see also [293,302,303,306]). The equation is shown thus to be fundamentally correct both for equilibrium and nonequilibrium states of the system independent of the size of the clusters considered.

In Eq. (3.1), $f(n,t)$ is the cluster-size distribution function obeying the condition

$$\int_0^\infty f(n,t)\,\mathrm{d}n = N^{(\mathrm{tot})}(t), \tag{3.2}$$

where $N^{(\mathrm{tot})}(t)$ is the total number of clusters per unit volume in the system with sizes $n > 1$, k_B is the Boltzmann constant and T is the absolute temperature. n is the number of building units in a cluster of the newly evolving phase, and t denotes the time.

In Eq. (3.1), $w^{(+)}_{n,n+1}$ is the probability that per unit time interval a primary building unit is added to a cluster consisting initially of n such units. The clusters are assumed to be of spherical shape with a radius R. $\Delta\Phi$ in Eq. (3.1) is the change of the (relevant for the given boundary conditions) thermodynamic potential resulting from the formation of a cluster consisting of n building units.

The cluster-size distribution function $f(n,t)$ obeys the boundary condition

$$f(n \to 0, t) = c(t). \tag{3.3}$$

The actual concentration $c(t)$ of single particles is given hereby by the mass-balance equation

$$c_0 = c(t) + \int_0^\infty n f(n,t)\,\mathrm{d}n. \tag{3.4}$$

Here c_0 is the concentration of single particles provided the segregating phase is totally dissolved in the solution, i.e., exists in the form of single particles only. It is assumed that such a state is realized at the beginning of the nucleation–growth process.

Equation (3.1) retains the same form also for more complex cases of phase formation, like bubble formation (in limiting cases, cf. [159, 160]), vacancy cluster evolution in metals under irradiation [301, 350], crystallization [94], and nucleation–growth in multicomponent solutions of a new phase with a given stoichiometric composition [298]. The results of the theory outlined are applicable thus with slight quantitative modifications also beyond the case of segregation in solid or liquid solutions considered as an example here.

The change of the cluster-size distribution function, $f(n,t)$, with time is connected with the flux, $J(n,t)$, in cluster-size space by

$$\frac{\partial f(n,t)}{\partial t} = -\frac{\partial J(n,t)}{\partial n}, \tag{3.5}$$

$$J(n,t) = -w^{(+)}_{n,n+1}\left\{\frac{\partial f(n,t)}{\partial n} + \frac{f(n,t)}{k_BT}\frac{\partial \Delta\Phi(n)}{\partial n}\right\}. \tag{3.6}$$

A comparison of Eq. (3.1) or Eqs. (3.5) and (3.6) with the respective relations describing mass transport processes in real space proves that the velocity of deterministic growth of a cluster in cluster-size space $v(n,t)$ is given by (cf. Section 2.4)

$$v(n,t) = \frac{\mathrm{d}n}{\mathrm{d}t} = -w^{(+)}_{n,n+1}\left\{\frac{1}{k_BT}\frac{\partial \Delta\Phi(n)}{\partial n}\right\}. \tag{3.7}$$

It follows furtheron that the kinetic coefficient $w^{(+)}_{n,n+1}$ has the meaning of a diffusion coefficient $D(n,t)$ in cluster-size space. We have

$$D(n,t) = w^{(+)}_{n,n+1} \geq 0. \tag{3.8}$$

In the initial stages of phase formation, the growth of the building units of the newly evolving phase is limited usually kinetically by processes of incorporation of monomers of the segregating phase into the cluster. In such cases, the kinetic parameters $w^{(+)}_{n,n+1}$ may be written in the form (cf. Eq. (2.97) and also Chapter 5)

$$w^{(+)}_{n,n+1} = \frac{4\pi R^2 D^{(*)} c}{a_m} = (4\pi D^{(*)} c a_m)\left(\frac{\omega_s}{\omega_m}\right)^{2/3} n^{2/3}. \tag{3.9}$$

Here the notations

$$\omega_s = \frac{4\pi}{3}a_s^3, \qquad \omega_m = \frac{4\pi}{3}a_m^3 \tag{3.10}$$

are introduced. ω_s and ω_m are the volumes of a monomeric building unit of the segregating particles in the newly evolving phase and the ambient phase, respectively, and a_s or a_m are their linear sizes. Moreover, the relation

$$\left(\frac{4\pi}{3\omega_s}\right) R^3 = n \tag{3.11}$$

is used. $D^{(*)}$ is the diffusion coefficient of the segregating particles in the ambient phase in the immediate vicinity of the aggregates of the new phase.

In certain cases like bubble formation in liquid gas solutions, vacancy formation in solids under irradiation, nucleation of gas-filled voids in solid solutions and, in particular, at later stages of the nucleation–growth process, the formation and growth of the clusters of the newly evolving phase may be determined by diffusional fluxes of the segregating particles to the aggregates of the newly evolving phase. In this case, we have (cf. Eq. (2.98))

$$w^{(+)}_{n,n+1} = 4\pi D a_s c n^{1/3}. \tag{3.12}$$

Here D is the coefficient of bulk diffusion of the segregating particles.

To carry out the subsequent derivations independently of the individual mechanism controlling the formation and growth of the clusters, we introduce two parameters α_1 and κ and a dimensionless time scale τ. These quantities are defined for kinetically limited growth as

$$\alpha_1 = \omega_s^{2/3}\omega_m^{1/3}, \qquad \kappa = \frac{2}{3}, \qquad \tau = \frac{D^{(*)}t}{a_m^2}. \tag{3.13}$$

For bulk diffusion-limited growth we have similarly

$$\alpha_1 = \omega_s, \qquad \kappa = \frac{1}{3}, \qquad \tau = \frac{Dt}{a_s^2}. \tag{3.14}$$

With these notations we get

$$w^{(+)}_{n,n+1} = \begin{cases} 3\alpha_1 c n^{\kappa} \dfrac{D^{(*)}}{a_m^2} & \text{kinetically limited growth} \\ 3\alpha_1 c n^{\kappa} \dfrac{D}{a_s^2} & \text{diffusion-limited growth} \end{cases} \tag{3.15}$$

It is evident that $w^{(+)}_{n,n+1}$ is, in general, a time-dependent quantity due to the dependence of the average concentration c of the segregating particles in the matrix on time. However, as will be shown in Section 3.5 (see also [298]), in the first stage of the nucleation–growth process, variations of the concentration are not significant. In these cases, c can be replaced by the initial concentration c_0.

As mentioned already, $\Delta\Phi$ in Eq. (3.1) is the change of the characteristic thermodynamic potential resulting from the formation of a cluster consisting of n building units. Assuming constancy of the pressure p and temperature T, Φ has to be identified with the Gibbs free energy and, in line with the approach commonly employed in classical nucleation theory, we may write

$$\Delta\Phi = -n\Delta\mu + \sigma A, \qquad \Delta\mu = \mu_\beta(p,T) - \mu_\alpha(p,T). \tag{3.16}$$

The subscript β specifies the chemical potential μ per building unit in the ambient phase, while α refers to the respective value in the newly evolving phase (both taken at a temperature T and a pressure p). The surface area of the cluster is A and σ is the surface tension or interfacial specific energy.

A number of generalizations of Eq. (3.16) have been developed in the past (see, e.g., Lothe and Pound [158], Fisher [69], Reiss et al. [207], Oxtoby et al. [198], Reiss et al. [208], Schmelzer et al. [239]). However, since the classical expression was shown to work well in a large variety of applications, we will use here Eq. (3.16) with a cluster size independent value of the surface tension (capillarity approximation) leaving possible generalizations to a future discussion (see also [249]).

The surface area A of the cluster, assumed, as mentioned, to be of spherical shape, can be expressed through the number n of particles in the cluster by (cf. Eq. (3.11))

$$A = 4\pi R^2 = 4\pi n^{2/3}\left(\frac{3\omega_s}{4\pi}\right)^{2/3}. \tag{3.17}$$

In terms of the number of monomers in the cluster, Eq. (3.16) can be reformulated thus to give

$$\Delta\Phi(n) = -n\Delta\mu + \alpha_2 n^{2/3}, \qquad \alpha_2 = 4\pi\sigma\left(\frac{3\omega_s}{4\pi}\right)^{2/3}. \tag{3.18}$$

With these notations, the critical cluster size n_c, corresponding to a maximum of the Gibbs free energy, is given by

$$\left.\frac{\partial\Delta\Phi(n)}{\partial n}\right|_{n=n_c} = -\Delta\mu + \frac{2}{3}\alpha_2 n_c^{-1/3} = 0, \qquad n_c^{1/3} = \frac{2\alpha_2}{3\Delta\mu}. \tag{3.19}$$

Moreover, for the second derivative of $\Delta\Phi$ we obtain

$$\left.\frac{\partial^2\Delta\Phi(n)}{\partial n^2}\right|_{n=n_c} = -\frac{2}{9}\alpha_2 n_c^{-4/3}. \tag{3.20}$$

These expressions will be used below for subsequent derivations.

In particular, the value of $\Delta\Phi$ at $n = n_c$ is given by

$$\Delta\Phi(n_c) = \frac{1}{3}\alpha_2 n_c^{2/3} = \left(\frac{4}{27}\right)\frac{\alpha_2^3}{(\Delta\mu)^2}. \tag{3.21}$$

Moreover, the range δn_c of n-values in the vicinity of the extremum of $\Delta\Phi$, where the inequality

$$\Delta\Phi(n_c) - \Delta\Phi(n) \le k_B T \tag{3.22}$$

holds, and thermal fluctuations determine the motion of the clusters in cluster-size space, can be written as

$$|n - n_c| \leq \delta n_c = \frac{1}{\sqrt{-\frac{1}{2k_BT}\left(\left.\frac{\partial^2 \Delta\Phi}{\partial n^2}\right|_{n=n_c}\right)}}. \tag{3.23}$$

With the above-derived equations we obtain

$$\frac{\Delta\Phi(n)}{k_BT} = -\beta n\left[\frac{1}{n_c^{1/3}} - \frac{1}{n^{1/3}}\right] + \frac{\beta}{2}n^{2/3}, \tag{3.24}$$

$$\beta = \frac{2\alpha_2}{3k_BT} = \frac{8\pi}{3}\left(\frac{\sigma a_s^2}{k_BT}\right), \tag{3.25}$$

$$\frac{1}{k_BT}\left(\frac{\partial \Delta\Phi}{\partial n}\right) = \beta\left[\frac{1}{n^{1/3}} - \frac{1}{n_c^{1/3}}\right], \quad n_c^{1/3} = \frac{\beta}{(\Delta\mu/k_BT)}, \tag{3.26}$$

$$-\frac{1}{k_BT}\left.\left(\frac{\partial^2 \Delta\Phi}{\partial n^2}\right)\right|_{n=n_c} = \frac{\beta}{3}\left(\frac{1}{n_c^{4/3}}\right) = \frac{2}{(\delta n_c)^2}, \tag{3.27}$$

$$v(n,\tau) = \frac{\mathrm{d}n}{\mathrm{d}\tau} = 3\alpha_1\beta c n^{\kappa}\left[\frac{1}{n_c^{1/3}} - \frac{1}{n^{1/3}}\right]. \tag{3.28}$$

Practically in the whole analysis, specific expressions for the chemical potential difference $\Delta\mu$ are not required. They have to be employed only in applications of the final results of the theory to specific systems. For completeness, the respective expressions for $\Delta\mu$ are given, however, already here.

In general, the chemical potential of the segregating component in a solution can be expressed as [97]

$$\mu_\beta(p,T,\phi) = \mu_\beta(p,T,\phi_{\text{eq}}^{(\infty)}) + k_BT\ln\left(\frac{\phi}{\phi_{\text{eq}}^{(\infty)}}\right). \tag{3.29}$$

Here ϕ is the activity of the respective component and $\phi_{\text{eq}}^{(\infty)}$ is its value in equilibrium with the newly evolving phase at a planar interface at the given values of pressure p and temperature T. This expression results in

$$\Delta\mu = k_BT\ln\left(\frac{\phi}{\phi_{\text{eq}}^{(\infty)}}\right). \tag{3.30}$$

For illustration purposes and comparison, we will also refer from time to time to segregation in a perfect solution. For perfect solutions, the chemical potential of the segregating particles in the ambient phase may be expressed in the form

$$\mu_\beta(p,T,c) = \mu_\beta(p,T,c_{\text{eq}}^{(\infty)}) + k_BT\ln\left(\frac{c}{c_{\text{eq}}^{(\infty)}}\right), \tag{3.31}$$

resulting in (cf. Eq. (3.16))

$$\Delta\mu = k_B T \ln\left(\frac{c}{c_{\text{eq}}^{(\infty)}}\right). \tag{3.32}$$

Similar to the case of real solutions, $c_{\text{eq}}^{(\infty)}$ is the concentration of the segregating particles in the ambient phase in equilibrium with the newly evolving phase at a planar interface at the given values of pressure p and temperature T. By definition of $c_{\text{eq}}^{(\infty)}$ (or $\phi_{\text{eq}}^{(\infty)}$), the relations $\mu_\alpha(p,T) = \mu_\beta(p,T,c_{\text{eq}}^{(\infty)})$ (or, in general, $\mu_\alpha(p,T) = \mu_\beta(p,T,\phi_{\text{eq}}^{(\infty)})$) hold. These relations were used, in addition to Eq. (3.16), in the derivation of Eqs. (3.30) and (3.32).

3.3 Nonsteady-State Effects in the Initial Stage of Nucleation

When an equilibrium system is brought into a metastable state by a rapid quench, the initial cluster-size distribution is determined by the spectrum of heterophase fluctuations which were present in the equilibrium state. This initial cluster size distribution $f(n,0)$ is a rapidly decreasing function of n, i.e., the relation $(\partial f(n,0)/\partial n) < 0$ holds. In the metastable state the negative gradient $(\partial f(n,0)/\partial n)$ in cluster-size space results in a flux into the positive direction of the n-axis compensated at part by the deterministic flow term proportional to $(\partial \Delta\Phi/\partial n)$ (cf. Eq. (3.1)).

Starting with the considered initial state it takes some time, the so-called time lag in nucleation, to establish a time-independent flux in an interval $0 < n \leq g$. The length of this time lag depends, in general, on the chosen value of g. We demand that g fulfills the condition $g > (n_c + \delta n_c)$ due to the following considerations: Once a cluster has reached this size, the further evolution in cluster-size space is dominated by the deterministic growth term in Eq. (3.1) and not by diffusion-like processes as it is the case in the interval $(n_c - \delta n_c, n_c + \delta n_c)$ or by stochastic processes below $n_c - \delta n_c$. Thus the clusters with sizes $n \geq g$, defined in the above-described way, may serve as centers for the evolution of the newly evolving phase. A more precise specification of the value of g will be given somewhat later.

By introducing the dimensionless time scale τ defined by Eqs. (3.13) and (3.14), the Frenkel–Zeldovich equation (3.1) gets the form

$$\frac{\partial f(n,\tau)}{\partial \tau} = \frac{\partial}{\partial n}\left\{ w_n^{(+)}\left[\frac{\partial f(n,\tau)}{\partial n} + \frac{f(n,\tau)}{k_B T}\frac{\partial \Delta\Phi(n)}{\partial n}\right]\right\}. \tag{3.33}$$

The reduced dimensionless coefficients of aggregation $w_n^{(+)}$ are given, now, by (cf. Eqs. (3.9), (3.12), (3.13), (3.14), and (3.15))

$$w_n^{(+)} = 3\alpha_1 c n^{\kappa}. \tag{3.34}$$

In the considered first initial stage of the nucleation–growth process the concentration of single particles is nearly constant and we may consider c in Eq. (3.34) as a time-independent quantity.

Equation (3.33) may be rewritten in a simpler form by introducing new functions

$$\widetilde{f}(n,\tau) = w_n^{(+)} f(n,\tau), \qquad \Delta\widetilde{\Phi} = \Delta\Phi - k_B T \ln w_n^{(+)}. \tag{3.35}$$

We obtain

$$\frac{\partial \widetilde{f}(n,\tau)}{\partial \tau} = w_n^{(+)} \left\{ \frac{\partial^2 \widetilde{f}(n,\tau)}{\partial n^2} + \frac{1}{k_B T}\frac{\partial \Delta\widetilde{\Phi}}{\partial n}\frac{\partial \widetilde{f}(n,\tau)}{\partial n} + \frac{1}{k_B T}\frac{\partial^2 \Delta\widetilde{\Phi}}{\partial n^2}\widetilde{f}(n,\tau)\right\}. \tag{3.36}$$

Possible analytical solutions of this equation will be discussed now for different regions in cluster-size space.

3.3.1 Approximative Solution in the Range $1 \lesssim n \lesssim n_c - \delta n_c$

First we consider the range $1 \leq n \leq n_c - \delta n_c$ in cluster-size space. In order to obtain analytic expressions for the evolution of the cluster-size distribution function to its steady-state shape $\widetilde{f}^{(\mathrm{st})}(n)$ and the characteristic time scale τ_{rel} of this process, we proceed in the following way. We write Eq. (3.36) as

$$\frac{\partial \widetilde{f}(n,\tau)}{\partial \tau} = a\frac{\partial \widetilde{f}(n,\tau)}{\partial n} + b\frac{\partial^2 \widetilde{f}(n,\tau)}{\partial n^2} - d\widetilde{f}(n,\tau). \tag{3.37}$$

The parameters a, b, and d in Eq. (3.37) do not change their sign in the considered interval. We take them as constants equal, in general, to the average values of the respective quantities, i.e.,

$$a \cong \left\langle w_n^{(+)} \frac{1}{k_B T}\frac{\partial \Delta\widetilde{\Phi}}{\partial n}\right\rangle \cong \left\langle w_n^{(+)} \frac{1}{k_B T}\frac{\partial \Delta\Phi}{\partial n}\right\rangle = -\left\langle \frac{\mathrm{d}n}{\mathrm{d}\tau}\right\rangle > 0, \tag{3.38}$$

$$b = \left\langle w_n^{(+)}\right\rangle > 0, \tag{3.39}$$

$$d \cong -\left\langle w_n^{(+)} \frac{1}{k_B T}\frac{\partial^2 \Delta\widetilde{\Phi}}{\partial n^2}\right\rangle \cong -\left\langle w_n^{(+)} \frac{1}{k_B T}\frac{\partial^2 \Delta\Phi}{\partial n^2}\right\rangle = \left\langle w_n^{(+)} \frac{\beta}{3n^{4/3}}\right\rangle. \tag{3.40}$$

In order to determine $\widetilde{f}(n,\tau)$, we first express this function as

$$\widetilde{f}(n,\tau) = \widetilde{f}^{(\mathrm{st})}(n) + \Delta\widetilde{f}(n,\tau). \tag{3.41}$$

Since $\widetilde{f}^{(\mathrm{st})}(n)$ is a solution of Eq. (3.37), $\Delta\widetilde{f}(n,\tau)$ is also determined by this equation, i.e.,

$$\frac{\partial \Delta\widetilde{f}(n,\tau)}{\partial \tau} = a\frac{\partial \Delta\widetilde{f}(n,\tau)}{\partial n} + b\frac{\partial^2 \Delta\widetilde{f}(n,\tau)}{\partial n^2} - d\Delta\widetilde{f}(n,\tau) \tag{3.42}$$

holds.

The initial condition for $\Delta\widetilde{f}(n,\tau)$ reads

$$\begin{aligned}\Delta\widetilde{f}(n,\tau=0) = &-\left(\widetilde{f}^{(\mathrm{st})}(n) - \widetilde{f}(n,\tau=0)\right)\\ &\times \theta\left(n_c - \delta n_c - n\right)\theta\left(n-1\right).\end{aligned} \tag{3.43}$$

Here $\theta(x)$ is determined by

$$\theta(x) = \begin{cases} 1 & \text{for} \quad x > 0 \\ 0 & \text{for} \quad x \le 0\,. \end{cases} \tag{3.44}$$

Starting with an initial distribution consisting of monomeric particles only, i.e.,

$$\widetilde{f}(n \ge 2, \tau = 0) = 0, \tag{3.45}$$

$\Delta\widetilde{f}(n,0)$ in the considered range of cluster sizes is equal (with a minus sign) to the steady-state cluster-size distribution in the considered part of cluster-size space.

We try to find the solution of Eq. (3.42) in the form

$$\Delta\widetilde{f}(n,\tau) = p(n,\tau)\exp[-d\tau + \chi(n,\tau)]. \tag{3.46}$$

Here $p(n,\tau)$ is some new unknown function while $\chi(n,\tau) = C_1\tau + C_2 n$ is, by assumption, a linear function of the variables τ and n. The constants C_1 and C_2 will be determined in such a way as to yield a relatively simple equation for the determination of $p(n,\tau)$.

A substitution of the ansatz (3.46) into Eq. (3.42) results with

$$\chi(\tau,n) = -\frac{a^2}{4b}\left(\tau + \frac{2n}{a}\right) \tag{3.47}$$

into the following partial differential equation for the determination of the unknown function $p(n,\tau)$:

$$\frac{\partial p(n,\tau)}{\partial \tau} = b\frac{\partial^2 p(n,\tau)}{\partial n^2}. \tag{3.48}$$

The solution of Eq. (3.48), obeying the required initial conditions $\Delta\widetilde{f}(n,\tau=0)$, is given by

$$p(n,\tau) = \frac{1}{\sqrt{4\pi b\tau}}\int\limits_{-\infty}^{+\infty} \Delta\widetilde{f}(n',\tau=0)\exp\left[\frac{an'}{2b} - \frac{(n-n')^2}{4b\tau}\right]\mathrm{d}n'. \tag{3.49}$$

Indeed, Eqs. (3.46) and (3.49) yield

$$\Delta\widetilde{f}(n,\tau) = \frac{1}{\sqrt{4\pi b\tau}} \exp\left\{-d\tau - \frac{a^2}{4b}\left[\tau + \frac{2n}{a}\right]\right\} \times \int_{-\infty}^{+\infty} \Delta\widetilde{f}(n',\tau=0) \exp\left[\frac{an'}{2b} - \frac{(n-n')^2}{4b\tau}\right] \mathrm{d}n' \tag{3.50}$$

Taking into account the relation

$$\frac{1}{\sqrt{\gamma\pi}} \int_{-\infty}^{+\infty} \exp\left(-\frac{x^2}{\gamma}\right) \mathrm{d}x = 1 \tag{3.51}$$

immediately the initial condition

$$\lim_{\tau\to 0} \Delta\widetilde{f}(n,\tau) = \Delta\widetilde{f}(n,0) \tag{3.52}$$

is re-established. Therefore, the behavior of the cluster-size distribution function in the initial stage of the nucleation–growth process is uniquely determined by Eq. (3.50).

Since $\Delta\widetilde{f}(n,\tau)$ is known now, $\widetilde{f}(n,\tau)$ can be established easily as

$$\widetilde{f}(n,\tau) = \widetilde{f}^{(\mathrm{st})}(n) + \frac{1}{\sqrt{4\pi b\tau}} \exp\left\{-d\tau - \frac{a^2}{4b}\left[\tau + \frac{2n}{a}\right]\right\} \times \int_{-\infty}^{+\infty} \Delta\widetilde{f}(n',\tau=0) \exp\left[\frac{an'}{2b} - \frac{(n-n')^2}{4b\tau}\right] \mathrm{d}n' \, . \tag{3.53}$$

With Eq. (3.35) we further obtain

$$f(n,\tau) = f^{(\mathrm{st})}(n) + \frac{1}{w_n^{(+)}\sqrt{4\pi b\tau}} \exp\left\{-d\tau - \frac{a^2}{4b}\left[\tau + \frac{2n}{a}\right]\right\} \times \int_{-\infty}^{+\infty} w_{n'}^{(+)} \Delta f(n',\tau=0) \exp\left[\frac{an'}{2b} - \frac{(n-n')^2}{4b\tau}\right] \mathrm{d}n' . \tag{3.54}$$

Moreover, by applying Eqs. (3.5) and (3.6), Eq. (3.54) may be employed to establish also an explicit analytic expression for $J(n,t)$ in the considered range of cluster sizes. We have

$$J(n,\tau) = -w_n^{(+)} \exp\left(-\frac{\Delta\Phi(n)}{k_B T}\right) \frac{\partial}{\partial n}\left\{f(n,\tau)\exp\left(\frac{\Delta\Phi(n)}{k_B T}\right)\right\} . \tag{3.55}$$

A detailed explanation of the way this equation can be obtained is given in the course of derivation of Eq. (3.73). Therefore, here the respective discussion may be omitted.

Finally, we would like to mention that the expressions obtained for $f(n,\tau)$ and $J(n,\tau)$ have to be considered as approximations, only, due to the assumptions made in the derivation of the respective results. It is believed, however, that the given expressions reflect the basic properties of both mentioned functions, at least, in a qualitatively correct way.

3.3.2 Time Scale of Establishment of Steady-State Cluster-Size Distributions in the Range $1 \lesssim n \lesssim n_c - \delta n_c$

In order to find an estimate for the characteristic time scale of establishment of the steady-state distribution in the range of cluster sizes $1 \leq n \leq n_c - \delta n_c$, we rewrite Eq. (3.50) as

$$\Delta\widetilde{f}(n,\tau) = \frac{1}{\sqrt{4\pi b\tau}} \exp(-d\tau) \times \int_{-\infty}^{+\infty} \Delta\widetilde{f}(n',\tau=0) \exp\left[-\frac{(n-n'+a\tau)^2}{4b\tau}\right] \mathrm{d}n' \,. \tag{3.56}$$

Equation (3.56) allows us to give an upper estimate of the characteristic time $\tau_{\mathrm{rel}}^{(1)}$ to reach steady-state conditions in the range of cluster sizes $1 \leq n \leq n_c - \delta n_c$. For this purpose, we assume that in the initial state all segregating particles are found in the system in the form of monomers, only. In this and other most typical cases, $\Delta\widetilde{f}(n',0)$ is a strongly decreasing function of n' and Eq. (3.56) gets the form

$$\Delta\widetilde{f}(n,\tau) \cong \frac{1}{\sqrt{4\pi b\tau}} \exp\left\{-d\tau - \frac{(n+a\tau)^2}{4b\tau}\right\} \int_{-\infty}^{+\infty} \Delta\widetilde{f}(n',\tau=0)\,\mathrm{d}n' . \tag{3.57}$$

The characteristic time scale $\tau_{\mathrm{rel}}^{(1)}$ for the establishment of the steady-state cluster-size distribution in the considered range of cluster sizes is determined by $\Delta\widetilde{f}(n,\tau) \propto \exp(-\tau/\tau_{\mathrm{rel}}^{(1)})$). From Eq. (3.57) we have

$$\tau_{\mathrm{rel}}^{(1)} \cong \frac{1}{d + \dfrac{a^2}{4b}} . \tag{3.58}$$

We may get the upper estimate for this quantity by replacing $a \to a_{\min}$, $b \to b_{\max}$ and $d \to d_{\min}$, i.e., by the respective lowest (min) or highest (max) values of the parameters in the considered range of cluster sizes. We find

$$a_{\min} = w_n^{(+)}(n_c)\frac{\beta}{3}\left(\frac{\delta n_c}{n_c^{4/3}}\right) = 2\left(\frac{w_n^{(+)}(n_c)}{\delta n_c}\right) , \tag{3.59}$$

$$b_{\max} = w_n^{(+)}(n_c) \,, \qquad d_{\min} = w_n^{(+)}(n_c)\left(\frac{\beta}{3n_c^{4/3}}\right) , \tag{3.60}$$

resulting in

$$\tau_{\mathrm{rel}}^{(1)} \cong 2\left(\frac{n_c^{4/3}}{\beta w_n^{(+)}(n_c)}\right) \cong \frac{1}{3}\left(\frac{(\delta n_c)^2}{w_n^{(+)}(n_c)}\right) \cong \frac{2}{3}\left(\frac{n_c^{[(4/3)-\kappa]}}{\alpha_1 c\beta}\right) . \tag{3.61}$$

3.3.3 Results for the Range $n_c - \delta n_c \lesssim n \lesssim n_c + \delta n_c$

The analysis can be carried out in the same way in the range of cluster sizes $n_c - \delta n_c \leq n \leq n_c + \delta n_c$. The only difference is that the parameter a in the considered here range is nearly equal to zero. We obtain thus from Eq. (3.50)

$$\Delta\widetilde{f}(n,\tau) = \frac{1}{\sqrt{4\pi b\tau}}\exp(-d\tau) \times \int_{-\infty}^{+\infty} \Delta\widetilde{f}(n',\tau=0)\exp\left[-\frac{(n-n')^2}{4b\tau}\right] dn'. \tag{3.62}$$

The relaxation time of the distribution to the steady state is now given by

$$\tau_{\text{rel}}^{(2)} \cong \frac{1}{d_{\min}} \cong 3\left(\frac{n_c^{4/3}}{\beta w_n^{(+)}(n_c)}\right) \cong \frac{1}{2}\left(\frac{(\delta n_c)^2}{w_n^{(+)}(n_c)}\right) = \frac{n_c^{[(4/3)-\kappa]}}{\alpha_1\beta c}. \tag{3.63}$$

A comparison with Eq. (3.61) shows that $\tau_{\text{rel}}^{(2)}$ is of the same order of magnitude as $\tau_{\text{rel}}^{(1)}$.

As an estimate for the total time $\tau_{\text{rel}}^{(\text{I})}$ of relaxation to the steady state in the range of cluster sizes $1 \leq n \leq n_c + \delta n_c$ we obtain the result

$$\tau_{\text{rel}}^{(\text{I})} = \tau_{\text{rel}}^{(1)} + \tau_{\text{rel}}^{(2)} \cong 5\left(\frac{n_c^{4/3}}{\beta w_n^{(+)}(n_c)}\right) \cong \frac{5}{6}\left(\frac{(\delta n_c)^2}{w_n^{(+)}(n_c)}\right) = \frac{5}{3}\left(\frac{n_c^{[(4/3)-\kappa]}}{\alpha_1\beta c}\right). \tag{3.64}$$

For both the considered modes of growth we have thus (in real-time scale; cf. Eqs. (3.13) and (3.14))

$$t_{\text{rel}}^{(\text{I})} = \begin{cases} \frac{5}{3}\left(\frac{\beta a_m^2}{\omega_s^{2/3}\omega_m^{1/3}D^{(*)}c}\right)\left(\frac{k_BT}{\Delta\mu}\right)^2 & \text{kinetically limited growth} \\ \frac{5}{3}\left(\frac{\beta^2 a_s^2}{\omega_s Dc}\right)\left(\frac{k_BT}{\Delta\mu}\right)^3 & \text{diffusion-limited growth.} \end{cases} \tag{3.65}$$

The functions $f(n,\tau)$ and $J(n,\tau)$ are given by Eqs. (3.54) and (3.55), again, where the parameter a has to be set equal to zero.

To complete the analysis of the behavior of the cluster-size distribution function in the region of cluster-size space $1 \leq n \leq n_c + \delta n_c$, the steady-state cluster-size distribution function $f^{(\text{st})}(n)$ and the steady-state flux $J(n)$ have to be determined. This task will be carried out in the next subsection.

3.3.4 Steady-State Nucleation Rate and Steady-State Cluster-Size Distribution in the Range $1 \lesssim n \lesssim n_c + \delta n_c$

Assuming that a steady state has been established in the system in the range of cluster sizes $1 \leq n \leq n_c + \delta n_c$, the cluster-size distribution and the flux in cluster-size space become independent of time. Applying the dimensionless time scale τ, again, both quantities are generally determined by

$$\frac{\partial f(n,\tau)}{\partial \tau} = -\frac{\partial J(n,\tau)}{\partial n}, \tag{3.66}$$

$$J(n,\tau) = -w_n^{(+)} \left\{ \frac{\partial f(n,\tau)}{\partial n} + \frac{f(n,\tau)}{k_B T}\frac{\partial \Delta\Phi(n)}{\partial n} \right\}. \tag{3.67}$$

Once $J(n,\tau) = J$ is a constant, the cluster-size distribution in the considered region does not depend on time (cf. Eq. (3.66)).

For the determination of J and of the steady-state cluster-size distribution function $f^{(\mathrm{st})}(n)$ in the considered range of n-values, we introduce, in addition, another so far unknown function $\Psi(n,\tau)$ via

$$f(n,\tau) = \Psi(n,\tau)\exp\left(-\frac{\Delta\Phi(n)}{k_B T}\right). \tag{3.68}$$

Taking into account the relation $\Delta\Phi(n \to 0) = 0$ we have

$$f(n,\tau)\big|_{n\to 0} = \Psi(n,\tau)\big|_{n\to 0} = c. \tag{3.69}$$

A substitution of the ansatz (3.68) into Eq. (3.67) yields

$$J(n,\tau) = -w_n^{(+)}\left(\frac{\partial \Psi(n,\tau)}{\partial n}\right)\exp\left(-\frac{\Delta\Phi(n)}{k_B T}\right) \tag{3.70}$$

or

$$\frac{\partial \Psi(n,\tau)}{\partial n} = -\frac{J(n,\tau)}{w_n^{(+)}}\exp\left(\frac{\Delta\Phi(n)}{k_B T}\right). \tag{3.71}$$

Integration of Eq. (3.71) with respect to n in the range from n to ∞ leads to

$$\Psi(n,\tau) = \int_n^\infty \frac{J(n',\tau)\exp\left(\frac{\Delta\Phi(n')}{k_B T}\right)}{w_{n'}^{(+)}}\,\mathrm{d}n'. \tag{3.72}$$

With Eq. (3.68) we get (cf. also [131])

$$f(n,\tau) = \exp\left(-\frac{\Delta\Phi(n)}{k_B T}\right)\int_n^\infty \frac{J(n',\tau)\exp\left(\frac{\Delta\Phi(n')}{k_B T}\right)}{w_{n'}^{(+)}}\,\mathrm{d}n'. \tag{3.73}$$

This equation can be reversed to express $J(n,t)$ via $f(n,t)$. As a result immediately Eq. (3.55) is obtained.

After a time interval $\tau \geq \tau_{\rm rel}^{(\rm I)}$ (cf. Eq. (3.64)), the flux in the range $1 \leq n \leq n_c + \delta n_c$ of cluster-size space becomes a constant. In this case we have instead of Eq. (3.73)

$$f^{(\rm st)}(n) = J \exp\left(-\frac{\Delta\Phi(n)}{k_B T}\right) \int_n^\infty \frac{\exp\left(\frac{\Delta\Phi(n')}{k_B T}\right)}{w_{n'}^{(+)}} \, dn' . \tag{3.74}$$

Taking into account, in addition, Eq. (3.69) we find

$$c = J \int_0^\infty \frac{\exp\left(\frac{\Delta\Phi(n')}{k_B T}\right)}{w_{n'}^{(+)}} \, dn' \tag{3.75}$$

and

$$J = \frac{c}{\int_0^\infty \frac{\exp\left(\frac{\Delta\Phi(n')}{k_B T}\right)}{w_{n'}^{(+)}} \, dn'} . \tag{3.76}$$

The function $\exp[\Delta\Phi(n)/(k_B T)]$ has a sharp maximum in the vicinity of $n = n_c$ and one obtains, in a good approximation, the following expression for the steady-state nucleation rate:

$$J = c w_n^{(+)}(n_c) \Upsilon_{(z)} \exp\left(-\frac{\Delta\Phi(n_c)}{k_B T}\right) , \tag{3.77}$$

$$\Upsilon_{(z)} = \sqrt{-\frac{1}{2\pi k_B T}\left(\left.\frac{\partial^2 \Delta\Phi}{\partial n^2}\right|_{n=n_c}\right)} . \tag{3.78}$$

In the derivation of the above-given equations, $\Delta\Phi(n)$ was expanded into a Taylor series including second-order terms. Moreover, $w_n^{(+)}$ was set equal to the respective value at $n = n_c$. $\Upsilon_{(z)}$ is the so-called Zeldovich factor [357].

With Eq. (3.27) we obtain for both considered modes of growth

$$J = \sqrt{\left(\frac{3}{2\pi}\beta\right)} \alpha_1 c^2 n_c^{[\kappa-(2/3)]} \exp\left(-\frac{\Delta\Phi(n_c)}{k_B T}\right) . \tag{3.79}$$

Equations (3.74) and (3.75) immediately yield

$$f^{(\rm st)}(n) = c \exp\left(-\frac{\Delta\Phi(n)}{k_B T}\right) \left[\frac{\Xi(n,\infty)}{\Xi(0,\infty)}\right] , \tag{3.80}$$

$$\Xi(a,b) = \int_a^b \frac{\exp\left(\frac{\Delta\Phi(n')}{k_B T}\right)}{w_{n'}^{(+)}} \, dn' . \tag{3.81}$$

With the same approximations as in the derivation of Eq. (3.77) (Taylor expansion of $\Delta\Phi$, $w_n^{(+)} \cong w_n^{(+)}(n_c)$) one gets in a good approximation

$$f^{(\mathrm{st})}(n) = \left(\frac{c}{2}\right) \exp\left(-\frac{\Delta\Phi(n)}{k_B T}\right) \mathrm{erfc}\left\{\Upsilon_{(z)}\sqrt{\pi}(n - n_c)\right\}. \tag{3.82}$$

The function erfc(ξ) is connected with the error function erf(ξ) by the relation erfc(ξ)=1−erf(ξ). We have

$$\mathrm{erf}(\xi) = \frac{2}{\sqrt{\pi}} \int_0^{\xi} \mathrm{d}z \, \exp(-z^2), \tag{3.83}$$

$$\mathrm{erfc}(\xi) = \frac{2}{\sqrt{\pi}} \int_{\xi}^{\infty} \mathrm{d}z \, \exp(-z^2). \tag{3.84}$$

Taking into account Eqs. (3.23) and (3.78), Eq. (3.82) may be rewritten in the form (cf. also [28, 329])

$$f^{(\mathrm{st})}(n) = \left(\frac{c}{2}\right) \exp\left(-\frac{\Delta\Phi(n)}{k_B T}\right) \mathrm{erfc}\left\{\frac{(n - n_c)}{\delta n_c}\right\}. \tag{3.85}$$

Note that in the limit $n_c \to \infty$, Frenkel's expression for heterophase fluctuations, Eq. (2.15), is retained in this equation as a special case.

In this way, the determination of the characteristics of the nucleation–growth process in the range $1 \leq n \leq n_c + \delta n_c$ is completed.

3.4 Flux and Cluster Distributions in the Range of Supercritical Cluster Sizes

For the subsequent analysis, in addition to Eq. (3.1), a similar partial differential equation determining the change of the flux $J(n,t)$ with time and number of particles in a cluster n is required. Once the flux $J(n,\tau)$ is known, the evolution of the cluster-size distribution can be determined via Eq. (3.73).

By taking the derivative of Eq. (3.67) with respect to dimensionless time, we get with Eq. (3.66)

$$\frac{\partial J(n,\tau)}{\partial \tau} = w_n^{(+)} \left\{\frac{\partial^2 J(n,\tau)}{\partial n^2} + \left[\frac{1}{k_B T}\frac{\partial \Delta\Phi(n)}{\partial n}\right]\left(\frac{\partial J(n,\tau)}{\partial n}\right)\right\}. \tag{3.86}$$

According to Eqs. (3.21) and (3.77) the nucleation rate decreases rapidly with a decrease of the supersaturation. Therefore, in the transient stage to a steady state (for $0 < n \leq n_c + \delta n_c$) and some additional time interval, where intensive nucleation proceeds, the supersaturation, i.e., the concentration, can be considered as nearly time independent. This result implies, in particular, that for the description of nucleation $w_n^{(+)}$ in Eq. (3.67) can be taken as a constant (cf. also [298]). For this reason, the term containing the derivative of $w_n^{(+)}$ with respect to time is omitted in Eq. (3.86).

As in the analysis of the cluster-size distribution in the range $n < n_c$, we will use different approximations of Eq. (3.86) appropriate for the respective range of cluster sizes.

3.4.1 Results in the Range $n_c \lesssim n \lesssim 8n_c$

Taking into account Eq. (3.26) we obtain, by a Taylor expansion of the term in the square brackets in Eq. (3.26) in the vicinity of $n = n_c$, the following result:

$$n^\kappa \frac{1}{k_B T}\frac{\partial \Delta\Phi}{\partial n} = n^\kappa \beta \left[\frac{1}{n^{1/3}} - \frac{1}{n_c^{1/3}}\right] \cong -\beta \frac{(n-n_c)}{3n_c^{(4/3)-\kappa}} . \tag{3.87}$$

It can be easily verified that in the considered region $n_c \lesssim n \lesssim 8n_c$ the Taylor expansion gives a quite correct fit of the difference in the square brackets. Indeed, we may write

$$n^\kappa \left[\frac{1}{n^{1/3}} - \frac{1}{n_c^{1/3}}\right] = n_c^{[\kappa-(1/3)]}\widehat{x}^{3\kappa-2}\widehat{x}(1-\widehat{x}), \quad \widehat{x} = \left(\frac{n}{n_c}\right)^{1/3} . \tag{3.88}$$

Moreover, the relation

$$\widehat{x}(1-\widehat{x}) = -(\widehat{x}-1)\frac{\widehat{x}^2+\widehat{x}+1}{\widehat{x}+(1/\widehat{x})+1} = -\frac{\widehat{x}^3-1}{\widehat{x}+(1/\widehat{x})+1} \tag{3.89}$$

holds. We find

$$n^\kappa \left[\frac{1}{n^{1/3}} - \frac{1}{n_c^{1/3}}\right] = -\frac{(n-n_c)}{3n_c^{(4/3)-\kappa}}\Lambda(\widehat{x}) , \tag{3.90}$$

$$\Lambda(\widehat{x}) = \frac{3}{\widehat{x}^{2-3\kappa}\left[\widehat{x}+(1/\widehat{x})+1\right]} . \tag{3.91}$$

In the considered range $n_c \le n \le 8n_c$ (or $1 \le \widehat{x} \le 2$) the factor $\Lambda(\widehat{x})$ has a value of the order of 1.

For the analysis of the behavior of the distribution and the flux in the region considered now, the concentration c is considered, again, as a constant. Therefore, we may introduce a modified reduced time scale $\widehat{\tau}$ via

$$\widehat{\tau} = (\alpha_1 c)\tau . \tag{3.92}$$

In the new time scale, the value $\widehat{\tau} = 0$ refers to the moment of time, when steady-state conditions in the range $1 \le n \le n_c + \delta n_c$ have been established.

Moreover, for the analysis in the region considered now, the replacement

$$w_n^{(+)}\frac{\partial^2 J(n,\widehat{\tau})}{\partial n^2} \Longrightarrow w_n^{(+)}(n_c)\frac{\partial^2 J(n,\widehat{\tau})}{\partial n^2} \tag{3.93}$$

is made. This replacement is quite correct near $n = n_c$. For larger values of n, this term loses its importance [304]. Therefore, the respective substitution can be considered as a satisfactory approximation in the whole considered range of cluster sizes $n_c \leq n \leq 8n_c$ (and also even for $n \geq 8n_c$).

With the above approximations, using Eq. (3.34), we get from Eq. (3.86)

$$\frac{\partial J(n,\widehat{\tau})}{\partial \widehat{\tau}} = -\beta n_c^{[\kappa-(4/3)]}(n-n_c)\frac{\partial J(n,\widehat{\tau})}{\partial n} + 3n_c^{\kappa}\left(\frac{\partial^2 J(n,\widehat{\tau})}{\partial n^2}\right). \tag{3.94}$$

Moreover, similar to the previous approach in the analysis of the Frenkel–Zeldovich equation, we set

$$\widehat{a} = \beta n_c^{[\kappa-(4/3)]}, \qquad \widehat{b} = 3n_c^{\kappa}, \qquad x = n - n_c. \tag{3.95}$$

Equations (3.94) and (3.95) yield

$$\frac{\partial J(x,\widehat{\tau})}{\partial \widehat{\tau}} = -\widehat{a}x\frac{\partial J(x,\widehat{\tau})}{\partial x} + \widehat{b}\frac{\partial^2 J(x,\widehat{\tau})}{\partial x^2}. \tag{3.96}$$

In order to find the solution of this equation, we introduce new variables $\widehat{p}(\widehat{\tau})$ and $\widehat{\chi}(x,\widehat{\tau})$. Provided $\widehat{\chi}$ and $\widehat{p}$ are chosen as

$$\widehat{\chi} = x\exp(-\widehat{a}\widehat{\tau}), \qquad \widehat{p}(\widehat{\tau}) = \frac{1}{2\widehat{a}}\left[1-\exp(-2\widehat{a}\widehat{\tau})\right], \tag{3.97}$$

Eq. (3.96) gets the simpler form

$$\frac{\partial J(\widehat{\chi},\widehat{p})}{\partial \widehat{p}} = \widehat{b}\frac{\partial^2 J(\widehat{\chi},\widehat{\tau})}{\partial \widehat{\chi}^2}. \tag{3.98}$$

Hereby the new variables fulfill the conditions

$$\widehat{p}(\widehat{\tau}=0) = 0, \qquad \chi(x=0,\widehat{\tau}=0) = 0. \tag{3.99}$$

For $\widehat{\tau} = 0$, by assumption, the relations $J(n = n_c, \widehat{\tau} = 0) = J(x = 0, \widehat{\tau} = 0) = J(\widehat{\chi} = 0, \widehat{p} = 0) = J =$ constant have to be fulfilled. Moreover, the condition $J(\widehat{\chi} > 0, \widehat{p} = 0) = 0$ holds. We have to find, therefore, a solution of Eq. (3.98) obeying the above-given initial and boundary conditions.

The appropriate solution of Eq. (3.96) can be written in the form

$$J(\widehat{\chi},\widehat{p}) = \frac{1}{\sqrt{4\pi\widehat{b}\widehat{p}}}\int_{-\infty}^{+\infty} J(\widehat{\chi}',\widehat{p}=0)\exp\left[-\frac{(\widehat{\chi}-\widehat{\chi}')^2}{4\widehat{b}\widehat{p}}\right]\mathrm{d}\widehat{\chi}'. \tag{3.100}$$

It is easily verified that this solution fulfills the initial conditions

$$\lim_{\widehat{p}\to 0} J(\widehat{\chi},\widehat{p}) = J(\widehat{\chi},0). \tag{3.101}$$

When $\widehat{p} = 0$ the flux is zero for $\widehat{\chi} > 0$ and equal to the constant value J for $\widehat{\chi} \leq 0$. Taking into account this dependence and returning to the original variables, we find after some straightforward transformations

$$J(n,\widehat{\tau}) = J(n_c)\,\mathrm{erfc}\left[\widehat{\xi}(\widehat{\tau})(n-n_c)\right], \qquad n_c \leq n \leq 8n_c, \tag{3.102}$$

$$\hat{\xi}(\hat{\tau}) = \frac{\exp(-\hat{a}\hat{\tau})}{\sqrt{\frac{2\hat{b}}{\hat{a}}[1 - \exp(-2\hat{a}\hat{\tau})]}}. \tag{3.103}$$

The evolution of the cluster-size distribution function with time is given by Eq. (3.73), again.

For $\hat{\tau} \to 0$, $\hat{\xi}$ tends to infinity and the flux is equal to zero. The approach to the steady state is governed by the relaxation of $\hat{\xi}$ to its steady-state value $\hat{\xi} = 0$. The respective relaxation time $\hat{\tau}_{\text{rel}}^{(\text{II})}$ is given by

$$\hat{\tau}_{\text{rel}}^{(\text{II})} = \frac{1}{\hat{a}} = \frac{n_c^{[(4/3)-\kappa]}}{\beta} \tag{3.104}$$

or (in terms of the time scale τ, cf. Eqs. (3.13), (3.14), and (3.92))

$$\tau_{\text{rel}}^{(\text{II})} = \frac{n_c^{[(4/3)-\kappa]}}{\alpha_1 c \beta} = \frac{3 n_c^{4/3}}{\beta w_n^{(+)}(n_c)}. \tag{3.105}$$

A comparison with Eq. (3.64) proves that the typical relaxation time is of the same order of magnitude as $\tau_{\text{rel}}^{(\text{I})}$, the time scale of the approach of steady-state conditions in the range of cluster sizes $1 \leq n \leq n_c + \delta n_c$. In real time $t_{\text{rel}}^{(\text{II})}$ is given by Eq. (3.65), again, where the factor $5/3$ has to be replaced by 1.

After the time $\tau_{\text{rel}} = \tau_{\text{rel}}^{(\text{I})} + \tau_{\text{rel}}^{(\text{II})}$ steady-state conditions are established thus in the whole range of cluster sizes $1 \leq n \leq 8n_c$. The flux equals $J = J(n_c)$ (Eq. (3.79)) while the steady-state cluster-size distribution is given by Eqs. (3.80) and (3.85), respectively.

The number of particles, segregated in clusters with sizes $n \leq n_c(0)$, is relatively small. Therefore, significant changes of the concentration do not occur in the initial stage of the nucleation–growth process for $\tau \leq \tau_{\text{rel}}$. Moreover, as will be shown subsequently, the characteristic time scale of steady-state nucleation τ_N is large as compared with the time lag in nucleation τ_{rel}. Therefore, taking into account variations of the state of the system due to the further nucleation–growth process at times $\tau_{\text{rel}} < \tau < \tau_N$, the characteristic time scales of variation of the state of the ambient phase are always large as compared with τ_{rel}. Therefore, as far as intensive nucleation processes take place in the system, the flux and the cluster-size distribution in the range of cluster sizes $1 \leq n \leq 8n_c$ are given by the steady-state expressions derived earlier with slowly varying values of the concentration c. By the mentioned reasons we will identify the parameter g, introduced at the beginning of Section 3.3, with $g \cong 8n_c$.

3.4.2 Results in the Range $n \gtrsim 8n_c$

For an adequate description of the flux in cluster-size space and the evolution of the cluster-size distribution function in the range $n \geq g \cong 8n_c$ the variation of $J(n_c)$ with time has to be taken into account from the very beginning. Otherwise stationary values of the flux and steady-state cluster-size distributions would develop up to $n \to \infty$. Such a result is meaningless from a physical point of view. In particular, it would require the existence of an infinite number of segregating particles in the system. The change of the nucleation rate is due to the decrease of the concentration of single particles from the initial value $c(t = 0) = c_0$

to the actual value $c(t)$. Such changes of the concentration of the single particles affect the nucleation rate mainly via its influence on the work of critical cluster formation $\Delta\Phi(n_c)$.

The work of critical cluster formation at some arbitrary moment of time $\Delta\Phi(n_c)$ (corresponding to a critical cluster size n_c and a concentration of single particles c or the activity $\phi(c)$) may be expressed through the respective expression in the initial state $\Delta\Phi[n_c(0)]$ (at a concentration $c = c_0$ or an activity $\phi(c_0)$ referring to a critical cluster size $n_c(0)$). In the subsequent analysis, we start with the identity

$$\frac{\Delta\Phi(n_c)}{k_B T} = \frac{\Delta\Phi[n_c(0)]}{k_B T} \frac{\Delta\Phi(n_c)}{\Delta\Phi[n_c(0)]} . \tag{3.106}$$

According to Eq. (3.21), the work of critical cluster formation depends on the actual value of concentration or activity mainly via its dependence on $\Delta\mu$. It is assumed here that, for the considered relatively small variations of the concentration in the initial stages of the transformation, the dependence of the specific interfacial energy on concentration is negligible in comparison with its effect on the variations of $\Delta\mu$. Since the considered variations of $\Delta\mu$ are also small, we may write

$$\Delta\Phi(n_c) = \Delta\Phi[n_c(0)] + \left.\frac{\partial \Delta\Phi(n_c)}{\partial \Delta\mu}\right|_{c=c_0} [\Delta\mu(c) - \Delta\mu(c_0)] . \tag{3.107}$$

Employing Eqs. (3.21), (3.24), (3.25), and (3.26), we have

$$\frac{\Delta\Phi(n_c)}{k_B T} = \frac{\Delta\Phi[n_c(0)]}{k_B T} - n_c(0) \left[\frac{(\Delta\mu(c) - \Delta\mu(c_0))}{k_B T} \right] . \tag{3.108}$$

With the notation

$$\varphi = \left[\frac{(\Delta\mu(c_0) - \Delta\mu(c))}{k_B T} \right] \tag{3.109}$$

and Eq. (3.77) we obtain

$$J(n_c) = J[n_c(0)] \exp\left[-n_c(0)\varphi\right] . \tag{3.110}$$

If we express the difference of the chemical potential of the particles in both phases via the respective expressions for real or perfect solutions (cf. Eqs. (3.30) and (3.32)), we may write, in particular,

$$\varphi = -\ln\left(\frac{\phi(c)}{\phi(c_0)}\right) \cong \left(1 - \frac{\phi(c)}{\phi(c_0)}\right) \tag{3.111}$$

for real solutions and

$$\varphi = -\ln\left(\frac{c}{c_0}\right) \cong \left(1 - \frac{c}{c_0}\right) \tag{3.112}$$

for perfect ones.

The time of active nucleation $\widehat{\tau}_N$ we define via

$$n_c(0)\varphi(\widehat{\tau}_N) \cong 1. \tag{3.113}$$

This definition corresponds to a decrease of the steady-state nucleation rate by a factor e^{-1} in comparison with its initial value. It follows as a consequence that during the time of active nucleation $\widehat{\tau} < \widehat{\tau}_N$ the inequality $\varphi \ll 1$ holds.

For a determination of the flux in the considered range of cluster sizes we start with Eq. (3.86), again. Hereby we go over from the variable n to the radius of the aggregates, R. Moreover, the radius will be expressed in units of a_s (cf. Eq. (3.10)) as $r = R/a_s$. Employing, again, the time scale $\widehat{\tau}$ (cf. Eq. (3.92)) we find

$$\frac{\partial J(r,\widehat{\tau})}{\partial \widehat{\tau}} = \beta r^{3\kappa-2}\left(\frac{1}{r} - \frac{1}{r_c}\right)\frac{\partial J(r,\widehat{\tau})}{\partial r} + \frac{r^{3\kappa-4}}{3}\left[\frac{\partial^2 J(r,\widehat{\tau})}{\partial r^2} - \frac{2}{r}\frac{\partial J(r,\widehat{\tau})}{\partial r}\right]. \tag{3.114}$$

The initial moment $\widehat{\tau} = 0$ in the analysis is to be identified, now, with a moment of time $\tau = \tau_{\text{rel}} = \tau_{\text{rel}}^{(\text{I})} + \tau_{\text{rel}}^{(\text{II})}$ after the initiation of the nucleation–growth process.

The further derivations have to be carried out for kinetically limited and diffusion-limited nucleation–growth in somewhat different but similar ways.

3.4.2.1 Kinetically Limited (or Ballistic) Growth

For kinetically limited growth the parameter κ in the expression for $w_n^{(+)}$ equals $\kappa = 2/3$ (cf. Eq. (3.13)). Moreover, for the considered range of cluster sizes the inequality $n \gg n_c$ holds, generally, and we may write in a good approximation

$$\frac{\partial J(r,\widehat{\tau})}{\partial \widehat{\tau}} = -\frac{\beta}{r_c}\frac{\partial J(r,\widehat{\tau})}{\partial r} + \frac{1}{3r_c^2}\frac{\partial^2 J(r,\widehat{\tau})}{\partial r^2}. \tag{3.115}$$

For the subsequent analysis, we introduce temporarily a new size variable $\widehat{r}$ as

$$\widehat{r} = r - r_g, \qquad r_g = \sqrt[3]{8n_c} = 2n_c^{1/3} = 2r_c. \tag{3.116}$$

Equation (3.115) then reads

$$\frac{\partial J(\widehat{r},\widehat{\tau})}{\partial \widehat{\tau}} = -\frac{\beta}{r_c}\frac{\partial J(\widehat{r},\widehat{\tau})}{\partial \widehat{r}} + \frac{1}{3r_c^2}\frac{\partial^2 J(\widehat{r},\widehat{\tau})}{\partial \widehat{r}^2}. \tag{3.117}$$

This equation has to be solved by applying the boundary conditions

$$J(\widehat{r}=0,\widehat{\tau}) = J[n_c(0)]\exp[-n_c(0)\varphi(\widehat{\tau})]. \tag{3.118}$$

We try to find the solution of Eq. (3.117) with the ansatz

$$J(\widehat{r},\widehat{\tau}) = \widetilde{p}(\widehat{r},\widehat{\tau})\exp(\widetilde{a}_k\widehat{r})\exp(-\widetilde{b}_k\widehat{\tau}). \tag{3.119}$$

If the parameters $\widetilde{a}_k$ and $\widetilde{b}_k$ are chosen as

$$\widetilde{a}_k = \frac{3\beta}{2}r_c, \qquad \widetilde{b}_k = \frac{3\beta^2}{4} \tag{3.120}$$

the differential equation for $\widetilde{p}(\widehat{r},\widehat{\tau})$ gets the form

$$\frac{\partial \widetilde{p}(\widehat{r},\widehat{\tau})}{\partial \widehat{\tau}} = b_k^{(*)} \frac{\partial^2 \widetilde{p}(\widehat{r},\widehat{\tau})}{\partial \widehat{r}^2}, \qquad b_k^{(*)} = \frac{1}{3r_c^2}. \tag{3.121}$$

Note that the parameters $\widetilde{a}_k$, $\widetilde{b}_k$, and $b_k^{(*)}$ obey the relation

$$\widetilde{b}_k = b_k^{(*)} \widetilde{a}_k^2. \tag{3.122}$$

We have to find the solution of Eq. (3.121) which fulfills the boundary condition (cf. Eqs. (3.118) and (3.119))

$$\widetilde{p}(0,\widehat{\tau}) = J(0,\widehat{\tau}) \exp\left(\widetilde{b}_k \widehat{\tau}\right). \tag{3.123}$$

This solution is given by

$$\widetilde{p}(\widehat{r},\widehat{\tau}) = \frac{\widehat{r}}{\sqrt{4\pi b_k^{(*)}}} \int_0^{\widehat{\tau}} \widetilde{p}(0,\widehat{\tau}') \left\{ \frac{\exp\left[-\frac{\widehat{r}^2}{4b_k^{(*)}(\widehat{\tau}-\widehat{\tau}')}\right]}{(\widehat{\tau}-\widehat{\tau}')^{3/2}} \right\} d\widehat{\tau}'. \tag{3.124}$$

Indeed, it can be verified easily that Eq. (3.124) is a solution of Eq. (3.121). Moreover, from Eq. (3.124) we have

$$\lim_{\widehat{r}\to 0} \widetilde{p}(\widehat{r},\widehat{\tau}) = \widetilde{p}(0,\widehat{\tau}) \lim_{\widehat{r}\to 0} \left\{ \frac{\widehat{r}}{\sqrt{4\pi b_k^{(*)}}} \int_0^{\widehat{\tau}} \left\{ \frac{\exp\left[-\frac{\widehat{r}^2}{4b_k^{(*)}(\widehat{\tau}-\widehat{\tau}')}\right]}{(\widehat{\tau}-\widehat{\tau}')^{3/2}} d\widehat{\tau}' \right\}\right\}. \tag{3.125}$$

By introducing a new variable of integration z via

$$z = \frac{\widehat{r}}{\sqrt{4b_k^{(*)}(\widehat{\tau}-\widehat{\tau}')}}, \qquad dz = \frac{\widehat{r}}{2\sqrt{4b_k^{(*)}}(\widehat{\tau}-\widehat{\tau}')^{3/2}} d\widehat{\tau}' \tag{3.126}$$

we find (cf. Eqs. (3.83) and (3.84))

$$\lim_{\widehat{r}\to 0} \widetilde{p}(\widehat{r},\widehat{\tau}) = \widetilde{p}(0,\widehat{\tau}) \lim_{\widehat{r}\to 0} \left\{ \frac{2}{\sqrt{\pi}} \int_{\widehat{r}/\sqrt{4b_k^{(*)}\widehat{\tau}}}^{\infty} \exp\left(-z^2\right) dz \right\} \tag{3.127}$$

or

$$\lim_{\widehat{r}\to 0} \widetilde{p}(\widehat{r},\widehat{\tau}) = \widetilde{p}(0,\widehat{\tau}). \tag{3.128}$$

This way, it is shown that the function given by Eq. (3.124) fulfills all requirements.
According to Eqs. (3.119) and (3.124), the general solution of Eq. (3.117) is given by

$$J(\widehat{r},\widehat{\tau}) = \frac{\widehat{r}}{\sqrt{4\pi b_k^{(*)}}} \exp(\widetilde{a}_k \widehat{r}) \int_0^{\widehat{\tau}} J(0,\widehat{\tau}') \exp\left[-\widetilde{b}_k(\widehat{\tau}-\widehat{\tau}')\right] \times \left\{ \frac{\exp\left[-\frac{\widehat{r}^2}{4b_k^{(*)}(\widehat{\tau}-\widehat{\tau}')}\right]}{(\widehat{\tau}-\widehat{\tau}')^{3/2}} \right\} \mathrm{d}\widehat{\tau}'. \tag{3.129}$$

By choosing, again, the variable z as the variable of integration we have (cf. Eq. (3.126))

$$J(\widehat{r},\widehat{\tau}) = \frac{2}{\sqrt{\pi}} \exp(\widetilde{a}_k \widehat{r}) \int_{z(\widehat{\tau}'=0)}^{\infty} J(0,\widehat{\tau}') \exp\left[-\frac{\widetilde{b}_k \widehat{r}^2}{4b_k^{(*)} z^2} - z^2\right] \mathrm{d}z, \tag{3.130}$$

$$z(\widehat{\tau}'=0) = \frac{\widehat{r}}{\sqrt{4b_k^{(*)}\widehat{\tau}}}. \tag{3.131}$$

With Eq. (3.122) we may write

$$\widetilde{a}_k \widehat{r} - \frac{\widetilde{b}_k \widehat{r}^2}{4b_k^{(*)} z^2} - z^2 = -\left(z - \frac{\widetilde{a}_k \widehat{r}}{2z}\right)^2 \tag{3.132}$$

and Eq. (3.130) gets the form

$$J(\widehat{r},\widehat{\tau}) = \frac{2}{\sqrt{\pi}} \int_{z(\widehat{\tau}'=0)}^{\infty} J(0,\widehat{\tau}') \exp\left[-\left(z - \frac{\widetilde{a}_k \widehat{r}}{2z}\right)^2\right] \mathrm{d}z. \tag{3.133}$$

The exponential function in Eq. (3.133) has a sharp maximum in the vicinity of the value $z = z_0$. This particular value of z is determined by

$$h(z=z_0) = \left(z - \frac{\widetilde{a}_k \widehat{r}}{2z}\right)^2\Bigg|_{z=z_0} = 0, \qquad z_0^2 = \frac{\widetilde{a}_k \widehat{r}}{2}. \tag{3.134}$$

In order to evaluate the integral (3.133), we replace the function $h(z)$ (defined by Eq. (3.134)) by its Taylor expansion in the vicinity of $z = z_0$. We get

$$h(z) \cong 4(z - z_0)^2. \tag{3.135}$$

Moreover, the variable $\widehat{\tau}'$ is connected with z by (cf. Eq. (3.126))

$$\widehat{\tau}' = \widehat{\tau} - \frac{\widehat{r}^2}{4b_k^{(*)} z^2}. \tag{3.136}$$

Since only values of $J(0, \widehat{\tau}')$ in the vicinity of $z = z_0$ contribute to the integral in Eq. (3.133), we replace in Eq. (3.136) the variable z by z_0. We have then

$$\widehat{\tau}' = \widehat{\tau} - \frac{\widehat{r}}{2b_k^{(*)}\widetilde{a}_k} = \widehat{\tau} - \frac{\widehat{r}r_c}{\beta} \equiv \widehat{\tau}_0(\widehat{r}, \widehat{\tau}). \tag{3.137}$$

This way, we may write with Eq. (3.110)

$$\begin{aligned} J(\widehat{r}, \widehat{\tau}) &= J[n_c(0)] \exp\left\{-n_c(0)\varphi\left[\widehat{\tau}_0(\widehat{r}, \widehat{\tau})\right]\right\} \\ &\times \frac{2}{\sqrt{\pi}} \int\limits_{z(\widehat{\tau}'=0)}^{\infty} \exp\left[-4\left(z - z_0\right)^2\right] \mathrm{d}z. \end{aligned} \tag{3.138}$$

Finally, with $\xi = 2(z - z_0)$, Eq. (3.138) yields

$$J(\widehat{r}, \widehat{\tau}) = J[n_c(0)] \exp\left\{-n_c(0)\varphi\left[\widehat{\tau}_0(\widehat{r}, \widehat{\tau})\right]\right\} \frac{1}{\sqrt{\pi}} \int\limits_{\xi(\widehat{\tau}'=0)}^{\infty} \exp\left(-\xi^2\right) \mathrm{d}\xi, \tag{3.139}$$

$$\xi(\widehat{\tau}' = 0) = 2\left[z(\widehat{\tau}' = 0) - z_0\right] = \sqrt{\frac{3r_c^2\widehat{r}}{\widehat{\tau}}}\left(\sqrt{\widehat{r}} - \sqrt{\frac{\beta\widehat{\tau}}{r_c}}\right). \tag{3.140}$$

The flux in cluster-size space has, according to Eq. (3.139), properties of a kink solution. It can be approximated by

$$J(\widehat{r}, \widehat{\tau}) = J[n_c(0)] \exp\left\{-n_c(0)\varphi\left[\widehat{\tau}_0(\widehat{r}, \widehat{\tau})\right]\right\} \theta(\widehat{r}_{\max} - \widehat{r}). \tag{3.141}$$

The theta function $\theta(x)$ is defined by Eq. (3.44), again.

The largest clusters formed at time $\widehat{\tau}$ have a size $\widehat{r}_{\max}$. Its value is determined by $\xi(\widehat{\tau}' = 0) = 0$ or

$$\widehat{r}_{\max} = \frac{\beta}{r_c}\widehat{\tau}. \tag{3.142}$$

Equation (3.137) shows that this value corresponds to $\widehat{\tau}_0 = 0$, i.e., $\varphi(\widehat{\tau}_0) = 0$.

It follows that steady-state conditions are established at a cluster size $\widehat{r}$ after a time interval $\widehat{\tau}$. Going over to the time scale τ (cf. Eq. (3.92)) and the size variable r (cf. Eq. (3.116)) we obtain the following expression for the cluster size-dependent time lag $\tau_{\mathrm{rel}}(r > 2r_c(0))$:

$$\tau_{\mathrm{rel}}(r > 2r_c) = \tau_{\mathrm{rel}}(r \le 2r_c) + \frac{r_c}{\alpha_1 c\beta}\left(r - 2r_c\right). \tag{3.143}$$

Here it is taken into account that $\widehat{\tau} = 0$ corresponds, by definition, to the moment of time when steady-state conditions have been established in the range of cluster sizes $n \le 8n_c$ (or $r \le 2r_c$). Due to such definition, in the transformation to the time scale τ in Eq. (3.143), $\tau_{\mathrm{rel}}(r \le 2r_c)$, the time interval for the establishment of steady-state conditions in the range of cluster sizes $n \leq 8n_c$ (or $r \leq 2r_c$) is added. This quantity is given by (cf. Eqs. (3.34), (3.64), and (3.105))

$$\tau_{\mathrm{rel}}(r \leq 2r_c) = \frac{8n_c^{4/3}}{\beta w_n^{(+)}(n_c)} = \frac{8n_c^{[(4/3)-\kappa]}}{3\alpha_1 c\beta}. \tag{3.144}$$

For kinetically limited growth $\kappa = 2/3$ holds. Equations (3.143) and (3.144) then yield

$$\tau_{\mathrm{rel}}(r > 2r_c) = \tau_{\mathrm{rel}}(r \leq 2r_c)\left\{1 + \frac{3}{4}\left[\left(\frac{r}{2r_c}\right) - 1\right]\right\}. \tag{3.145}$$

Once the flux $J(\widehat{r}, \widehat{\tau})$ (or $J(n, \tau)$) is determined, the cluster-size distribution function can be obtained based on Eq. (3.73). We have, in general,

$$f(n,\tau) = \exp\left(-\frac{\Delta\Phi(n)}{k_B T}\right)\int_n^{\infty}\frac{J(n',\tau)\exp\left(\dfrac{\Delta\Phi(n')}{k_B T}\right)}{w_{n'}^{(+)}}\,\mathrm{d}n'. \tag{3.146}$$

In order to get an analytical expression for $f(n, \tau)$ in the considered range of cluster sizes, we realize that $J(n, \tau)$, $w_n^{(+)}$, and $v(n, \tau)$ are slowly varying functions of n in comparison with the term $\exp(\Delta\Phi(n)/(k_B T))$. By introducing the function η via

$$\eta = -\frac{\Delta\Phi(n')}{k_B T}, \qquad \mathrm{d}\eta = -\frac{1}{k_B T}\frac{\partial\Delta\Phi(n')}{\partial n'}\,\mathrm{d}n', \tag{3.147}$$

we get

$$f(n,\tau) = -\exp\left(-\frac{\Delta\Phi(n)}{k_B T}\right) \times \int_{-\Delta\Phi(n)/k_B T}^{\infty}\frac{J(n',\tau)}{\left[w_{n'}^{(+)}\dfrac{1}{k_B T}\dfrac{\partial\Delta\Phi(n')}{\partial n'}\right]}\exp(-\eta)\,\mathrm{d}\eta. \tag{3.148}$$

Taking into account the remarks made above and Eqs. (3.7) and (3.28), we have, finally,

$$f(n,\tau) \cong \frac{J(n,\tau)}{\left(\dfrac{\mathrm{d}n}{\mathrm{d}\tau}\right)}, \qquad \frac{\mathrm{d}n}{\mathrm{d}\tau} = \frac{3\alpha_1\beta c n^{2/3}}{n_c^{1/3}}, \qquad n \geq 8n_c. \tag{3.149}$$

$J(n, \tau)$ is given here either by Eq. (3.139) or by Eq. (3.141).

3.4.2.2 Diffusion-Limited Growth

The method outlined in detail for kinetically limited growth can be applied with minor modifications also to diffusion limited aggregation processes. We start with Eq. (3.114), again, and set the parameter κ equal to $\kappa = 1/3$ (cf. Eq. (3.14)), now. We obtain with $r \gg r_c$

$$\frac{\partial J(r,\widehat{\tau})}{\partial\widehat{\tau}} = -\frac{\beta}{r_c r}\frac{J(r,\widehat{\tau})}{\partial r} + \frac{1}{3r^2}\left(\frac{\partial^2 J(r,\widehat{\tau})}{\partial r^2} - \frac{2}{r}\frac{\partial J(r,\widehat{\tau})}{\partial r}\right). \tag{3.150}$$

Taking into account the identity

$$\frac{\partial}{\partial r}\left[\frac{1}{r}\frac{\partial J}{\partial r}\right] = -\frac{1}{r^2}\frac{\partial J}{\partial r} + \frac{1}{r}\frac{\partial^2 J}{\partial r^2}, \tag{3.151}$$

we first have

$$\frac{\partial J(r,\widehat{\tau})}{\partial \widehat{\tau}} = -\frac{\beta}{r_c r}\frac{J(r,\widehat{\tau})}{\partial r} + \frac{1}{3r^2}\left[\frac{\partial}{\partial r}\left(\frac{1}{r}\frac{\partial J(r,\widehat{\tau})}{\partial r}\right) - \frac{1}{r^2}\frac{\partial J(r,\widehat{\tau})}{\partial r}\right]. \tag{3.152}$$

With the new variable

$$y = r^2, \tag{3.153}$$

Eq. (3.152) gets the form

$$\frac{\partial J(y,\widehat{\tau})}{\partial \widehat{\tau}} = -\frac{2\beta}{r_c}\frac{\partial J(y,\widehat{\tau})}{\partial y} + \frac{4}{3r_c}\frac{\partial^2 J(y,\widehat{\tau})}{\partial y^2}. \tag{3.154}$$

Similar to the previous case, we set

$$\widehat{y} = y - y_g = r^2 - r_g^2, \qquad r_g = \sqrt[3]{8n_c} = 2n_c^{1/3} = 2r_c. \tag{3.155}$$

With this notation, Eq. (3.154) reads

$$\frac{\partial J(\widehat{y},\widehat{\tau})}{\partial \widehat{\tau}} = -\frac{2\beta}{r_c}\frac{\partial J(\widehat{y},\widehat{\tau})}{\partial \widehat{y}} + \frac{4}{3r_c}\frac{\partial^2 J(\widehat{y},\widehat{\tau})}{\partial \widehat{y}^2}. \tag{3.156}$$

In this way, the partial differential equation for $J(\widehat{y},\widehat{\tau})$ is of the same form as Eq. (3.117) for $J(\widehat{r},\widehat{\tau})$. As a consequence, we can apply the same methods of solution with similar results. Hereby the following replacements have to be made in the final expressions:

$$\widehat{r} \Longrightarrow \widehat{y} \Longrightarrow \widehat{r}^2 = r^2 - r_g^2, \qquad \widetilde{a}_k \Longrightarrow \widetilde{a}_d = \frac{3}{4}\beta, \tag{3.157}$$

$$\widetilde{b}_k \Longrightarrow \widetilde{b}_d = \frac{3}{4}\frac{\beta^2}{r_c}, \qquad b_k^{(*)} \Longrightarrow b_d^{(*)} = \frac{4}{3r_c}. \tag{3.158}$$

Note that the relation $\widetilde{b}_d = b_d^{(*)}\widetilde{a}_d^2$ is fulfilled, again.

A similar relation was used in the analysis of kinetically limited nucleation–growth processes. We have, thus (cf. Eqs. (3.137), (3.139), (3.140), (3.141), (3.142), (3.148), and (3.149))

$$\widehat{\tau}' = \widehat{\tau} - \frac{\widehat{y}}{2b_d^{(*)}\widetilde{a}_d} = \widehat{\tau} - \frac{\widehat{r}^2 r_c}{2\beta} \equiv \widehat{\tau}_0(\widehat{r},\widehat{\tau}), \tag{3.159}$$

$$J(\widehat{r},\widehat{\tau}) = J[n_c(0)]\exp\left\{-n_c(0)\varphi\left[\widehat{\tau}_0(\widehat{r},\widehat{\tau})\right]\right\}\frac{1}{\sqrt{\pi}}\int\limits_{\xi(\widehat{\tau}'=0)}^{\infty}\exp\left(-\xi^2\right)\,d\xi, \tag{3.160}$$

$$\xi(\widehat{\tau}'=0)=2\left\{\frac{\widehat{y}}{\sqrt{4b_d^{(*)}\widehat{\tau}}}-\sqrt{\frac{\widetilde{a}_d\widehat{y}}{2}}\right\}=\sqrt{\frac{3r_c^2\widehat{r}^2}{4\widehat{\tau}}}\left(\sqrt{\widehat{r}^2}-\sqrt{\frac{2\beta\widehat{\tau}}{r_c}}\right), \tag{3.161}$$

$$J(\widehat{r},\widehat{\tau})=J[n_c(0)]\exp\left\{-n_c(0)\varphi\left[\widehat{\tau}_0(\widehat{r},\widehat{\tau})\right]\right\}\theta(\widehat{r}_{\max}-\widehat{r}), \tag{3.162}$$

$$\widehat{r}^2_{\max}=\frac{2\beta}{r_c}\widehat{\tau}, \tag{3.163}$$

$$\tau_{\text{rel}}(r>2r_c)=\tau_{\text{rel}}(r\le 2r_c)\left\{1+\frac{3}{4}\left[\left(\frac{r}{2r_c}\right)^2-1\right]\right\}, \tag{3.164}$$

$$f(n,\tau)\cong\frac{J(n,\tau)}{\left(\frac{\mathrm{d}n}{\mathrm{d}\tau}\right)},\qquad \frac{\mathrm{d}n}{\mathrm{d}\tau}=\frac{3\alpha_1\beta c n^{1/3}}{n_c^{1/3}},\qquad n\ge 8n_c. \tag{3.165}$$

This way the respective analysis for the case of diffusion-limited nucleation–growth processes for the considered region in cluster-size space is also completed.

3.5 Time Interval for Steady-State Nucleation

Based on the results concerning the evolution of the flux and the cluster-size distribution function in the first stages of the nucleation–growth process, an estimate of the time interval for steady-state nucleation is developed, now. This derivation is based on the mass-balance equation. It may be written in the form

$$\Delta_0=\Delta+\int_0^\infty f(n,\tau)n\,\mathrm{d}n\,, \tag{3.166}$$

$$\Delta_0=c_0-c_{\text{eq}}^{(\infty)},\qquad \Delta=c-c_{\text{eq}}^{(\infty)}. \tag{3.167}$$

Here c_0 is the initial and c the actual concentration of segregating particles in the ambient phase ($\Delta(\tau=0)=\Delta_0$, $\Delta(\tau\to\infty)\to 0$).

A derivation of Eq. (3.166) with respect to time yields

$$\frac{\mathrm{d}\Delta(\tau)}{\mathrm{d}\tau}=-\int_0^\infty\left(\frac{\partial f(n,\tau)}{\partial\tau}\right)n\,\mathrm{d}n=\int_0^\infty\left(\frac{\partial J(n,\tau)}{\partial n}\right)n\,\mathrm{d}n \tag{3.168}$$

and after partial integration

$$\frac{\mathrm{d}\Delta(\tau)}{\mathrm{d}\tau}=-\int_0^\infty J(n,\tau)\,\mathrm{d}n. \tag{3.169}$$

For the range of cluster sizes $n \leq g \cong 8n_c$ we assume that steady-state conditions are fulfilled ($J(n,\tau) = J(\Delta)$ =const), while for $n > g$ the flux has to be considered, in general, as a function of n and τ. It follows

$$\frac{\mathrm{d}\Delta(\tau)}{\mathrm{d}\tau} = -\int_0^g J(\Delta)\,\mathrm{d}n - \int_g^\infty J(n,\tau)\,\mathrm{d}n = -J(\Delta)g - \int_g^\infty J(n,\tau)\,\mathrm{d}n. \tag{3.170}$$

The first term on the right-hand side of Eq. (3.170) accounts for the influence of the newly formed clusters of size g on the supersaturation Δ while the second term reflects the influence of the clusters with sizes $n > g$. Both terms on the right-hand side of Eq. (3.170) contribute to the change in the supersaturation. Therefore, as a next step the second term in Eq. (3.170) has to be evaluated. The respective derivations differ to some extent, again, for diffusion and kinetically limited nucleation–growth processes. They have to be carried out separately.

3.5.1 Kinetically Limited Growth

Since for a given moment of time τ the flux is different from zero only in the range of cluster sizes $n \leq n_{\max}$ (cf. Eq. (3.142)), we may write

$$\int_g^\infty J(n,\tau)\,\mathrm{d}n = \int_g^{n_{\max}} J(n,\tau)\,\mathrm{d}n. \tag{3.171}$$

In order to evaluate this integral, we realize that r can be expressed generally as (cf. Eqs. (3.92), (3.116), and (3.137))

$$r = r_g + \frac{\beta\alpha_1 c_0}{r_c}\left[\tau - \tau_0(r,\tau)\right]. \tag{3.172}$$

In terms of the number of particles n in the cluster, we thus have

$$n = \left\{ g^{1/3} + \frac{\beta\alpha_1 c_0}{r_c}\left[\tau - \tau_0(n,\tau)\right] \right\}^3. \tag{3.173}$$

The characteristic curve $\tau_0(n,\tau)$ of the growth equation fulfills hereby the conditions

$$\tau_0(n,\tau) = \begin{cases} \tau & \text{for} \quad n = g \\ 0 & \text{for} \quad n = n_{\max}. \end{cases} \tag{3.174}$$

In addition, according to Eq. (3.139) or (3.141), the flux $J(n,\tau)$ depends on the variables n and τ only via $\tau_0(n,\tau)$. With Eqs. (3.171), (3.173), and (3.174), we obtain

$$\int_g^\infty J(n,\tau)\,\mathrm{d}n = \int_\tau^0 J[\tau_0(n,\tau)]\frac{\mathrm{d}n}{\mathrm{d}\tau_0}\,\mathrm{d}\tau_0 = \int_0^\tau J[\tau_0(n,\tau)]\frac{\mathrm{d}n(\tau-\tau_0)}{\mathrm{d}\tau}\,\mathrm{d}\tau_0. \tag{3.175}$$

Here $[\mathrm{d}n(\tau-\tau_0)/\mathrm{d}\tau] \equiv v(n,\tau-\tau_0)$ is the growth rate of a cluster of size n at a moment of time $(\tau-\tau_0)$.

A substitution of Eq. (3.175) into Eq. (3.170) yields

$$\frac{\mathrm{d}\widetilde{\varphi}(\tau)}{\mathrm{d}\tau} = \frac{J(\tau)g}{c_0} + \int_0^{\tau} \left[\frac{J(\tau_0)}{c_0}\right] \frac{\mathrm{d}n(\tau-\tau_0)}{\mathrm{d}\tau}\,\mathrm{d}\tau_0, \tag{3.176}$$

$$\widetilde{\varphi} = 1 - \frac{c(t)}{c_0}. \tag{3.177}$$

Here $J(\tau)$ is given by $J(\tau) = J(0)\exp[-n_c(0)\varphi(\tau)]$ (cf. Eq. (3.118)).

A partial integration of the second term on the right-hand side of Eq. (3.176) results in

$$\frac{\mathrm{d}\widetilde{\varphi}(\tau)}{\mathrm{d}\tau} = \frac{J(0)g}{c_0}\left(1 + \frac{\beta\alpha_1 c_0}{2n_c^{2/3}}\tau\right)^3. \tag{3.178}$$

Small terms of the order $[J(0)/c_0]^2$ are omitted in Eq. (3.178). The solution of Eq. (3.178) with the initial condition $\widetilde{\varphi}(0) = 0$ gives

$$\widetilde{\varphi}(\tau) = \frac{4J(0)n_c^{7/3}}{\beta\alpha_1 c_0^2}\left[\left(1 + \frac{\beta\alpha_1 c_0}{2n_c^{2/3}}\tau\right)^4 - 1\right]. \tag{3.179}$$

The time interval of steady-state nucleation is determined by $\varphi(\tau_N)n_c(0) \cong 1$ (cf. Eqs. (3.109)–(3.113)). With this expression and Eq. (3.179) we obtain

$$\frac{\widetilde{\varphi}(\tau_N)}{\varphi(\tau_N)}\varphi(\tau_N)n_c = \frac{\widetilde{\varphi}(\tau_N)}{\varphi(\tau_N)} = \frac{4J(0)n_c^{10/3}}{\beta\alpha_1 c_0^2}\left(\frac{\beta\alpha_1 c_0}{2n_c^{2/3}}\tau_N\right)^4. \tag{3.180}$$

Here it was employed that the prefactor to the brackets in Eq. (3.179) is a small quantity.

Moreover, as it is evident from a comparison of Eqs. (3.112) and (3.176), for perfect solutions $\widetilde{\varphi}$ and φ coincide. Therefore, we may introduce a parameter of nonideality as

$$\Omega = \frac{\widetilde{\varphi}[c(\tau_N)]}{\varphi[c(\tau_N)]}. \tag{3.181}$$

Here $c(\tau_N)$ is the value of the concentration at the moment of time $\tau = \tau_N$. It is determined by Eq. (3.113). For perfect solutions, the parameter of nonideality Ω equals 1.

Equations (3.180) and (3.181) yield

$$\tau_N^{(k)} = \left(\frac{c_0}{J(\tau=0)}\right)^{1/4}\left(\frac{4\Omega}{n_c^{2/3}\beta^3(\alpha_1 c_0)^3}\right)^{1/4}. \tag{3.182}$$

The superscript (k) in Eq. (3.182) specifies that the respective results hold for kinetically limited nucleation–growth processes. In real time, this relation reads

$$t_N^{(k)} \cong \frac{3}{2}\frac{a_m^2\Omega^{1/4}}{\omega_s^{2/3}\omega_m^{1/3}c_0\beta^{7/8}n_c^{1/6}D^{(*)}}\exp\left(\frac{1}{4}\frac{\Delta\Phi(n_c)}{k_B T}\right). \tag{3.183}$$

3.5.2 Diffusion-Limited Growth

Instead of Eq. (3.173), we have in the case of diffusion-limited nucleation–growth processes the relation (cf. Eqs. (3.92), (3.157), and (3.159))

$$n = \left\{ g^{2/3} + \frac{2\beta\alpha_1 c_0}{r_c} \left[\tau - \tau_0(n,\tau) \right] \right\}^{3/2} . \tag{3.184}$$

The characteristic curve $\tau_0(n,\tau)$ of the growth equation obeys condition (3.174), again. Moreover, the dependence $\widetilde{\varphi} = \widetilde{\varphi}(\tau)$ is also given by Eq. (3.176). The solution of this equation for diffusion-limited growth reads

$$\widetilde{\varphi}(\tau) = \frac{32 J(0) n_c^2}{5\beta\alpha_1 c_0^2} \left[\left(1 + \frac{\beta\alpha_1 c_0}{2 n_c} \tau \right)^{5/2} - 1 \right] . \tag{3.185}$$

The time interval of steady-state nucleation for diffusion-limited nucleation–growth processes $\tau_N^{(d)}$ is then obtained as

$$\tau_N^{(d)} = \left(\frac{c_0}{J(\tau = 0)} \right)^{2/5} \left(\frac{25\Omega^2}{32 n_c \beta^3 (\alpha_1 c_0)^3} \right)^{1/5} . \tag{3.186}$$

In real time, the respective expression reads

$$t_N^{(d)} \cong 1.4 \frac{a_s^2 \Omega^{2/5} n_c^{2/15}}{(\omega_s c_0) \beta^{4/5} D} \exp\left(\frac{2}{5} \frac{\Delta\Phi(n_c)}{k_B T} \right) . \tag{3.187}$$

Since the order of the nonideality parameter is expected not to differ considerably from unity, $\Omega^{1/4}$ and $\Omega^{2/5}$ can be set equal to 1 in a good approximation. For this reason, nonideality effects enter the expression for the time of steady-state nucleation mainly via the work of critical cluster formation, $\Delta\Phi$.

3.5.3 Nonsteady-State Time Lag and the Time Scale of Steady-State Nucleation

According to Eqs. (3.13), (3.14), and (3.144), the time required to establish steady-state conditions in the range of cluster sizes $n \leq 8n_c$ is given by

$$t_{\mathrm{rel}}(n \leq 8n_c) \cong \begin{cases} \dfrac{8}{3} \dfrac{a_m^2 n_c^{2/3}}{\omega_s^{2/3} \omega_m^{1/3} D^{(*)} c_0 \beta} & \text{kinetically limited growth} \\[2ex] \dfrac{8}{3} \dfrac{a_s^2 n_c}{\omega_s D c_0 \beta} & \text{diffusion-limited growth.} \end{cases} \tag{3.188}$$

With Eqs. (3.183) and (3.187), we obtain the following results for the ratio (t_N/t_{rel}):

$$\frac{t_N^{(k)}}{t_{\mathrm{rel}}^{(k)}} \cong 0.34 \frac{\Omega^{1/4} \beta^{1/8}}{n_c^{5/6}} \exp\left[\frac{1}{4} \frac{\Delta\Phi(n_c)}{k_B T} \right] \gg 1 , \tag{3.189}$$

$$\frac{t_N^{(d)}}{t_{\mathrm{rel}}^{(d)}} \cong 0.5 \frac{\Omega^{2/5}\beta^{1/5}}{n_c^{13/15}} \exp\left[\frac{2}{5}\frac{\Delta\Phi(n_c)}{k_B T}\right] \gg 1. \tag{3.190}$$

In both the cases, the time interval of steady-state nucleation exceeds the time lag considerably.

Once the characteristic time scales for steady-state nucleation are determined, a number of additional basic characteristics of the nucleation–growth process may be derived in a straightforward way.

3.6 Further Basic Characteristics of Nucleation–Growth Processes

3.6.1 Number of Clusters Formed by Nucleation

The maximum number, N_{max}, of supercritical clusters formed by nucleation processes and capable to a further deterministic growth may be determined via

$$N_{\mathrm{max}} = \int_0^{\tau_N} J(g,\tau)\, \mathrm{d}\tau. \tag{3.191}$$

Rewriting this equation in the form

$$N_{\mathrm{max}} = \int_0^{\tau_N} J(g,0)\left(1 + \frac{[J(g,\tau) - J(g,0)]}{J(g,0)}\right) \mathrm{d}\tau \tag{3.192}$$

we get

$$N_{\mathrm{max}} = J(g,0)\tau_N \left[1 - \frac{1}{\tau_N}\int_0^{\tau_N} \frac{(J(g,0) - J(g,\tau))}{J(g,0)}\right] \mathrm{d}\tau. \tag{3.193}$$

Taking into account the relations

$$J(\tau) = J(0)\exp[-n_c(0)\varphi(\tau)], \tag{3.194}$$

$$n_c(0)\varphi(\tau) \leq n_c(0)\varphi(\tau_N) \cong 1, \tag{3.195}$$

we have

$$\frac{1}{\tau_N}\int_0^{\tau_N}\left[\frac{J(g,0) - J(g,\tau)}{J(g,0)}\right] \mathrm{d}\tau \leq \frac{1}{\tau_N}\int_0^{\tau_N}\left[\frac{J(g,0) - J(g,\tau_N)}{J(g,0)}\right] \mathrm{d}\tau < \frac{2}{3}. \tag{3.196}$$

We may formulate, therefore, in a good approximation a very simple estimate for the maximum number of supercritical clusters capable to a further deterministic growth:

$$N_{\mathrm{max}} \cong J(g,0)\tau_N. \tag{3.197}$$

With Eqs. (3.182) and (3.186), we have for the both considered modes of growth

$$N_{\max}^{(k)} = J^{3/4}(0)\left(\frac{4c_0\Omega}{n_c^{2/3}\beta^3(\alpha_1 c_0)^3}\right)^{1/4}, \tag{3.198}$$

$$N_{\max}^{(d)} = J^{3/5}(0)\left(\frac{25c_0^2\Omega}{32n_c\beta^3(\alpha_1 c_0)^3}\right)^{1/5}. \tag{3.199}$$

Substituting $J(0)$ with Eqs. (3.13), (3.14), (3.15), (3.77), and (3.78) one gets, finally,

$$N_{\max}^{(k)} \cong \frac{c_0\Omega^{1/4}}{n_c^{1/6}\beta^{3/8}}\exp\left(-\frac{3}{4}\frac{\Delta\Phi(n_c)}{k_BT}\right), \tag{3.200}$$

$$N_{\max}^{(d)} \cong \frac{3c_0\Omega^{1/5}}{4n_c^{2/5}\beta^{3/10}}\exp\left(-\frac{3}{5}\frac{\Delta\Phi(n_c)}{k_BT}\right). \tag{3.201}$$

Hereby the quantities n_c, β, and $\Delta\Phi(n_c)$ are given by (cf. Eqs. (3.21), (3.24), (3.25), and (3.26))

$$n_c^{1/3} = \frac{\beta}{[\Delta\mu/(k_BT)]}, \qquad \beta = \frac{8\pi}{3}\left(\frac{\sigma a_s^2}{k_BT}\right), \tag{3.202}$$

$$\frac{\Delta\Phi(n_c)}{k_BT} = \frac{1}{2}\frac{\beta^3}{[\Delta\mu/(k_BT)]^2}. \tag{3.203}$$

As is evident from Eqs. (3.200) and (3.201), the number of clusters formed in the system is determined mainly by the initial concentration of the segregating particles c_0 and the value of the work of critical cluster formation. It does not depend on the diffusion coefficient D or $D^{(*)}$. In the calculation of the work of critical cluster formation nonideality effects have to be taken into account, of course. This is the main factor where deviations from ideality enter the description.

3.6.2 Average Size of the Clusters

The stage of dominating nucleation is followed, in general, by a stage of dominating independent growth of the supercritical clusters present in the system [28, 230]. Independent growth means that the supercritical clusters do not influence each other directly but grow at the expense of the excess monomers and the supersaturation tends to zero [$\Delta(t) = (c - c_{\text{eq}}^{(\infty)}) \to 0$]. After this process is finished, the transformation reaches a certain degree of completion (cf. [28]) and goes over into a third, late, stage of competitive growth denoted usually as coarsening or Ostwald ripening (cf. Chapter 4 and Lifshitz and Slezov [155], Slezov and Sagalovich [289]).

In nucleation–growth experiments, the earliest states commonly observed correspond to the end of the stage of independent growth. Therefore, in the calculation of the average sizes of the clusters to compare them with experiment one has to take into account the deterministic growth of the supercritical clusters. This growth is stopped after the supersaturation reaches values near zero (for a more detailed and precise formulation see, e.g., [220, 230]).

Since we have already determined the number of clusters, $N_{\max}$, formed in the course of nucleation, by a purely thermodynamic argumentation it is possible to give estimates also of the average size, $\langle R \rangle$, of the clusters and its dependence on the initial supersaturation. Approximately, we may write the mass-balance equation for the final state of independent growth in the form

$$\left(c_0 - c_{\mathrm{eq}}^{(\infty)}\right) = \frac{4\pi \langle R \rangle^3}{3\omega_s} N_{\max} . \tag{3.204}$$

This relation is equivalent to

$$\langle R \rangle^3 = \left(\frac{3\omega_s c_{\mathrm{eq}}^{(\infty)}}{4\pi} \right) \left(\frac{1}{N_{\max}} \right) \left[\left(\frac{c_0}{c_{\mathrm{eq}}^{(\infty)}} \right) - 1 \right] . \tag{3.205}$$

A substitution of the expressions for $N_{\max}$ (cf. Eqs. (3.200) and (3.201)) into Eq. (3.205) yields

$$\langle R \rangle \cong \begin{cases} a_s \left\{ \left[\dfrac{c_0 - c_{\mathrm{eq}}^{(\infty)}}{c_0} \right] \left(\dfrac{n_c^{1/6} \beta^{3/8}}{\Omega^{1/4}} \right) \right\}^{1/3} \exp\left(\dfrac{1}{4} \dfrac{\Delta\Phi(n_c)}{k_B T} \right) \\ \qquad \text{for kinetically limited growth} \\ a_s \left\{ \left[\dfrac{c_0 - c_{\mathrm{eq}}^{(\infty)}}{c_0} \right] \left(\dfrac{4 n_c^{2/5} \beta^{3/10}}{3\Omega^{1/5}} \right) \right\}^{1/3} \exp\left(\dfrac{1}{5} \dfrac{\Delta\Phi(n_c)}{k_B T} \right) \\ \qquad \text{for diffusion-limited growth} \end{cases} \tag{3.206}$$

or in a good approximation

$$\langle R \rangle \cong \begin{cases} a_s \exp\left(\dfrac{1}{4} \dfrac{\Delta\Phi(n_c)}{k_B T} \right) & \text{kinetically limited growth} \\ a_s \exp\left(\dfrac{1}{5} \dfrac{\Delta\Phi(n_c)}{k_B T} \right) & \text{diffusion-limited growth.} \end{cases} \tag{3.207}$$

3.6.3 Time Interval of Independent Growth

For segregation processes in solutions, the deterministic growth of the clusters (in the second stage of the phase transformation) is limited commonly by diffusion processes of the segregating component to the newly evolving phase. Neglecting surface effects, we have in this case (cf. also Eqs. (3.12), (3.26), (3.28), and (3.30))

$$w_{n,n+1}^{(+)} = 4\pi R D c , \tag{3.208}$$

$$\frac{\mathrm{d}n}{\mathrm{d}t} = -w_{n,n+1}^{(+)} \left\{ \frac{1}{k_B T} \frac{\partial \Delta\Phi(n)}{\partial n} \right\} = 4\pi R D c \left[\ln \left(\frac{\phi}{\phi_{\mathrm{eq}}^{(\infty)}} \right) \right] . \tag{3.209}$$

In terms of the radius of the clusters this equation reads (cf. Eq. (3.11))

$$\frac{\mathrm{d}R}{\mathrm{d}t} = \frac{(D\omega_s)}{R} c \left[\ln \left(\frac{\phi}{\phi_{\mathrm{eq}}^{(\infty)}} \right) \right] . \tag{3.210}$$

From the mass-balance equation, furthermore we obtain, assuming that all clusters are nearly of equal sizes,

$$c = c_0 - N R^3 \left(\frac{4\pi}{3\omega_s} \right) . \tag{3.211}$$

Substituting into Eq. (3.210) yields

$$\frac{\mathrm{d}R}{\mathrm{d}t} = \frac{D\omega_s}{R} \left[c_0 - \left(\frac{4\pi N}{3\omega_s} \right) R^3 \right] \left[\ln \left(\frac{\phi}{\phi_{\mathrm{eq}}^{(\infty)}} \right) \right] . \tag{3.212}$$

From this equation, as an estimate for the duration t_{growth} of the stage of independent growth, we obtain for diffusion limited growth

$$t_{\mathrm{growth}}^{(d)} \cong \frac{\langle R \rangle^2}{2 D \omega_s c_0} . \tag{3.213}$$

Substituting the mean radius $\langle R \rangle$, with Eq. (3.207), we thus have

$$t_{\mathrm{growth}}^{(kd)} \cong \frac{a_s^2}{2 D \omega_s c_0} \exp \left(\frac{1}{2} \frac{\Delta\Phi(n_c)}{k_B T} \right) , \tag{3.214}$$

if the nucleation process proceeds by kinetically limited growth, and

$$t_{\mathrm{growth}}^{(dd)} \cong \frac{a_s^2}{2 D \omega_s c_0} \exp \left(\frac{2}{5} \frac{\Delta\Phi(n_c)}{k_B T} \right) , \tag{3.215}$$

for diffusion-limited nucleation.

The ratio of independent growth time to that of steady-state nucleation $(t_{\mathrm{growth}}/t_N)$ can be written as (cf. Eqs. (3.183), (3.187), (3.214), and (3.215))

$$\frac{t_{\mathrm{growth}}^{(kd)}}{t_N^{(k)}} \cong \frac{\beta^{7/8} n_c^{1/6}}{3\Omega^{1/4}} \left[\frac{D^{(*)}}{D} \right] \exp \left(\frac{1}{4} \frac{\Delta\Phi(n_c)}{k_B T} \right) , \tag{3.216}$$

$$\frac{t_{\mathrm{growth}}^{(dd)}}{t_N^{(d)}} \cong \frac{\beta^{4/5}}{3\Omega^{2/5} n_c^{2/15}} . \tag{3.217}$$

In the case when both nucleation and the stage of independent growth are limited kinetically, we have instead of Eq. (3.213)

$$t_{\mathrm{growth}}^{(k)} \cong \frac{a_m \langle R \rangle}{D^{(*)} \omega_s c_0} . \tag{3.218}$$

With Eqs. (3.183) and (3.207) we arrive at

$$\frac{t_{\text{growth}}^{(kk)}}{t_N^{(k)}} \cong \frac{2\beta^{7/8} n_c^{1/6}}{3\Omega^{1/4}} . \tag{3.219}$$

In all considered cases, t_{growth} noticeably exceeds t_N. It follows that the time t_{compl},

$$t_{\text{compl}} = t_{\text{rel}} + t_N + t_{\text{growth}} , \tag{3.220}$$

required to reach the stage of competitive growth, is determined mainly by the value of t_{growth}.

3.7 Time of Steady-State Nucleation and Induction Time

In the kinetics of phase transformation processes, a less detailed description is widely employed dealing exclusively with the time evolution of the total mass or volume fraction of the newly evolving phase [13, 49, 129, 191]. In such type of descriptions, this induction period or latent time of the transformation, till the appearance of the new phase can be observed experimentally, plays a major role. In order to give a comparison of both approaches, their advantages and limitations, such a method of description is described here briefly as well.

Assume that the nucleation process proceeds with some given rate $J(t')$ starting with some moment of time $t = 0$. The number of supercritical clusters, formed in the time range $t', t' + \mathrm{d}t'$, is then given by

$$\mathrm{d}N(t') = J(t')\,\mathrm{d}t' . \tag{3.221}$$

The clusters, once formed, grow and give at time t a contribution $\mathrm{d}V(t,t')$

$$\mathrm{d}V(t,t') = v(t,t')\,\mathrm{d}N(t') \tag{3.222}$$

to the total volume $V(t)$ of the newly evolving phase. Here $v(t,t')$ denotes the volume of a cluster at time t is denoted which has been formed originally at time t'. It is then commonly assumed that this quantity is determined mainly by the time of growth $t - t'$.

Denoting further by G_R the growth rate of the linear dimensions of the aggregates of the newly evolving phase and by ω_n a geometrical shape factor, the cluster volume $v(t,t')$ may be expressed as

$$v(t,t') = \omega_n \left\{ \int_{t'}^{t} G_R(t'' - t')\,\mathrm{d}t'' \right\}^n . \tag{3.223}$$

Here the parameter n specifies the number of independent directions of cluster growth in space.

Going temporarily over to the variable $x = t'' - t'$, we obtain

$$v(t,t') = \omega_n \left\{ \int_{0}^{t-t'} G_R(x)\,\mathrm{d}x \right\}^n . \tag{3.224}$$

We then find immediately

$$V(t) = \omega_n \int_0^t J(t')\,\mathrm{d}t' \left\{ \int_0^{t-t'} G_R(t'')\,\mathrm{d}t'' \right\}^n . \tag{3.225}$$

For spherical clusters and three-dimensional phase formation, we have

$$\frac{\mathrm{d}R}{\mathrm{d}t} = \gamma_1, \qquad \gamma_1 = \text{constant}, \qquad G_R(t) = \gamma_1 t \tag{3.226}$$

for kinetically limited growth and

$$\frac{\mathrm{d}R}{\mathrm{d}t} = \frac{\gamma_2}{2R}, \qquad \gamma_2 = \text{constant}, \qquad G_R(t) = \frac{\sqrt{\gamma_2}}{2t^{1/2}} \tag{3.227}$$

for diffusion-limited growth. Assuming, in addition, constancy of the nucleation rate, we arrive at

$$V(t) = \Gamma_1 J t^{n+1}, \qquad \Gamma_1 = \frac{\omega_n \gamma_1^n}{(n+1)} \tag{3.228}$$

for kinetically limited growth and

$$V(t) = \Gamma_2 J t^{(n+2)/2}, \qquad \Gamma_2 = \left(\frac{2}{n+2}\right) \omega_n \gamma_2^{n/2} \tag{3.229}$$

for diffusion-limited growth.

Remember that the induction time of the transformation is usually defined as follows: it is the time from the beginning of the nucleation–growth process till the moment when the new phase is observed first experimentally. Therefore, if we select some special value for the total amount of the new phase as the lower limit for experimental observation, we then obtain the following dependences for the induction times:

$$t_{\mathrm{ind}}^{(k)} \propto J^{1/(n+1)}, \tag{3.230}$$

$$t_{\mathrm{ind}}^{(d)} \propto J^{2/(n+2)}, \tag{3.231}$$

respectively.

The induction times, determined in the sketched way, are characterized by similar dependences from the steady-state nucleation rate as the expressions for the time of steady-state nucleation, Eqs. (3.183) and (3.187), derived here earlier. Note, however, that the induction times do not coincide, in general, with the times of steady-state nucleation. Their values depend on the choice of the volume of the newly evolving phase, which is experimentally measurable. Note that in our approach the length of the interval of time, where steady-state nucleation with constant rates occurs, is determined as a result of the theory and not taken as an assumption. Taking into account that the time of steady-state nucleation is, in general, much shorter as compared with the time of independent growth, the applicability of the assumption of constancy of the nucleation rate for the estimation of the induction times is, in general, obviously not correct. It could be correct only if the detectable volumes of the newly evolving phase are quite small, so that the new phase is observed already in time scales comparable with the time of steady-state nucleation.

For an accurate description of the time evolution of the total volume of the newly evolving phase one has to employ first the above method assuming steady-state nucleation. But after the time of steady-state nucleation has passed, the further growth of the new phase is governed by the simultaneous independent growth of a nearly constant number of supercritical clusters. Thus, the respective $V(t)$-dependences may differ from the results derived by assuming constancy of the nucleation rate. Indeed, for times $t \geq t_N$ we have instead of Eq. (3.225)

$$V(t) = \omega_n \int_0^{t_N} J(t')\,\mathrm{d}t' \left\{ \int_0^{t-t'} g(t'')\,\mathrm{d}t'' \right\}^n , \qquad t \geq t_N . \tag{3.232}$$

For the case of kinetically limited growth we then obtain with $J =$const

$$V(t) = \frac{\omega_n \gamma_1^n}{(n+1)} J t^{n+1} \left[1 - \left(1 - \frac{t_N^{(k)}}{t} \right)^{n+1} \right] . \tag{3.233}$$

For large times ($t_N/t \ll 1$) this equation is reduced, approximately, to

$$V(t) \cong \omega_n \gamma_1^n \left(J t_N^{(k)} \right) t^n . \tag{3.234}$$

Similarly, we have for diffusion- limited growth

$$V(t) = \omega_n \gamma_2^{n/2} \left(\frac{2}{n+2} \right) J t^{(n+2)/2} \left[1 - \left(1 - \frac{t_N^{(d)}}{t} \right) \right] \tag{3.235}$$

and (in the range $t_N/t \ll 1$)

$$V(t) \cong \omega_n \gamma_2^{n/2} \left(J t_N^{(d)} \right) t^{n/2} . \tag{3.236}$$

Thus, one has to take care in deducing conclusions concerning possible growth mechanisms from the time dependence of the $V(t)$-curves, only, if it is not established by an independent analysis whether the condition of constancy of the nucleation rate is fulfilled or not. A more detailed analysis of related problems can be found in Ref. [115].

3.8 Formation of a New Phase with a Given Stoichiometric Composition

3.8.1 The Model

In the description of the kinetics of phase transformations, mainly two kinds of processes are well studied [51, 94, 110]

- processes of phase formation when the composition of the ambient and the newly evolving phases are practically the same;
- segregation processes of one of the components in the ambient phase.

The general case that several components segregate and form a new phase of more or less arbitrary composition (nucleation in multicomponent systems) is intensively studied as well. However, due to the enormous theoretical difficulties involved in such a task the problem remains, at present, far from being satisfactorily solved.

Nevertheless, in the limiting case that the newly evolving phase has a definite stoichiometric composition, the situation is less complicated. As will be shown (cf. also Refs. [293, 304, 309]), in this limiting case of multicomponent nucleation–growth processes, the kinetic equations can be reduced to the respective expressions valid for nucleation and growth in one-component systems. However, the effective diffusion coefficients and the supersaturation of the system are expressed in a fairly complicated way via the kinetic and thermodynamic properties of the different components involved in the phase formation process. The development of the respective expressions and their possible application in the further study of the course of the phase separation is the aim of the present section.

3.8.2 Basic Equations

By assumption, the newly evolving multi-component phase has a definite stoichiometric composition. This assumption implies that the aggregates of the new phase consist of n primary units (or groups of particles). The composition of each of such a group of particles is given by the stoichiometric coefficients ν_i.

The process of formation and growth of aggregates of a given stoichiometric composition can be understood thus as mediated by incorporation or emission of such primary building units each consisting of ν_i particles of the different components i. The state of the system is characterized then by the cluster-size distribution function $f(n,t)$, which is the number (or number density) of clusters containing n primary units. Its evolution is governed by a kinetic equation of the standard form (Frenkel–Zeldovich equation) [303]

$$\frac{\partial f(n,t)}{\partial n} = \frac{\partial}{\partial n}\left\{ w^{(+)}_{n,n+1}\left[\frac{\partial f(n,t)}{\partial n} + \frac{f(n,t)}{k_B T}\frac{\partial \Delta\Phi(n)}{\partial n}\right]\right\}. \tag{3.237}$$

The kinetic coefficients $w^{(+)}_{n,n+1}$ have the meaning of the average number of primary building units of the new phase incorporated into a cluster of size n per unit time interval. k_B is the Boltzmann constant and T the absolute temperature.

The peculiarities of the considered particular process enter the description via the determination of the expressions for the work of cluster formation, $\Delta\Phi(n)$, and the coefficients of aggregation, $w^{(+)}_{n,n+1}$. For the work of cluster formation, $\Delta\Phi(n)$, we may write

$$\Delta\Phi(n) = -n\left[\mu_s - \mu_s^{(n)}\right] \tag{3.238}$$

or, applying the concept of a specific interfacial energy

$$\Delta\Phi(n) = -n\Delta\mu + \alpha n^{2/3}, \qquad \alpha = 4\pi\sigma\left(\frac{3\omega_s}{4\pi}\right)^{2/3} = 4\pi\sigma a_s^2. \tag{3.239}$$

The thermodynamic potentials, μ_s, of the primary building units of the newly evolving phase are connected with the respective values, μ_i, of the single components by the following dependences:

$$\Delta\mu = \left[\mu_s - \mu_s^{(\infty)}\right], \qquad \mu_s^{(\infty)} = \sum_i \nu_i \mu_i^{(\infty)}, \tag{3.240}$$

$$\mu_s^{(n)} = \sum_i \nu_i \mu_i^{(n)}, \qquad \mu_s = \sum_i \nu_i \mu_i. \tag{3.241}$$

Here μ_s is the chemical potential of a primary building unit in the ambient phase for the given values of pressure p, temperature T and molar fractions x_i, $i = 1, 2, \ldots, k-1$, of the k different components, $\mu_s^{(\infty)}$ is its value in the case of stable coexistence of both phases at a planar interface while $\mu_s^{(n)}$ refers to the respective equilibrium value required for a stable existence of a cluster of size n in the ambient phase. Similarly, μ_i are the chemical potentials of the different components in the ambient phase, while $\mu_i^{(\infty)}$ and $\mu_i^{(n)}$ denote the respective values for a stable coexistence of both phases at a planar interface or for a cluster of size n. The primary units in the cluster phase are characterized by their volume ω_s and radius a_s; σ is their surface tension or specific interfacial energy. Since the composition and state of the newly evolving phase is assumed to be fixed (independent of cluster size), σ has to be taken also as independent of the size of the cluster of the newly evolving phase.

According to Eq. (3.238), the critical cluster size n_c and the work of formation of critical clusters $\Delta\Phi(n_c)$ have the same form as for one-component systems, i.e.,

$$n_c^{1/3} = \frac{2\alpha}{3\Delta\mu}, \qquad \Delta\Phi(n_c) = \frac{1}{3}\alpha n_c^{2/3}. \tag{3.242}$$

In contrast to the one-component case, the thermodynamic driving force of the transformation is, however, a function of the chemical potentials μ_i of the different components. We may express the chemical potentials of the k different components in the ambient phase via their activities φ_i, $i = 1, 2, \ldots, k$, or the coefficients of activity f_i as [97]

$$\mu_i(p, T, x_1, x_2, \ldots, x_{k-1}) = \mu_{i0}(p, T) + k_B T \ln\varphi_i, \quad \varphi_i = f_i x_i. \tag{3.243}$$

For perfect solutions ($f_i = 1$, $i = 1, 2, \ldots, k$), the activities of the different components φ_i are equal to the molar fractions x_i.

In terms of the activities of different components, the difference of the chemical potential $\Delta\mu$ may be thus written as

$$\frac{\Delta\mu}{k_B T} = \frac{1}{k_B T} \sum_i \nu_i \left[\mu_i - \mu_i^{(\infty)}\right] = \ln \left[\frac{\prod_i \varphi_i^{\nu_i}}{\prod_i \left[\varphi_i^{(\infty)}\right]^{\nu_i}}\right]. \tag{3.244}$$

The superscript (∞) specifies, again, the respective values of the activities of the different components for the case of a stable coexistence of both phases near a planar interface.

In order to apply Eq. (3.237), the coefficients of aggregation $w_{n,n+1}^{(+)}$ have to be specified as a next step. Generally, they are connected with the macroscopic growth rates, v_n, of an aggregate of given stoichiometric composition and thermodynamic characteristics via [298]

$$v_n = \frac{\mathrm{d}n}{\mathrm{d}t} = -4\pi R_n^2 j_n = -w_{n,n+1}^{(+)} \left[\frac{1}{k_B T}\left(\frac{\partial \Delta\Phi}{\partial n}\right)\right]. \tag{3.245}$$

Here j_n is the density of fluxes of primary building units to the aggregate of the newly evolving phase near and perpendicular to the surface of the aggregate. Similar to the thermodynamic properties, the effective kinetic parameters of the primary building units have to be expressed, now, via the respective characteristics of the different components.

The rate of incorporation of particles of the component i through the interface of an aggregate of size n or a radius R_n can be written similar to Eq. (3.245) as [298]

$$\frac{\mathrm{d}n_i}{\mathrm{d}t} = -4\pi R_n^2 j_i = -w_{n_i,n_i+1}^{(+)} \left[\frac{1}{k_B T}\left(\frac{\partial \Delta\Phi}{\partial n_i}\right)\right]. \tag{3.246}$$

Here j_i is the density of fluxes of particles of the ith component in the immediate vicinity of the aggregate of the new phase perpendicular to its surface.

The coefficients of aggregation can be expressed from microscopic considerations (number of jumps per unit time) in the form [293]

$$w_{n_i,n_i+1}^{(+)} = \left(\frac{D_i^{(s)}}{a_m^2}\right)\left(\frac{4\pi R_n^2 a_m}{\omega_m}\right) x_i, \qquad \omega_m = \frac{4\pi}{3} a_m^3. \tag{3.247}$$

In the above equations, ω_m is the average volume a particle of one of the components occupies in the solid solution, and $D_i^{(s)}$ is the partial diffusion coefficient of the respective component in the immediate vicinity of the cluster of the newly evolving phase. It may be equal but generally less than the respective bulk value, D_i. Thus, we may replace $D_i^{(s)}$ by $D_i\alpha_i$ with $\alpha_i \le 1$. The term $(D_i^{(s)}/a_m^2)$ has the meaning of the average frequency of a jump of a particle of the component i into the direction of the aggregate perpendicular to the interface, and a_m is a measure of the average length for such a jump. The remaining term in Eq. (3.247) gives the number of particles of the respective component capable to perform such an elementary process.

Since we consider here the case of formation of a new phase with a given stoichiometric composition, the aggregation rates of the different components are connected by the constraints

$$n_i = \nu_i n , \qquad n = \frac{4\pi}{3} \frac{R_n^3}{\omega_s} = \left(\frac{R_n}{a_s} \right)^3 . \tag{3.248}$$

With these relations, Eqs. (3.245) and (3.246), the rate of change of the number of structural units n may be written as

$$\frac{\mathrm{d}n}{\mathrm{d}t} = - \frac{4\pi R_n^2 \sum_i \omega_i j_i}{\omega_s} = -4\pi R_n^2 \left(\frac{j_i}{\nu_i} \right) . \tag{3.249}$$

In the above equation, the identities

$$\frac{j_1}{\nu_1} = \frac{j_2}{\nu_2} = \cdots = \frac{j_k}{\nu_k} , \qquad \omega_s = \sum_i \nu_i \omega_i \tag{3.250}$$

have been employed. ω_i is the volume of the component i in the newly evolving phase.

A substitution of the expressions for (j_i/ν_i) (cf. Eqs. (3.246) and (3.247)) into the right-hand side of Eq. (3.249) yields

$$\left(\frac{\nu_i^2}{D_i^{(s)} x_i} \right) \frac{\mathrm{d}n}{\mathrm{d}t} = \frac{3n^{2/3}}{a_m^2} \left(\frac{\omega_s}{\omega_m} \right)^{2/3} \frac{1}{k_B T} \left\{ \nu_i \left(\mu_i - \mu_i^{(n)} \right) \right\} . \tag{3.251}$$

By taking the sum over all components, we have

$$\frac{\mathrm{d}n}{\mathrm{d}t} = \frac{3D_{\text{eff}}^{(s)} n^{2/3}}{a_m^2} \left(\frac{\omega_s}{\omega_m} \right)^{2/3} \frac{1}{k_B T} \sum_{i=1}^{k} \nu_i \left[\mu_i - \mu_i^{(n)} \right] \tag{3.252}$$

or with Eqs. (3.238)–(3.240)

$$\frac{\mathrm{d}n}{\mathrm{d}t} = - \frac{3D_{\text{eff}}^{(s)} n^{2/3}}{a_m^2} \left(\frac{\omega_s}{\omega_m} \right)^{2/3} \frac{1}{k_B T} \left(\frac{\partial \Delta\Phi(n)}{\partial n} \right) . \tag{3.253}$$

$D_{\text{eff}}^{(s)}$ denotes the effective diffusion coefficient; it is given according to the above derivations by

$$D_{\text{eff}}^{(s)} = \frac{1}{\sum_i \frac{\nu_i^2}{x_i D_i^{(s)}}} = \frac{1}{\sum_i \frac{\nu_i^2}{(x_i D_i \alpha_i)}} . \tag{3.254}$$

Going over from the partial diffusion coefficients, D_i, to the values $D_i^{(0)}(p,T)$ for the case of the absence of interactions of the solute particles (perfect solution) we may replace in Eq. (3.254) $D_i x_i$ by $D_i^{(0)} \varphi_i$ [304].

It follows as a consequence from Eqs. (3.245) and (3.253) that the coefficient of aggregation $w^{(+)}_{n,n+1}$ can be expressed as

$$w^{(+)}_{n,n+1} = \frac{3D^{(s)}_{\text{eff}}n^{2/3}}{a^2_m}\left(\frac{\omega_s}{\omega_m}\right)^{2/3} = 4\pi D^{(s)}_{\text{eff}}n^{2/3}a_m c\left(\frac{\omega_s}{\omega_m}\right)^{2/3} . \tag{3.255}$$

Here the relation $c = 1/\omega_m$ has been employed, in addition. Since ω_m is, by definition, equal to the average volume a particle occupies in the solid solution, c has the meaning of the total volume concentration of the particles in the solution. Latter replacement transfers the expression for $w^{(+)}_{n,n+1}$ thus also formally into the same form as obtained for segregation in a one-component solid solution. Consequently, almost any of the results obtained earlier for this case can be employed, now, for the analysis of the problem considered here.

Denoting by $\left\{\varphi^{(n)}_i\right\}$ the values of the activities of the different components in the ambient phase required for an equilibrium coexistence of a cluster of size n in the ambient phase, we may rewrite Eq. (3.253) in the form

$$\frac{dn}{dt} = \frac{3D^{(s)}_{\text{eff}}n^{2/3}}{a^2_m}\left(\frac{\omega_s}{\omega_m}\right)^{2/3}\ln\left[\frac{\prod\limits_i \varphi_i^{\nu_i}}{\prod\limits_i \left[\varphi^{(n)}_i\right]^{\nu_i}}\right] . \tag{3.256}$$

The particular values of the activities $\left\{\varphi^{(n)}_i\right\}$ can be determined from

$$\mu^{(n)}_i = \mu^{(\infty)}_i + \frac{2\alpha}{3\nu_i n^{1/3}} . \tag{3.257}$$

Indeed, to prove this equation we may write down according to Eqs. (3.239), (3.240), and (3.248) the change of the thermodynamic potential in the formation of an aggregate of size n as

$$\Delta\Phi = -\sum_i n_i\left[\mu_i - \mu^{(\infty)}_i\right] + \alpha\left(\frac{n_i}{\nu_i}\right)^{2/3} . \tag{3.258}$$

A derivation of this equation with respect to n_i yields

$$\frac{\partial\Delta\Phi}{\partial n_i} = -\left[\mu_i - \mu^{(\infty)}_i\right] + \frac{2\alpha}{3\nu_i n^{1/3}} = 0. \tag{3.259}$$

The cluster of a given size is in equilibrium with the ambient phase if the chemical potential of the different components in the cluster $\mu^{(n)}_i$ is equal to the respective values μ_i in the ambient phase. This way, Eq. (3.257) is verified.

So far we have considered the case of kinetically limited growth. In this case, the concentration in the vicinity of the growing or dissolving aggregates coincides with the average composition of the ambient phase far away from the growing or dissolving clusters. Analogous expressions may be established similarly for other mechanisms of cluster growth. In particular, for diffusion-limited growth we obtain [304]

$$w^{(+)}_{n,n+1} = \frac{3D^{(b)}_{\text{eff}} n^{1/3}}{a_m^2} \left(\frac{\omega_s}{\omega_m}\right)^{1/3} = 4\pi D^{(b)}_{\text{eff}} a_s c n^{1/3}, \tag{3.260}$$

$$D^{(b)}_{\text{eff}} = \frac{1}{\sum\limits_i \dfrac{\nu_i^2}{x_i D_i}}. \tag{3.261}$$

While nucleation is usually dominated by kinetically limited growth, Eq. (3.260) has to be employed for the description of the growth of the supercritical clusters and for the description of Ostwald ripening (see Chapter 4 and [155]). Note as well that the expressions for the effective diffusion coefficients in the case of formation of a phase with a well-defined stoichiometric composition – as discussed here – can be obtained straightforwardly from the more general relations, Eqs. (2.96)–(2.98), derived in Chapter 2 by setting the derivative of cluster composition with respect to cluster size equal to zero.

Summarizing the first part of this section, the specification of both the thermodynamic and kinetic quantities required for the theoretical description of formation of a phase of a given stoichiometric composition is completed and the kinetics of phase formation may be considered, now.

3.8.3 Applications

As a first consequence from the analysis of Eq. (3.237), we obtain the expression for the steady-state nucleation rate J in the form (see, e.g., [298, 303, 304, 306])

$$J = \left[c \prod_i x_i^{\nu_i}\right] w^{(+)}_{n,n+1}(n_c) \Upsilon_{(z)} \exp\left(-\frac{\Delta\Phi(n_c)}{k_B T}\right), \tag{3.262}$$

$$\Upsilon_{(z)} = \sqrt{-\frac{1}{2\pi k_B T}\left(\left.\frac{\partial^2 \Delta\Phi}{\partial n^2}\right|_{n=n_c}\right)} = \sqrt{\frac{1}{6\pi}\left(\frac{\beta}{n_c^{4/3}}\right)}. \tag{3.263}$$

Here $\Delta\Phi(n_c)$ and n_c are the work of critical cluster formation and the critical cluster size in nucleation determined by Eq. (3.242) or

$$n_c^{1/3} = \frac{2\alpha}{3\Delta\mu} = \frac{\beta}{(\Delta\mu/k_B T)}, \qquad \beta = \frac{2\alpha}{3k_B T}. \tag{3.264}$$

The term in the square brackets in Eq. (3.262) is equal to the number of configurations in the initial state of the melt which may act as centers for the evolution of the new phase (cf. Chapter 2). $\Upsilon_{(z)}$ is the Zeldovich factor, again. Steady-state conditions are established in the range $0 \le n \le 8n_c$ after a time lag τ equal to [298, 309]

$$\tau \cong \frac{8n_c^{4/3}}{\beta w^{(+)}_{n,n+1}(n_c)}. \tag{3.265}$$

Following the methods of the above analysis (for details see [298, 309]), a number of further characteristics of the nucleation–growth process may be established like the number of clusters, N, formed in the nucleation process and their average size, $\langle R \rangle$, at the end of the stage of independent growth of the already formed supercritical clusters, etc. The change of the nucleation rate in the course of the transformation is due to the decrease of the concentration of single particles from the initial values $\{c_i(t=0)\} = \{c_{i0}\}$ to the actual values $\{c_i(t)\}$. Such changes of the concentration of the single particles affect the nucleation rate mainly via its influence on the work of critical cluster formation $\Delta\Phi(n_c)$.

The work of critical cluster formation at some arbitrary moment of time $\Delta\Phi(n_c)$ (corresponding to a critical cluster size n_c and a concentration of single particles $\{c_i\}$ or the activities $\{\varphi_i\}$) may be expressed through the respective expression in the initial state $\Delta\Phi[n_c(0)]$ (at concentrations $\{c_i = c_{i0}\}$ or activities $\{\varphi_{i0}\}$ referring to a critical cluster size $n_c(0)$). In the subsequent analysis, we start with the identity

$$\frac{\Delta\Phi(n_c)}{k_BT} = \frac{\Delta\Phi[n_c(0)]}{k_BT}\frac{\Delta\Phi(n_c)}{\Delta\Phi[n_c(0)]}. \tag{3.266}$$

According to Eq. (3.244), the work of critical cluster formation depends on the actual value of concentration or activity mainly via its dependence on $\Delta\mu$. It is assumed here that for the considered relatively small variations of the concentration in the initial stages of the transformation, the dependence of the specific interfacial energy on concentration is negligible in comparison with its effect on the variations of $\Delta\mu$. Since the considered variations of $\Delta\mu$ are also small, we may write

$$\Delta\Phi(n_c) = \Delta\Phi[n_c(0)] + \left.\frac{\partial\Delta\Phi(n_c)}{\partial\Delta\mu}\right|_{\{c_i=c_{i0}\}}\big[\Delta\mu(\{c_i\}) - \Delta\mu(\{c_{i0}\})\big]. \tag{3.267}$$

Employing Eqs. (3.242), we have

$$\frac{\Delta\Phi(n_c)}{k_BT} = \frac{\Delta\Phi[n_c(0)]}{k_BT} - n_c(0)\left[\frac{(\Delta\mu(\{c_i\}) - \Delta\mu(\{c_{i0}\}))}{k_BT}\right]. \tag{3.268}$$

With the notation

$$\widetilde{\varphi} = \left[\frac{(\Delta\mu(\{c_{i0}\}) - \Delta\mu(\{c_i\}))}{k_BT}\right] \tag{3.269}$$

and Eq. (3.262) we obtain

$$J(n_c) = J[n_c(0)]\exp\left[-n_c(0)\widetilde{\varphi}\right], \tag{3.270}$$

again. In the initial state, $\widetilde{\varphi} = 0$ holds and $J(n_c)$ equals $J[n_c(0)]$.

As a next step, we have to determine the value of $\widetilde{\varphi}$, at which the process of active steady-state nucleation is finished. For such purposes, we express the difference of the chemical potentials of the components in both phases via the respective expressions for real or perfect solutions (cf. Eq. (3.243)). We may write, in particular,

$$\widetilde{\varphi} = -\ln\left(\frac{\prod_i \varphi_i^{\nu_i}(\{c_i\})}{\prod_i \varphi_i^{\nu_{i0}}(\{c_{i0}\})}\right) \cong 1 - \frac{\prod_i \varphi_i^{\nu_i}(\{c_i\})}{\prod_i \varphi_{i0}^{\nu_i}(\{c_{i0}\})} \tag{3.271}$$

for real solutions and

$$\widetilde{\varphi} = -\ln\left(\frac{\prod_i c_i^{\nu_i}}{\prod_i c_{i0}^{\nu_i}}\right) \cong 1 - \frac{\prod_i c_i^{\nu_i}}{\prod_i c_{i0}^{\nu_i}} \tag{3.272}$$

for perfect ones. We define the time of active nucleation, τ_N, via

$$n_c(0)\widetilde{\varphi}(\tau_N) \cong 1. \tag{3.273}$$

This definition corresponds to a decrease of the steady-state nucleation rate by a factor e^{-1} in comparison with its initial value. It follows as a consequence that during the time of active steady-state nucleation $t < t_N$ the inequality $\widetilde{\varphi} \ll 1$ holds.

The value of $\widetilde{\varphi}(\tau_N)$, respectively τ_N, may be determined in the following way. From mass balance, we obtain

$$c_{i0} = c_i + \int_0^\infty n_i f(n,t)\,\mathrm{d}n = c_i + \nu_i \int_0^\infty n f(n,t)\,\mathrm{d}n \tag{3.274}$$

or

$$\frac{\mathrm{d}c_i}{\mathrm{d}t} = -\nu_i \int_0^\infty n \frac{\partial f(n,t)}{\partial t}\,\mathrm{d}n. \tag{3.275}$$

On the other hand, Eq. (3.272) gives

$$\frac{\mathrm{d}\widetilde{\varphi}}{\mathrm{d}t} = -\sum_i \frac{\nu_i}{c_{i0}} \frac{\mathrm{d}c_i}{\mathrm{d}t}. \tag{3.276}$$

Equations (3.275) and (3.276) yield

$$\frac{\mathrm{d}\widetilde{\varphi}}{\mathrm{d}t} = \left(\sum_i \frac{\nu_i^2}{c_{i0}}\right) \int_0^\infty n \frac{\partial f(n,t)}{\partial t}\,\mathrm{d}n. \tag{3.277}$$

The integral in latter equation can be evaluated analytically. We get similar to the above-derived expressions

$$\frac{\mathrm{d}\widetilde{\varphi}}{\mathrm{d}t} = \left(\sum_i \frac{\nu_i^2}{c_{i0}}\right) J[n_c(0)]n(t) = A\left\{1 + Bt\right\}^3 \tag{3.278}$$

with

$$A = \left(\sum_i \frac{\nu_i^2}{c_{i0}}\right) J[n_c(0)]g, \qquad g = 8n_c(0), \tag{3.279}$$

$$B = \frac{D_{\text{eff}}^{(s)}}{g^{1/3}a_m^2}\left(\frac{\omega_s}{\omega_m}\right)^{2/3} \ln\left[\frac{\prod_i \varphi_{i0}^{\nu_i}}{\prod_i \left(\varphi_i^{(\infty)}\right)^{\nu_i}}\right]. \tag{3.280}$$

The solution of Eq. (3.278) reads

$$\widetilde{\varphi} = \frac{A}{4B}\left[(1+Bt)^4 - 1\right]. \tag{3.281}$$

With Eq. (3.262), (3.263), and (3.273), τ_N is then obtained as

$$\tau_N \cong \left(\frac{a_m^2}{D_{\text{eff}}^{(s)}}\right)\left(\frac{\omega_m}{\omega_s}\right)^{4/3}\left[\frac{2^{5/2}}{\beta^{7/2}\left[\sum_i\left(\frac{c\nu_i^2}{c_{i0}}\right)\right]\left[\prod_i (x_i^{\nu_i})\right]}\right]^{1/4}$$

$$\times \exp\left(\frac{1}{4}\frac{\Delta\Phi(n_c)}{k_BT}\right). \tag{3.282}$$

It can be shown that the account of nonideality effects does not lead to significant changes of the value of τ_N as expressed by Eq. (3.282). Moreover, $\tau \ll \tau_N$ also holds in this more general situation. With $N = J\tau_N$, we obtain

$$N \cong \left[c\prod_i (x_i^{\nu_i})\right]\left\{\frac{(\omega_m/\omega_s)^{2/3}}{\beta^{3/8}\left[\sum_i\left(\frac{c\nu_i^2}{c_{i0}}\right)\prod_i (x_i^{\nu_i})\right]^{1/4}}\right\}\exp\left(-\frac{3}{4}\frac{\Delta\Phi(n_c)}{k_BT}\right). \tag{3.283}$$

The further evolution of the ensemble of clusters is connected with a consumption of free monomeric particles of the different components. This process proceeds until for one of the components the equilibrium concentration $c_i^{(\infty)}$ is not reached. From mass balance we have, for the highest number of particles of any of the components which may be incorporated into the newly evolving phase, the following relation:

$$c_{i0} - c_i^{(\infty)} = \langle n_i\rangle N = \nu_i\langle n\rangle N, \qquad i = 1, 2, \ldots, k \tag{3.284}$$

or

$$\langle n\rangle = \frac{\left[c_{i0} - c_i^{(\infty)}\right]}{\nu_i}\frac{1}{N}, \qquad i = 1, 2, \ldots, k. \tag{3.285}$$

Obviously, the size of the clusters is determined by the lowest of the values $[(c_{i0} - c_i^{(\infty)})/\nu_i]$ of the terms on the right-hand side of Eq. (3.285), i.e.,

$$\langle n\rangle = \frac{1}{N}\min\left\{\frac{\left[c_{i0} - c_i^{(\infty)}\right]}{\nu_i}\right\}\Bigg|_{i=1}^{i=k}. \tag{3.286}$$

With Eqs. (3.248) and (3.283), we have, consequently,

$$\langle n\rangle \propto \exp\left(\frac{3}{4}\frac{\Delta\Phi(n_c)}{k_BT}\right) , \tag{3.287}$$

$$\langle R\rangle \propto a_s \exp\left(\frac{1}{4}\frac{\Delta\Phi(n_c)}{k_BT}\right) , \tag{3.288}$$

again, in full qualitative coincidence with the one-component case.

The growth of the supercritical clusters is limited, in the stage of independent growth, by diffusion processes of the different components to the aggregates of the newly evolving phase. The growth equation is obtained in this case from Eqs. (3.242), (3.245), and (3.260) as

$$\frac{\mathrm{d}n}{\mathrm{d}t} = \frac{3\beta D_{\mathrm{eff}}^{(b)}}{a_m^2}\left(\frac{\omega_s}{\omega_m}\right)^{1/3} n^{1/3}\left[\frac{1}{n_c^{1/3}} - \frac{1}{n^{1/3}}\right] . \tag{3.289}$$

In terms of the radius of the clusters (cf. Eq. (3.248)), this equation reads

$$\frac{\mathrm{d}R}{\mathrm{d}t} = \frac{2\sigma D_{\mathrm{eff}}^{(b)}}{k_BT}\left(\frac{\omega_s^2}{\omega_m}\right)\frac{1}{R}\left[\frac{1}{R_c} - \frac{1}{R}\right] . \tag{3.290}$$

In the stage of dominating independent growth, we generally have $(1/R) \ll (1/R_c)$ and Eqs. (3.289) and (3.290) yield

$$\frac{\mathrm{d}}{\mathrm{d}t}R^2 = 2D_{\mathrm{eff}}^{(b)}\left(\frac{a_s}{a_m}\right)^3\left(\frac{\Delta\mu}{k_BT}\right) , \tag{3.291}$$

$$\frac{\Delta\mu}{k_BT} = \ln\left[\frac{\prod_i \varphi_{i0}^{\nu_i}}{\prod_i \left[\varphi_i^{(\infty)}\right]^{\nu_i}}\right] . \tag{3.292}$$

The time of independent growth τ_{growth} may thus be estimated as

$$\tau_{\mathrm{growth}} \propto \frac{a_s^2}{2D_{\mathrm{eff}}^{(b)}\left(\frac{a_s}{a_m}\right)^3\left(\frac{\Delta\mu}{k_BT}\right)}\exp\left(\frac{1}{2}\frac{\Delta\Phi(n_c)}{k_BT}\right) . \tag{3.293}$$

As a rule, it exceeds τ_N, given by Eq. (3.282), considerably.

In addition to the mentioned characteristics of the nucleation–growth process, also the evolution of the cluster-size distribution function in different regions of cluster-size space may be determined. The method is fully identical to the respective approach described in detail for the one-component case.

For the description of the late stages of coarsening (Chapter 4), interfacial effects have to be accounted for, again. In the description of this process, thus Eqs. (3.289) or (3.290) have to be employed. These equations are of the same form as the respective relations governing coarsening in the one-component case. However, the effective diffusion coefficient is, now, a nontrivial combination of the partial diffusion coefficients of the different components. Note that in the expression for the diffusion coefficients, Eqs. (3.254) and (3.260), only one of the components may have reached a concentration close to the respective equilibrium value. The concentrations of all other components are determined then by Eqs. (3.284).

3.9 Summary of Results

In the present chapter, the first stages of nucleation–growth processes are analyzed. The method is based on the reduction of the basic set of kinetic equations of nucleation theory to a Fokker–Planck-type equation and the approximative solution of this equation. In the initial state the segregating particles are assumed to exist in the form of single monomers only.

The respective derivations are outlined above in detail to make the analysis repeatable. However, for those interested in the results only, a summary of the basic theoretical predictions is believed to be of use. Such a summary will be given below.

3.9.1 Results for the Range of Cluster Sizes $n \lesssim n_c$

The time evolution of the cluster-size distribution function and the flux in the range of cluster sizes $1 \leq n \leq n_c + \delta n_c$ is given by (cf. Eqs. (3.54) and (3.55))

$$f(n,\tau) = f^{(\mathrm{st})}(n) + \frac{1}{w_n^{(+)}\sqrt{4\pi b\tau}} \exp\left\{-d\tau - \frac{a^2}{4b}\left[\tau + \frac{2n}{a}\right]\right\}$$

$$\times \int_{-\infty}^{+\infty} w_{n'}^{(+)} \Delta f(n', \tau = 0) \exp\left[\frac{an'}{2b} - \frac{(n-n')^2}{4b\tau}\right] \mathrm{d}n' \,. \tag{3.294}$$

$$J(n,\tau) = -w_n^{(+)} \exp\left(-\frac{\Delta\Phi(n)}{k_B T}\right) \frac{\partial}{\partial n}\left\{f(n,\tau)\exp\left(\frac{\Delta\Phi(n)}{k_B T}\right)\right\} . \tag{3.295}$$

The dimensionless time scale τ is determined by Eqs. (3.13) and (3.14); the other parameters $w_n^{(+)}$, a, b, and d are determined by Eqs. (3.34), (3.38)–(3.40).

A steady-state cluster-size distribution and a time-independent flux in cluster-size space are established in the considered region after an interval (cf. Eqs. (3.64) and (3.65))

$$\tau_{\mathrm{rel}}^{(\mathrm{I})} \cong 5\left(\frac{n_c^{4/3}}{\beta w_n^{(+)}(n_c)}\right) \cong \frac{5}{6}\left(\frac{(\delta n_c)^2}{w_n^{(+)}(n_c)}\right) = \frac{5}{3}\left(\frac{n_c^{[(4/3)-\kappa]}}{\alpha_1 \beta c}\right) . \tag{3.296}$$

The mentioned characteristics, steady-state nucleation rate, and steady-state cluster-size distribution can be generally expressed as (cf. Eqs. (3.79) and (3.80))

$$J = \sqrt{\left(\frac{3}{2\pi}\beta\right)} \alpha_1 c^2 n_c^{[\kappa-(2/3)]} \exp\left(-\frac{\Delta\Phi(n_c)}{k_B T}\right) , \tag{3.297}$$

$$f^{(\mathrm{st})}(n) = c \exp\left(-\frac{\Delta\Phi(n)}{k_B T}\right)\left[\frac{\Xi(n,\infty)}{\Xi(0,\infty)}\right] , \tag{3.298}$$

where

$$\Xi(a,b) = \int_a^b \frac{\exp\left(\frac{\Delta\Phi(n')}{k_B T}\right)}{w_{n'}^{(+)}} \mathrm{d}n' , \tag{3.299}$$

or, in a good approximation (cf. Eq. (3.85)), as

$$f^{(\mathrm{st})}(n) = \left(\frac{c}{2}\right) \exp\left(-\frac{\Delta\Phi(n)}{k_B T}\right) \mathrm{erfc}\left\{\frac{(n - n_c)}{\delta n_c}\right\}. \tag{3.300}$$

To obtain the expression, e.g., for the steady-state nucleation rate in real time scale, one has to multiply the right-hand side of Eq. (3.297) with the ratio (τ/t) (cf. Eqs. (3.13) and (3.14)). One has to proceed similarly to obtain the expression for the time lag in real time scale.

3.9.2 Results for the Range of Cluster Sizes $n \gtrsim n_c$

3.9.2.1 Results for the Range of Cluster Sizes $n_c \lesssim n \lesssim 8n_c$

Once steady-state conditions are established in the range $n \lesssim n_c$, the further evolution is governed by (cf. Eqs. (3.92), (3.102), and (3.103))

$$J(n, \tau) = J(n_c)\, \mathrm{erfc}[\widehat{\xi}(\tau)(n - n_c)], \qquad n_c \le n \le 8n_c, \tag{3.301}$$

$$\widehat{\xi}(\tau) = \frac{\exp\left[-\widehat{a}\left(\tau - \tau_{\mathrm{rel}}^{(\mathrm{I})}\right)(\alpha_1 c)\right]}{\sqrt{\dfrac{2\widehat{b}}{\widehat{a}}\left\{1 - \exp\left[-2\widehat{a}\left(\tau - \tau_{\mathrm{rel}}^{(\mathrm{I})}\right)(\alpha_1 c)\right]\right\}}}, \qquad \text{for} \qquad \tau \ge \tau_{\mathrm{rel}}^{(\mathrm{I})}. \tag{3.302}$$

The parameters $\widehat{a}$ and $\widehat{b}$ are given by Eq. (3.95). The evolution of the cluster-size distribution function with time is governed by Eq. (3.73), again.

Steady-state conditions are established in the considered range in a time interval $\tau_{\mathrm{rel}}^{(\mathrm{II})}$. It is determined by (cf. Eq. (3.105))

$$\tau_{\mathrm{rel}}^{(\mathrm{II})} = \frac{n_c^{[(4/3)-\kappa]}}{\alpha_1 c \beta} = \frac{3n_c^{4/3}}{\beta w_n^{(+)}(n_c)}. \tag{3.303}$$

Its value is of the same order of magnitude as $\tau_{\mathrm{rel}}^{(\mathrm{I})}$ (cf. Eq. (3.296)). The total time required to establish steady-state conditions in the range of cluster sizes $n \le 8n_c$ (or $R \le 2R_c$) is given thus by (cf. Eq. (3.144))

$$\tau_{\mathrm{rel}}(R \le 2R_c) = \frac{8n_c^{4/3}}{\beta w_n^{(+)}(n_c)} = \frac{8n_c^{[(4/3)-\kappa]}}{3\alpha_1 c \beta}. \tag{3.304}$$

Again, to express the time lag in real time scale, one has to multiply Eq. (3.304) with (t/τ).

3.9.2.2 Results for the Range of Cluster Sizes $n \gtrsim 8n_c$

In the now considered range of cluster sizes, the flux in cluster-size space has, according to Eq. (3.139), properties of a kink solution. It can be approximated for $\tau \le \tau_N$ by

$$\begin{aligned} J(R, \tau) &= J\big[n_c(0)\big] \exp\big\{-n_c(0)\varphi\left[\widehat{\tau}_0(\widehat{r}, \widehat{\tau})\right]\big\}\theta(R_{\max} - R) \\ &\cong J\big[n_c(0)\big]\theta(R_{\max} - R) \end{aligned} \tag{3.305}$$

The theta function $\theta(x)$ is defined by Eq. (3.44), again.

The position of the kink or the size of the largest clusters formed at time τ is given by $R_{\max}$. This quantity is determined for kinetically limited growth by (cf. Eq. (3.142))

$$R_{\max} = 2R_c + \frac{\beta a_s^2}{R_c}(\alpha_1 c)\left[\tau - \tau_{\text{rel}}(R \leq 2R_c)\right], \quad \text{for} \quad \tau - \tau_{\text{rel}}(R \leq 2R_c) \geq 0. \tag{3.306}$$

It follows that the time lag for the establishment of steady-state conditions for clusters of sizes $R \geq 2R_c$ may be written in the form

$$\tau_{\text{rel}}(R > 2R_c) = \tau_{\text{rel}}(R \leq 2R_c)\left\{1 + \frac{3}{4}\left[\left(\frac{R}{2R_c}\right) - 1\right]\right\}, \quad \text{for} \quad R \geq 2R_c. \tag{3.307}$$

Based on this equation, one can determine the cluster-size-dependent time lag for any appropriate value of $R \geq 2R_c$. This possibility is of considerable importance for a comparison of theoretical and experimental results, since in experiments commonly only such clusters are detected that exceed the size of the critical clusters by several times. For kinetically limited growth we have further the following expression for the cluster-size distribution:

$$f(n,\tau) \cong \frac{J(n,\tau)}{\left(\dfrac{dn}{d\tau}\right)}, \qquad \frac{dn}{d\tau} = \frac{3\alpha_1 \beta c n^{2/3}}{n_c^{1/3}}. \tag{3.308}$$

Similarly, we obtained for diffusion-limited growth the following results (cf. Eqs. (3.162)–(3.165)):

$$J(R,\tau) = J[n_c(0)]\exp\left\{-n_c(0)\varphi\left[\hat{\tau}_0(\hat{r},\hat{\tau})\right]\right\}\theta(R_{\max} - R) \cong J[n_c(0)]\theta(R_{\max} - R), \tag{3.309}$$

$$R_{\max}^2 = (2R_c)^2 + \frac{2\beta}{R_c}a_s^3(\alpha_1 c)\left[\tau - \tau_{\text{rel}}(R \leq 2R_c)\right], \quad \text{for} \quad \tau \geq \tau_{\text{rel}}(R \leq 2R_c), \tag{3.310}$$

$$\tau_{\text{rel}}(R > 2R_c) = \tau_{\text{rel}}(R \leq 2R_c)\left\{1 + \frac{3}{4}\left[\left(\frac{R}{2R_c}\right)^2 - 1\right]\right\}, \quad \text{for} \quad R \geq 2R_c, \tag{3.311}$$

$$f(n,\tau) \cong \frac{J(n,\tau)}{\left(\dfrac{\mathrm{d}n}{\mathrm{d}\tau}\right)}, \qquad \frac{\mathrm{d}n}{\mathrm{d}\tau} = \frac{3\alpha_1\beta c n^{1/3}}{n_c^{1/3}}, \qquad n \geq 8n_c. \tag{3.312}$$

3.9.3 Integral Characteristics of the Nucleation–Growth Process

3.9.3.1 Time Interval of Steady-State Nucleation

Due to the depletion of the system (decrease of the concentration of the segregating particles in the course of the nucleation–growth process), steady-state nucleation can be established for a finite interval of time only. For kinetically limited nucleation, this time interval can be written as (cf. Eqs. (3.182) and (3.183))

$$\tau_N^{(k)} = \left(\frac{c_0}{J(\tau=0)}\right)^{1/4}\left(\frac{4\Omega}{n_c^{2/3}\beta^3(\alpha_1 c_0)^3}\right)^{1/4}, \tag{3.313}$$

$$t_N^{(k)} \cong \frac{3a_m^2\Omega^{1/4}}{2\omega_s^{2/3}\omega_m^{1/3}c_0\beta^{7/8}n_c^{1/6}D^{(*)}}\exp\left(\frac{1}{4}\frac{\Delta\Phi(n_c)}{k_BT}\right). \tag{3.314}$$

For diffusion-limited nucleation processes the relations

$$\tau_N^{(d)} = \left(\frac{c_0}{J(\tau=0)}\right)^{2/5}\left(\frac{25\Omega^2}{32n_c\beta^3(\alpha_1 c_0)^3}\right)^{1/5}, \tag{3.315}$$

$$t_N^{(d)} \cong 1.4\frac{a_s^2\Omega^{2/5}n_c^{2/15}}{\omega_s c_0\beta^{4/5}D}\exp\left(\frac{2}{5}\frac{\Delta\Phi(n_c)}{k_BT}\right) \tag{3.316}$$

hold (cf. Eqs. (3.186) and (3.187)). Since $\Omega^{1/4}$ and $\Omega^{2/5}$ can be set equal to 1 in a good approximation, nonideality effects enter the expression for the time of steady-state nucleation mainly via the work of critical cluster formation $\Delta\Phi(n_c)$.

3.9.3.2 Number of Clusters Formed by Nucleation–Growth Processes

As a result of depletion effects only a finite number of supercritical clusters may be formed. This number is given by

$$N_{\max}^{(k)} \cong \frac{c_0\Omega^{1/4}}{n_c^{1/6}\beta^{3/8}}\exp\left(-\frac{3}{4}\frac{\Delta\Phi(n_c)}{k_BT}\right) \tag{3.317}$$

for kinetically limited nucleation–growth processes (cf. Eq. (3.200)) and by

$$N_{\max}^{(d)} \cong \frac{3c_0\Omega^{1/5}}{4n_c^{2/5}\beta^{3/10}}\exp\left(-\frac{3}{5}\frac{\Delta\Phi(n_c)}{k_BT}\right) \tag{3.318}$$

for diffusion-limited nucleation (cf. Eq. (3.201)). In both the cases, the number of clusters is determined mainly by the work of critical cluster formation; these numbers are independent of the value of the diffusion coefficients.

3.9.3.3 Average Size of the Clusters at the End of the Stage of Independent Growth

After completion of the nucleation process ($\tau \geq \tau_N$), the supercritical clusters grow at the expense of excess monomers existing in the system. The average size of the clusters at the end of this stage is given by (cf. Eqs. (3.207))

$$\langle R \rangle \cong \begin{cases} a_s \exp\left(\frac{1}{4}\frac{\Delta\Phi(n_c)}{k_B T}\right) & \text{kinetically limited growth} \\ \\ a_s \exp\left(\frac{1}{5}\frac{\Delta\Phi(n_c)}{k_B T}\right) & \text{diffusion-limited growth.} \end{cases} \quad (3.319)$$

It can be easily recognized that the number of clusters increases while the average size of the clusters decreases with an increase of the initial supersaturation.

3.9.3.4 Time Interval of Independent Growth

Provided the growth of the supercritical clusters is limited by diffusion processes of the segregating particles the time interval for the stage of independent growth is given by

$$t_{\text{growth}}^{(kd)} \cong \frac{a_s^2}{2D\omega_s c_0} \exp\left(\frac{1}{2}\frac{\Delta\Phi(n_c)}{k_B T}\right) \quad (3.320)$$

if the nucleation process proceeds by kinetically limited growth, and

$$t_{\text{growth}}^{(dd)} \cong \frac{a_s^2}{2D\omega_s c_0} \exp\left(\frac{2}{5}\frac{\Delta\Phi(n_c)}{k_B T}\right) \quad (3.321)$$

for diffusion-limited nucleation.

This way, the basic characteristics of the initial stages of first-order phase transitions are established analytically. Hereby both time-lag effects are reconsidered as well as time dependences connected with the depletion of the state of the ambient phase in the course of the nucleation–growth processes. Moreover, simultaneously the respective expressions have been derived for two of the most important mechanisms of kinetically and diffusion-limited growth of the aggregates of the newly evolving phase. An application of these methods to bubble formation in liquid–gas solutions is given in Chapter 9; the particular case, when the growth of the bubbles is governed by Rayleigh's equation, is analyzed in the project report [241].

In comparison with previous investigations [298], the method of analysis employed here is applicable to condensation or segregation processes not only in ideal but equally well in any kind of real systems. It was shown further that nonideality effects enter the description mainly in the final expressions of the theory via the appropriate formulas for the work of critical cluster formation. Moreover, in comparison with Ref. [298] the accuracy of some of the estimates could be improved.

As shown also the respective results for nucleation and growth of a phase with a given stoichiometric composition can be obtained immediately by an appropriate replacement of some of the parameters of the theory.

3.10 Conclusions

In the present chapter, a theoretical description of the basic characteristics of the first stages of nucleation–growth processes has been developed. The results can be compared both with experimental findings and the results of computer calculations [164,247] (see Chapter 5). Such a detailed comparison between the theoretical predictions as outlined here and the results of experimental research as well as computer calculations requires an extensive discussion which goes beyond the scope of the present monograph. To some extent, it will be given in Chapter 5. Moreover, as it has been shown recently the results may be employed also for an analysis of nucleation–growth processes at time-dependent external conditions. For an applications of the methods outlined here to such boundary conditions, cf., e.g., [90, 244].

As already mentioned, once the first stages of nucleation–growth processes, analyzed here, are passed, the system goes over continuously into a stage of competitive growth denoted commonly as Ostwald ripening or coarsening. A satisfactory theory of this process has been developed first by the author in cooperation with Lifshitz [155]. The theory has been extended later considerably [289, 290], but its basic conclusions remain unchanged. This theory is presented in Chapter 4. This way, a complete description of the whole course of first-order phase transitions is available, now.

Finally, we would like to note that the results obtained may be used also as a means to test the accuracy of different expressions for the work of critical cluster formation. Quantities like the number of clusters formed in nucleation–growth and their average size may be determined easily experimentally. Therefore, by applying the respective theoretical expressions for these quantities, conclusions concerning the appropriate expression for the work of critical cluster formation can be drawn straightforwardly.

4 Theory of the Late Stages of Nucleation–Growth Processes: Ostwald Ripening

4.1 Coarsening

4.1.1 Introduction: Formulation of the Problem

As was already discussed in detail in Chapter 3, first-order phase transitions which result in the formation of a new phase via decomposition of a supersaturated solid (or liquid) solution, proceed in several stages. In the first stages, described in detail in Chapter 3, new-phase aggregates arise in a stochastic manner via thermal fluctuations. These precipitates grow then in the supersaturated solution consuming monomers and reducing the supersaturation considerably. In the subsequent final stage of the process, when the precipitates are large enough but the supersaturation is low, the further evolution of the ensemble of clusters is mainly determined by coarsening processes, i.e., larger clusters grow at the expense of the smaller ones which are dissolved. In English literature, in order to describe these late stages of the decomposition process, the term "ripening" or "Ostwald ripening" is commonly applied while in Russian literature the term "coalescence" is often used instead. As shown in Chapter 3 already at the intermediate stage of independent growth and, in particular, at the late stages considered now, the stochastic generation of new supercritical nuclei is an unlikely event since they would have to be macroscopic in size.

In the present chapter, the theoretical description of the kinetics of precipitate growth, the coarsening process, at this later stage is developed. In formulating the problem, we make first the simplest assumptions concerning the system under consideration reflecting the most essential features, for example, we neglect a possible anisotropy and assume the precipitates to be spherical. Deviations from such simplifying assumptions, for example, from a spherical shape result in modifications of some numerical constants in the subsequent equations [153, 155], however, not changing the basic results. Going over to the theoretical description, we derive first in the subsequent analysis some basic dependences required to develop the theory.

We consider a solid solution consisting of two components. One of the components with the volume concentration c is assumed to segregate and to form clusters consisting exclusively of this component (generalizations of this assumption will be considered in more detail in Chapter 8). The equilibrium conditions for a stable coexistence of the evolving macroscopic phase with the chemical potential $\mu_\alpha(p,T)$ with the macroscopic solution with the chemical potential $\mu_\beta(p,T,c)$ then read

$$\mu_\alpha(p,T) = \mu_\beta(p,T,c_\infty). \tag{4.1}$$

Here c_∞ is the concentration of the saturated solution.

Kinetics of First-order Phase Transitions. Vitaly V. Slezov

ISBN: 978-3-527-40775-0

For a critical cluster of finite size, being in equilibrium with the solution, we have instead

$$\mu_\alpha(p_\alpha, T) = \mu_\beta(p, T, c_R). \tag{4.2}$$

Here c_R is the concentration of segregating particles in the solution required to fulfill the necessary conditions for thermodynamic equilibrium.

The pressure in the critical cluster is connected with the pressure in the surrounding solution via the relation (Young–Laplace equation)

$$p_\alpha - p = \frac{2\sigma}{R}. \tag{4.3}$$

By a Taylor expansion of the chemical potential of the cluster, we then obtain

$$\mu_\alpha(p_\alpha, T) \cong \mu_\alpha(p, T) + \left(\frac{\partial \mu_\alpha}{\partial p}\right)_{p_\alpha = p} \frac{2\sigma}{R} = \mu_\alpha(p, T) + \frac{2\sigma\omega}{R}. \tag{4.4}$$

Here ω is the volume of an atom of the segregating component. Assuming that the solution can be described as a perfect solution

$$\mu(p, T, c) = \mu_0 + k_B T \ln c \tag{4.5}$$

(k_B is the Boltzmann constant) and taking into account the equilibrium conditions for planar interfaces, Eq. (4.1), we arrive at the well-known Gibbs–Thomson equation [144]

$$c_R \cong c_\infty + \frac{\alpha}{R}, \tag{4.6}$$

where α is related to the interfacial tension σ via

$$\alpha = \frac{2\sigma\omega c_\infty}{k_B T}. \tag{4.7}$$

Therefore, the equilibrium concentration of a dissolved substance at the surface of small-sized precipitates exceeds that at the surface of large precipitates, thus inducing a flow of the dissolved component, from small precipitates into the matrix and to large precipitates out of the matrix. Neglecting the diffusional interaction between the precipitates in a polydispersed ensemble (assuming that the ratio of the mean size of precipitates, $\overline{R}$, to the mean distance, $\overline{l}$, between them is small, $\overline{R} \ll \overline{l}$), we get the following equation for the diffusional flux of dissolved particles per unit area of the precipitate surface:

$$j_R = -D \left(\frac{\partial c}{\partial r}\right)\bigg|_{r=R}. \tag{4.8}$$

To determine the concentration, we use the self-consistent mean-field approximation, which is valid when $Q_0 \ll 1$, where Q_0 is the volume fraction of the precipitating species. Note that the form of the equation for the diffusion flux is generally preserved even with $Q_0^{1/3} \lesssim 1$.

The variation of the precipitate volume is determined by the flux of the dissolved atoms per unit time onto the precipitate surface

$$\frac{\mathrm{d}}{\mathrm{d}t}\left(\frac{4\pi}{3}R^3\right) = -4\pi R^2 j_R = 4\pi R^2 D \left(\frac{\partial c}{\partial r}\right)\bigg|_{r=R}, \tag{4.9}$$

and hence the rate of change of the precipitate radius is given by

$$\frac{\mathrm{d}R}{\mathrm{d}t} = D\left(\frac{\partial c}{\partial r}\right)\bigg|_{r=R}. \tag{4.10}$$

The partial derivative, $(\partial c/\partial r)$, can be determined by solving the corresponding diffusion problem. We get

$$\frac{\partial c}{\partial t} = D\triangle c, \qquad c|_{r=R} = c_R, \qquad c|_{r\to\infty} = \overline{c}. \tag{4.11}$$

It is easily seen that for small initial supersaturations,

$$\triangle_0 = (\overline{c}_0 - c_\infty) \ll 1, \tag{4.12}$$

the flux can be obtained by solving the steady-state diffusion problem. In fact, the ratio of the characteristic time of establishment of a steady-state diffusion flux of the dissolved component at the surface of a precipitate,

$$\tau_{\text{diff}} \sim \left(\overline{R}^2/D\right), \tag{4.13}$$

to the characteristic time of variation of the precipitate size,

$$\tau_{\text{ch}} \sim \frac{\overline{R}}{D\left(\partial c/\partial r\right)} \sim \frac{\overline{R}^2}{D\triangle_0}, \tag{4.14}$$

is given by

$$\frac{\tau_{\text{diff}}}{\tau_{\text{ch}}} \sim \triangle_0 \ll 1. \tag{4.15}$$

Solving the steady-state diffusion equation ($(\partial c/\partial t) = 0$) assuming that the supersaturation, $\Delta = c - c_\infty$, all the time remains low, $\Delta \ll 1$, and the diffusional interactions between the precipitates can be neglected, we arrive at an equation for the diffusion flux of dissolved matter normalized per unit area of the precipitate surface in the form

$$-j = D\left(\frac{\partial c}{\partial r}\right)\bigg|_{r=R} = \frac{D}{R}(c - c_R) = \frac{D}{R}\left(\Delta - \frac{\alpha}{R}\right), \tag{4.16}$$

and, accordingly, at an equation for the rate of change of the precipitate radius

$$\frac{\mathrm{d}R}{\mathrm{d}t} = -j = \frac{D}{R}\left(\Delta - \frac{\alpha}{R}\right). \tag{4.17}$$

Thus, for any given supersaturation, Δ, a critical radius $R_{\text{cr}} = (\alpha/\Delta)$ exists such that the precipitate is in equilibrium with the solution, with the precipitate growing when $R > R_{\text{cr}}$ and dissolving when $R < R_{\text{cr}}$. The existence of such critical cluster radius R_{cr} implies the dissolution of smaller precipitates by greater ones. The supersaturation Δ itself and R_{cr} are time dependent.

From here on we shall use the dimensionless variables $\rho = (R/R_{\text{cr}})$, $t' = (t/T)$, $R_{\text{cr}_0} = (\alpha/\Delta_0)$, and $T = (R^3_{\text{cr}_0}/\alpha D)$, Δ_0 being the initial supersaturation and R_{cr_0} the initial critical radius. By omitting the prime in t' we get

$$\frac{\mathrm{d}\rho^3}{\mathrm{d}t} = 3\left(\frac{\rho}{x} - 1\right), \qquad x(t) = \frac{R_{\text{cr}}}{R_{\text{cr}_0}}, \tag{4.18}$$

where $x(t)$ is the dimensionless critical radius, $x(0) = 1$. Introducing further the distribution function $f(\rho, t)$ of the precipitates with respect to size, and considering $v_\rho = (\mathrm{d}r/\mathrm{d}t)$ as the rate of change of the precipitate in cluster-size space, we can formulate the equations for the determination of the unknown functions $f(\rho, t)$ and $x(t)$. The first of these equations is the continuity equation in cluster-size space

$$\frac{\partial f}{\partial t} + \frac{\partial}{\partial \rho}(f v_\rho) = 0. \tag{4.19}$$

The second one is the law of conservation of matter

$$Q_0 = \Delta_0 + q_0 = \Delta + q, \qquad q = \frac{4\pi}{3} R^3_{\text{cr}_0} \int_0^\infty f\rho^3\,\mathrm{d}\rho, \tag{4.20}$$

or, taking into account that $x = (\Delta_0/\Delta)$, we have

$$1 = \frac{\Delta_0}{Q_0 x} + \kappa \int_0^\infty f\rho^3\,\mathrm{d}\rho, \qquad \kappa = \frac{4\pi}{3}\frac{R^3_{\text{cr}_0}}{Q_0}. \tag{4.21}$$

In these equations, Q_0 is the total initial supersaturation that includes the initial volume of the material in the precipitates, q_0. In Eqs. (4.19) and (4.21), f is normalized to a unit volume; therefore, $n = \int_0^\infty f\,\mathrm{d}\rho$ is the number of precipitates in a unit volume. Equations (4.19) and (4.21) provide a complete set of equations for obtaining an asymptotic solution for the given initial conditions in the case when stochastic formation of new phase nucleation centers is completed or can be ignored.

Note that Eq. (4.19) includes the hydrodynamic term $\frac{\partial}{\partial\rho}(f v_\rho)$ only. In fact, this is an approximate equation that can be refined by writing a Fokker–Planck-type equation involving next-order terms of expansion in small gradients in size space. These corrections may seem considerable in a region where $\mathrm{d}R/\mathrm{d}t$ is small. However, with the supersaturation monotonically decreasing, this region is in continuous motion and the precipitates of a given size enter the region for only a very short period of time. Under this condition, the diffusion term in cluster-size space is insignificant. This conclusion will become evident upon going over to relative variables. Our aim is to find an asymptotic solution to Eqs. (4.19) and (4.21), with the initial condition $f(\rho, 0) = f_0(\rho)$.

4.1.2 Asymptotic Behavior of the Critical Cluster Size

In order to solve the problem posed, the knowledge is required first of all of the asymptotic behavior of the critical cluster size, $x(t)$ (or alternatively the supersaturation, $\Delta(t) =$

$(\Delta_0/x(t)))$. The pattern of motion of the point ρ representing the precipitate radius along the size axis is as follows: the points to the left of $x(t)$ are accelerated to the left and disappear on reaching the origin (complete dissolution of precipitates); the points located initially to the right of $x(t)$ move further to the right (precipitates are growing) but, with the further decrease of the supersaturation, the critical size $x(t)$ increases, and catches up, one after another, the points moving toward the right-hand side of $x(t)$. The latter points then start moving into the opposite direction and also disappear on reaching the origin. The motion proceeds in an ordered fashion so that the original sequence of points is preserved.

Both the form of Eq. (4.18) and the physical meaning of the variable $x(t)$ make it natural for us to choose, as an independent variable in Eqs. (4.18)–(4.21), the radius divided by the critical radius, rather than the precipitate radius ρ, i.e.,

$$u = \frac{\rho}{x(t)}. \tag{4.22}$$

Since the supersaturation tends to zero ($\Delta \to 0$) with times approaching infinity ($t \to \infty$), so that $x(t) \to \infty$, the value of $x(t)$ can therefore be used as a measure of time.

Formally, the case when the original distribution is described by a δ-function ($f_0 = A\delta(\rho - \rho_0)$), or by a sum of δ functions, is an exception. In this case, $x(t) \to \mathrm{const}$ when the size of the clusters reaches the equilibrium value. However, this situation corresponds, in general, to an unstable state since any arbitrarily small perturbations of the original distribution lead to $x(t) \to \infty$.

A detailed analysis shows that the canonical form of the equation of motion (4.18) is obtained if the time is measured in terms of

$$\tau = \ln x^3. \tag{4.23}$$

Substitution of Eqs. (4.22) and (4.23) into Eq. (4.18) leads to

$$\frac{\mathrm{d}u^3}{\mathrm{d}\tau} = \gamma(u-1) - u^3, \tag{4.24}$$

$$\gamma = \gamma(\tau) = 3\frac{\mathrm{d}t}{\mathrm{d}x^3}. \tag{4.25}$$

Let us denote by $u(v,\tau)$ the solution of Eq. (4.24) with the initial condition $u|_{\tau=0} = v$. Then, taking into account that $\rho(v,\tau) = xu(v,\tau)$, $x(0) = 1$, and $\tau|_{t=0} = 0$, the total amount, q, of matter in the precipitates can be expressed in terms of the original distribution function $f_0(\rho)$ as

$$q = \kappa Q_0 \int_{v_0(\tau)}^{\infty} f_0(v) u^3(v,\tau)\,\mathrm{d}v. \tag{4.26}$$

Here $v_0(\tau)$ is the solution of the equation $u(v_0(\tau),\tau) = 0$, i.e., the lower limit of the initial size of the precipitates that remained undissolved at time τ. Noticing that $x^3 = e^{\tau}$, we can rewrite Eq. (4.21) as

$$1 - e^{-\tau/3} = \kappa e^{\tau} \int_{v_0(\tau)}^{\infty} f_0(v) u^3(v,\tau)\,\mathrm{d}v. \tag{4.27}$$

Equations (4.24) and (4.27) provide a complete system of equations with $\gamma(t) = 3(\mathrm{d}t/\mathrm{d}x^3)$ being an unknown function. They can be used to determine it and to then obtain $x(t)$.

There exist three possible types of asymptotic behavior for $\gamma(\tau)$ when $\tau \to \infty$: (i) $\gamma(\tau) \to \infty$; (ii) $\gamma(\tau) \to \mathrm{const}$ and (iii) $\gamma(\tau) \to 0$. We begin the detailed analysis of these possibilities with the case $\gamma(\tau) \to \mathrm{const}$. Depending on the value of γ, the plot of the velocity, $\mathrm{d}u^3/\mathrm{d}r$, vs u may either touch the abscissa axis (when $\gamma = \gamma_0 = 27/4$) or pass below this axis (with $\gamma < \gamma_0$) or have a segment of positive values (when $\gamma > \gamma_0$) (see Figure 4.1).

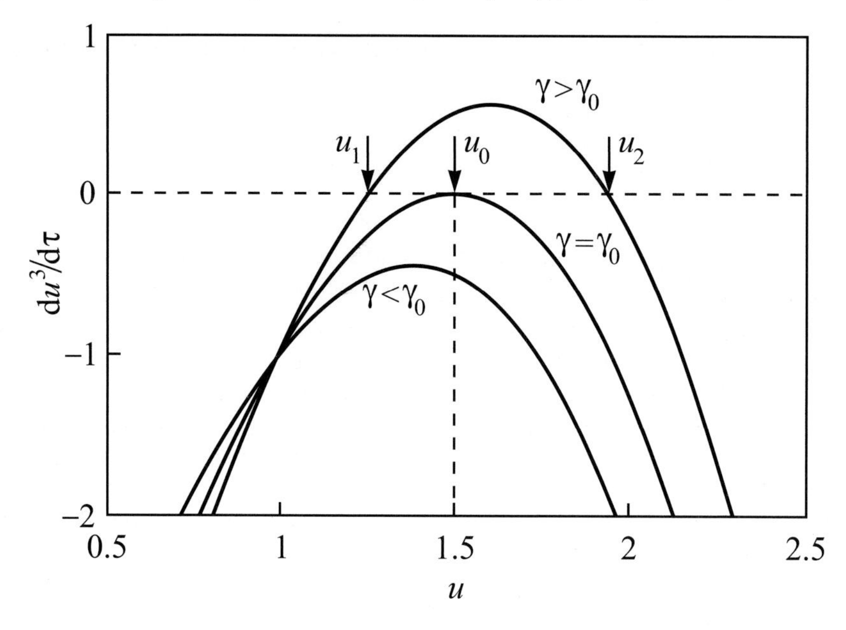

Figure 4.1: $\mathrm{d}u^3/\mathrm{d}\tau$ as a function of u for different values of γ.

Let us now consider several different situations: (a) When $\gamma > \gamma_0$, all the points to the left of u_1 are moving leftwards and disappear on reaching the origin; all the points to the right of u_1 are moving to the point u_2 asymptotically approaching it from the left or from the right. The integral on the right-hand side of Eq. (4.27) with $\tau \to \infty$ approaches the constant value

$$I_0 = u_2^3 \int_{u_1}^{\infty} f_0(v)\,\mathrm{d}v, \tag{4.28}$$

while the total precipitate volume q on the right-hand side increases as e^{τ}, i.e.,

$$q = \kappa I_0 e^{\tau} \to \infty\,, \qquad \tau \to \infty, \tag{4.29}$$

so Eq. (4.27) is not satisfied. The fact that a constant value of $\gamma > \gamma_0$ is attained only asymptotically does not jeopardize the above statements: we should only move the time origin and relate the expression $f_0(v)$ to a moment when $\gamma(\tau)$ approaches its asymptotic value.

(b) When $\gamma < \gamma_0$, all the points are moving leftward and reach the origin in a finite time. By the time, τ, according to Eq. (4.24), all the precipitates with an original size smaller than $v_0(\tau)$ dissolve, and the value of $v_0(\tau)$ is given by the equation

$$\int_0^{v_0(\tau)} \frac{3u^2\,\mathrm{d}u}{u^3 - \gamma(u-1)} = \tau. \tag{4.30}$$

With $\tau \to \infty$ we have $v_0(\tau) = e^{\tau/3}$. The total precipitate volume is therefore determined by the tail of the original distribution, $f_0(v)$,

$$q(\tau) = \kappa e^{\tau} \int_{e^{\tau/3}}^{\infty} f_0(v) u^3(v,\tau)\,\mathrm{d}v \approx \kappa \int_{e^{\tau/3}}^{\infty} f_0(v) v^3\,\mathrm{d}v. \tag{4.31}$$

When $v \to \infty$ then $f_0(v) \geq v^{-n}$ with $n > 4$ (since $u(v,\tau) \approx v e^{-\tau/3}$). Thus, $q(\tau)$ in this case approaches zero and Eq. (4.27) has no solution.

The above considerations with respect to the cases when $\gamma > \gamma_0$ and $\gamma < \gamma_0$ are further strengthened for the cases when $\gamma \to \infty$ and $\gamma \to 0$, respectively. Thus, only the case when $\gamma(\tau) \to \gamma_0 = 27/4$ should be considered as a physically reasonable solution.

First of all it should be noted that once the exact identity, $\gamma = \gamma_0$ is fulfilled all the points to the right of the point of contact, $u_0 = 3/2$, in their motion toward the left cannot pass the point of contact u_0 and become embedded in it. In view of this, when $\gamma > \gamma_0$, Eq. (4.27) cannot be satisfied ($q(\tau) \sim e^{\tau} \to \infty$ when $\tau \to \infty$). This result suggests that the expression $\gamma(\tau)$ should approach γ_0 from below, i.e., as

$$\gamma(\tau) = \gamma_0 \left(1 - \varepsilon(\tau)\right). \tag{4.32}$$

As shown below, $\varepsilon^2(\tau) \sim \tau^{-2}$. The points approaching u_0 from the right leak through the blocking point, $u_0 = 3/2$, with decreasing rates. The leakage rate is determined by the value of $\varepsilon(\tau)$ that is to be evaluated along with $\gamma(\tau)$ from Eq. (4.27) and the equation of motion (Eq. (4.24)). This form of the function $\gamma(\tau)$ is obligatory for an ordered motion of the precipitates from right to left in the relative size coordinates, which corresponds to $\mathrm{d}u/\mathrm{d}\tau < 0$ for all u. Otherwise there would be no leakage, in the precipitate size space, from the right into the region to the left of the blocking point $u_0 = 3/2$, and as $R_{\mathrm{cr}} = (\alpha/\Delta) \to \infty$ with $\tau \to \infty$, $\Delta \to 0$, the amount of matter would be increasing infinitely, which is impossible. The leakage should proceed in such a way that the size distribution ensuring the balance of matter has enough time to evolve.

Note that, for $\gamma > \gamma_0 = \mathrm{const}$, the specific initial conditions nullifying $\mathrm{d}u/\mathrm{d}\tau$ at the point $u = u_0$ can be formally found in an infinite variety of ways. Accordingly, an infinite number of solutions satisfying both the continuity equation and the balance of matter can be obtained. However, taking fluctuations into account immediately leads to a nonzero probability of the precipitates arising to the right of u_0, which violates the balance of matter. All such solutions (for $\gamma > \gamma_0 = \mathrm{const}$) are therefore unstable and physically meaningless. As is readily seen, the amount of matter to the right of the blocking point u_0 is negligible. Neglecting this amount of matter means that $\varepsilon^2(\tau)$ is set equal to zero along with f at $u > u_0$. We call this approximation the "zeroth-order hydrodynamic approximation."

A unique solution corresponding to $\gamma = \gamma_0$ really exists in this zeroth-order approximation. The velocity $(\mathrm{d}u/\mathrm{d}r)$ must have a second-order zero in this approximation; then, taking into account $\varepsilon^2(\tau) = 0$, we immediately arrive at the asymptotic values of u_0 and γ_0, and hence at the asymptotic values of critical sizes and supersaturation

$$\left(\frac{\mathrm{d}u}{\mathrm{d}\tau}\right)\bigg|_{u=u_0} = 0, \qquad \frac{\partial}{\partial \tau}\left(\frac{\mathrm{d}u}{\mathrm{d}\tau}\right)\bigg|_{u=u_0} = 0, \tag{4.33}$$

from which we get the already known values of $u_0 = 3/2$ and $\gamma_0 = 27/4$. Accordingly, taking into account that $R_{\mathrm{cr}} = (\alpha/\Delta)$, and $\Delta = (\alpha/R_{\mathrm{cr}})$, we have equations for the critical size from Eq. (4.25)

$$R_{\mathrm{cr}}^3 = R_{\mathrm{cr}_0}^3 + \frac{4}{9} D\alpha t, \tag{4.34}$$

$$R_{\mathrm{cr}}^3 = \frac{4}{9} D\alpha t, \qquad \Delta = \left(\frac{4}{9}\frac{D}{\alpha^2} t\right)^{-1/3} \qquad \text{when} \qquad \tau \to \infty, \tag{4.35}$$

where R_{cr_0} is the critical size of the system at a stage when the decomposition obeys the asymptotic equations (unlike the initial critical size introduced in Eq.(4.18)).

4.1.3 Asymptotic Behavior of the Distribution Function

In accordance with the results obtained in the previous section, we shall look now for the distribution function depending on the new variables u and τ. The distribution function $\varphi(u, \tau)$ with respect to relative cluster sizes, $u = (\rho/x)$, is related to $f(\rho, t)$ through the obvious equation

$$\varphi(u, \tau)\, \mathrm{d}u = f(\rho, t)\, \mathrm{d}\rho = x f(\rho, t)\, \mathrm{d}u, \tag{4.36}$$

where

$$f = \frac{\varphi(u, \tau)}{x}. \tag{4.37}$$

The continuity equation for $\varphi(u, \tau)$, with $\tau \gg 1$, assumes the following form everywhere except in the vicinity of the point u_0:

$$\frac{\partial \varphi}{\partial \tau} - \frac{\partial}{\partial u}(\varphi g(u)) = 0, \tag{4.38}$$

$$\frac{\mathrm{d}u}{\mathrm{d}\tau} = -g(u)|_{\gamma=\gamma_0} = \frac{1}{3u^2}\left(u - \frac{3}{2}\right)^2 (u+3), \qquad 0 \le u \le u_0 = \frac{3}{2}, \tag{4.39}$$

$$\frac{\mathrm{d}u}{\mathrm{d}\tau} = 0, \qquad u \ge u_0. \tag{4.40}$$

The solution to this equation in the region to the left of u_0 is

$$\varphi = \frac{\chi(\tau+\psi)}{g(u)}, \qquad 0 \le u \le u_0, \tag{4.41}$$

$$\varphi = 0, \qquad u \ge u_0, \tag{4.42}$$

where we introduced the notation

$$\psi = \int_0^u \frac{du}{g(u)} = \frac{4}{3}\ln(u+3) + \frac{5}{3}\ln\left(\frac{3}{2}-u\right) + \frac{1}{1-(2/3)u} - \ln\frac{3^3 e}{2^{5/3}}, \tag{4.43}$$

and χ is an arbitrary function to be specified.

The above analysis of the equation of motion (Eq. (4.24)), i.e., of the characteristics of Eq. (4.40), suggests that the vicinity of the point u_0 can be considered as a sink for all the points $u > u_0$ and as a source for the region $u < u_0$ (in the region $u < u_0$, the origin $u = 0$ is a sink). While moving from left to right, all the points pass through the vicinity of the blocking point u_0; the later they arrive in this vicinity, the longer they stay there. The distribution function in a region to the right of the point u_0 where $\tau \to \infty$ is determined by an infinitely distant part of the tail of the original distribution, and its cumulative contribution, as shown below, rapidly approaches zero in terms of both absolute and relative values. As shown below, the relative contribution from the vicinity of the point u_0 also vanishes when $\tau \to \infty$.

Thus, precipitates with $u < u_0$ contribute predominantly to the law of conservation of matter. Accordingly, the law of conservation of matter can be employed as an integral equation defining the zeroth-order asymptotic behavior of the distribution function for $u < u_0$ (the distribution function for $u > u_0$ is equal to zero in this approximation). Taking this into account, we substitute the solution of Eq. (4.40) into the law of conservation of matter to get an asymptotic equation for χ in the form

$$1 = \kappa e^{\tau} \int_0^{3/2} \chi(\tau+\psi)\frac{u^3}{g(u)}\,du. \tag{4.44}$$

The functions $\chi(\tau+\psi)$ and $\varphi(u,\tau)$ are easily inferred from these equations as

$$\chi(\tau+\psi) = A e^{-\tau-\psi}, \tag{4.45}$$

$$\varphi(u,\tau) = A\frac{e^{-\tau-\psi}}{g(u)} = \frac{3^4 e}{2^{5/3}} a e^{-\tau} \frac{u^2 \exp\left\{-\dfrac{1}{1-(2/3)u}\right\}}{(u+3)^{7/3}((3/2)-u)^{11/3}}, \qquad u \le \frac{3}{2}, \tag{4.46}$$

where

$$A = \left[\kappa \int_0^{3/2} e^{-\psi}\frac{u^3}{g(u)}\,du\right]^{-1} \approx \frac{3Q_0}{4\pi R^3_{\mathrm{cr}_0} 1.11} \approx 0.22\frac{Q_0}{R^3_{\mathrm{cr}_0}}. \tag{4.47}$$

It is evident that the asymptotic behavior of the distribution function for $u < u_0$ is independent of the original distribution function.

The number of particles in unit volume, according to Eq. (4.46), is given by

$$n(\tau) = \int_0^{3/2} \varphi(u,\tau)\,\mathrm{d}u = Ae^{-\tau} = \left(\frac{3}{2}\right)^{2/3} \frac{A}{t}. \tag{4.48}$$

Let, now, $P(u)\,\mathrm{d}u$ be the probability of a particle having a size between u and $u + \mathrm{d}u$. Then

$$\psi(u,\tau) = n(\tau)\,P(u), \tag{4.49}$$

where

$$P(u) = \begin{cases} \dfrac{3^4 e}{2^{5/3}} \dfrac{u^2 \exp\left\{-\dfrac{1}{1-(2/3)\,u}\right\}}{(u+3)^{7/3}\left((3/2)-u\right)^{11/3}}, & u < u_0 = \frac{3}{2}, \\ 0, & u > u_0 = \frac{3}{2}. \end{cases} \tag{4.50}$$

The function $P(u)$ is shown in Figure 4.2.

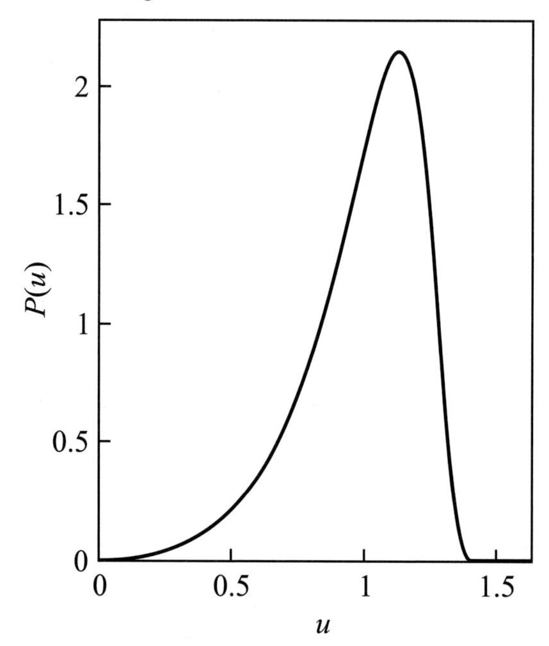

Figure 4.2: Dependence of the distribution function density $P(u)$ on the reduced radius $u = r/r_c$.

This probability $P(u)$ can be conveniently expressed in terms of a slightly different reduced variable by introducing the maximum size

$$\rho_m = \frac{3}{2}x, \qquad v = \frac{\rho}{\rho_m} = \frac{u}{u_m} = \frac{2}{3}u. \tag{4.51}$$

Then we get

$$P(u)\,\mathrm{d}u = P_1(v)\,\mathrm{d}v, \tag{4.52}$$

where

$$P_1(v) = \begin{cases} e2^{7/3}v^2 \exp\left\{-\dfrac{1}{1-v}\right\} \dfrac{1}{(v+2)^{7/3}(1-v)^{11/3}}, & v < 1, \\ 0, & v > 1. \end{cases} \tag{4.53}$$

These equations completely define the asymptotic distribution of particles with respect to size and time.

When measurements are performed at some etched microsection of the system under investigation, the distribution function $\Phi(r,\tau)$ with respect to the circular sections of the precipitates r is given, as can be easily shown, by

$$F(l,\tau)\,\mathrm{d}l = \Phi(r,\tau)\,\mathrm{d}r, \qquad \Phi(r,\tau) = x^{-1}F\left(\frac{r}{x(\tau)},\tau\right), \tag{4.54}$$

$$F(l,\tau) = 2x(\tau)\int\limits_l^{3/2} \frac{\varphi(u,\tau)\,\mathrm{d}u}{\left[(u^2/l^2)-1\right]^{1/2}}, \tag{4.55}$$

where

$$l = \frac{r}{x(\tau)}, \qquad r = \frac{R}{R_{\mathrm{cr}_0}}. \tag{4.56}$$

These dependences are illustrated in Figure 4.3.

Now note that

$$\int\limits_0^{3/2} e^{-\psi}\frac{u-1}{g(u)}\,\mathrm{d}u = \int\limits_0^{\infty} e^{-\psi}\left[u(\psi)-1\right]\mathrm{d}\psi = e^{-\psi}u^3(\psi)\Big|_0^{\infty} = 0. \tag{4.57}$$

We have used here the relationship

$$\frac{\mathrm{d}u^3}{\mathrm{d}\psi} = u^3 - \frac{4}{27}(u-1). \tag{4.58}$$

It suggests that $\overline{u} = 1$; hence

$$\overline{\rho} = \overline{u}x = x(\tau). \tag{4.59}$$

The distribution function in absolute variables ρ can be obtained by simply replacing u in Eq. (4.46) by ρ/x and dividing by x, where

$$x^3 = \frac{4}{9}t. \tag{4.60}$$

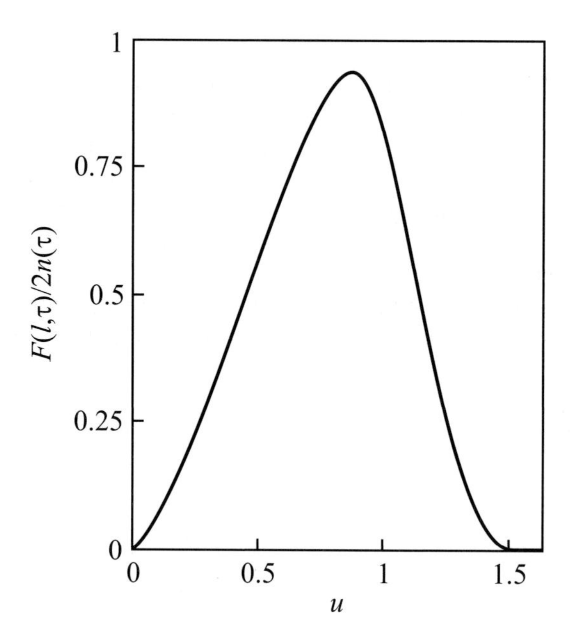

Figure 4.3: Distribution function density in dependence on the size u of the precipitates as revealed on the surface of the microsection.

Bearing in mind that, according to Eq. (4.59), $u = R/\overline{R}$, and turning back to the original dimensional variables, we get

$$f(R,t) = n(t)\,P\left(\frac{R}{\overline{R}}\right)\frac{1}{\overline{R}}, \tag{4.61}$$

$$n(t) = \beta Q_0(\overline{R})^{-3}, \qquad \beta \approx 0.22, \qquad \overline{R}^3 = \frac{4}{9}D\alpha t, \tag{4.62}$$

and $P(u)$ is given by Eq. (4.49). The supersaturation at a given time is given by the following expression:

$$\Delta(t) = \frac{\Delta_0}{x(t)} = \left(\frac{3}{2}\right)^{2/3}\Delta_0\left(\frac{T}{t}\right)^{1/3} = \left(\frac{3}{2}\right)^{2/3}\left(\frac{\alpha^2}{Dt}\right)^{1/3}. \tag{4.63}$$

Let us finally consider the limits of applicability of the above equations. It follows from the above analysis that the asymptotic formulas are valid provided that the inequalities

$$\tau^2 = (3\ln x)^2 = 9\left(\ln\frac{\overline{R}}{R_{\mathrm{cr}_0}}\right)^2 \gg 1, \qquad \overline{R} \gg R_{\mathrm{cr}_0}, \tag{4.64}$$

are fulfilled. Here $R_{\mathrm{cr}_0} = (\alpha/\Delta_0)$ is the initial critical radius for the coarsening process (with Δ_0 being the initial supersaturation).

It should be remembered in this connection that, with the initial mean size of precipitates being of the order of the critical size ($R_0 \approx R_{\mathrm{cr}_0}$), it is the latter size that appears in the above estimate. If, however, $R_0 \gg R_{\mathrm{cr}_0}$, then, at the early stage, the precipitates start growing from the solution directly, with the process going on until the supersaturation drops to so low values that the mean size approaches the critical size ($\overline{R}_1 \sim R_{\mathrm{cr}}$) thus initiating the coarsening process. These size values should appear as the initial ones. They are largely determined not only by the initial supersaturation but also by the initial number of nucleation centers (provided that this number can be considered as being fixed).

If, say, the initial supersaturation is Δ_0, the initial number of precipitates is n_0 and $\overline{R}_0 \gg R_{\mathrm{cr}_0} = (\alpha/\Delta_0)$, then the precipitates will go on growing from the solution until they reach the size

$$\frac{4\pi}{3}\overline{R}_1^3 \approx \frac{\Delta_0}{n_0}, \tag{4.65}$$

while at the early stage

$$\frac{\mathrm{d}\overline{R}^2}{\mathrm{d}t} = 2D\left(\Delta_0 - \frac{4\pi}{3}\overline{R}^3 n_0\right) = 2D\Delta_0\left(1 - \frac{\overline{R}^3}{R_{\mathrm{cr}_0}^3}\right). \tag{4.66}$$

The duration of the early stage of independent growth is $t_1 \sim \left(\overline{R}_1^2/D\Delta_0\right)$.

The later stage (coarsening) begins at the characteristic time

$$t_0 \sim \frac{\overline{R}_1^3}{D\alpha} \sim t_1\frac{\overline{R}_1}{R_{\mathrm{cr}_0}}, \tag{4.67}$$

i.e., in the case under consideration when $\overline{R}_1 \gg R_{\mathrm{cr}_0}$, $t_0 \gg t_1$. The time dependence of the mean cluster size in both stages of independent growth and coarsening is shown schematically in Figure 4.4. A more detailed numerical analysis will be given in Chapter 5 (for the respective analytic estimates of the number of clusters evolving in nucleation-growth processes and their average size at the end of the stage of independent growth, being the initial state of coarsening, see Chapter 3).

4.1.4 Boundary Effects and Theory of Sintering

In the foregoing section we discussed coarsening in an infinite homogeneous space. Accordingly, macroscopic diffusion flows of the dissolved material were not included in the formulation of the problem. The situation changes dramatically if spatial homogeneity is lacking; the most important inhomogeneous case is associated with the presence of a border between the solution and another phase. If the second phase (or, more generally, the phase constituting the precipitates under consideration) is formed by a purely dissolved material, then the supersaturation at this border is given by $\Delta|_{z=0}$ and it induces a macroscopic diffusion flow toward the boundary.

We shall now consider here the important case where vacancies act as the atoms of a dissolved material while the pores (voids) produced by coagulation of vacancies inside a crystal supersaturated with vacancies plays the role of the precipitates. In the presence of a free

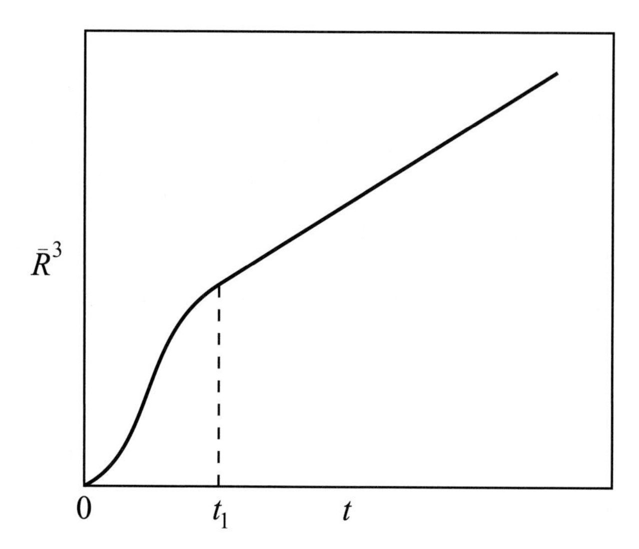

Figure 4.4: Time dependence of $\overline{R}^3$.

surface, two competing processes will occur in such a crystal: growth of pores and their coarsening in accordance with the above scheme in regions distant from the border, and dissolution of pores and repulsion of vacancies to the boundary (considered as a pore of an infinite radius) by a diffusion flow. The latter process of repulsion of pores is the physical origin of the sintering phenomenon. It is this particular process that is discussed below although all the results obtained in the course of the analysis apply, of course, also to more general cases of decomposition of supersaturated solutions.

The equations defining the precipitate growth (Eq. (4.17)) and the distribution function (Eq. (4.19)) remain valid in the half-space case, too. However, the law of conservation ($Q_0 = \Delta + q$) of mass that corresponds to Eq. (4.21) must be replaced by the diffusion equation. Every point z is a source (or a sink) with a strength, $(\mathrm{d}q/\mathrm{d}t)$, $q = q(z, t)$, determined by the dissolution of pores. Thus, the exact equations take on the following form:

$$\frac{\partial\left(\Delta + q\right)}{\partial t} = D\frac{\partial^2 \Delta}{\partial z^2}, \qquad \Delta|_{z=0} = 0, \tag{4.68}$$

$$q = \kappa x^3 \int\limits_{v_0(\tau)}^{\infty} f_0\left(v\right) u^3\left(v, \tau\right) \mathrm{d}v, \qquad x = \frac{\Delta_0}{\Delta\left(z, t\right)}, \tag{4.69}$$

where $u(v, \tau)$ satisfies Eq. (4.24) and is determined by the form of the function $\Delta(z, t)$.

The above analysis showed a kind of stability of the resulting asymptotic variation of the supersaturation $\Delta = \lambda t^{-1/3}$: namely, an arbitrarily small slowing down of the asymptotic decrease of the supersaturation potentially stimulates an unlimited growth of the precipitate material ($q \to \infty$), which is impossible. On the other hand, even a small acceleration of the asymptotic decrease of the supersaturation would result in a relatively fast and complete dissolution of the precipitates ($q \to 0$). The asymptotic behavior of the sintering process proceeds therefore as described below.

There are three distinct regions in a crystal: (a) The most distant region ($z_2 < z < \infty$) is that where the effects of the border are negligible and the supersaturation is given by Eq. (4.63) for an infinite space. (b) The region $z_1 < z < z_2$ is the range where dissolution of pores (precipitates) is observed. (c) The region $0 < z < z$ is the range where pores are completely lacking and a purely diffusive repulsion of vacancies takes place. The boundaries of these regions $z_1(t)$ and $z_2(t)$ are moving into the sample interior, so the pore-free crust is continuously thickening. In fact, the set of Eqs. (4.68) and (4.69) has to be solved only for the intermediate region $z_1 < z < z_2$.

However,

$$\frac{z_2 - z_1}{z_2} \ll 1, \tag{4.70}$$

as shown below. By taking

$$\frac{z_1 + z_2}{2} = \xi \tag{4.71}$$

as a first-order approximation, and replacing this region by a corresponding boundary condition, we derive the position of the boundary $\xi(t)$ and the concentration variation in the surface layer $0 < z < \xi$ (in the crust). Thus, we have

$$\frac{\partial \Delta}{\partial t} = D \frac{\partial^2 \Delta}{\partial z^2}, \tag{4.72}$$

$$\Delta|_{z=0} = 0, \qquad \Delta|_{z=\xi} = \lambda t^{-1/3}, \qquad D\left(\frac{\partial \Delta}{\partial z}\right)\bigg|_{z=\xi} = Q_0 \frac{\mathrm{d}\xi}{\mathrm{d}t}. \tag{4.73}$$

The latter condition replaces the intermediate region and represents the fact that the boundary $\xi(t)$ is a source with the strength $Q_0(\mathrm{d}\xi/\mathrm{d}t)$ (since all the excess vacancies on the right are concentrated inside the pores). We shall seek a solution to Eq. (4.69), with the boundary condition $\Delta|_{z=0} = 0$, in the form $\Delta = \sum_n \alpha_n(t) z^n$. From Eq. (4.72) we get

$$\Delta = \sum_{n=0}^{\infty} \frac{\hat{P}^n \alpha_1(t)}{(2n+1)!} z^{2n+1}, \qquad \hat{P} = \frac{1}{D} \frac{\mathrm{d}}{\mathrm{d}t}. \tag{4.74}$$

Before we go over to satisfy the other two boundary conditions we note that asymptotically $\alpha_1(t)$ is a decreasing function of time since it represents the flow at the boundary, $z = 0$ ($\alpha_1(t) = D(\mathrm{d}\Delta/\mathrm{d}z)|_{z=0}$). Accordingly, we shall seek the asymptotic value of $\alpha_1(t)$ as $\alpha_1 = \beta t^{-s}$, $s > 0$. It follows from Eq. (4.72) that

$$\Delta = \beta \frac{z}{t^s} \sum_{n=0}^{\infty} (-1)^n \left(\frac{z^2}{Dt}\right)^n \frac{s(s+1)\cdots(s+n)}{(2n+1)!}. \tag{4.75}$$

As shown below, asymptotically $\xi^2(t)$ varies more slowly than t, i.e., $\left(\xi^2/t\right) \to 0$ as $t \to \infty$. Thus, the asymptotic behavior is mainly determined by a term with $n = 0$ resulting in

$$\Delta = \frac{\beta z}{t^s} = \alpha_1 z. \tag{4.76}$$

Substituting Eq. (4.75) into Eq. (4.72), we have

$$\alpha_1 \xi = \lambda t^{-1/3}, \qquad \alpha_1 = \frac{Q_0}{D} \frac{\mathrm{d}\xi}{\mathrm{d}t}. \tag{4.77}$$

This results in

$$\xi = \sqrt{3} \left(\frac{3}{2} \right)^{2/3} \overline{R} Q_0^{-1/2} = m \left(D\alpha t \right)^{1/3} Q_0^{-1/2}, \tag{4.78}$$

$$m = \left(\frac{3}{2} \right) \sqrt{2} \approx 2, \tag{4.79}$$

$$\Delta = \lambda t^{-1/3} \frac{z}{\xi} = \frac{m}{3} Q_0^{1/2} \frac{\alpha^{1/3} z}{(Dt)^{2/3}}. \tag{4.80}$$

It only remains for us to compute the relative width of the intermediate region. According to Eq. (4.80), we have

$$\frac{\delta \xi}{\xi} = \frac{\delta t}{3t} = \frac{T_d}{(t)\, 3t}, \tag{4.81}$$

where T_d is the dissolution time of the largest precipitates residing on the boundary by the time t, $R_{\mathrm{max}} = (3/2)\, R_{\mathrm{cr}}$. A simple calculation shows that

$$\gamma = 3 \frac{\mathrm{d}t}{\mathrm{d}x^3} \leq \frac{27}{8} < \gamma_0 \tag{4.82}$$

in the equation of motion. This equation yields the dissolution time of the biggest precipitates as

$$u_0 = \frac{3}{2}, \qquad T_d \approx \frac{t}{3}. \tag{4.83}$$

For the relative width we have

$$\frac{\delta \xi}{\xi} \approx \frac{1}{9}. \tag{4.84}$$

Incidentally, note that

$$\frac{\delta \xi}{R} = \frac{1}{3} Q_0^{-1/2} \gg 1. \tag{4.85}$$

Furthermore,

$$\frac{\delta \xi}{l} \approx Q_0^{1/3} \frac{1}{3} Q_0^{-1/2} = \frac{1}{3} Q_0^{-1/6} \gg 1, \qquad Q_0 \to 0, \tag{4.86}$$

where l is the mean distance between precipitates. The inequalities (4.85) and (4.86) suggest that $\delta\xi \gg 1 \gg \overline{R}$, i.e., the width of the intermediate region is considerably larger than the mean interpore distance and the mean pore (precipitate) size, and at the same time it is considerably smaller than the width of the first region (the crust). All our assumptions made when evaluating the asymptotic behavior are thus valid. The asymptotically valid relation (4.80) is applicable to the same times as those obtained for the asymptotic behavior of coarsening in an infinite volume.

A peculiar situation occurs when we consider a sample of a finite size (rather than a sample with one boundary). To illustrate this, a plate of thickness a manifests three characteristic times: the time it takes the vacancies to diffuse from the interior of the plate onto its surface, $T_0 \sim (a^2/D)$; the time it takes the pore to grow from the solution, $T_1 \sim (\overline{R}^2/D)\Delta_0$, and the characteristic time of coarsening, $T_0 \sim (\overline{R}^3/D\alpha)$. To allow for the crust mechanism of sintering described above, the time of the diffusion outflow, T_0, must by far exceed T_{cr}, i.e., $a^2 \gg \left(\overline{R}^3/\alpha\right)$; the mean pore sizes appearing in this inequality must correspond to a well-developed coarsening process. In another extreme case (at $T_0 \ll T_{\text{cr}}$) the vacancies flow outward having no time to form pores. In the intermediate case, $a^2 \sim \left(\overline{R}^3/\alpha\right)$, the process kinetics is determined by several factors associated with the original distribution function and cannot be described in a general way.

4.1.5 Diffusive Decomposition Involving Different Mass-transfer Mechanisms

The late stage of first-order phase transitions, or coarsening, studied in this chapter, is essentially observed in experiments as the diffusive decomposition of supersaturated solid (liquid) solutions. In what follows we shall often use the term "diffusive decomposition" as an equivalent to the notion "coarsening."

Hitherto we have discussed the diffusive decomposition of a supersaturated solid solution under the condition where a transfer of mass between macrodefects was mediated by volume diffusion. Sometimes mass transfer is not primarily attributable to volume diffusion but, depending on the external conditions and the structure of the material, to other mechanisms such as diffusion along the dislocation lines or dislocation network, and diffusion along the grain boundaries. In addition, one can imagine growth processes where the decomposition kinetics is completely determined by the rate at which point defects cross the interfacial boundary or by the rate at which bonds are formed on the growing surface of a macrodefect [342]. In all these cases the basic canonical equations remain the same as those in the case of the volume diffusion transfer of mass. The only difference is found in the equation for the rate of growth due to the particular mass-transfer mechanism.

The rates of diffusional growth of macrodefects for different mass-transfer mechanisms are specified by the equation

$$\frac{\mathrm{d}R}{\mathrm{d}t} = \frac{D_n a^{n-3}}{R^{n-2}}\left(\Delta - \frac{\alpha}{R}\right), \qquad n \geq 2. \tag{4.87}$$

It can be expressed in terms of the equations given above when specific values are assigned to the parameters n and $D_n a^{n-3}$. When $n = 3$, we obtain Eq. (4.17) for the growth due to mass transfer through the volume, and when $n = 4$, we obtain the equation for the growth governed by diffusion along the grain boundaries. Let us consider the diffusive decomposition of a solid solution with the mass transfer given by this general equation.

The approach developed above for the case of diffusive decomposition of a solid solution, under the condition where mass is transferred through the volume from one precipitate to another, is obviously applicable to the diffusive decomposition of a solid solution with other mechanisms of mass transfer as given by Eq. (4.87), the volume mass transfer being a special case of this equation. We introduce again $u = (R/R_{\text{cr}})$, $R_{\text{cr}} = (\alpha/\Delta)$, and $\tau = n \ln(R_{\text{cr}}/R_{\text{cr}0})$. Then we get

$$\frac{\mathrm{d}u^n}{\mathrm{d}\tau} = \gamma(\tau)(u-1) - u^2, \qquad n \geq 2, \tag{4.88}$$

$$\gamma(\tau) = D_n \alpha n a^{n-3} \frac{\mathrm{d}t}{\mathrm{d}R_{\text{cr}}^n}. \tag{4.89}$$

The notations here are the same as those previously used.

By using the above-outlined approach we can easily obtain equations for the blocking point

$$u_0 = \frac{n}{n-1}, \tag{4.90}$$

$$\gamma_0 = n u_0^{n-1} = \frac{n^n}{(n-1)^{n-1}}. \tag{4.91}$$

It follows from Eq. (4.91) that in the case when the above mass-transfer mechanisms are operating, the blocking point can only be located in the interval $1 < u_0 < 2$.

The time dependence of the critical size of a precipitate immediately follows from the definition of $\gamma(r)$ (see Eq. (4.89)) as

$$R_{\text{cr}}^n = R_{\text{cr}_0}^n + \frac{D_n \alpha a^{n-3}}{u_0^{n-1}} t. \tag{4.92}$$

Solving the continuity equation and taking into consideration the law of conservation of mass and the equation of motion using the above method, we get an equation for the distribution function in the form

$$\varphi_n(u,\tau) = \begin{cases} \dfrac{A}{g_n(u)} \exp\{-\tau - \psi_n(u)\}, & u \leq u_0, \\ 0, & u \geq u_0, \end{cases} \tag{4.93}$$

$$\psi_n(u) = \int_0^u \frac{\mathrm{d}u'}{g_n(u')}, \tag{4.94}$$

where

$$-g_n(u) = \frac{\mathrm{d}u}{\mathrm{d}\tau} = \frac{1}{n u^{n-1}} \left[\gamma(u-1) - u^n\right], \tag{4.95}$$

$$\varphi(u,\tau) = N(\tau)\, P_n(u) \left(\overline{R}\right)^{-1}, \tag{4.96}$$

$$\overline{R} = R_{\mathrm{cr}} \frac{\int\limits_0^{u_0} P_n(u)\, u\, \mathrm{d}u}{\int\limits_0^{u_0} P_n(u)\, \mathrm{d}u}. \tag{4.97}$$

The values of the parameters, u_0, γ_0, R_{cr}, $P_n(u)$, and $\overline{R}$, for different specific mass-transfer mechanisms (for specific values of n) are summarized in Tables 4.1 and 4.2. There K_{s} is the rate of chemical bond creation on the precipitate surface.

Table 4.1: Coarsening characteristics for different mass-transfer mechanisms: size parameters.

n	R_{cr}^n	$\overline{u} = \frac{\overline{R}}{R_{\mathrm{cr}}}$
2	$R_{\mathrm{cr}0}^2 + \frac{1}{2}\frac{2\sigma n_0 V^2 c_\infty K_{\mathrm{s}} t}{k_B T}$	$\frac{8}{9}$
3	$R_{\mathrm{cr}0}^3 + \frac{4}{9}\frac{2\sigma n_0 V^2 c_\infty D t}{k_B T}$	1
4	$R_{\mathrm{cr}0}^4 + \frac{27}{64}\frac{2\sigma n_0 V^2 c_\infty D_0 t}{k_B T}$	1.03

Let us compare the results obtained for different mass-transfer mechanisms. It is evident from Tables 4.1 and 4.2 that the time dependence of the growth rate of the precipitate decreases with increasing n as $\left(R \sim t^{1/n}\right)$. To illustrate this statement, for example, the growth governed by diffusion along the grain boundaries ($n = 4$) is slower than growth due to diffusional mass transfer through the volume ($n = 3$), although usually $D_2 \gg D_0$ holds. In fact, there is no controversy here, as in the case of mass transfer along the grain boundary, a small cross-sectional area of the boundary leads to a large resistance to diffusion.

The blocking point $u_0 = [n/(n-1)]$ specifies the initial point of the range of sizes where the distribution function is identically equal to zero in the zeroth-order approximation. With increasing n, $u_0 \to 1$, and a part of the distribution function to the right of the peak becomes narrower. In the range of small values of $u_0 \to 0$, the distribution function behaves as $P_n(u) \sim u^{n-1}$, and the width of the distribution function decreases with increasing n.

As is readily seen from Eq. (4.89), the growth rate $\mathrm{d}R/\mathrm{d}t$ has its maximum at a precipitate size, $u = [(n-1)/(n-2)])$, located to the right of the blocking point u_0. The relation

Table 4.2: Coarsening characteristics for different mass-transfer mechanisms: distribution functions.

n	$P_n(u)$	
2	$\dfrac{24e^3 u \exp\left\{-\dfrac{6}{2-u}\right\}}{(2-u)^5}$	$u \leq 2$
	0	$u \geq u_0 = 2$
3	$\dfrac{3^4 e}{2^{5/3}} \dfrac{u^2 \exp\left\{-\dfrac{1}{1-(2/3)\,u}\right\}}{(u+3)^{7/3}\left(\dfrac{3}{2}-u\right)^{11/3}}$	$u \leq \frac{3}{2}$
	0	$u \geq u_0 = \frac{3}{2}$
4	$\dfrac{3^4 \exp\left\{-\dfrac{2}{4-3u} - \dfrac{1}{6\sqrt{2}} \arctan^{-1}\left(\dfrac{u+(4/3)}{(4/3)\sqrt{2}}\right)\right\}}{4\left(\dfrac{4}{3}-u\right)^{19/6}\left(u^2+\dfrac{8}{3}u+\dfrac{16}{3}\right)^{23/12}}$	$u \leq \frac{4}{3}$
	0	$u \geq u_0 = \frac{4}{3}$

between R_{cr} and R follows from the definition of u and $\overline{u} = (\overline{R}/R_{\text{cr}})$, where

$$\overline{u} = \int_0^{u_0} u P_n(u)\, \mathrm{d}u. \tag{4.98}$$

Differentiating the integral equation of conservation of matter with respect to time (under the assumption that $P(u)$ and $\overline{c}$ vary slowly with time) and using Eq. (4.89) for the growth rate, the following relation can be easily obtained:

$$R_{\text{cr}} = \frac{\overline{u^{4-n}}}{\overline{u^{3-n}}}, \tag{4.99}$$

which leads to an equality of the first moments of the function $P_n(u)$ of the order of $(4-n)$ and $(3-n)$, $\overline{u^{4-n}} = \overline{u^{3-n}}$. Then for $n = 3$ we immediately get $\overline{u} = 1$, i.e., $\overline{R} = R_{\text{cr}}$. The values of $\overline{u}$ for different mass-transfer mechanisms are given in Table 4.1, and the distribution function density $P_n(u)$ for different n is plotted in Figure 4.5.

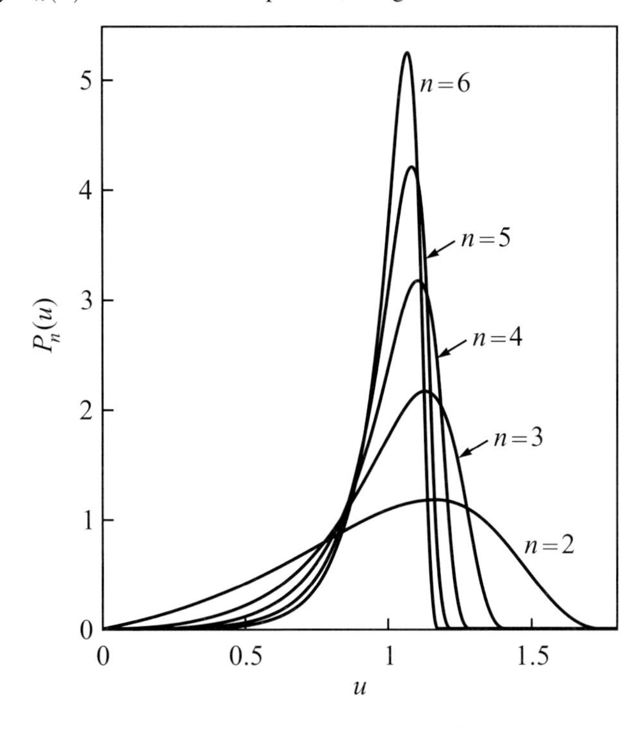

Figure 4.5: Dependence of the distribution function density $P_n(u)$ on u for different values of n.

4.1.6 Effects of Competition of Several Mass-Transfer Mechanisms

The theoretical description of the kinetics of diffusive decomposition of supersaturated solid solutions involving only one mass-transfer mechanism has been developed in the previous sections. In a real situation, however, several such mechanisms can be involved simultaneously. Thus, for example, a transfer of mass always occurs along the grain and block boundaries and along the dislocation lines concurrently with volume diffusion [285]. All these mechanisms operate concurrently for all new-phase precipitates. Due to these mechanisms only the precipitates with the fastest growth rate survive at a later stage of decomposition. The technique to solve the equations determining diffusive decomposition when several mass-transfer mechanisms are operating concurrently, discussed in the present section, has been developed for the first time in [285].

Let us consider the process of decomposition in the intermediate region. Generally, the growth rate of a spherical particle is specified by the equation

$$\frac{\mathrm{d}R}{\mathrm{d}t} = \sum_n J_n = J, \tag{4.100}$$

where J_n is the point defect flux density (per unit area of a macrodefect surface), which is determined by the nth mass-transfer mechanism. The equations for the respective fluxes are given above.

Suppose that a transfer of mass occurs under the concurrent operation of volume diffusion, diffusion along the block boundaries and dislocations with the corresponding diffusivities D_0, D_s, and D_d. Summing up the expressions given earlier for the respective flow densities, we obtain

$$\frac{\mathrm{d}R}{\mathrm{d}t} = \frac{D_{\mathrm{eff}}\,(R)}{R}\left(\Delta - \frac{\alpha}{R}\right), \tag{4.101}$$

where

$$D_{\mathrm{eff}} = \frac{D\,(R)}{1 + [D\,(R)\,/\,(K_{\mathrm{s}}R)]}, \tag{4.102}$$

$$D\,(R) = D_0\left[1 + \frac{\alpha}{R}\left(\frac{ND_s}{2D_0\ln\left(2b/R\right)} + \frac{M\,(D_d)^{1/2}}{[8D_0\ln\left(l/a\right)]^{1/2}}\right)\right], \tag{4.103}$$

and N and M are the numbers of block surfaces and dislocation lines that intersect the precipitate. When $K_{\mathrm{s}} \to \infty$, $D_{\mathrm{eff}} \to D_0$. When $K_{\mathrm{s}} \to 0$, then $D(R)$ drops out from Eq. (4.101), and we have a growth process determined by the rate of formation of the chemical bonds.

Since at sufficiently long times the mass transfer is attributable primarily to volume diffusion mechanism, we choose dimensionless canonical variables τ and u corresponding to the canonical form of this mechanism:

$$\tau = 3\ln\frac{R_{\mathrm{cr}}}{R_{\mathrm{cr}0}}, \qquad R_{\mathrm{cr}} = \frac{\alpha}{\Delta}, \qquad u = \frac{R}{R_{\mathrm{cr}}}, \tag{4.104}$$

$$R_{\mathrm{cr}0} = \frac{\alpha}{\Delta_0}, \qquad \Delta_0 = \Delta\,(t)|_{t=0}\,. \tag{4.105}$$

For the sake of clarity let us assume that the mass transfer is accomplished through volume diffusion and through diffusion along the surface of the blocks and along the dislocation lines, i.e., $K_{\mathrm{s}} \to \infty$. Then, using the chosen variables, a complete set of equations can be written as follows:

$$\frac{\partial\varphi}{\partial\tau} + \frac{\partial\,(\varphi g)}{\partial u} = 0, \tag{4.106}$$

$$\frac{\mathrm{d}u}{\mathrm{d}t} = g\,(u,\tau)\,, \qquad g\,(u,\tau) = \frac{1}{3u^2}\left[\gamma\,(u-1)\left(1+\frac{\beta}{u}\right) - u^3\right], \tag{4.107}$$

$$1 - \frac{\Delta_0}{Q_0}e^{-\tau/3} = \kappa e^{\tau}\int\limits_{v_0(\tau)}^{\infty} f_0\,(v)\,u^3\,(v,\tau)\;\mathrm{d}v, \tag{4.108}$$

where $f_0(v)$ is the initial distribution function of the new-phase precipitates with respect to the sizes $v = (R/R_{\mathrm{cr}})$, $u\,(v,\tau)$ is the solution of Eq. (4.105) with the initial condition $u\,(v,\tau)|_{\tau=0} = v$, Δ_0 is the initial supersaturation,

$$\kappa = \frac{4\pi}{3}\frac{R_{\mathrm{cr}0}^3}{Q_0}, \qquad \beta = \frac{D_{\mathrm{eff}}a}{DR_{\mathrm{cr}}}, \qquad \gamma = 3D\alpha\frac{\mathrm{d}t}{\mathrm{d}R_{\mathrm{cr}}^2}, \tag{4.109}$$

and $v_0(\tau)$ is the root of the equation $u(v,\tau) = 0$; in other words, $v_0(\tau)$ is the initial size of the precipitate dissolved by time τ.

The above-given detailed analysis shows that the functions $\gamma(\tau)$ and $\beta(\tau)$ for large τ should vary in such a way that, at each moment of time, the plot of $g(u,\tau)$ as a function of u would lie below the u axis, and in the zeroth-order approximation should touch it at one point. The form of the $g(u,\tau)$ function suggests that there is only one such point $u = u_0$. This tangency condition is given by the following equations:

$$g(u,\tau)|_{u=u_0} = 0, \qquad \left.\left(\frac{\partial g(u,\tau)}{\partial u}\right)\right|_{u=u_0} = 0. \tag{4.110}$$

Introducing the notation

$$\gamma(u,\tau) = \gamma(\tau)\left(1 + \frac{\beta}{u}\right), \tag{4.111}$$

we obtain from the first equation

$$\gamma(u_0,\tau) = \frac{u_0^3}{u_0 - 1}, \tag{4.112}$$

and from the definition of $\gamma(\tau)$,

$$\frac{u_0^3}{u_0 - 1} = 3D\alpha\left(1 + \frac{\beta}{u_0}\right)\frac{\mathrm{d}t}{\mathrm{d}R_{\mathrm{cr}}^3}. \tag{4.113}$$

The second equation together with Eq. (4.112) is then reduced to

$$u_0(3 - 2u_0) = (3u_0 - 4)\beta. \tag{4.114}$$

Solving this equation for u_0, we obtain

$$u_0 = \frac{3}{4}(1-\beta) + \left(\frac{9}{16}(1-\beta)^2 + 2\beta\right)^{1/2}. \tag{4.115}$$

We can represent β as

$$\beta(\tau) = \beta(0)\frac{R_{\mathrm{cr}0}}{R_{\mathrm{cr}}}, \tag{4.116}$$

and since $(R_{\mathrm{cr}0}/R_{\mathrm{cr}}) = e^{-\tau/3}$ holds also as $\beta(\tau) = \beta(0)\,e^{-\tau/3}$ and

$$u_0 = \frac{3}{4}\left(1 - \beta(0)\,e^{-\tau/3}\right) + \left(\frac{9}{16}\left(1 - \beta(0)\,e^{-\tau/3}\right)^2 + 2\beta(0)\,e^{-\tau/3}\right)^{1/2}. \tag{4.117}$$

It can be seen from Eq. (4.115) that when $\tau \to 0$ ($\beta \to 0$), then $u_0 \to 3/2$. When $\beta \to \infty$, we obtain from the same equation $u_0 = 4/3$. $\beta \to \infty$ corresponds to the case $D \to 0$. By rewriting Eq. (4.114) in the form

$$\beta = \frac{u_0(3 - 2u_0)}{3u_0 - 4}, \tag{4.118}$$

we can see that u_0 has to have a value in the range $(4/3 < u_0 < 3/2)$ because β is positive. It can be shown that for all the values of u_0 located in the range $(4/3 < u_0 < 3/2)$, the derivative $(\mathrm{d}u_0/\mathrm{d}r) > 0$, i.e., the $u_0\,(\tau)$ function is an increasing function of τ. It varies from its initial value $u_0(0)$, which is also in this range, to the value 3/2 when $\tau \to \infty$.

Equation (4.113) makes it possible to determine the dependence of the critical cluster radius R_{cr} on time t. In fact, u_0 is a function of β, and β is related to R_{cr} through Eq. (4.109). Substituting for u_0 in Eq. (4.113) its expression through β (Eq. (4.115)), and expressing β through R_{cr} according to Eq. (4.109), we obtain a differential equation relating R_{cr} to t. The solution of this equation, which has separable variables, yields a rather complex dependence of R_{cr} on t in the form

$$t = \frac{27}{16}\frac{R_{\mathrm{cr}}^3\beta^3}{\alpha D} \tag{4.119}$$

$$\times \int\limits_{\beta(0)\frac{R_{\mathrm{cr}0}}{R_{\mathrm{cr}}}}^{\beta(0)} \frac{\mathrm{d}x}{x^4}\left\{\frac{\left(1+\frac{14}{9}x+x^2\right)^{1/2}-x-\frac{7}{9}}{\left[\left(1+\frac{14}{9}x+x^2\right)^{1/2}-x-\frac{1}{3}\right]^2}\left[\left(1+\frac{14}{9}x+x^2\right)^{1/2}-x+1\right]^3\right\}.$$

When t is large, we have

$$R_{\mathrm{cr}}^3 = \frac{4}{9}D\alpha t, \tag{4.120}$$

consistent with the volume diffusion mechanism. If $\beta(0) > 1$, but the times are such that $\beta(0)(R_{\mathrm{cr}_0}/R_{\mathrm{cr}}) \gg 1$, we can easily obtain the dependence

$$R_{\mathrm{cr}}^4 - R_{\mathrm{cr}_0}^4 = \left(\frac{3}{4}\right)^3 D_{\mathrm{eff}}\alpha a t, \tag{4.121}$$

$$D_{\mathrm{eff}} = \frac{N D_{\mathrm{s}}}{2\ln\left(2b/R_{\mathrm{cr}_0}\right)} + \frac{M\left(D D_{\mathrm{d}}\right)^{1/2}}{\left(8\ln(l/a)\right)^{1/2}}, \tag{4.122}$$

characteristic of the mass transfer along the block surfaces and dislocation lines.

The dependence of R_{cr} on t is cumbersome because of the difficulty of expressing u_0 in terms of β and using the exact relation (4.115). We introduce into Eq. (4.115) a new unknown variable $x = 3/2 - u_0$, which can be expressed as

$$x = \frac{\beta}{6(1+\beta)}\frac{2}{[1-\beta/(9(1+\beta)^2)]^{1/2}+1}. \tag{4.123}$$

This equation can be approximated by a more simple expression

$$x \approx \frac{\beta}{6(1+\beta)}, \tag{4.124}$$

which can be used for small and large values of β. Then for β we obtain

$$u_0 = \frac{3}{2}\left(1-\frac{1}{9}\frac{\beta}{1+\beta}\right). \tag{4.125}$$

When $\beta = 0$, $u_0 = 3/2$, and when $\beta \to \infty$, $u_0 = 3/4$. We substitute u_0, as expressed through β, into Eq. (4.113) and replace $R_{\rm cr}$ in that equation by its expression through β according to Eq. (4.109). In this way, we obtain a differential equation that can be solved resulting in

$$\int_{\beta}^{\beta(0)} \frac{{\rm d}x}{x^4} \frac{\left(x + \frac{9}{8}\right)^4}{(x+1)^2 \left(x + \frac{3}{2}\right)\left(x^2 + \frac{7}{3}x + \frac{3}{2}\right)} = \frac{3^3}{2^8} D\alpha t \left(\frac{D}{\alpha D_{\rm eff}}\right)^3 . \tag{4.126}$$

When $\beta \ll 1$, expressing β in terms of $R_{\rm cr}$, we obtain

$$R_{\rm cr}^3 = \frac{4}{9} D\alpha t, \tag{4.127}$$

characteristic of mass transfer through volume diffusion. When $\beta \gg 1$, we obtain

$$R_{\rm cr}^4 - R_{\rm cr_0}^4 = \left(\frac{3}{4}\right)^3 \alpha D_{\rm eff} a t, \tag{4.128}$$

characteristic of mass transfer through diffusion along the boundaries of blocks and dislocations.

To determine the kinetics of motion of the blocking point in the general case, for an arbitrary number of mass-transfer mechanisms, we must determine its location when the growth rate is described by Eq. (4.109). Introducing the reduced variable $z = (u/u_0)$ instead of u, the equation for $({\rm d}z/{\rm d}\tau)$ becomes universal and gets the form

$$\frac{{\rm d}z^n}{{\rm d}\tau} = -(z-1)^2 \sum_{l=1}^{n-1} (n-1) z^{l-1}, \qquad n \geq 2. \tag{4.129}$$

Since the average radius $\overline{R}$, as shown above, increases with time, the relative contribution of the volume diffusion mechanism increases and becomes dominant at large times, and the blocking point approaches $3/2$. The growth rate of the precipitate is proportional to the sum of the flow densities for the different mechanisms and each of them, in the dimensionless coordinates, is analogous to the right-hand side of Eq. (4.129), i.e., has an essentially negative value. Therefore, u_0 monotonically approaches the point $3/2$; in fact, u_0 is located in a much narrower range depending on the initial conditions, either $1 < u_0 < (3/2)$ or $(3/2) < u_0 < 2$. Since the variation of u_0 is very small, its time derivative can be ignored in the zeroth-order approximation when determining the characteristics of the continuity equation. The solution is the same as that for the volume diffusion because τ was chosen for this particular mechanism. We get

$$\varphi(u,\tau) = \begin{cases} \dfrac{A}{g(u,u_0)} \exp\left\{-\tau - \psi(u,u_0)\right\}, & u \leq u_0 \\ 0, & u \geq u_0 \end{cases} \tag{4.130}$$

$$\psi(u,u_0) = \int_0^u \frac{{\rm d}u'}{g(u',u_0)}. \tag{4.131}$$

The growth rate of the precipitate determines $g(u,u_0)$.

If mass transfer along the boundaries of the blocks and dislocations and the volume mass-transfer mechanism are in operation, $g(u, u_0)$ can be determined from Eq. (4.108), and $u_0(\tau)$ from Eqs. (4.115) or (4.125). By determining the change in $u_0(\tau)$ more precisely, we obtain an additional term in the expression for $\varphi(u, \tau)$, which is close to unity because of the smallness of the derivative $(\mathrm{d}u_0/\mathrm{d}\tau)$.

We now determine the number of maxima in the distribution function with respect to the variable u for a given τ by solving the equation $(\mathrm{d}\varphi/\mathrm{d}u) = 0$ with respect to u. If there is only one mass-transfer mechanism, we have $(\mathrm{d}g/\mathrm{d}u) = 1$. Because the left-hand side of the latter equation decreases monotonically, it has only one time-independent root corresponding to one maximum of the φ function. This result has a simple physical explanation: the distribution peak is attributable to the presence of a single blocking point. For the same reason there is only one maximum when several mass-transfer mechanisms are in operation: there is only one blocking point, although in this case it is not stationary with respect to time. Several maxima may be observed in the distribution function as a result of experimental preparation of the samples. This is attributable to spatial inhomogeneity of the solutions investigated, which results in a mass transfer in different regions via different mechanisms. A composite curve with several maxima can be obtained by superimposing the corresponding distribution curves. Apparently this was done in [322] where two peaks were observed.

Consider now the decomposition process determined by the formation rate of the chemical bonds, and by mass transfer through the volume, to illustrate the evolution of the critical size of precipitates and the distribution function in the intermediate asymptotic region. In this case,

$$\beta = \frac{u_0(u_0 - 3)}{2 - u_0}, \tag{4.132}$$

where $\beta = [D/(K_{\mathrm{ph}} R_{\mathrm{cr}})]$. Using the definition of $\gamma(\tau)$, we obtain, for the asymptotic critical sizes, an equation in the dimensional variables

$$\int_{\beta}^{\beta_0} \frac{\mathrm{d}x}{x^4} \frac{(x - 9/4)^3}{(x + 3)^4} = f(\beta) - f(\beta_0) = \frac{3}{2^4} D\alpha \left(\frac{K}{D}\right)^3 t, \tag{4.133}$$

where

$$f(\beta) = \frac{1}{3\beta^3} \frac{(\beta + 9/4)^3}{(\beta + 3)^2} + \frac{1}{6\beta^2} \frac{(\beta + 9/4)^2(\beta + 9/2)}{(\beta + 3)^3} + \frac{9}{64\beta} \frac{7\beta + 9}{(\beta + 3)^4}$$
$$+ \frac{1}{72} \left\{ \ln \frac{\beta + 3}{\beta} + \frac{\beta^2 + 12\beta}{(\beta + 3)^2} + \frac{4\beta^3}{3(\beta + 3)^3} - \frac{\beta^4}{4(\beta + 3)^4} \right\}, \tag{4.134}$$

and

$$f(\beta_0) = f(\beta)|_{\beta = \beta_0}. \tag{4.135}$$

The distribution function is finally given by

$$\varphi(u, \tau) = N(\tau) P(u, u_0)/R_{\mathrm{cr}}, \tag{4.136}$$

where

$$N(\tau) = A\exp(-\tau) = AR_{\rm cr}^3, \qquad A = \frac{3}{4\pi}Q_0\left[\int_0^{u_0} P(u,u_0)u^3\,{\rm d}u\right]^{-1}, \tag{4.137}$$

$$P(u,u_0) = \begin{cases} \dfrac{3uu_0^{c+d}\{u(2-u_0)+u_0(2u_0-3)\}}{(2-u_0)^{1+d}\left(u+\dfrac{u_0}{2-u_0}\right)^{1+b}(u_0-u)^{2+c}} & \\ \times\exp\left\{-\dfrac{3u_0(u_0-1)}{(3-u_0)(u_0-u)}+\dfrac{3u_0-1}{3-u_0}\right\}, & \frac{3}{2}\le u\le 2, \\ 0\,, & u\ge u_0. \end{cases} \tag{4.138}$$

Here

$$b = 6\frac{2-u_0}{3-u_0}, \qquad c = 3\frac{u_0^2-4u_0+5}{(3-u_0)^2}, \qquad d = 6\frac{2-u_0}{(3-u_0)^2}. \tag{4.139}$$

By determining the motion of the blocking point u_0 we can easily obtain the evolution of the critical size and the distribution function in the intermediate region.

When $\beta\to\infty$, $u_0 = 2$, and we immediately arrive at expressions for the critical size and the distribution function that are characteristics of the decomposition determined by the formation of chemical bonds ($n = 2$). When $\beta\to 0$, $u_0 = 3/2$ and the equations define the decomposition process when the mass transfer is accomplished through volume diffusion ($n = 3$).

4.1.7 Asymptotic Stability of Solid Solutions

As already mentioned, the set of equations defining coarsening (diffusive decomposition) corresponds to the hydrodynamic approximation. The equations for the growth rate of particles involve, in fact, only the average supersaturation. It means that local fluctuations of concentration near a growing or dissolving particle are not taken into account. These fluctuations result in fluctuations of the particle growth rate. This effect can be taken into account by the use of a mechanism of collisions between particles which corresponds to the so-called kinetic approximation in the basic equations [286]. It is this approximation that enables us to choose, from all formal solutions, the only one which is stable against various disturbances. The term collision refers here to either the direct fusion or diffusive interaction of particles separated by a distance smaller than their size (in English literature this type of fusion process is commonly called coalescence).

The mechanism by which an asymptotic distribution of particles with respect to their sizes is established is largely influenced by the passage of the particles through the blocking point $u_0 = 3/2$, moving with the hydrodynamic velocity $v = -g(u,\tau)$. As a result of these collisions particles cross this point, thus changing the form of the equations. Therefore, the collisions greatly affect the stable asymptotic distribution even when the total supersaturation is low and the collisions are rare.

We now introduce the relative variables $\xi = r/R_{\text{cr}}$ in ordinary three-dimensional space. If we put the origin of the coordinate system at a certain particle, then all other particles, while approaching each other, will be moving toward the chosen particle in these variables (as the coordinates of the particles remain unchanged whereas $R_{\text{cr}}(t) \to \infty$). Most of the particles will dissolve before reaching the center and without experiencing a collision.

The distribution function and the density of particles will be normalized to a unit of relative volume as well. They are related to the previously used quantities φ and n by the following equations:

$$f = R_{\text{cr}}^3 \varphi = e^{\tau} \varphi, \tag{4.140}$$

$$N = R_{\text{cr}}^3 n = e^{\tau} n. \tag{4.141}$$

As shown above, in the asymptotic state, when all the excess volume is practically concentrated within the particles, $n \sim e^{\tau}$, i.e., the density N is constant. Accordingly, the probability of an interparticle collision, in unit time in a unit of this relative volume, is time independent. When two particles collide, the larger one absorbs the smaller one by diffusion and their total volume remains unchanged. In fact, the volume is kept only approximately constant, as some amount of matter can pass directly into the solution. By taking this effect into account, it will obviously not result in considerable changes.

The dimensionless collision time is of the order of unity ($\tau_{\text{col}} \approx 1$) because all the parameters in the equation defining the dissolution of a smaller particle,

$$\frac{\mathrm{d}u^3}{\mathrm{d}\tau} \approx -\gamma - u^3, \tag{4.142}$$

and in the initial condition $u|_{\tau=0} = u_0 \approx 1$, are of the order of unity. Incorporation of collisions into the system of equations defining coarsening affects only Eq. (4.19). An expression corresponding to an ordinary collision integral appears on the right-hand side of Eq. (4.19).

Since the volume of particles is conserved during the collisions, we can conveniently use the distribution function with argument $z = u^3$. This distribution function is defined by the following equation:

$$\kappa e^{\tau} \varphi \mathrm{d}u = f(z, \tau)\, \mathrm{d}z, \tag{4.143}$$

where the factor

$$\kappa = \frac{4\pi}{3} \frac{1}{Q_0} R_{\text{cr}0}^3 \tag{4.144}$$

is introduced for convenience.

Before writing the collision integral it should be noted that the number of collisions of particles, between u' and $u' + \mathrm{d}u'$ in size, with particles of size u, in unit time, is given by

$$\nu = \varphi(u, \tau)\varphi(u', \tau) \frac{V_{\text{eff}}}{\tau_{\text{col}}}\, \mathrm{d}u', \tag{4.145}$$

where τ_{col} is the collision time; $V_{\text{eff}} = W(z, z')e^{\tau} R_{\text{cr}0}^3$ is the effective volume within which the particles' centers can interact with other particles around them, and $W(z, z')$ is the relative effective volume. Obviously, $W(z, z')$ is of the order of the total relative volume of the particles; $W(z, z') = z + z'$. It is evident that the effective volume increases asymptotically along with the mean size of the particle. Since at long times $\varphi \sim e^{-\tau}$ asymptotically, $f(z, \tau) = f(z)$ and the number of collisions can be written as

$$\nu = \frac{1}{\kappa^2} \frac{1}{\tau_{\mathrm{col}}} f(z) W(z, z') f(z') e^{-\tau} \, \mathrm{d}z'. \tag{4.146}$$

For sufficiently long times, the system of Eqs. (4.17)–(4.19) rewritten for $f(z)$ in the new variable z with allowance for collisions assumes the following form:

$$\frac{\mathrm{d}}{\mathrm{d}z}(f\beta) + f = -I_{\mathrm{col}}, \qquad -\beta = \left(z^{1/3} - 1\right)\gamma - z, \tag{4.147}$$

$$\int_0^\infty z f(z) \, \mathrm{d}z = 1, \qquad \gamma = \mathrm{const}, \tag{4.148}$$

where

$$I_{\mathrm{col}} = \frac{3}{4\pi} Q_0 \frac{1}{\tau_{\mathrm{col}}} \left\{ \frac{1}{2} \int_0^z W(z - z', z') f(z - z') f(z') \, \mathrm{d}z' - f(z) \int_0^\infty W(z, z') f(z') \, \mathrm{d}z' \right\}. \tag{4.149}$$

The factor $1/2$ in the first term is needed to avoid taking the collisions between two groups of particles into account twice. It follows from the symmetry of $W(z, z')$, with respect to interchanging arguments, that the number of particles decreases with the total volume held constant as the collisions occur.

Since the collision integral I_{col} is small, we can solve the resulting set of equations by means of successive approximations. It should be stressed that from physical considerations and from the analysis at the beginning of this section we should expect the value of γ to decrease as compared to γ_0 because of the collisions (since the particles crossing the point $z_0 = (3/2)^3$ due to the collisions should be able to leak back). Thus, in the zeroth-order approximation, with γ approaching γ_0 from below, the solution reads

$$P(u, u_0) = \begin{cases} \left(\frac{A}{\beta}\right) \exp(-\psi), & z < z_0, \\ \left(\frac{A}{\beta}\right) \exp(-\psi - \Delta), & z > z_0. \end{cases} \tag{4.150}$$

Here again

$$\psi = \int_0^z \frac{\mathrm{d}z}{\beta} = \frac{4}{3} \ln\left(z^{1/3} + 3\right) + \frac{5}{3} \ln\left|z^{1/3} - z_0^{1/3}\right| + \frac{1}{1 - \left(\frac{z}{z_0}\right)^{1/3}} - \ln\frac{3^3 e}{2^{5/3}}. \tag{4.151}$$

The constant

$$\Delta = \frac{3\pi}{(3\Delta\gamma/\gamma_0)^{1/2}}, \qquad \Delta\gamma = \gamma_0 - \gamma > 0, \tag{4.152}$$

reflects a rapid change of ψ in the region of transition from $z < z_0$ to $z > z_0$. The integral $\int_{\Delta z} \beta^{-1}\,\mathrm{d}z$ over the transition region is equal to Δ.

It follows that when $\gamma = \gamma_0$, f_0 reduces to zero along with all its derivatives, i.e., it matches zero in the smoothest possible way,

$$f_0(z,\gamma_0) = \begin{cases} \left(\frac{A}{\beta}\right) e^{-\psi(z,\gamma_0)}, & z < z_0, \\ 0, & z > z_0, \end{cases} \tag{4.153}$$

where the constant A is again specified by the normalization condition

$$\int_0^z z f(z)\,\mathrm{d}z = \frac{1}{\kappa}, \tag{4.154}$$

and κ is determined by Eq. (4.144).

When $\gamma < \gamma_0$, there is no stationary solution ($\int_0^\infty z f(z)\,\mathrm{d}z$ is logarithmically diverging). When $\gamma > \gamma_0$, $\beta(z,\gamma) = 0$ at two points, z_1 and z_2 ($z_1 < z_2$), and, near z_1, the function $f_0(z,\gamma)$ decreases as $|z - z_1|^b$, where $b = \left(\partial\beta/\partial z|_{z=z_1}\right)^{-1} - 1$. When $\gamma - \gamma_0 \ll 1$, $f_0 \approx |z - z_1|^c$, where $c = 2(3\gamma)^{-1}(\gamma - \gamma_0)^{-1/2} - 1$. A zeroth-order stationary solution can therefore exist for $\gamma > \gamma_0$ and is given by

$$f_0(z,\gamma_0) = \begin{cases} \left(\frac{A}{\beta}\right) e^{-\psi(z,\gamma_0)}, & z < z_1, \\ 0, & z > z_1. \end{cases} \tag{4.155}$$

This solution is continuous together with its derivative at the point $z = z_1$; however, it is unstable against an appearance of the right-hand side in the equation.

In fact, in the next-order approximation the distribution function $f(z)$ for $\gamma > \gamma_0$ is

$$f(z) = \frac{e^{-\psi}}{\beta}\left[\text{const} - \int I_{\text{col}}\{f\}\, e^{\psi}\,\mathrm{d}z'\right], \tag{4.156}$$

and diverges at the point z_2. This instability reflects the fact that the particles arriving at the region $z_1 \le z < z_2$ are unable to leak back, and the volume of the particles in this region would increase with time, which is impossible. Thus, $f_0(z,\gamma_0)$ should be taken as the zeroth-order approximation. Then, in this approximation,

$$I_{\text{col}}\{f\} \approx I_{\text{col}}\{f_0\} = \frac{3Q_0}{4\pi} I_0(z), \tag{4.157}$$

$$f_0(z,\gamma_0) = \begin{cases} \frac{1}{2}\int_0^z W(z-z',z')f_0(z-z')f_0(z')\,\mathrm{d}z' \\ -f_0(z)\int_0^\infty W(z,z')f_0(z')\,\mathrm{d}z', & z < 2z_0, \\ 0, & z > 2z_0. \end{cases} \tag{4.158}$$

Successive approximations involve incorporation of the right-hand part with simultaneous refinement of the γ value like

$$\frac{\mathrm{d}f_n\beta_n}{\mathrm{d}z} + f_n = -I_{\mathrm{col}}\{f_{n-1}\} \ , \tag{4.159}$$

$$\int_0^z zf(z)\,\mathrm{d}z = 1, \qquad \beta_n \equiv \beta(z,\gamma_n). \tag{4.160}$$

The solution satisfying Eq. (4.160) in the first-order approximation is eventually given by

$$f_1(z) = \begin{cases} \frac{3Q_0}{4\pi}\frac{1}{\tau_{\mathrm{col}}}e^{\Delta}\frac{A}{\beta}e^{-\psi}, & z < z_0, \\ \frac{3Q_0}{4\pi}\frac{1}{\tau_{\mathrm{col}}}\frac{1}{\beta}e^{-\psi}\int_z^{2z_0} I_0(z')e^{-\psi}\,\mathrm{d}z', & z_0 \le z \le 2z_0, \\ 0, & z > 2z_0. \end{cases} \tag{4.161}$$

$$A = \int_z^{2z_0} I_0(z')e^{\psi}\,\mathrm{d}z'. \tag{4.162}$$

Hence $f(z) \sim Q_0$ when $z > z_0$, and the function varies smoothly in the vicinity of $z = z_0$. The next-order approximation of the distribution function can be obtained by substituting the above function into the collision integral. The distribution function can be calculated on any interval of the z values with an arbitrary accuracy by successive iterations.

If we keep only the lowest order terms in Q_0 for every interval $nz_0 \le z \le (n+1)z_0$, the resulting solution can be written as

$$f(z) = Q_0^{z/z_0}\varphi(z), \tag{4.163}$$

where $\varphi(z)$ can be determined for every interval ($\varphi(z)$ for the interval $0 \le z \le 2z_0$ is given above in Eq. (4.138)). Physically, this Q_0-dependence is attributable to the fact that the particles of size z are created due to the collisions of the particles of smaller size, $z = z_1 + z_2$.

It is evident from Eq. (4.162) that the distribution function for the interval $z < z_0$ contributes primarily to the law of conservation of volume with an accuracy to terms of the order of Q_0. Let us now derive $\Delta\gamma$ as a function of Q_0 with the same accuracy

$$1 = \frac{3}{4\pi}\frac{Q_0}{\tau_{\mathrm{col}}}e^{\Delta}A\int_0^{z_0} e^{\psi}\frac{z}{\beta(z,\gamma)}\,\mathrm{d}z. \tag{4.164}$$

Hence, as $\Delta\gamma = \gamma_0 - \gamma \ll 1$, $Q_0 \ll 1$, and all the other factors are of the order of unity, we obtain

$$\frac{\Delta\gamma}{\gamma_0} = \frac{3\pi^2}{(\ln Q_0)^2}. \tag{4.165}$$

Thus, the allowance for collisions results in an insignificant variation of the distribution function defined by a term of the order of Q_0 in the vicinity of the $z = z_0$ point, and in the appearance of an exponentially decreasing tail of the order of Q_0 in the region beyond the blocking point. Qualitatively, incorporation of collisions between pores enables us to choose the only stable solution.

Finding the exact solution for a specific initial distribution is of great interest [286]. This exact solution can be derived for the case when $\gamma = \text{const} = \gamma_0$ at all times, i.e., when the point $u = u_0 = 3/2$ is precisely a blocking point, and $\varphi(u, \tau)$ is nonzero only for the interval $0 \le u \le u_0$. The initial distribution function in this case is

$$\varphi(u, 0) = \varphi_0(v) = \begin{cases} AP(R)\left(1 - Ce^{-\psi(R)/3}\right), & R < 3/2, \\ 0, & R \ge 3/2. \end{cases} \tag{4.166}$$

$$u|_{\tau=0} = v = R, \qquad P(R) = \frac{1}{g(R)} \exp\{-\psi(R)\}, \tag{4.167}$$

$$A = \left(\kappa \int_0^{3/2} e^{-\psi} \frac{u^3}{g(u)} \,\mathrm{d}u\right)^{-1},$$

$$C = \frac{\Delta_0}{Q_0}\left(\int_0^{3/2} e^{-\psi} \frac{u^3}{g(u)} \,\mathrm{d}u\right)\left(\int_0^{3/2} e^{-4\psi/3} \frac{u^3}{g(u)} \,\mathrm{d}u\right)^{-1}. \tag{4.168}$$

The exact distribution function with respect to the absolute sizes that holds for all times is

$$f_0(R, t) = \varphi\left(u, \tau(t)\right) R_{\mathrm{cr}}^{-1}(t) \tag{4.169}$$

$$= \begin{cases} AP\left(\frac{R}{R_{\mathrm{cr}}(t)}\right)\left(1 + \frac{3t}{\gamma_0}\right)^{-4/3}\left[1 - C\left(1 + \frac{3t}{\gamma_0}\right)^{-1/3} e^{-\psi/3}\right], & R < 3/2, \\ 0, & R \ge 3/2. \end{cases}$$

Its asymptotic behavior is, of course, similar to that obtained above.

4.2 Rigorous Analysis of the Transformation of an Arbitrary Initial Distribution Function into a Universal One

4.2.1 Introduction

A theory of first-order phase transitions at its later stage (coarsening) has been developed above in the zeroth-order approximation. A qualitative analysis of the equations defining the process of coarsening (or solution diffusive decomposition) in this approximation suggested that at sufficiently long times the contribution to the balance of matter from new-phase particles of sizes, exceeding the blocking point, was negligible. At the same time, it was shown that the process is characterized by a definite self-consistent behavior of the supersaturation with time.

As far as the growth rate of particles (expressed in relative sizes) is concerned, this behavior suggests that the growth rate touches the relative size axis at the blocking point, u_0, from below, or, otherwise, both the rate and its derivative with respect to relative size vanish at this point. This approximation unambiguously leads to a universal solution for the size distribution function for $u \leq u_0$ which is independent of the initial conditions. In the region where $u \geq u_0$, the distribution function is zero in the zeroth-order approximation. This approximation, however, does not allow us to describe the evolution of the universal distribution function with time and its relation to the original distribution. It is also impossible to obtain the distribution function within and outside the region of the blocking point u_0, as well as to calculate more precisely the asymptotic behavior of supersaturation of the solution with time, and to specify the time τ_0 at which we can start using asymptotic equations to describe the time course of diffusive decomposition with a sufficient accuracy. All these values are closely related to the form of the original distribution function, and their calculation requires the development of a systematic approach to solving the set of equations that define the diffusive decomposition of a supersaturated solution starting with some certain initial condition [342].

4.2.2 Canonical Form of the Basic System of Equations

We consider a single-component solid or liquid solution characterized by the supersaturation $\Delta(t) = \overline{c} - c_\infty$ (where $\overline{c} = \overline{c}(t)$ is the average concentration of the solution at a given time and c_∞ is the equilibrium concentration at a planar boundary), and by the time-dependent size distribution function $f(R, t)$ of the new-phase particles. This distribution is normalized to unit volume; $f(R, t)\,\mathrm{d}R = \mathrm{d}N$ is the number of particles in unit volume within the size interval $(R, R + \mathrm{d}R)$.

We introduce again the more convenient relative size of particles $u = (R/R_{\mathrm{cr}}(t))$ and reduced time

$$\tau = \ln \frac{R_{\mathrm{cr}}(t)}{R_{\mathrm{cr}}(0)} = 3 \ln \frac{\Delta(0)}{\Delta(t)}, \tag{4.170}$$

$$R_{\mathrm{cr}}(t) = \frac{\alpha}{\Delta(t)}, \tag{4.171}$$

where $\alpha = (2\sigma\omega c_\infty/k_B T)$, σ is the interfacial tension; ω is the volume of an atom of the dissolved component; k_B is the Boltzmann constant, and T is the absolute temperature.

The distribution functions in absolute and relative variables are interrelated by an obvious equation

$$f(R,t)\,\mathrm{d}R = \varphi(u,\tau)\,\mathrm{d}u. \tag{4.172}$$

The introduction of the relative sizes of particles is of great importance because, as described by these variables, from a certain time onward the particles dissolve in an orderly fashion, and their distribution function at a given time is determined by the asymptotic behavior of the initial distribution function.

Note that, if the particles contain excess material, a limited time interval can exist when they partially dissolve temporarily increasing the supersaturation of the solution (this corresponds to $\mathrm{d}\tau/\mathrm{d}t < 0$) with the supersaturation monotonically decreasing later on, $\mathrm{d}\tau/\mathrm{d}t > 0$. The diffusive decomposition when $\mathrm{d}\tau/\mathrm{d}t > 0$ holds is discussed below. This is the case when the supersaturation is sufficiently low, independent of the initial conditions which affect only the time when $\mathrm{d}\tau/\mathrm{d}t$ begins rising monotonically with time.

Furthermore, when $\Delta \ll 1$, as shown above, the equilibrium concentration and the stationary diffusion flux have time to evolve at the boundaries of particles. Under these conditions, the rate of particle growth due to the mass transfer through the volume is

$$\frac{\mathrm{d}R}{\mathrm{d}t} = \frac{D}{R}\left(\Delta - \frac{\alpha}{R}\right), \tag{4.173}$$

where D is the diffusion coefficient of the dissolved component. The growth rate in the variables u and τ is given by

$$\frac{\mathrm{d}u}{\mathrm{d}\tau} = \frac{1}{3u^2}\left[\gamma(\tau)(u-1) - u^3\right], \tag{4.174}$$

$$\gamma(\tau) = 3\frac{\mathrm{d}t}{\mathrm{d}R_{\mathrm{cr}^3(t)}}\frac{R^3_{\mathrm{cr}}(0)}{t^*} = \frac{3D}{\alpha^2}\left(\frac{\mathrm{d}}{\mathrm{d}t}\frac{1}{\Delta^3}\right)^{-1} = \gamma_0\left(1 - \varepsilon^2(\tau)\right), \tag{4.175}$$

$$\gamma_0 = \frac{27}{4}, \qquad t^* = \frac{R^3_{\mathrm{cr}}(0)}{\alpha D}. \tag{4.176}$$

This form of the $\gamma(\tau)$ function is characteristic of an orderly motion of the particles from right to left, in the relative size variables, that corresponds to $\mathrm{d}u/\mathrm{d}\tau < 0$ for all u. Otherwise the particles located to the right of the blocking point, $u_0 = 3/2$, in the size space will be unable to pass into the region to the left of this point, and since $R_{\mathrm{cr}}(t) = \alpha/\Delta(t)$, when $\tau \to \infty$ and $\Delta \to 0$, the amount of the material contained in the particles would increase infinitely, which is impossible. The leakage should proceed in such a way that the size distribution function satisfying the condition of the balance of matter has time to evolve.

The equations defining coarsening are the continuity equation in the size space for a certain mass-transfer mechanism together with the equation of balance of matter. In our case they are the equations of the volume diffusion of a dissolved component and the conservation of material present in the solution and in the particles:

$$\frac{\partial\varphi}{\partial\tau} + \frac{\partial}{\partial u}\left(\varphi\frac{\mathrm{d}u}{\mathrm{d}\tau}\right) = 0, \tag{4.177}$$

$$\varphi|_{\tau=0} = f_0(v), \qquad u|_{\tau=0} = v. \tag{4.178}$$

When considering the coarsening as a late stage of diffusive decomposition, when $\Delta \ll 1$, we can ignore stochastic production of the new-phase particles with $R > \alpha/\Delta$ thus setting the right-hand side of the continuity equation equal to zero. This corresponds to the hydrodynamic approximation. Its general solution is

$$\varphi(u,\tau) = f_0(v(u,\tau))\frac{\partial v}{\partial u}, \tag{4.179}$$

where $v = v(u,\tau)$ is the characteristics of the equation

$$\frac{\mathrm{d}u}{\mathrm{d}\tau} = \frac{\gamma(u-1)-u^3}{3u^2}, \qquad u\big|_{\tau=0} = v. \tag{4.180}$$

The balance of a dissolved component is given by the equation

$$1 - \frac{\Delta(0)}{Q_0}e^{-\tau/3} = \kappa e^{\tau}\int_0^\infty \varphi(u,\tau)u^3\,\mathrm{d}u = \kappa e^{\tau}\int_{v(0,\tau)}^\infty f_0(v)u^3(v,\tau)\,\mathrm{d}v. \tag{4.181}$$

Here $\Delta(0)$ is the initial supersaturation, Q_0 is the total excess amount of the material, and

$$\kappa = \frac{4\pi}{3}\left(\frac{R_{\text{cr}}^3(0)}{Q_0}\right). \tag{4.182}$$

Taking into account the form of $\gamma(\tau)$ and the fact that with $\gamma = \gamma_0$, $\mathrm{d}u/\mathrm{d}\tau$ has a second-order zero at the point $u_0 = 3/2$, we can rewrite Eq. (4.180) as

$$\frac{\mathrm{d}u}{\mathrm{d}\tau} = -g\left(u,\varepsilon^2(\tau)\right) = -\frac{1}{3u^2}\left[(u+3)(u-u_0)^2 + \frac{27}{4}(u-1)\varepsilon^2(\tau)\right]. \tag{4.183}$$

When $\tau \to \infty$, $\varepsilon^2(\tau) \to 0$. The $\varepsilon^2(\tau)$ function behaves in such a way because with increasing τ during an orderly motion from right to left, the increasingly distant parts of the initial distribution function (which is a decreasing function, $f_0(v) \le v^{-n}$ $(n > 4)$) contribute to the distribution function, at a given time, to the left of the blocking point, which primarily contributes to the balance of material. Since, in fact, $\mathrm{d}u/\mathrm{d}\tau < 0$, as the time increases, fewer and fewer particles remain to the right of the blocking point so they have to leak more slowly through the transformation region around the blocking point to absorb all the excess material. Accordingly, the distribution function in relative sizes will narrow in a self-consistent manner.

The continuity equation possesses a general solution so the equation of the balance of matter, Eq. (4.181), combined with the characteristics, Eq. (4.180), represents a nonlinear equation for the $\varepsilon^2(\tau)$ function and the characteristics $v(u,\tau)$ as well as the distribution function. Taking into account that $\varepsilon^2(\tau) \to 0$ when $\tau \to \infty$, we can write the equation for the characteristic as

$$\frac{\mathrm{d}u}{\mathrm{d}\tau} = \begin{cases} -g(u), & u < u_2 < u_0, \\ -\frac{2}{3}(u-u_0)^2 - \frac{1}{2}\varepsilon^2(\tau), & u_2 \le u \le u_1, \\ -g(u), & u > u_1 > u_0, \end{cases} \tag{4.184}$$

where

$$g(u) \equiv g(u,0) = (3u)^{-2}(u+3)(u-u_0)^2. \tag{4.185}$$

Here we isolated a region around the point u_0 where the function $\varepsilon^2(\tau)$ contributes significantly to the overall value. The boundaries of the transformation region, u_1 and u_2, obey the obvious conditions

$$\varepsilon^2(\tau)(u_1-u_0)^{-2} \ll 1, \qquad \varepsilon^2(\tau)(u_2-u_0)^{-2} \ll 1. \tag{4.186}$$

Let us now introduce a function $\psi(u)$ for $u > u_0$ and $u < u_0$, such that the difference between its values at any two u points is equal to the time it takes to move from the larger u value to the smaller,

$$\psi(u) = \int \frac{\mathrm{d}u}{g(u)} = \frac{4}{3}\ln(u+3) + \frac{5}{3}\ln|u-u_0| + \frac{u_0}{u_0-u}. \tag{4.187}$$

The function

$$\psi(u) = \int\limits_0^u \frac{\mathrm{d}u}{g(u)} = \psi(u) - \psi(0) \tag{4.188}$$

was derived previously only for $u < u_0$. We can now write down the exact solution of the continuity equation based on the form of its characteristics in these three regions,

$$\varphi(u,\tau) = f_0(v(u,\tau))\frac{\partial v}{\partial u}, \tag{4.189}$$

where $v(u,\tau)$, with $u \geq u_1$, can be obtained from the solution of the equation for the characteristics that is obeyed in this region at all times,

$$\psi(v) = \psi(u) + \tau, \qquad u \geq u_1, \tag{4.190}$$

$$\varphi(u,\tau) = f_0(v(u_1,\tau_1(u,\tau,u_1)))\frac{\partial v}{\partial \tau_1}\frac{\partial \tau_1}{\partial u}, \tag{4.191}$$

$$\psi(v) = \psi(u_1) + \tau_1(u,\tau,u_1), \qquad u_2 \leq u \leq u_1, \tag{4.192}$$

$$\varphi(u,\tau) = f_0(v(u_1,\tau_1(u_2,\tau_2(u,\tau,u_2),u_1)))\frac{\partial v}{\partial \tau_1}\frac{\partial \tau_1}{\partial \tau_2}\frac{\partial \tau_2}{\partial u}, \tag{4.193}$$

$$\psi(v) = \psi(u_1) + \tau_1(u_2,\tau_2(u,\tau,u_2),u_1), \qquad u \leq u_2. \tag{4.194}$$

Here $\tau_1 = \tau_1(u,\tau,u_1)$ can be determined from the equation for the characteristics in the region $u_2 \leq u \leq u_1$

$$\frac{\mathrm{d}u}{\mathrm{d}\tau} = -\frac{2}{3}(u-u_0)^2 - \frac{1}{2}\varepsilon^2(\tau), \qquad u|_{\tau=\tau_1} = u_1. \tag{4.195}$$

The function $\tau_2 = \tau_2(u,\tau,u_2)$ can be determined from the equation for the characteristics in the region $u \leq u_2$ and at once written for all times

$$\tau_2 = \tau + \psi(u) - \psi(u_2). \tag{4.196}$$

The function $\varepsilon^2(\tau)$, and accordingly the distribution function, can be evaluated, with sufficient accuracy, from the moment when essentially all the excess material is concentrated in the particles of size $u < u_2$.

From that time onward the balance of material which is present in excess over the thermal equilibrium concentration is specified by

$$\kappa e^{\tau} \int_0^{u_2} \varphi(u) u^3 \, \mathrm{d}u = \kappa e^{\tau} \int_0^{u_2} f_0(v) \frac{\partial v}{\partial u} u^3 \, \mathrm{d}u = 1. \tag{4.197}$$

Here we neglect the amount of material in the solution and in particles of size $u > u_2$ which can be incorporated in the next-order approximation and which can introduce an exponentially small correction in τ.

Substituting the solution for φ at $u < u_2$, Eq. (4.191), into Eq. (4.193), we get

$$\kappa \int_0^{u_2} e^{\tau} f_0(v(\tau_1(\tau_2))) \frac{\partial v}{\partial \tau_1} \frac{\partial \tau_1}{\partial \tau_2} \frac{\partial \tau_2}{\partial u} u^3 \, \mathrm{d}u = 1. \tag{4.198}$$

Using the explicit form of τ_2, Eq. (4.196), we have

$$\kappa \int_0^{u_2} e^{\tau_2} f_0(v(\tau_2)) \frac{\partial v}{\partial \tau_2} \exp\left[-\psi(u) - \psi(u_2)\right] \frac{u^3}{g(u)} \, \mathrm{d}u = 1. \tag{4.199}$$

Equation (4.199) is obeyed when

$$e^{\tau_2} f_0(v(\tau_2)) \frac{\partial v}{\partial \tau_2} = C = e^{-\psi(u_2)} \left[\kappa \int_0^{u_2} e^{-\psi(u)} \frac{u^3}{g(u)} \, du \right]^{-1}. \tag{4.200}$$

This enables us to calculate the values of $\varphi(u, \tau)$, for $u < u_2$, that coincide with the above given zeroth-order approximation

$$\varphi(u, \tau) = f_0(v) \frac{\partial v}{\partial \tau_1} \frac{\partial \tau_1}{\partial \tau_2} \frac{\partial \tau_2}{\partial u} = f_0(v) \frac{\partial v}{\partial \tau_2} \frac{1}{g(u)} = C e^{\psi(u_2)} \frac{1}{g(u)} e^{-\psi(u) - \tau}. \tag{4.201}$$

We find now for $u < u_2$ the relation of the integral of motion $v = v(\tau_2)$ to the asymptotic distribution function, taking into account that $v(\tau_2)$ is the initial relative size of a particle which takes the value u at the time τ, with u rising infinitely when $\tau \to \infty$.

To this end, let us integrate Eq. (4.200) over τ_2 multiplying it previously by $e^{-\tau_2}$

$$\int_{v(\tau_2)}^{\infty} f_0(v) \, \mathrm{d}v = C \int_{\tau_2}^{\infty} e^{-\tau_2} \, \mathrm{d}\tau_2 = C e^{-\tau_2}. \tag{4.202}$$

Evaluating $v(\tau_2)$ from Eq. (4.202) and using Eq. (4.194), we obtain the dependence $\tau_1 = \tau_1(\tau_2)$ for $u \leq u_2$. Physically, the integral τ_1 represents the time it takes a particle to attain the point u_1 in the size space, provided that it is at the point u at the time τ.

By letting $u = u_2$ and, accordingly, $\tau_2 = \tau$ we evaluate the time it takes a particle to arrive at the boundary u_1 of the transformation region, provided that the particle is at the opposite boundary u_2 at the time τ,

$$\psi(v(\tau)) = \tau_1(\tau) + \psi(u_1). \tag{4.203}$$

Equation (4.203) holds for those times which obey the condition $v(\tau) \geq u_1$. This equation combined with the equation for the characteristics for the interval $u_2 \leq u \leq u_1$ (Eq. (4.195)) enables us to deduce $\varepsilon^2(\tau)$. The resulting equation can be conveniently written in a canonical form upon substituting $2(u - u_0)/3^{1/2} \to u$ and $\tau/3^{1/2} \to \tau'$. We obtain

$$\frac{\mathrm{d}u}{\mathrm{d}\tau'} = -u^2 - \varepsilon^2(\tau'), \qquad u|_{\tau'=\tau_1=\tau_1(\tau)} = u_1, \qquad u|_{\tau'=\tau} = u_2. \tag{4.204}$$

Equations (4.202) (where we should put $\tau_2 = \tau$), (4.203), and (4.204) provide a canonical system for $\varepsilon^2(\tau)$ and, accordingly, for the characteristics as well as for the distribution function in the region $u \geq u_2$.

Generally, we should solve a Riccati-like equation (4.204) for an arbitrary function $\varepsilon^2(\tau)$, and then derive a functional equation for $\varepsilon^2(\tau)$ from the boundary conditions. This procedure cannot be approached analytically in the general case, although the asymptotic behavior of physically meaningful initial distributions, decreasing with size, leads to the $\tau_1 = \tau_1(\tau)$ dependences that reduce the problem, with good accuracy, to ordinary differential equations. As it is readily seen, the time it takes to cross the region $u_2 \leq u \leq u_1$, $\Delta\tau = \tau - \tau_1(\tau)$, increases with τ, while $\varepsilon^2(\tau)$ decreases monotonically with τ increasing. This, in turn, suggests that $\varepsilon^2(\tau) = \varepsilon^2(\tau + \tau_0)$, where the parameter τ_0 is determined by the initial ($\tau = 0$) value of $\varepsilon_0 = \varepsilon(\tau_0)$.

Equation (4.204) has a group of functional transformations preserving its general form. Indeed, substituting

$$u \to \xi = u(\tau + \tau_0) - \frac{1}{2} \to \eta = \xi \ln(\tau + \tau_0) - \frac{1}{2} = \cdots, \tag{4.205}$$

$$\mathrm{d}\tau \to \frac{\mathrm{d}\tau}{\tau + \tau_0} \to \frac{\mathrm{d}}{\ln(\tau + \tau_0)} \ln(\tau + \tau_0) = \cdots, \tag{4.206}$$

$$\tau + \tau_0 \to \ln(\tau + \tau_0) \to \ln\ln(\tau + \tau_0) = \cdots, \tag{4.207}$$

we obtain

$$\varepsilon^2(\tau + \tau_0) \to \delta^2 = \varepsilon^2(\tau + \tau_0) - \frac{1}{4} \to e^2 = \delta^2\left(\ln(\tau + \tau_0)\right)^2 - \frac{1}{4}, \text{ etc.}, \tag{4.208}$$

and the form of Eq. (4.204) remains unchanged. Note that the equations always involve the ratios $\frac{\mathrm{d}}{\ln(\tau+\tau_0)} \ln(\tau + \tau_0)$, etc., so the absolute magnitudes of the logarithms should be taken in the region where they are negative.

Furthermore, when the initial distribution function is of infinite extent, the right-hand side of Eq. (4.204) cannot be equal to zero at any values of u (or ξ, η, etc.) as a result of these transformations. Otherwise the leakage through the blocking point would cease, thus violating the balance of matter. It suggests positivity of ε^2, δ^2, and l^2 or else, an obvious restriction on

ε^2, δ^2, and l^2 from below. Given the dependence $\tau_1 = \tau_1(\tau)$, it is natural to attempt to find a functional transformation that renders the right-hand side of Eq. (4.204) independent of or weakly dependent on its argument over a certain time interval. After such a transformation Eq. (4.204) can be easily solved.

4.2.3 Coarsening in the Case of Power-Dependent Initial Cluster Size Distributions

In the case assumed now, we have $f_0(R) = AR^{-n}$, or using the relative variables, $f_0(v) = Bv^{-n}$, where $B = AR_{\rm cr}^{-n}(0)$. Equation (4.202) can be rearranged to give for $v(\tau) \gg 1$

$$v(\tau) = \left[\frac{B}{C(n-1)}e^{-\tau}\right]^{1/(n-1)} = \exp\left[\frac{\tau + \ln D}{n-1}\right], \tag{4.209}$$

$$D = \frac{B}{C(n-1)}. \tag{4.210}$$

Further substituting Eq. (4.210) into Eq. (4.203), we have

$$\psi\left\{\exp\left[\frac{\tau + \ln D}{n-1}\right]\right\} = \tau_1(\tau) + \psi(u_1). \tag{4.211}$$

Since $\tau_1(\tau) \geq 0$, Eq. (4.211) holds for $\tau \geq (n-1)\ln w - \ln D$.

When $\tau \to \infty$, using the asymptotic function $\psi(v) = 3\ln u - \ln D$ for $v \gg 1$, we obtain

$$\tau_1(\tau) = d\tau + \beta\,, \qquad d = \frac{3}{n-1} < 1, \qquad \beta = d\ln D - \psi(u_1). \tag{4.212}$$

Since τ_0 and β depend on the zero moment of time, it can be conveniently chosen to obey the following relation:

$$\beta = -(1-d)\tau_0. \tag{4.213}$$

Hence

$$\tau_1(\tau) + \tau_0 = d(\tau + \tau_0), \tag{4.214}$$

and Eq. (4.204) reads

$$\frac{\mathrm{d}u}{\mathrm{d}\tau'} = -u^2 - \varepsilon^2(\tau' + \tau_0), \tag{4.215}$$

$$u|_d\,(\tau + \tau_0) = u_1, \qquad u|_{\tau+\tau_0} = u_2. \tag{4.216}$$

Introducing

$$\xi = u(\tau + \tau_0) - \frac{1}{2}, \tag{4.217}$$

we have

$$\frac{d\xi}{d\ln(\tau+\tau_0)} = -\xi^2 - \delta^2, \tag{4.218}$$

where

$$\delta^2 = \xi^2(\tau+\tau_0) - \frac{1}{4}, \tag{4.219}$$

$$\xi_1 = u_1(\tau+\tau_0) - \frac{1}{2} \to \infty \quad \text{when} \quad \ln(\tau_1+\tau_0) = \ln(\tau+\tau_0) - \ln\frac{1}{d}, \tag{4.220}$$

$$\xi_2 = u_2(\tau+\tau_0) - \frac{1}{2} \to -\infty \quad \text{when} \quad \ln(\tau_1+\tau_0), \qquad \tau \to \infty. \tag{4.221}$$

In order to obey these boundary conditions, δ must be set equal to constant. Then

$$\frac{1}{\delta}\left(\arctan\frac{\xi_2}{\delta} - \arctan\frac{\xi_1}{\delta}\right) = -\ln\frac{\tau+\tau_0}{\tau_1(\tau)+\tau_0} = \ln\frac{1}{d}, \tag{4.222}$$

and for $\xi_2 \to -\infty$, $\xi_1 \to \infty$ we get

$$\delta = \frac{\pi}{\ln(1/d)} = \frac{\pi}{\ln((n-1)/3)}, \tag{4.223}$$

$$\varepsilon^2(\tau+\tau_0) = \frac{1}{4(\tau+\tau_0)^2}\left(1+4\delta^2\right). \tag{4.224}$$

A more exact solution is given below.

We now return back to the previous variables via

$$\tau+\tau_0 \to \frac{1}{\sqrt{3}}(\tau+\tau_0), \qquad \tau_1+\tau_0 \to \frac{1}{\sqrt{3}}(\tau_1+\tau_0), \tag{4.225}$$

$$u \to \frac{2}{\sqrt{3}}(u-u_0), \qquad u \to \frac{2}{3}(u-u_0)(\tau+\tau_0) - \frac{1}{2}. \tag{4.226}$$

Then

$$\varepsilon^2(\tau+\tau_0) = \frac{3}{4(\tau+\tau_0)^2}\left(1+4\delta^2\right). \tag{4.227}$$

Knowing $\varepsilon^2(\tau)$, we can now obtain the integral $\tau_1 = \tau_1(\xi,\tau)$ over the region $u_2 \le u \le u_1$, and hence the distribution function. To finally perform this, we should replace ξ_2 by ξ in Eq. (4.222) and make the substitution

$$\ln\frac{1}{d} \to \ln\frac{\tau+\tau_0}{\tau_1(\xi,\tau)+\tau_0} \tag{4.228}$$

on the right-hand side since the boundary ξ_1 is now attained at the time $\tau_1 = \tau_1(\xi,\tau)$, provided that the ξ point is attained at the time τ. It is obvious that $\tau_1(\xi,\tau)|_{\xi\to\infty} = \tau \to \infty$. It follows that

$$\ln\frac{\tau+\tau_0}{\tau_1(\xi,\tau)+\tau_0} = \frac{1}{\delta}\left(\frac{\pi}{2} - \arctan\frac{\xi}{\delta}\right). \tag{4.229}$$

Hence

$$\tau_1(\xi,\tau)+\tau_0=d^P(\tau+\tau_0), \qquad P=\frac{1}{2}\left(1-\frac{2}{N}\arctan\frac{\xi}{\delta}\right), \tag{4.230}$$

$$\frac{\partial\tau_1}{\partial u}=\frac{\partial\tau_1}{\partial\xi}\frac{\partial\xi}{\partial u}=\frac{d^P\left(\frac{\ln(1/d)}{\pi}\right)^2\frac{2}{3}(\tau+\tau_2)}{1+\left[\frac{2}{3}(u-u_0)(\tau+\tau_2)-\frac{1}{2}\right]^2\left(\frac{\ln(1/d)}{\pi}\right)^2}. \tag{4.231}$$

By using the asymptotically valid relations

$$\psi(v)=3\ln v, \quad v\gg 1, \quad B=\left(\frac{3}{d}\right)Ce^{\ln D}, \quad \frac{dv}{d\tau_1}=g(v)=\frac{v}{3}, \tag{4.232}$$

and explicit expressions for τ_0 and D in the constant factor, we obtain from Eqs. (4.191) and (4.192) for sufficiently large τ

$$\begin{aligned}\varphi(u,\tau) &= f_0(v(\tau_1(\xi,\tau)))g(v(\tau_1(\xi,\tau)))\frac{\partial\tau_1}{\partial u}\\ &= \frac{B}{3}\frac{\partial\tau_1}{\partial u}\exp\left\{-\frac{1}{d}\left(\tau_1\left(\xi,\tau\right)+\psi\left(u_1\right)\right)\right\}\\ &= C\exp\left\{-\tau\right\}\frac{1}{d}\frac{\partial\tau_1}{\partial u}\exp\left\{-\left(d^{-q}-1\right)\left(\tau+\tau_0\right)\right\}, \qquad u_2\le u\le u_1,\end{aligned}$$

$$q=\frac{1}{2}\left(1+\frac{2}{\pi}+\arctan\frac{\xi}{\delta}\right). \tag{4.233}$$

For $u\ge u_1$, Eqs. (4.189) and (4.190) with the asymptotic relation $\psi(v)=3\ln v$, $v\gg 1$, yield

$$\varphi(u,\tau)=Bv^{-n}\frac{g(v)}{g(u)}=\frac{B}{3g(u)}\exp\left\{-\frac{\tau+\psi(u)}{d}\right\}, \qquad u>u_1. \tag{4.234}$$

Thus, Eqs. (4.201), (4.233), and (4.234) define the size-distribution function over the whole range of variation of u at sufficiently long times, $\tau\gg 1$. Clearly, for $\tau=0$ and $u\to\infty$ we get $\varphi(u,0)=Bu^n$. The resulting solution in the region $u_2\le u\le u_1$ connects the solutions for $u\le u_2$ and $u\ge u_1$ smoothly, while the joining points asymptotically drop out of the resulting solution. In fact we get

$$\left.\left(\frac{\partial\tau_1}{\partial u}\right)\right|_{\xi=\xi_2\to-\infty;\tau\to\infty}=\frac{d}{g\left(u_2\right)}, \tag{4.235}$$

$$\left.\left(\frac{\partial\tau_1}{\partial u}\right)\right|_{\xi=\xi_1\to-\infty;\tau\to\infty}=\frac{1}{g\left(u\right)}. \tag{4.236}$$

Substituting these results into Eq. (4.233) and using τ_0 in explicit form, we obtain

$$\left.\varphi\left(u,\tau\right)\right|_{u\to u_2}=C\exp\left\{-\tau\right\}\frac{1}{g\left(u\right)}, \tag{4.237}$$

$$\varphi(u,\tau)|_{u\to u_1} = \frac{B}{3}\exp\left\{-\frac{1}{d}(\tau+\psi(u))\right\}\frac{1}{g(u)}. \tag{4.238}$$

The mean particle size $\overline{R}(t)$, as shown above for the zeroth-order approximation and for the mass-transfer mechanism under consideration, coincides with $R_{cr}(t)$. By using a more exact value of $\gamma(t)$ we obtain the supersaturation $\Delta(t)$ and critical radius $R_{cr}(t)$,

$$\Delta(t) = \frac{\alpha}{R_{cr}(t)}, \tag{4.239}$$

$$R_{cr}^3(t) = R_{cr}^3(0) + \frac{4}{9}D\alpha t\left\{1 - \frac{1+4\pi^2\left(\ln\frac{n-1}{3}\right)^{-2}}{4\left[\ln\left(1+\frac{4D\alpha t}{R_{cr}^3(0)}\right)+\tau_0\right]^2}\right\}^{-1}. \tag{4.240}$$

Let us now find the time at which we can start using the above asymptotic equations. The parameter τ_0 is a function of $R_{cr}(0) = \alpha/\Delta_0$ by definition and thus depends on the time zero we choose. The explicit form of τ_0 can be deduced from Eq. (4.213) by substituting the values of the constants B and C into D. We get

$$\begin{aligned}\tau_0 &= \frac{3(n-3)}{n-4}\ln\frac{1}{\Delta(0)} - \frac{3}{n-4}\ln\left(\frac{A}{\alpha^{n-3}(n-1)Q_0}\right)\\ &- \frac{3}{n-4}\psi(u_2) + \frac{n-1}{n-4}\psi(u_1),\end{aligned} \tag{4.241}$$

where we take into account that

$$\frac{4\pi}{3}R_{cr}^3(0)\int_{R_{cr}}^{u_2}\exp\{\psi(u)\}\frac{u^3}{g(u)}\,du \approx 1. \tag{4.242}$$

Although the original equations define $\varepsilon^2(\tau)$ for $\tau \gg 1$, the initial conditions $\varepsilon^2(\tau_0) \ll 1$ make these equations also valid for $\tau \geq 0$ since $\varepsilon^2(\tau+\tau_0)$ is a monotonic function of time and the time zero is chosen arbitrarily.

For the time interval $\Delta\tau = [(1-d)/d])]\tau_0$, we have $\tau_1(\tau) < 0$. It means that the particles which are of size u_2 at the time $\tau = 0$ were of the size u_1 at the time $\tau = -(1-d)\tau_0$. When moving the time zero (the $\tau = 0$ point), $\varepsilon_0\tau_0$ becomes constant with time

$$\begin{aligned}\varepsilon_0\tau_0 &= \left[1 - \frac{3D}{\gamma_0\alpha^2}\left(\frac{\mathrm{d}}{\mathrm{d}t}\frac{1}{\Delta^3}\right)^{-1}\right]^{1/2}\frac{3(n-3)}{n-4}\ln\frac{1}{\Delta}\\ &\to \frac{3^{1/2}}{2}\left[1+\frac{4\pi^2}{(\ln[(n-1)/3])^2}\right]^{1/2}.\end{aligned} \tag{4.243}$$

Here minor terms are omitted as, when essentially all the excess material is concentrated inside the particles, we have asymptotically

$$\frac{\psi(u_2)}{\tau_0} \sim \frac{1}{u_2-u_0}\frac{1}{\tau_0} \ll 1, \quad \frac{\psi(u_1)}{\tau_0} \sim \frac{1}{u_1-u_0}\frac{1}{\tau_0} \ll 1, \quad \Delta(0) \ll Q_0. \tag{4.244}$$

The approach of a stationary value of this expression determines the time when we can start using asymptotic equations. Any moment from that time on can be chosen as the time zero ($\tau = 0$) and all the dimensional variables can be related to the value of $R_{\text{cr}}(0) = \alpha/\Delta(0)$ at a given time. The boundary conditions in this case assume a canonical form (4.216). The investigation of the distribution function in the region beyond the blocking point allows us to evaluate n for the asymptotic distribution function and compare it with Eq. (4.243). It should be noted that unless we choose the time zero in a specific way (Eq. (4.243)) but seek the solution for an arbitrary time zero with the only constraint, $\varepsilon^2(\tau) \ll 1$, the resulting solution for $u > u_2$ will coincide with Eq. (4.233) up to terms as small as $\beta/d\tau \to 0$, $\tau \to \infty$, i.e., with the precision with which we evaluate the characteristics in this case.

The amount of excess material in the region around the blocking point and beyond it is given by

$$Q' = \kappa u_0^3 C \approx Q_0 \exp\{-\psi(u_2)\} \ll Q_0. \tag{4.245}$$

Since our analysis holds true when the region around the blocking point where a universal distribution is formed becomes asymptotically small (i.e., $\varphi(u_2) \gg 1$) so $Q' \ll 1$, as it must. Thus, the asymptotic time behavior of the distribution function is specified by Eqs. (4.201), (4.233), (4.234), and (4.240) when the initial size distribution function has a power-dependent asymptotics.

4.2.4 Coarsening in the Case of Exponentially Decaying Initial Cluster-Size Distributions

Generally we now have

$$f_0(R) = A \exp\left\{-\left(\frac{R - R_0}{r_0}\right)^m\right\} \frac{1}{R_n}, \qquad m > 0, \tag{4.246}$$

and in the relative variables

$$f_0(v) = B v^{-n} \exp\{-(v - v_0)^m P^m\}, \tag{4.247}$$

where

$$B = \frac{A}{R_{\text{cr}}^n(0)} = \frac{A\Delta^n(0)}{\alpha^n}, \qquad P = \frac{R_{\text{cr}}(0)}{r_0} = \frac{\alpha}{r_0 \Delta(0)}. \tag{4.248}$$

Substituting $f_0(v)$ into Eq. (4.202), we get, for $\tau \gg 1$,

$$v(\tau_2) = v_0 + \frac{1}{P}\left\{(\tau_2 + \ln D) - \frac{m-1}{m}\ln(\tau_2 + \ln D) - \frac{n}{m}\ln(\tau_2 + \ln D + P v_0)\right\}^{1/m}, \tag{4.249}$$

$$D = \frac{B P^{n-1}}{mC}. \tag{4.250}$$

Logarithmic terms are significant only when $v(\tau)$ appears in the exponent and can be neglected otherwise when $\tau \to \infty$.

By substituting Eq. (4.250) into Eq. (4.203) and again setting $u = u_2$ and $\tau = \tau_2$, we obtain at the upper boundary $u = u_1$ of the transformation region

$$\tau_1(\tau) = \psi(v(\tau)) - \psi(u_1), \tag{4.251}$$

where $v(\tau)$ is given by Eq. (4.250). In this case, no choice of the time zero enables us to rearrange these conditions, at the boundaries of the transformation region, u_1 and u_2, to the canonical form that yields an exact solution of the equation for the characteristics in this region. The latter characteristics can be evaluated with sufficient accuracy only for specific time intervals. To this end, Eq. (4.204) must be rearranged to make the right-hand side time independent with sufficient accuracy for these intervals.

First we can use the solution of Eq. (4.222) by substituting $\tau_1(\tau)$ into it from Eq. (4.251). Then Eq. (4.218) where we replace $1/d$ by $(\tau+\tau_0)/(\tau_1(\tau)+\tau_0)$ leads to

$$\delta = \pi\left(\ln\frac{\tau+\tau_0}{\tau_1(\tau)+\tau_0}\right)^{-1}, \qquad \varepsilon^2(\tau+\tau_0) = \frac{3}{4}\frac{1+4\delta^2}{(\tau+\tau_0)^2}, \tag{4.252}$$

and

$$\tau_1 = (\xi,\tau) + \tau_0 = \left(\frac{\tau_1(\tau)+\tau_0}{\tau+\tau_0}\right)^b(\tau+\tau_0), \tag{4.253}$$

where

$$b = \frac{1}{2}\left(1 - \frac{2}{\pi}\arctan\frac{\xi}{\delta}\right). \tag{4.254}$$

Hence

$$\frac{\partial v}{\partial u} = g(v)\frac{\partial \tau_1}{\partial u} = g(v)\frac{\partial \tau_1}{\partial \xi}\frac{\partial \xi}{\partial u}, \tag{4.255}$$

$$\varphi(u,\tau) = f_0(v(\tau,\xi))\,g(v)\frac{\partial \tau_1}{\partial \xi}\frac{\partial \xi}{\partial u}, \qquad u_1 \le u \le u_2, \tag{4.256}$$

and

$$\varphi(u,\tau) = f_0(v(u,\tau))\frac{\partial v}{\partial u}, \tag{4.257}$$

$$\psi(v) = \tau + \psi(u), \qquad u \ge u_1. \tag{4.258}$$

In this case, the distribution function in the range of reduced cluster sizes $u > u_2$ is defined completely by Eqs. (4.252)–(4.258).

The derivative

$$\frac{\partial \tau_1}{\partial u} = \frac{\partial \tau_1}{\partial \xi}\frac{\partial \xi}{\partial u} \tag{4.259}$$

at this time interval is given by Eq. (4.231) where we replace $1/d$ by $(\tau+\tau_0)/(\tau_1(\tau)+\tau_0)$. Note that the distribution function in the region $u_2 \le u < u_1$ always matches the distribution function in the region $u \ge u_1$ smoothly, and is independent of the accuracy in evaluating the characteristics. This is due to the fact that the boundary condition, $\tau_1(\tau,\xi) \to \tau \to \infty$ when $\xi \to \infty\,(u \to u_1)$, is obeyed exactly and the characteristics to the left of the point u_1 smoothly goes over into the characteristics to the right of u_1, $\tau_1(\tau,\xi)+\psi(u_1) \to \tau+\psi(u)$ when $u \to u_1$.

The requirement of smooth matching at the point u_2 imposes a condition for the time interval where our solution is acceptable. Substituting into Eq. (4.256), $v(\tau,\xi)$ when $u \to u_2, \xi \to -\infty$, we get

$$\varphi(u,\tau) = Ce^{-\tau}\frac{g(v)}{g(u)}mP(\tau+\ln D)^{(m-1)/m}\frac{\tau_1(\tau)+\tau_0}{\tau+\tau_0}, \qquad u \to u_2. \tag{4.260}$$

Thus, our solution holds true when

$$\begin{aligned} f(\tau) &= mP(\tau+\ln D)^{(m-1)/m}g(v)\frac{\tau_1(\tau)+\tau_0}{\tau+\tau_0} \\ &\approx g(v)\left(\frac{\partial v}{\partial \tau}\right)^{-1}\frac{\psi(v(\tau))-\psi(u_1)+\tau_0}{\tau+\tau_0} \approx 1, \end{aligned} \tag{4.261}$$

where $v(\tau))$ is given by Eq. (4.250).

The accuracy of the above solution is independent of the initial distribution function but is entirely determined by the accuracy in evaluating the characteristics. If we substitute the approximate solution, Eq. (4.222), with $\tau_1(\tau)$ as given by Eq. (4.251) into the equation for the characteristics, this equation is accurate to a term which must be small for this interval,

$$\frac{\pi d}{\delta\ln(\tau+\tau_0)} = 1-\frac{1}{f(\tau)} \approx 0. \tag{4.262}$$

This condition coincides closely with the requirement of Eq. (4.261). We evaluate this time interval by noting that, for $\tau \gg 1$ and $v \gg 1$,

$$g(v) = \frac{v}{3}, \tag{4.263}$$

$$g(v)\left(\frac{\partial v}{\partial \tau}\right)^{-1}\frac{1}{\tau+\tau_0} \approx \frac{1}{3}\frac{\mathrm{d}\ln v}{\mathrm{d}\ln(\tau+\tau_0)} \approx \frac{1}{3}, \tag{4.264}$$

$$\psi(v) = 3\ln v \approx \frac{1}{3}\ln\frac{\tau}{P^m}, \tag{4.265}$$

and $\psi(u_1) = C$, $u_1 \approx 2$. Thus, the above solution is correct for the time

$$\frac{1}{m}\ln\frac{\tau}{P^m}+\frac{\tau_0}{3} \approx 1. \tag{4.266}$$

Let us now evaluate the parameter τ_0. Equation (4.251) is obviously valid when $\tau_1(\tau) \ge 0$ or $v(\tau) \ge u_1$. The condition $\tau_1(\tau_{\min}) = 0$ or $v_1(\tau_{\min}) = u_1$ specifies the time interval measured from the chosen time zero, when the distribution function for the interval $u_2 \le u \le u_1$

is determined by the distribution function for the interval $u \geq u_1$. The parameter τ_0 can be obtained from the following equation:

$$\varepsilon(\tau_{\min}+\tau_0) = \left[1 - \frac{3}{\gamma_0}\frac{D}{\alpha}\left(\frac{\mathrm{d}}{\mathrm{d}t}\frac{1}{\Delta^3}\right)^{-1}\right]^{1/2}\Bigg|_{\Delta(t)=\Delta(0)\exp\{-\tau_{\min}(0)\}}$$

$$= \frac{3^{1/2}}{2}\frac{1}{\tau_{\min}+\tau_0}\left[1 + \frac{4\pi^2}{\left(\ln\dfrac{\tau_{\min}+\tau_0}{\tau_0}\right)^2}\right]^{1/2}. \tag{4.267}$$

At a sufficiently late stage

$$\tau_{\min}(\Delta_0) = (Pu_0 - v_0)^m - \ln D \gg 1. \tag{4.268}$$

Thus, evaluation of τ_0 requires experimental measurement of the derivative $\left(\mathrm{d}\Delta^{-3}/\mathrm{d}t\right)^{-1}$ at the time when $\Delta(t) = \Delta(0)e^{-\tau_{\min}(\Delta_0)}$. For the exponential asymptotic behavior of the initial distribution function, in contrast to the distribution asymptotic to a power function, Eq. (4.251), which is generally nonlinear, does not allow us to evaluate the time $\tau_1(0)$ in the past, when particles of size u_2 had been of size u_1. This is due to the fact that solutions of Eqs. (4.252)–(4.258) which are valid for $\tau \geq \tau_{\min}$ change considerably with time and thus cannot be extended to the region $\tau < \tau_{\min}$.

For $m^{-1}\ln[\tau/(pm)] \gg 1$, $\ln\ln\tau \approx 1$, we perform another functional rearrangement of Eq. (4.218) by changing to a new argument (time scale) $\ln\ln(\tau+\tau_0)$ and a new function $\eta = \xi\ln(\tau+\tau_0) - (1/2)$. Then we obtain

$$\frac{\mathrm{d}\eta}{\mathrm{d}\ln\ln(\tau'+\tau_0)} = -\eta^2 - l^2, \tag{4.269}$$

$$l^2 = \left[\varepsilon^2(\tau+\tau_0)(\tau+\tau_0)^2 - \frac{1}{4}\right](\ln(\tau+\tau_0))^2 - \frac{1}{4}, \tag{4.270}$$

$$\eta|_{\tau'=\tau\to\infty} = -\infty, \quad \eta|_{\tau'=\tau_1\to\infty} = \infty. \tag{4.271}$$

For the time interval when $l^2 \approx$ const, the solution of Eq. (4.267) coincides with the obtained solution (Eq. (4.222)), in which $\ln(\tau+\tau_0)/\ln(\tau_1(\tau)+\tau_0)$ should be substituted for $1/d$

$$l = \pi\left[\ln\frac{\ln(\tau+\tau_0)}{\ln(\tau_1(\tau)+\tau_0)}\right]^{-1}, \tag{4.272}$$

where $\tau_1(\tau)$ is given by Eq. (4.251),

$$\ln(\tau_1(\xi,\tau)+\tau_0) = \left[\frac{\ln(\tau_1(\tau)+\tau_0)}{\ln(\tau+\tau_0)}\right]^c \ln(\tau+\tau_0), \tag{4.273}$$

$$c = \frac{1}{2}\left(1 - \frac{2}{\pi}\arctan\frac{\eta}{l}\right),$$

$$\varepsilon^2(\tau+\tau_0) = \frac{3}{4}\frac{1}{(\tau+\tau_0)^2}\left[1+\frac{1+4l^2}{(\ln(\tau+\tau_0))^2}\right] \approx \frac{3}{4}\frac{1}{\tau^2}, \qquad \ln\tau \gg 1. \tag{4.274}$$

Inserting the time $\tau_1(\xi,\tau)$ from Eq. (4.273) into Eqs. (4.256) and (4.258), we obtain the distribution function. At such long times we can assume $\tau_1(\tau) \approx (3/m)\ln(\tau/P^m)$ and neglect τ_0. As already mentioned, the matching of the distribution functions in the region u_1 is smooth. In the region $u = u_2$, $\eta \to -\infty$, substitution of Eq. (4.273) into Eq. (4.256) yields

$$\varphi(u,\tau) = Ce^{-\tau} mP\tau^{(m-1)/m} g(v)\frac{\partial\tau_1}{\partial u} = Ce^{-\tau}\frac{1}{g(u)}\ln\ln\tau, \tag{4.275}$$

$$\eta \approx u\tau\ln\tau, \tag{4.276}$$

$$g(v) = \frac{v}{3} = \frac{\tau^{1/m}}{3P}, \tag{4.277}$$

$$\frac{\partial\tau_1}{\partial u} = \tau_1\frac{\partial\ln\tau_1}{\partial\eta}\frac{\partial\eta}{\partial\xi}\frac{\partial\xi}{\partial u} = \frac{3}{m}\frac{1}{g(u)}\frac{\ln\ln\tau}{\tau}. \tag{4.278}$$

It follows that the obtained solution holds for the time interval when $\ln\ln\tau \approx 1$.

Equation (4.267) for the characteristics is accurate to a small term:

$$\frac{\mathrm{d}}{\mathrm{d}\ln\ln\tau}\left(\frac{\pi}{l}\right) = 1 - \frac{1}{\ln\ln\tau} \approx 0. \tag{4.279}$$

Thus, for the time interval $\ln\ln\tau \simeq 1$ the accuracy in evaluating the distribution function for $u > u_2$ depends only on the accuracy in evaluating the characteristics, as it must be the case. For $u > u_1$ the distribution function

$$\varphi(u,\tau) \sim \exp\{-m\exp[\tau+\psi(u)]\} \tag{4.280}$$

is very small and does not contribute to the balance of matter significantly. Note that if the initial size distribution function has an end point, i.e., equal to zero when $R \geq R^0$, we can easily evaluate the time when it transforms into an universal distribution function, and $\varepsilon^2(\tau) = 0$. This takes place when

$$\frac{R^0}{R_{\mathrm{cr}}} = \frac{R^0}{\alpha}\Delta(t) = \frac{3}{2}. \tag{4.281}$$

From this time on, under these initial conditions, we obtain

$$\varphi(u,\tau) = \begin{cases} \dfrac{A}{g(u)}e^{-\psi(u)-\tau}, & 0 \leq u < u_0, \\ 0, & u \geq u_0, \end{cases} \tag{4.282}$$

$$A = \left[\kappa\int_0^{u_0} e^{-\psi(u)}\frac{u^3}{g(u)}\,\mathrm{d}u\right]^{-1}. \tag{4.283}$$

Let us solve Eq. (4.218) with higher accuracy. Introducing

$$x = \ln(\tau + \tau_0), \qquad y = \ln(\tau' + \tau_0) \tag{4.284}$$

for convenience, we can rewrite Eq. (4.218) as

$$\frac{d\xi}{dy} = -\xi^2 - \delta^2(y), \tag{4.285}$$

$$\xi|_{y=\eta(x)=\ln(\tau_1(\tau)+\tau_0)} = u_1 e^{\eta(x)} - \frac{1}{2}, \tag{4.286}$$

$$\xi|_{y=x=\ln(\tau+\tau_0)} = -|u_2| e^x - \frac{1}{2}. \tag{4.287}$$

Equation (4.287) with its boundary conditions is a canonical equation since it holds true for any mass-transfer mechanism and any initial distribution function. They affect only the form of $\eta(x)$. If $f_0 = AR^{-n}$ ($n > 4$), then $\eta(x) = x - x_0$, $x_0 = \ln[(n-1)/3]$, and, accordingly, $(x - x_0) < y < x$.

We shall seek the solution of Eq. (4.287) in the form

$$\xi = \delta_0 \cot[\delta_0(y - x + \psi(x, y))], \tag{4.288}$$

$$\delta_0 = \frac{\pi}{x_0}, \qquad -\pi < \delta_0[y - x + \psi(x, y)] < 0, \qquad |\xi| < \infty. \tag{4.289}$$

Substitution into Eq. (4.287) leads to

$$\frac{d\psi}{dy} = -f(y), \tag{4.290}$$

$$f(y) = 1 - \delta^2/\delta_0^2. \tag{4.291}$$

For sufficiently large x the boundary conditions are

$$\psi(x, y)|_{y=x-x_0} = \frac{1}{u_1} e^{-(x-x_0)} + O\left(e^{-2x}\right), \tag{4.292}$$

$$\psi(x, y)|_{y=x} = -\frac{1}{|u_2|} e^{-x} + O\left(e^{-2x}\right). \tag{4.293}$$

Equation (4.291) can be written in an integral form

$$\psi(x, y) = \frac{1}{u_1} e^{-(x-x_0)} - \int_{x-x_0}^{y} f(y') \sin^2[\delta_0(y' - x + \psi(x, y'))] \, dy'. \tag{4.294}$$

Using the second boundary condition, we obtain

$$\int_{x-x_0}^{y} f(y') \sin^2[\delta_0(y' - x + \psi(x, y'))] \, dy' = \alpha e^{-x}, \tag{4.295}$$

$$\alpha\left(\frac{1}{u_1}e^{x_0}+\frac{1}{|u_2|}\right)>0. \tag{4.296}$$

Equations (4.294) and (4.296) represent a complete system of equations for $f(y)$ and $\psi(x,y)$.
Let us replace $z=y-x$, and $f(y)=A(y)e^{-y}$. Then

$$\psi(z+x,x)=\frac{1}{u_1}e^{-(x-x_0)} - e^{-x}\int\limits_{-x_0}^{z} A(z'+x)e^{-z}\sin^2\left[\delta_0(z'\psi(z'+x,x))\right]\mathrm{d}z', \tag{4.297}$$

$$\int\limits_{-x_0}^{0} A(z'+x)e^{-z}\sin^2\left[\delta_0(z'+\psi(z'+x,x))\right]\mathrm{d}z'=\alpha. \tag{4.298}$$

It follows from Eq. (4.298) that if A has a fixed sign (that is physically meaningful), it is limited and, according to Eq. (4.297), $\psi\sim e^{-x}\to 0$ when $x\to\infty$. Then we can neglect ψ in the argument of $\sin^2\left[\delta_0(z'+\psi)\right]\approx\sin^2(\delta_0 z')$ with an accuracy to the next-order terms in e^{-x}. We obtain

$$\delta^2(y)=\delta_0\left(1-Ae^{-y}\right),\qquad A=\frac{2\alpha}{e^{x_0}-1}\left(1+\frac{x_0}{4\pi^2}\right), \tag{4.299}$$

$$\psi(y,x)=\frac{1}{u_1}e^{-(x-x_0)}+A\left\{-\frac{1}{2}e^{-(x-x_0)}\left(1+\frac{x_0}{4\pi^2}\right)^{-1} -\frac{1}{2}e^{-y}\left[-1+\left(\cos\frac{2\pi(y-x)}{x_0}-\frac{2\pi}{x_0}\sin\frac{2\pi(y-x)}{x_0}\right)\left(1+\frac{x_0}{4\pi^2}\right)^{-1}\right]\right. . \tag{4.300}$$

These equations obviously hold true for $Ae^{-y}<1$, and it is physically clear that the smaller x_0 ($n\to 4$), the larger the amount of material in the tail of the distribution function and, accordingly, the later diffusive decomposition begins.

In a more exact way, Eq. (4.218) can be solved for an exponentially decreasing tail of the distribution function in a similar way by substituting $x\to\eta(x)$ in the expression for ξ, replacing δ by $\pi/y-\eta(y)$ and taking into account slow variations of this function with respect to τ on different intervals.

4.2.5 Generalizations

The above solutions obtained for mass transfer via volume diffusion can be easily generalized to other mass-transfer mechanisms. The basic canonical system of equations remains unchanged, however, with the only difference which resides in the form of the function $\psi(u)$, determined by the specific mass-transfer mechanism. An iterative procedure can be developed to evaluate the corrections to these solutions.

The first correction to the characteristics can be derived by taking into account the overall amount of material in the solid solution, $\Delta\sim e^{-\tau/3}$, and in the tail of the distribution function for $u \geq u_2$, as evaluated in the zeroth-order approximation. This leads to the substitution $e^{-\tau} \to e^{-\tau}(1 + O(e^{-\tau/3}))$. To incorporate the correction δv, the integrand in Eq. (4.197) of the balance of matter should be rewritten as

$$f_0(v_0 + \delta v) = f_0(v_0) + \frac{\mathrm{d}f(v_0)}{\mathrm{d}v_0}\delta v\,. \tag{4.301}$$

By retaining the terms of the next order of smallness, it can be easily shown that the corrections to the characteristics and hence to $\varepsilon^2(\tau)$ are of the order of $O\left(e^{-\tau/3}\right)$, i.e., exponentially small.

In general, further refinement of the "hydrodynamic" approximation does not make much sense because of the contribution of the local fluctuations of concentration near the particles, which increases with time. These fluctuations are primarily attributable to direct diffusive interaction (collisions) between the new-phase particles that may be located at a distance smaller than their size. Allowance for this interaction, as shown above, gives rise to a supplementary tail of the distribution function, in the region $u > u_0$ which is mainly determined by the distribution function in the region $u < u_0$, and makes $\varepsilon^2(\tau) = \Delta\gamma/\gamma_0$ tend to a constant value, $\varepsilon^2(\tau) \sim (\ln Q_0)^{-2}$. This result suggests that the universal distribution function evolves in the region $u < u_0$ in the hydrodynamic mode, while the collisions between particles in the region $u < u_0$ contribute primarily to the value of the tail of the distribution function in the region $u > u_0$. The contribution of the collisions apparently becomes significant once $\varepsilon^2(\tau)$, as evaluated in the hydrodynamic approximation, attains a value of the order of $(\ln Q_0)^{-2}$. Note that the distribution function has a peak at a point to the left of the point of lowest velocity. The reason is that the distribution function decreases in the region $u > u_0$ in any approximation. Upon leaking through the blocking point region, u_0, it has time to take the maximum value which is practically independent of the initial conditions with $\tau \to \infty$ only at some distance from u_0.

The above results can be easily extended to incorporate sources of a dissolved component. The expression for the growth rate of the new-phase particles is independent of the presence of sources, while in the equation of the balance of matter we have $e^{\tau} \to e^{n\tau}$, where the power $n \leq 1$ depends on the strength of the source of a component. Depending on the particular mass-transfer mechanism, n is constrained from below to ensure $\Delta(t) \to 0$ with $t \to \infty$, thus enabling the universal distribution function to evolve for any initial distribution.

These results can also be extended to the multicomponent, multiphase case. For such systems in their later stages, the continuity equations, as shown in the next section, separate into independent equations for each phase, to an exponential accuracy. Only the algebraic equations defining the regions of coexistence of phases and the distribution of components over phases remain involved. The corrections to the concentration of a component in such systems are clearly determined by the initial distribution functions of the phases involving this component. These problems will be analyzed in the subsequent sections.

4.3 Theory of Diffusive Decomposition of Multicomponent Solutions

4.3.1 Introduction

The diffusive decomposition and coarsening of multicomponent solid solutions represents the most general case of the processes considered. This situation is of greatest interest for practical purposes, as most of the materials under consideration are in fact supersaturated multicomponent solid solutions, where various phases can precipitate under certain operating conditions, thus determining the properties of such materials. A theory describing the evolution of the precipitates of these phases has been developed in [279–283]. As we shall see below, the quasithermodynamic equilibrium conditions produce the most efficient distribution of the components among the phases and determine the regions of their coexistence, while the surface tension leads in the zeroth-order approximation to an universal distribution of particle sizes in the coexistent phases (here $\alpha^s/\overline{R}^s \ll 1$, where $\overline{R}^s$ is the average size of the particles of the sth phase, and α^s is proportional to the surface tension σ^s of the sth phase).

The growth of precipitates at the early stage of diffusive decomposition depends on the history of the sample. At sufficiently long times it is independent of the initial size distribution function for the precipitates, which, because of the nonlinear processes proceeding in the system, becomes universal in the zeroth-order approximation. The corrections, which depend on the initial conditions, diminish with increasing duration of the decomposition process. The initial conditions also determine the time for establishing the asymptotic behavior. The latter behavior can be determined by comparing the theoretical and experimental distribution functions, or from the saturation time of any material property sensitive to the impurity concentration in the solid solution (using dilatometry, and measuring lattice parameters, electrical resistance, etc.).

Let us examine an N-component solid solution in which k different phases (chemical compounds) can be formed from the components contained in the solution as a consequence of decomposition. The coexistence of different phases in a matrix is determined by the ratios of the initial concentrations of the components producing the given phases and by the thermodynamic efficiency of these phases. Of all the chemical compounds occurring in a given multicomponent system, only those precipitates can be stable whose constituents are present in the solution in a concentration corresponding to a certain supersaturation. This necessary condition for decomposition, at low concentration of the components $(c_i^0 \ll 1)$ when the law of mass action applies to the chemical reactions, can be written as

$$\sum_i \nu_i^s \mu_i^s = \ln\left(\frac{\prod_i (c_i^0)^{\nu_i^s}}{K_\infty^s}\right) > 0, \tag{4.302}$$

where c_i^0 is the initial concentration of the ith component in the solid solution, K_∞^s is the equilibrium constant of the sth phase chemical reaction at the precipitate surface, and ν_i^s and μ_i^s are respectively the stoichiometric coefficient and the chemical potential of the ith component of the sth phase.

If the phases formed in the system do not have any common components, the condition given by Eq. (4.302) is sufficient, since the growth of precipitates of the different phases occurs independently. If, however, the phases share common components, we notice that, although the solution initially has been supersaturated with respect to some phases, the material can be redistributed during decomposition and the solution in the phases will no longer be saturated. Therefore, in this case Eq. (4.302) is merely a necessary condition for selecting the phases whose precipitates are capable of further growth during diffusive decomposition.

4.3.2 Basic Equations and Their Solution

The set of equations which describes the diffusive decomposition or coarsening process, in an N-component system producing k phases of stoichiometric composition, including pure solute components and compounds with the same material as that of the matrix, consists of k continuity equations (Eq (4.303)), plus N laws of conservation of the components (Eq. (4.304)) plus $\sum_s n^s - k$ stoichiometric ratios for the diffusion flux J^s_{ik} (Eq. (4.305)), and plus k laws of mass action (Eq. (4.306)):

$$\frac{\partial f^s}{\partial t} + \frac{\partial}{\partial R}\left(f^s v^s_R\right) = 0, \tag{4.303}$$

$$\frac{\mathrm{d}\bar{c}_i}{\mathrm{d}t} + \sum_{s,i} \frac{4\pi}{n_0} \int_0^\infty f^s R^2 J^s_{i,R}\,\mathrm{d}R = 0, \tag{4.304}$$

$$\frac{J^s_{i,R}}{\nu^s_i} = \frac{J^s_{i',R}}{\nu^s_{i'}}, \tag{4.305}$$

$$\prod_i^k \left(c^s_{i,R}\right)^{\nu^s_i} = K^s_R. \tag{4.306}$$

Here $f^s(R,t)$ is the precipitate size distribution function of the sth chemical compound (phase) in the matrix; $\sum_{s,i}$ denotes summation over all the phases containing the ith component; $\bar{c}_i(t)$ is the average concentration of the ith component in the solid solution at a given time; $c^s_{i,R}$ is the equilibrium concentration of the ith component at the surface of a particle of the sth phase; n_0 is the number of matrix sites per unit volume; K^s_R is the equilibrium constant of the sth chemical reaction at the surface of a particle of the radius R, and n^s is the number of components in the sth phase.

Since the supersaturation Δ_i of the components at a later stage of decomposition is much less than unity, we can, similar to single-component solutions, use the expressions for quasi-stationary diffusion flux $J^s_{i,R}$ of atoms of the ith component to the particles of the sth phase, normalized per unit surface area of the particles of radius R, which were obtained in the approximation of a self-consistent diffusion field:

$$J^s_{i,R} = \frac{D_i n_0}{R}\left(\bar{c}_i - c^s_{i,R}\right). \tag{4.307}$$

Here D_i is the diffusion coefficient of the ith component in the matrix. In this case, the ratio of the characteristic time for establishing the diffusion flux for the slowest moving component

$\tau_{\rm dif} \sim \overline{R}_i/(6D_i)$ to that for the change in the precipitate size,

$$\tau_{\rm ch} \sim \frac{\overline{R}_i}{D_i}\left(\frac{{\rm d}c_i}{{\rm d}r}\right)^{-1} \sim \frac{\overline{R}_i^2}{D_i\Delta_i}, \tag{4.308}$$

is small $\tau_{\rm dif}/\tau_{\rm ch} \sim \Delta_i \ll 1$. The smallness of the parameter Δ_i makes it possible to use the quasistationary conditions at the surface of the precipitates. Since $\Delta_i \to 0$, the law of mass action can be used at a later stage, and the interaction of the components in the solution can be ignored. Since

$$c^s_{i,R} = c^s_{i,\infty}\exp\left\{\delta^s_{i,R}\right\}, \tag{4.309}$$

$$\begin{aligned}(K^s_R)^{1/N_s} &= (K^s_\infty)^{1/N_s}\exp\left\{\frac{\alpha^s}{R}\right\} = \prod_i \left(c^s_{i,R}\right)^{P^s_i} \\ &= \prod_i \left(c^s_{i,\infty}\right)^{P^s_i}\prod_{i'}\exp\left\{P^s_{i'}\delta^s_{i',R}\right\} = \prod_i \left(c^s_{i,\infty}\right)^{P^s_i}\exp\left\{\sum_{i'} P^s_{i'}\delta^s_{i',R}\right\},\end{aligned} \tag{4.310}$$

then

$$\prod_i \left(c^s_{i,\infty}\right)^{P^s_i} = (K^s_\infty)^{1/N_s}, \qquad \sum_i P^s_i\delta^s_{i,R} = \frac{\alpha^s}{R}, \tag{4.311}$$

$$N_s = \sum_i \nu^s_i, \qquad \alpha^s = \frac{2\sigma^s\overline{V}^s}{k_BT}, \qquad \overline{V}^s = \sum_i P^s_i V^s_i. \tag{4.312}$$

Here $\overline{V}^s$ is the mean volume per atom of the compound; V^s_i is the atomic volume of the ith component in the sth phase; $i = 1, 2, \ldots$ are the numbers of the components comprising the given phase, and $P^s_i = \nu^s_i/N_s$ are the normalized stoichiometric coefficients.

Let us consider the following equation:

$$\begin{aligned}\sum_i \frac{P^s_i J^s_{i,R}}{D_i c^s_{i,\infty}} &= \frac{n_0}{R}\sum_i\left[P^s_i\left(\frac{\bar{c}_i}{c^s_{i,\infty}} - 1\right) - P^s_i\left(\frac{c^s_{i,R}}{c^s_{i,\infty}} - 1\right)\right] \\ &= \frac{1}{D_s}\frac{J_{i,R^s}}{P^s_i} = \frac{n_0}{R}\left(\Delta^s - \gamma\right).\end{aligned} \tag{4.313}$$

Here,

$$\frac{1}{D_s} = \sum_i \frac{(P^s_i)^2}{D_i c^s_{i,\infty}}, \tag{4.314}$$

and

$$\Delta^s = \sum_i P^s_i\frac{\bar{c}_i - c^s_{i,\infty}}{c^s_{i,\infty}} \tag{4.315}$$

is a time-dependent function representing the effective supersaturation at a planar surface of the compound, and

$$\gamma^s = \sum_i P_i^s \frac{c_{i,R}^s - c_{i,\infty}^s}{c_{i,\infty}^s} = \sum_i P_i^s \left(\exp\left\{\delta_{i,R}^s\right\} - 1\right) \tag{4.316}$$

is the deviation of the effective equilibrium concentration at the surface of a compound of curvature radius R from that at a planar interface ($R \to \infty$). Using the above equation and Eq. (4.305) we obtain

$$v_R^s = \left(\frac{\mathrm{d}R}{\mathrm{d}t}\right)^s = \sum_i^{n^s} V_i^s J_{i,R}^s = \frac{D^s \overline{V}^s n_0}{R}\left(\Delta^s - \gamma^s\right) , \tag{4.317}$$

$$\bar{c}_i - c_{i,\infty}^s = \frac{P_i^s D^s}{D_i}\Delta^s, \tag{4.318}$$

$$c_{i,R}^s - c_{i,\infty}^s = \frac{P_i^s D^s}{D_i}\gamma^s. \tag{4.319}$$

For sufficiently large $\overline{R}^s \gg \alpha^s$ and, accordingly, $\delta_{i,R}^s \ll 1$, we have

$$\gamma^s = \sum\nolimits_i P_i^s \delta_{i,R}^s = \frac{\alpha^s}{R}. \tag{4.320}$$

The precipitate growth rate is finally given by

$$\left(\frac{\mathrm{d}R}{\mathrm{d}t}\right)^s = \frac{D^s \overline{V}^s n_0}{R}\left(\Delta^s - \frac{\alpha^s}{R}\right). \tag{4.321}$$

Using Eq. (4.319), and integrating Eq. (4.304) with the help of Eq. (4.303), we can easily rearrange the law of conservation of matter (Eq. (4.304)) in the following manner:

$$\sum_{s,i} P_i^s J^s = \sum_{s,i} \frac{P_i^s}{\overline{V}^s n_0} \frac{4\pi}{3} \int_0^\infty f^s R^3 \,\mathrm{d}R = Q_i - \bar{c}_i. \tag{4.322}$$

Here

$$Q_i = c_i^0 + \sum_{s,i} \frac{P_i^s}{\overline{V}^s n_0} \frac{4\pi}{3} \int_0^\infty f_0^s R^3 \,\mathrm{d}R \tag{4.323}$$

is the total amount of the material of the ith component at zero time, and

$$J^s = \frac{1}{\overline{V}^s n_0} \frac{4\pi}{3} \int_0^\infty f^s R^3 \,\mathrm{d}R \tag{4.324}$$

is the relative number of molecules of the sth phase in unit volume of the precipitate. The law of conservation of matter as given by Eq. (4.322) is a natural extension of the law of conservation for a single-component solution. Using Δ^s and substituting $\bar{c}_i$ from Eq. (4.322), we get

$$\Delta^s = \sum_{s,i} \frac{P_i^s}{c_{i,\infty}^s} \left(Q_i - \sum_{s',i'} P_i^{s'} J^{s'} - c_{i,\infty}^s \right), \tag{4.325}$$

$$\Delta^s + \sum_{s,i} \frac{P_i^s}{c_{i,\infty}^s} \sum_{s',i'} P_i^{s'} J^{s'} = \sum_{s,i} \frac{P_i^s}{c_{i,\infty}^s} \left(Q_i - c_{i,\infty}^s \right). \tag{4.326}$$

Introducing dimensionless variables, as in the case of a single-component solution, and analyzing the law of conservation of matter as we have done above in Section 4.2, we can show that a solution exists when Δ^s approaches zero as $t^{-1/3}$:

$$\Delta^s = \left[\frac{4}{9} \frac{D^s \overline{V}^s n_0 t}{(\alpha^s)^2} \right]^{-1/3}. \tag{4.327}$$

Since $\Delta^s > 0$ and $J^s > 0$, the values of $\bar{c}_i$, $c_{i,\infty}^s$, and J^s which are generally functions of time, approach certain limiting values when the time is large, $t \to \infty$, and $\Delta^s \to 0$. The system behaves in this manner because of its tendency to achieve thermodynamic equilibrium. Thus, in the zero-order approximation (with accuracy to terms of the order of $t^{-1/3}$) the system of Eqs. (4.303)–(4.306) can be separated asymptotically into k independent subsystems. The above statement is a very important implication of the theory, as a complicated system of nonlinear differential equations, in the canonical variables, can be rearranged into a set of independent equations, defining the distribution functions and the parameters of decomposition of all the precipitated phases to a time-asymptotic accuracy.

The continuity equation (4.303), the law of size variation of precipitates of the sth phase, Eq. (4.321), and the tendency of the volume of the precipitates of a given phase to achieve a constant value, $J^s \overline{V}^s n_0$, which is determined by the law of conservation of matter (Eq. (4.322)), comprise k complete, independent sets of equations, corresponding to the set of equations for a single-component solution that has been obtained and analyzed in detail above in Section 4.2. Thus, the above analysis applies to each independent subsystem and makes it possible to write the asymptotic solutions of the given set of equations for the sth phase, after making the appropriate substitution. We get

$$D \to D^s \overline{V}^s n_0, \qquad Q_0 \to J^s \overline{V}^s n_0, \qquad \alpha \to \alpha^s = \frac{2\sigma^s \overline{V}^s}{k_B T}, \tag{4.328}$$

$$R_k^s = \frac{\alpha^s}{\Delta^s}, \tag{4.329}$$

$$f^s(R,t) = \frac{\varphi(u^s, \tau^s)}{\overline{R}^s} = \frac{N^s(t) P(u^s)}{\overline{R}^s}, \tag{4.330}$$

$$P^s(u^s) = \begin{cases} \dfrac{3^4}{2^{5/3}}, & 0 \le u^s < \dfrac{3}{2}, \\[2ex] 0, & u^s = \dfrac{R}{\overline{R}^s} \ge \dfrac{3}{2}, \end{cases} \tag{4.331}$$

$$N^s(t) = 0.22\frac{J^s\overline{V}^s n_0}{\left(\overline{R}^s\right)^3}, \tag{4.332}$$

$$\left(\overline{R}^s\right)^3 = (R^s_{\rm cr})^3 = \left(\frac{\alpha^s}{\Delta^s}\right)^3 = (R^s_{\rm cr}(0))^3 + \frac{4}{9}D^s\overline{V}^s n_0\alpha^3 t. \tag{4.333}$$

This result is an important implication of the theory that predicts the formation of a universal (in the appropriate relative variables, $u^s = R/\overline{R}^s$) time-asymptotic size distribution, for the precipitates of arbitrary phases produced during the diffusive decomposition of multicomponent solid solutions. Physically, this asymptotic behavior is clear: interfacial tension gives rise to a universal (in the appropriate variables) distribution function, which is the same for all the coexistent phases irrespective of their initial distributions. Thus, in the most general case of a multicomponent dispersed system there exists to a good accuracy a unique (in the appropriate variables) stable asymptotic state achieved by the system upon forgetting the initial distribution. The evolution of the mean sizes of the precipitates from all the phases also obeys a universal "$t^{1/3}$-law," which defines the coarsening (decomposition) controlled by the volume diffusion of the components.

Although the general solution is the same, the specific solution depends on the set of $(N + \sum_s n^s + k)$ limiting parameters $\overline{c}_i$, $c^s_{i,\infty}$, and J^s. They can be evaluated from the limiting form of Eqs. (4.305), (4.312), and (4.322), as $R \to \infty$. An additional set of equations (with accuracy to terms of the order of $t^{1/3}$) should use the condition for existence of a chemical solution $\Delta^s = 0$, which actually replaces, in the later stage, the initial conditions for the size distribution function of the phases. It follows from Eq. (4.319) that asymptotically as $\Delta^s \to 0$,

$$\overline{c}_i - c^s_{i,\infty} = \frac{P^s_i D^s}{D_i}\Delta^s \to 0, \tag{4.334}$$

$$c^{s1}_{i,\infty} = c^{s2}_{i,\infty} = c^{s3}_{i,\infty} = \cdots = c^{sN}_{i,\infty} = \overline{c}_i. \tag{4.335}$$

Physically, this condition is clear: only the precipitates of the chemical compounds, which have the same equilibrium concentration of the common components, can coexist. Thus, asymptotically the concentrations of the components at the surfaces of the precipitates are independent of the specific kind of the phase. The kinetic condition for the coexistence of phases, which is asymptotic with respect to time, corresponds during diffusive decomposition to the thermodynamic condition for the coexistence of phases. In fact, we obtain from Eq. (4.335) the equality of the chemical potentials of the ith component in the coexistent phases ($\mu^s_i = \psi_i + k_B T \ln c^s_{i,\infty}$).

Introducing $x_i = \overline{c}_i/Q_i$, and taking into account Eq. (4.319), we obtain from Eqs. (4.312) and (4.322) a set of equations for the asymptotic time behavior of the limiting parameters x_i

and J^s

$$\prod_i \left(x_i - \frac{P_i^s D^s \Delta^s}{D_i Q_i}\right)^{P_i^s} = \prod_i \left(1 - \frac{1}{Q_i}\sum_{s',i'} P_i^{s'} J^{s'} - \frac{P_i^s D^s \Delta^s}{D_i Q_i}\right)^{P_i^s}$$
$$= \frac{(K_\infty^s)^{1/N^s}}{\prod_i Q_i^{P_i^s}} \equiv \widetilde{K}_\infty^s. \tag{4.336}$$

These equations, with accuracy to terms $(\Delta^s)^2$, are

$$\prod_i (x_i)^{P_i^s} = \prod_i \left(1 - \frac{1}{Q_i}\sum_{s',i'} P_i^{s'} J^{s'} - \frac{P_i^s D^s \Delta^s}{D_i Q_i}\right). \tag{4.337}$$

The constant values of the limiting parameters x_i and J^s can be obtained, with accuracy to $(\Delta^s)^2$, by solving the system of equations (4.337).

Physically, it is clear that the roots x_i and J^s of the system of equations (4.337) must satisfy the inequalities $0 \leq x_i \leq 1$ and $J^s \geq 0$ ($x_i > 1$ indicates that the final quantity of the material in the solution is greater than the given quantity). It can easily be shown that this set of roots is unique. In fact, the left-hand side of Eq. (4.337) involves polynomials which are monotonically varying functions in the region of the physical roots. In this region, in the J^s coordinates, they form open hypersurfaces with a curvature of the same sign, and their intersection gives a unique set of physical roots. If the concentration of some of the components is higher than their solubility limit in the matrix, the precipitates of these pure components may be the coexistent phases, and the chemical reaction constant coincides with the solubility limit of these components $\bar{c}_i$; here $\nu_i = 1$ and $\nu_{i' \neq i} = 0$. This chemical reaction corresponds to the equation $c_{i,\infty} = \bar{c}_i$.

Note that all the phases for which the chemical solution is supersaturated should be taken into account in Eq. (4.337). The solution of these equations automatically selects the phases which asymptotically survive as a result of the competitive growth during coarsening. If, in Eq. (4.337), $J^s < 0$ for some roots, they should be set equal to zero, and the laws of mass action (i.e., relevant equations in Eqs. (4.337)) should be ignored, because the solution becomes unsaturated for these phases during the decomposition process. If, however, it turns out that $x_i > 1$, then the solid solution is unsaturated for all the chemical reactions containing the ith component. We can see from Eq. (4.337) that all the J^{s_i}'s should be set equal to zero then, the laws of mass action should be ignored and the system of equations (4.337) should be solved again after reducing the number of equations by S_i. If the system of equations (4.337) is separated into several subsystems without common components, this indicates that for them the decomposition process occurs independently.

It should be emphasized that the main conclusions of the theory of diffusive decomposition of multicomponent solid solutions in the later stage imply the formation of a universal distribution function; the separation of the complete system of equations describing this process into k similar subsystems of equations for the size distribution functions of each phase, and k sets of algebraic equations for determining the limiting parameters contained in these functions, are insensitive to the mass-transfer mechanism and are determined solely by the laws of conservation. The specific mechanism of atomic mass transfer, i.e., $(\mathrm{d}R/\mathrm{d}t)^s$, affects only the distribution function. The Brownian motion and particle coagulation, precipitation, and the random appearance of particles at small distances can change the equation for the distribution function. These variations can be taken into account by using the collision integral, which corresponds to incorporating the corrections for the volume fractions of the precipitates. Since the basic system of equations is separated into independent subsystems for each phase, all the conclusions, concerning the effect of fluctuations of the precipitate positions, and of their diffusive interactions on the stability of the universal distribution function, that have been drawn from the analysis of single-component solutions are applicable to multicomponent systems as well.

The set of equations (4.337) relates the external parameters, i.e., the parameters determined by the external conditions, with the internal, adjusting parameters. In the case under consideration, Q_i and K^s_∞ are the external parameters, and $\overline{c}_i$ and J^s are the internal parameters. In the general case, the external conditions divide the limiting parameters into external and internal parameters. If the concentration of a given component is constant, it is considered as an external parameter, and the corresponding relative quantity of the material is the internal parameter to be determined from Eqs. (4.337). Note that since the formation of compounds with the matrix material during the decomposition process is analogous to the production of a pure component, we can formally set $D_{\mathrm{matr}} \to \infty$ in the expressions obtained and consider the matrix concentration specified.

As an example, we derive in an explicit form some equations for the simplest case of diffusive decomposition with the formation of the stoichiometric precipitates $A^{(1)}_{\nu_1} A^{(2)}_{\nu_2}$ dispersed in the matrix M (see also Chapter 8). This is often the case for materials of practical importance, containing such inclusions as oxides, carbides, nitrides, etc. In this case, the system of equations (4.337) assumes the form

$$1 - x_i = \frac{P_i}{Q_i} J \qquad (i = 1, 2), \qquad P_i = \frac{\nu_i}{\nu_1 + \nu_2}, \tag{4.338}$$

$$x_1^{P_1} x_2^{P_2} = \frac{K_\infty^{1/(\nu_1+\nu_2)}}{Q_1^{P_1} Q_2^{P_2}} \equiv \widetilde{K}_\infty. \tag{4.339}$$

Note that the inequality $\widetilde{K}_\infty < 1$ must always be obeyed; otherwise the solution is unsaturated. An analytical solution, even of this simple system, is generally impossible. We shall find the solution for some practical limiting cases which are of great interest physically. In specific cases, the systems of equations (4.338) and (4.339) can be solved with a computer to any desired accuracy.

1. Let us first consider a weakly supersaturated solution ($1 - x_i \ll 1$). By expanding Eqs. (4.338) and (4.339) in the small parameter $(P_i/Q_i)J$ we obtain

$$J = \left(1 - \widetilde{K}_\infty\right) \left(\frac{P_1^2}{Q_1} + \frac{P_2^2}{Q_2}\right)^{-1}, \quad x_i = 1 - \left(1 - \widetilde{K}_\infty\right) \frac{P_i}{Q_i} \left(\frac{P_1^2}{Q_1} + \frac{P_2^2}{Q_2}\right)^{-1}. \tag{4.340}$$

Taking into account that $c_{i,\infty} = Q_i$ and using Eq. (4.323), we get

$$\overline{R}^3 = \frac{8}{9} \frac{\sigma \overline{V}^2 n_0}{k_B T} t \left(\frac{P_1^2}{D_1 Q_1} + \frac{P_2^2}{D_2 Q_2}\right)^{-1}, \tag{4.341}$$

$$n(t) = \frac{0.22\left(1-\widetilde{K}_\infty\right)\overline{V}n_0}{\overline{R}^3}\left(\frac{P_1^2}{Q_1}+\frac{P_2^2}{Q_2}\right)^{-1}. \tag{4.342}$$

2. If the initial concentrations of the components correspond to the stoichiometric composition ($Q_1/P_1 = Q_2/P_2$), then solving Eqs. (4.338) and (4.339), we have

$$\overline{R}^3 = \frac{8}{9}\frac{\sigma\overline{V}^2 n_0}{k_B T}t\left[P_2\left(\frac{P_1^2}{Q_1}+\frac{P_2^2}{Q_2}\right)\right]^{-1}\left(\frac{P_2}{P_1}\right)^{P_1} K_\infty^{1/(\nu_1+\nu_2)}, \tag{4.343}$$

$$n(t) = \frac{0.22\overline{V}n_0}{\overline{R}^3}\left[\frac{Q_2}{P_2}-\frac{1}{P_2}\left(\frac{P_2}{P_1}\right)^{P_1} K_\infty^{1/(\nu_1+\nu_2)}\right]. \tag{4.344}$$

3. If the initial concentration of one component by far exceeds that of the other component, e.g.,

$$x_i \ll \frac{Q_2}{Q_1}\frac{P_1}{P_2} - 1, \tag{4.345}$$

then

$$\overline{R}^3 = \frac{8}{9}\frac{\sigma\overline{V}^2 n_0}{k_B T}t\left\{\frac{P_1^2}{D_1 Q_1}\left[\frac{1}{\widetilde{K}_\infty}\left(1-\frac{Q_1 P_2}{Q_2 P_1}\right)^{P_2}\right]^{1/P_1} + \frac{P_2^2}{D_2 Q_2}\left(1-\frac{Q_1 P_2}{Q_2 P_1}\right)^{-1}\right\}^{-1}, \tag{4.346}$$

$$n(t) = \frac{0.22\overline{V}n_0}{\overline{R}^3}\frac{Q_1}{P_1}. \tag{4.347}$$

If the mass-transfer coefficient of one component is much larger than that of the other component, then we can easily obtain the results [153] by passing to the limit ($D_1 \to \infty$) in Eqs. (4.342)–(4.347). Thus, the diffusive decomposition of multicomponent solid solutions differs from the decomposition of single-component solutions, in that the decomposition depends on the initial concentrations of the dissolved components and the constants of the chemical reactions at the precipitate boundaries through Eq. (4.342). Physically, it is clear that the concentrations of the components at the precipitate surfaces are tightly bound by the law of mass action. In the multicomponent case, the additional parameters, Q_1, determined by the previous history of the sample, appear. They make it possible to manipulate the growth rate by optimizing the relative amounts of the dissolved materials.

The results obtained for diffusive decomposition under the condition of simultaneous operation of several mass-transfer mechanisms have been extended to the case of two-component precipitates of stoichiometric composition [285]. The expressions obtained are formally similar to those derived for the growth of single-component precipitates. The difference resides in that the effective coefficients, D_{eff} and K_{eff}, in the resulting equations, involve the concentrations, $c_{i,\infty}$, which are not specified in this case but are defined self-consistently by the asymptotic equations for the laws of conservation of matter and the law of mass action. It should be stressed that these coefficients depend on the external parameters Q_i and K_∞^s. It is this dependence that enables us, by changing the external parameters, to manipulate the diffusive decomposition, in the intermediate asymptotic region, with the goal of improving the thermal and phase stability of the newly created materials.

It should be emphasized that the theory solves the inverse problem of determining the characteristics of multicomponent systems in experiments on diffusive decomposition. It allows us to formulate an innovative approach to the experimental measurement of important and yet not easily measurable characteristics of multicomponent systems such as the constants of chemical reactions in the solids, K_∞^s, the specific surface energy of the phases, σ^s, and the partial diffusion coefficients of the precipitate components, D_i. They can be evaluated by comparing the time variation of the experimental size distribution functions for the phases (the histograms) with the predicted distribution function which depends on these parameters. The constant parameters in the theoretical size distribution function for the precipitates can be chosen to fit the experimental data to any desired accuracy, with the help of a computer.

The parameters K_∞^s, σ^s, and D_i can also be evaluated by measuring the decomposition rate constants. In the single-component case, the values D, σ, and c_∞ enter into the expression for the rate constant in a linear combination, so the measurements of the rate constants should be supplemented by independent experimental measurements of the supersaturation, as was done in [8]. For decomposition of multicomponent solutions, the rate constant, even in the simplest cases, is a nonlinear function of these values and of the relative amounts of the materials, Q_i, thus enabling us to determine D_i, σ^s, and K_∞^s by measuring the decomposition rate constant for different Q_i and solving Eq. (4.337). Assuming a weakly supersaturated solution with the decomposition specified by $A_{\nu_1}^{(1)} A_{\nu_2}^{(2)}$ (Eq. (4.342)), D_1, D_2, σ, and K_∞ are the unknown parameters. Accordingly, the knowledge of four values of the decomposition rate constants, with different Q_1 and Q_2 (satisfying, of course, the condition of weak supersaturation), would generally suffice to obtain from Eq. (4.342) a set of equations which, being solved numerically, yield the required parameters.

4.3.3 Regions of Phase Coexistence in Composition Space

In order to determine the regions of coexistence of the different phases, it is necessary to determine the boundaries of all k phases in the space $\{Q_i\}$ of relative contents Q_i of the N-component supersaturated solid solution. These boundaries are $(N-1)$-dimensional hypersurfaces. In the general case, the equations for these hypersurfaces can be obtained by determining J^s from Eq. (4.337) and setting them equal to zero:

$$J^s(Q_1, Q_2, \ldots, Q_N) = 0. \tag{4.348}$$

This hypersurface divides the N-dimensional space $\{Q_i\}$ into two regions. In the region where $J^s < 0$ (under the surface) the solid solution for the sth phase is unsaturated and therefore this phase does not exist. In the region where $J^s > 0$ (above the surface) the solid solution for the sth phase is supersaturated and the phase consists of precipitates. The subspace, defined by $Q_i > 0$ and $\sum_{i=1}^N Q_i < 1$, is divided into regions in the space $\{Q_i\}$ by the hypersurfaces of all the phases. Apparently only the phases for which $J^s(Q_1, Q_2, \ldots, Q_N) > 0$ can coexist for the Q_i's situated in any one of these sections. Since the algebraic equations for determining J^s are not linear, and only the solutions at each point in the space $\{Q_i\}$ that satisfy the physical conditions indicated above should be selected, in the general case the hypersurface $J^s = 0$ is piecewise continuous. If it is necessary to determine the hypersurface which separates the regions in which the phases containing the ith component are present from those in which they are missing, we must determine $x_i = x_i(Q_1, Q_2, \ldots, Q_N)$ from Eq. (4.337) and set it equal to zero, which corresponds to a saturated solution for the ith component.

To illustrate the above-made statements, we solve Eq. (4.337) for a two-phase ($s = 1, 2$), three-component ($i = 1, 2, 3$) system in which the component $i = 2$ is contained in both phases. Following the above procedure, we obtain an equation for the surface in a three-dimensional space $\{Q_1 Q_2 Q_3\}$ which separates the region where phase I exists (we say above the surface) and the region (below the surface) where phase I is absent ($J^{\text{I}} = 0$):

$$Q_2^{\text{I}} = \frac{K_1^{1/\nu_2}}{Q_1^{\nu_1/\nu_2}} + \frac{\nu_4}{\nu_3} Q_3 - \frac{\nu_4}{\nu_3} \frac{K_2^{1/\nu_2}}{K_1^{\nu_4/(\nu_2 \nu_3)}} Q_1^{\nu_1 \nu_4/(\nu_2 \nu_3)}, \tag{4.349}$$

and an equation for the surface which has phase II above it and no phase II ($J^{\text{II}} = 0$) below it:

$$Q_2^{\text{II}} = \frac{K_2^{1/\nu_1}}{Q_3^{\nu_3/\nu_4}} + \frac{\nu_2}{\nu_1} Q_1 - \frac{\nu_2}{\nu_1} \frac{K_1^{1/\nu_2}}{K_2^{\nu_2/(\nu_1 \nu_4)}} Q_3^{\nu_2 \nu_3/(\nu_1 \nu_4)}. \tag{4.350}$$

These surfaces intersect at $Q_2^{\text{I}} = Q_2^{\text{II}}$, which, as can easily be seen, coincides with the line for which the solution containing all the components is saturated, i.e., $\widetilde{K}_1 = \widetilde{K}_2 = 1$ or

$$\frac{K_1^{1/\nu_2}}{Q_1^{\nu_1/\nu_2}} = \frac{K_2^{1/\nu_1}}{Q_3^{\nu_3/\nu_4}}. \tag{4.351}$$

The intersection of the surfaces of Eqs. (4.349) and (4.350) forms four regions (see Figure 4.6). Phases I and II occur above these surfaces but not below them, where the solution is unsaturated. Only phase I exists in the region situated between the surfaces determined by Eqs. (4.349) and (4.350), when the surface equation (4.349) is below Eq. (4.350). However, if the surface (Eq. (4.350)) is below Eq. (4.349), then the region between the surfaces has only phase II.

As an example, we consider the phase diagram of decomposition, shown in Figure 4.6(b), as projected onto the plane $Q_1 Q_2$ ($Q_3 = \text{const}$). Here lines 1 and 2 correspond to Eqs. (4.349) and (4.350), while point A corresponds to Eq. (4.351). The state of the system under given external conditions is determined by point B with the initial concentrations Q_1^0 and Q_2^0.

A change of phase relations or a transition from one phase region to another, in the system under consideration, may be due to either a change of the systems initial state related to the change of the initial concentrations (Figure 4.7(a)) or a change of the boundaries of the region of coexistence of phases (Figure 4.7(b)). As shown in Figure 4.7(a), the system can be transferred from state B (phase I) to state B′ (phase II) by changing the initial concentrations ($Q_1^0, Q_2^0 \to Q_1^{\text{I}}, Q_2^{\text{I}}$). Changing the external conditions, such as temperature, pressure, etc., or in other words changing only K_∞^s and leaving the position of point B unchanged (Figure 4.7(b)), we can change the boundaries of the region of coexistence of phases in such a way that the system transfers to another region.

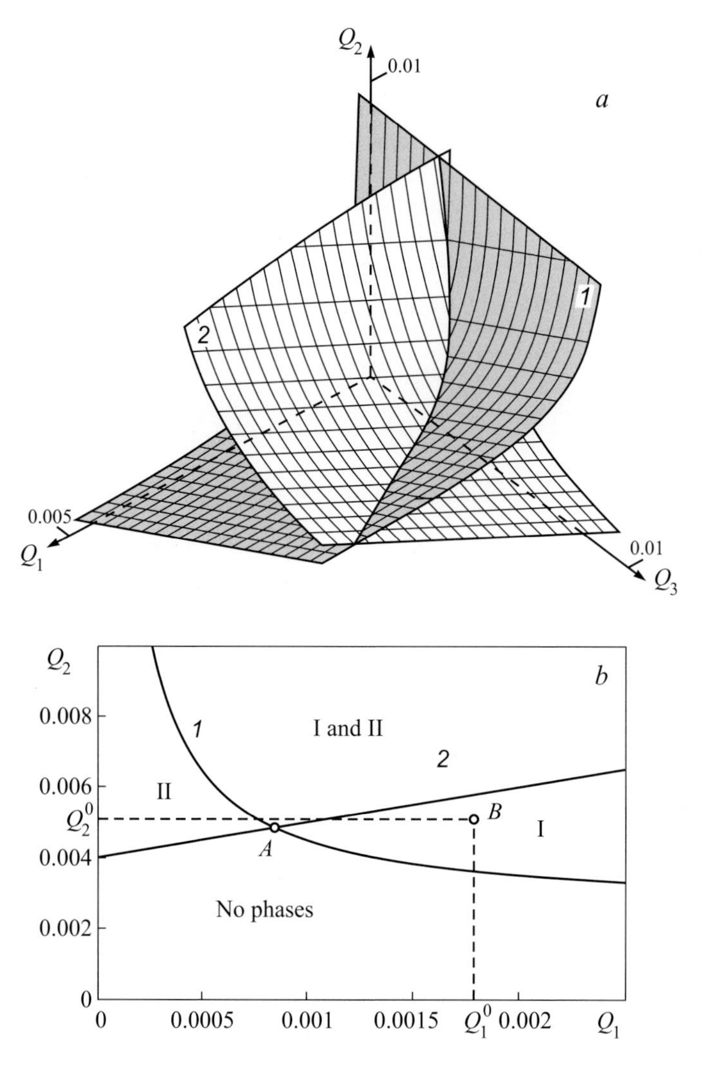

Figure 4.6: Phase diagram for an illustration of the decomposition process ($K_1 = 2 \times 10^{-6}$, $K_2 = 10^{-7}$, $\nu_1 = \nu_2 = \nu_4 = 1$, $\nu_3 = 2$) (a) in three-dimensional space $\{Q_1 Q_2 Q_3\}$ and (b) as projected onto the plane $Q_1 Q_2$, $Q_3 = 0.005$.

Let us now consider a solution which is weakly supersaturated in all of the components ($Q_i^{-1} \sum_{s,i} \nu_i^s J^s \ll 1$ and $K_\infty^s \approx 1$). The system of equations (4.337) can be linearized and solved for this specific case. In other words, it determines the hypersurfaces in the region near their common line of intersection $\widetilde{K}_\infty^1 = \widetilde{K}_\infty^2 = \cdots = \widetilde{K}_\infty^s = 1$. For example, setting first J^{I} and then J^{II} equal to zero, exactly as before, we obtain equations for the boundary surfaces of the phases for a two-phase, three-component system. Note that at $J^{\mathrm{I}} = 0$ the solution is sought near $\widetilde{K}_\infty^s = 1$, i.e., where $Q_2^{\mathrm{I}} = K_1^{1/\nu_2} Q_1^{-\nu_1/\nu_2}(1 + \varepsilon)$, and at $J^{\mathrm{II}} = 0$,

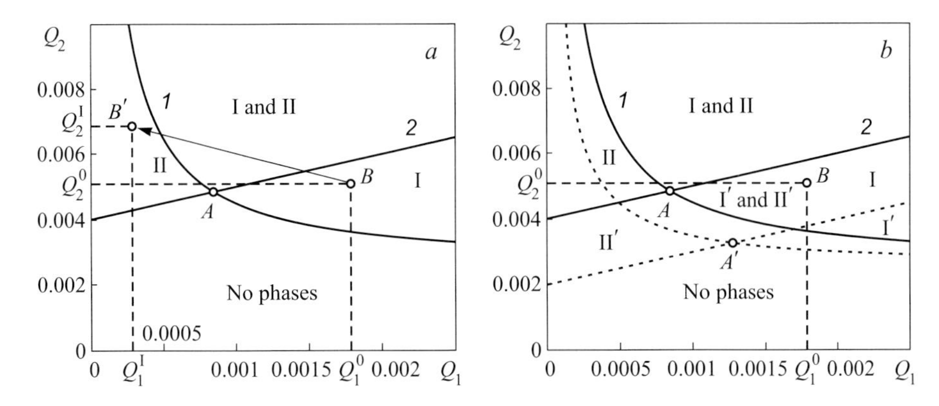

Figure 4.7: Scheme of phase change due to (a) change of initial conditions and (b) change of interphase position ($K_1' = 10^{-6}$, $K_2' = 5 \times 10^{-8}$); the other parameters are the same as in Figure 4.6.

$Q_2^{\mathrm{II}} = K_2^{1/\nu_1} Q_3^{-\nu_3/\nu_4}(1+\eta)$, where $\varepsilon \ll 1$ and $\eta \ll 1$. Thus we obtain equations for the surfaces near the line $\widetilde{K}_\infty^1 = \widetilde{K}_\infty^2 = 1$, which correspond to Eqs. (4.349) and (4.350) in the region Q_i. If the kinetic hindrance for phase formation is large enough, then the phase may fail to be formed in real time, though it is asymptotically stable. The transformation of the phase diagram with time should be taken into account. Solving the system of equations (4.337), we obtain the kinetic phase diagrams of the decomposition, enabling us to determine the stability of phases based on their kinetic characteristics.

For a solution weakly supersaturated in all the components, we have

$$J^s = \left[1 - \widetilde{K}_\infty^s - D^s \Delta^s(t) \sum_i \frac{(P_i^s)^2}{D_i Q_i}\right] \left[\sum_i \frac{(P_i^s)^2}{Q_i}\right]^{-1}, \tag{4.352}$$

$$c_{i,\infty} = Q_i - P_i^s \frac{1 - \widetilde{K}_\infty^s}{\sum\limits_i \frac{(P_i^s)^2}{Q_i}} - P_i^s \Delta^s(t) \left[\frac{1}{\sum\limits_i \frac{(P_i^s)^2}{Q_i}} + \frac{1}{D^s \sum\limits_i \frac{(P_i^s)^2}{D_i Q_i}}\right]. \tag{4.353}$$

Here $\bar{c}_i$ can be determined from Eq. (4.319) and $\Delta^s(t) = \mathrm{const}\cdot t^{-1/3}$. Equation (4.353) defines the dynamics of the relative quantities of matter in the precipitates, i.e., J^s, and the equilibrium concentrations $c_{i,\infty}^s$ in the asymptotic region. It is seen from Eq. (4.353) that these values vary with time as $t^{-1/3}$ and approach certain constant values at sufficiently long times.

The phases have diffuse boundaries because of the nonuniform distribution of the material and fluctuations of temperature and concentration of the components. As can be seen from Eqs. (4.349) and (4.350), the regions of coexistence of the phases depend on the relative amounts of the dissolved materials Q_i and on the constants of the chemical reactions K_∞^s. Thus, the surface tension during the diffusive decomposition process in a multicomponent solid solution gives rise to a universal size distribution of the precipitates of the phases, and the quasithermodynamic equilibrium conditions determine the phases produced in the

N-component solution. The phases can be changed by either varying the Q_i ratio or the temperature, since K^s_∞ vary differently with temperature.

The law of mass action in Eq. (4.306) can be easily replaced by a general equation for the phase equilibrium at the boundaries of the precipitates. We get

$$\prod_i \frac{\left(c^s_{i,R}\right)^{\nu^s_i}}{K^s_R} = 1 \to F\left(\left\{c^s_{i,R}\right\}\right) = 1. \tag{4.354}$$

Theoretically, this does not affect the conclusions concerning the separation of the system of equations (4.303)–(4.306) into s independent subsystems, as well as the formation of a universal distribution function, and the law of evolution of the critical size. It affects only the phase diagrams of the decomposition, i.e., the regions of coexistence of phases and the quantities of the material in the phases, which are determined by the system of algebraic equations (4.337) and depend on the specific equation for the phase balance $F\left(\left\{c^s_{i,R}\right\}\right) = 1$. Note that all the above conclusions hold true if $k < N$, i.e., the number of the coexisting phases is not larger than the number of the components. If we take into account the external parameters p and T, then the number of the coexisting phases is specified by the Gibbs phase rule $k \leq N + 2$.

To determine all the coexisting phases for $k > N$, it is necessary to separate the system of equations (4.337) into groups containing N phases. The overall number of such groups is C^N_k. Then each of the C^N_k groups of equations should be solved and only one group will have the roots (for given p and T) meeting the physical requirements of the coexistence of the phases that were discussed above ($J^s > 0$; $0 < x_i < 1$). It is the solution of this system of equations that selects those N phases that survive in the process of diffusive decomposition.

4.3.4 Competition of Different Phases in Coarsening

In the course of diffusive decomposition the amount of material available to generate particles becomes small, and competition arises between different phases. As a result, only some of those phases that precipitate at early stages of the decomposition can survive [305].

4.3.4.1 Phases with Simple Stoichiometry

Consider a three-component solid solution of atomic impurities A, B, and C in a chemically inert matrix, which can precipitate into two stoichiometric two-component phases containing a common component: $A_\mu C_\nu$ and $B_\eta C_\xi$. To start with, let us consider the simplest stoichiometric phases, where $\mu = \nu = \eta = \xi = 1$. The state of the system is determined by the law of mass action (see Eq. (4.306)),

$$c_a c_c = K_1, \qquad c_b c_c = K_2, \tag{4.355}$$

and by the equation of stoichiometry (Eq. (4.305))

$$q_a + q_b = q_c, \tag{4.356}$$

where c_i ($i = a, b, c$) is the concentration of component i averaged over the volume, $K_{1,2}$ are chemical reaction constants for the first and second phases, respectively, and q_i is the relative number of impurities of type i in particles per unit volume. Moreover, the first phase will precipitate out when the following inequality is satisfied:

$$q_a > 0, \tag{4.357}$$

and the second when the condition

$$q_b > 0 \tag{4.358}$$

is satisfied. Taking Eq. (4.356) into account, we see that the condition $q_c > 0$ is also necessary for the first and second phases to precipitate, whereas this takes place automatically when Eq. (4.357) or (4.358) are satisfied.

The equation of balance for component type i takes the form

$$Q_i = q_i + c_i, \tag{4.359}$$

where Q_i is the total number of impurities in solution and in particles. Taking this relation into account, it is convenient to write Eq. (4.356) in the form

$$c_a + c_b - c_c + \delta Q = 0, \tag{4.360}$$

where

$$\delta Q = Q_a + Q_b - Q_c. \tag{4.361}$$

The solution the system of Eqs. (4.355) and (4.360) gives the values of the component concentration at the late stage of the decomposition:

$$c_{a,b} = \frac{K_{1,2}}{2\left(K_1 + K_2\right)} \left[\delta Q^2 + 4\left(K_1 + K_2\right) + \delta Q\right]^{1/2}, \tag{4.362}$$

$$c_c = 2\left(K_1 + K_2\right)\left[\delta Q^2 + 4\left(K_1 + K_2\right) + \delta Q\right]^{-1/2}. \tag{4.363}$$

These expressions combined with Eqs. (4.357) and (4.358) determine the conditions for existence of the phases. Thus, the first phase (AC) exists when

$$Q_c > Q_c^{\mathrm{I}} = Q_b - \frac{1}{Q_a}\left(\frac{K_1}{K_2}Q_a^2 - K_1\right), \tag{4.364}$$

while the second (BC) exists when

$$Q_c > Q_c^{\mathrm{II}} = Q_a - \frac{1}{Q_b}\left(\frac{K_2}{K_1}Q_b^2 - K_1\right). \tag{4.365}$$

A situation of interest in applications is the one where the component precipitates strongly into particles, i.e., when $\delta Q \gg K_{1,2}$. Therefore, we will analyze the results for this case in particular. From Eqs. (4.362) and (4.363), we obtain for $\delta Q > 0$

$$c_{a,b} \approx \delta Q \frac{K_{1,2}}{K_1 + K_2}, \qquad c_c \approx \frac{K_1 + K_2}{\delta Q}, \tag{4.366}$$

while for $\delta Q < 0$

$$c_{a,b} \approx \frac{K_{1,2}}{|\delta Q|}, \qquad c_c \approx |\delta Q| \, . \tag{4.367}$$

The conditions for existence of phases (4.364) and (4.365) take the form

$$Q_c > Q_c^{\mathrm{I}} = Q_b - \frac{K_2}{K_1} Q_a , \tag{4.368}$$

$$Q_c > Q_c^{\mathrm{II}} = Q_a - \frac{K_1}{K_2} Q_b . \tag{4.369}$$

It is easy to see that when $\delta Q > 0$ (i.e., the number of components of type C is less than the numbers of types A and B taken together) the conditions given by Eqs. (4.368) and (4.369) cannot be satisfied at the same time, i.e., the phases cannot coexist. In fact, from Eq. (4.366) it follows that in the final state $c_c < c_a, c_b$, i.e., component C is almost completely absorbed, while impurities A and B remain in the solution. A competition then begins between particles of the first (AC) and second (BC) phases for the insufficient component C, as a result of which only one of the phases survives. For $\delta Q < 0$ there is enough of component type C to prevent competition between the phases, and they exist independently of each other. All excess impurities A and B precipitate into particles, while the type C component remains partially in solution.

4.3.4.2 Separation of Three Phases

Consider the case where, in addition to the two binary compounds, it is possible for a third phase to precipitate consisting of the pure components C, A, or B. As was shown above, when $\delta Q < 0$ the type C component partially remains in solution, and consequently it can precipitate in the form of a pure phase. In this case, system (1) must satisfy the equation

$$c_c = c_{\infty,c} . \tag{4.370}$$

From this we find

$$c_{a,b} = \frac{K_{1,2}}{c_{\infty,c}} . \tag{4.371}$$

Here $c_{\infty,c}$ is the equilibrium concentration at the planar boundary.

Stoichiometry (Eq. (4.356)) in this case determines the amount of precipitate of the third phase C:

$$q_c = (Q_c - c_c) - (Q_a - c_a) - (Q_b - c_b) . \tag{4.372}$$

Taking into account that three phases precipitate for $q_c > 0$, it is easy to obtain the conditions for coexistence of all three phases:

$$Q_c > Q_a + Q_b + c_{\infty,c} - \frac{K_{1,2}}{c_{\infty,c}} , \tag{4.373}$$

$$Q_{a,b} > K_{1,2}/c_{\infty,c}. \tag{4.374}$$

When $\delta Q > 0$, the excess of type C component is completely absorbed by particles of the new phases, and as we noted above, competition will allow only one of them to survive. The competition disappears if it is possible to precipitate a pure phase A (the case where phase B precipitates is entirely analogous). Adding the equation $c_a = c_{\infty,a}$ to Eq. (4.355), we obtain, as before, the condition for coexistence of all three phases:

$$Q_a > Q_c - Q_b + c_{\infty,c}\left(1 + \frac{K_2}{K_1}\right) - \frac{K_1}{c_{\infty,a}}, \tag{4.375}$$

$$Q_b > \frac{K_1}{K_2}c_{\infty,a}, \quad Q_c > \frac{K_1}{c_{\infty,a}}. \tag{4.376}$$

4.3.4.3 Phases with Arbitrary Stoichiometry

Let us consider precipitation of two phases with arbitrary stoichiometries: $A_\mu C_\nu$ and $B_\eta C_\xi$. Equations (4.355) and (4.356) in this case can be written in the form

$$c_a^\mu c_c^\nu = K_1, \tag{4.377}$$

$$c_b^\eta c_c^\xi = K_2, \tag{4.378}$$

$$\frac{q_a}{\mu} + \frac{q_b}{\eta} = \frac{q_c}{\nu + \xi}. \tag{4.379}$$

Let us introduce

$$\delta Q = \frac{Q_a}{\mu} + \frac{Q_b}{\eta} - \frac{Q_c}{\nu + \xi} \tag{4.380}$$

and, as before, consider the case where the chemical reaction constant is small compared with δQ. Taking Eqs. (4.378) and (4.379) into account, let us rewrite Eq. (4.380) in the form

$$\delta Q = \frac{c_a}{\mu} + \frac{1}{\eta}\left(c_a\right)^{\frac{\mu+\nu}{\eta+\xi}}\left(\frac{K_2^\nu}{K_1^\xi}\right)^{\frac{1}{\eta+\xi}}. \tag{4.381}$$

This equation can be solved rather simply in two cases: when $\mu + \nu = \eta + \xi$ and $\mu + \nu = 2(\eta + \xi)$ (the case where $\mu + \nu = (\eta + \xi)/2$ obviously reduces to the previous cases by making the substitution $A \leftrightarrow B$).

Note that this case, although restrictive, still includes a rather wide class of compounds: in the first case, AC_2 and BC_2, AC_2 and B_2C, A_2C_3 and B_2C_3, A_2C_3 and B_3C_2, etc., while in the second case, A_2C_2 and BC, A_3C and BC, etc. In the first case

$$\delta Q = c_a\left[\frac{1}{\mu} + \frac{1}{\eta}\left(\frac{K_2^\nu}{K_1^\xi}\right)^{\frac{1}{\mu+\nu}}\right]. \tag{4.382}$$

From this we obtain the condition for existence of the phase $A_\mu C_\nu$:

$$\kappa Q_a > Q_b - \frac{\eta}{\nu+\xi} Q_c, \tag{4.383}$$

and of the phase $B_\eta C_\xi$:

$$\frac{1}{\kappa} Q_b > Q_a - \frac{\mu}{\nu+\xi} Q_c, \tag{4.384}$$

where

$$\kappa = \frac{1}{\eta} \left(\frac{K_2^\nu}{K_1^\xi} \right)^{\frac{1}{\mu+\nu}} . \tag{4.385}$$

Analogously, in the second case where $\mu + \nu = 2(\eta + \xi)$, we find that phase $A_\mu C_\nu$ exists when the condition

$$\kappa Q_a^2 > Q_b - \frac{\eta}{\nu+\xi} Q_c \tag{4.386}$$

is satisfied, and phase $B_\eta C_\xi$ when the condition

$$\frac{1}{\sqrt{\kappa}} Q_b^2 > Q_a - \frac{\mu}{\nu+\xi} Q_c \tag{4.387}$$

holds.

So, at the initial stage of the decay all phases whose supersaturation is positive and sufficiently large precipitate. As the supersaturations decrease, competition begins both between particles of the same phase but different sizes and between different phases, as a result of which one or several phases survive, depending on the conditions of survival or extinction given by the inequalities (4.364), (4.365), (4.383)–(4.387), and a universal size distribution of particles is generated.

We note the following interesting feature of the behavior of this system with time. The rate of precipitation of a phase depends not only on its supersaturation but also on the conditions for its nucleation, and the diffusion coefficients of the reagents. Therefore, a situation is possible where, at the initial stage of the decomposition, the phase that precipitates most strongly is one that does not satisfy inequalities (4.364), (4.365), (4.383)–(4.387), while the phase that satisfies these inequalities precipitates more slowly. Then (at the late stages), the first phase dissolves due to the competition, while the second phase survives. Thus, in this case, a replacement of precipitating phases takes place in the course of diffusive decomposition of a supersaturated solution.

Note as well that a competition is possible, not only between different phases but also between particles of the same phase under different conditions (for example, at a boundary and within the body of a grain). Although the location where the phase precipitates (in the body of a grain, at its boundaries, or at some nucleation centers) can affect the kinetics of the initial stage of the decay and the expression for the chemical reaction constant $K_{1,2}$, it does not alter the analysis given above qualitatively, nor the results obtained, since the only important factor is the diffusive exchange of material between the phases.

4.3.5 Formation of Precipitates of Nonstoichiometric Composition

It is well established [68] that many compounds used as strengthening agents constitute in fact phases of variable composition, and deviations from stoichiometry greatly affect the properties of existing materials. In examining diffusive decomposition (coarsening), which produces precipitates of nonstoichiometric composition, we should first note that, physically, nonstoichiometricity of the compound (phase) is determined by the components that can change their presence in the compound to some extent. In stoichiometric compounds components do not change their stoichiometric coefficients ν_j. Although the ν_j can change in the homogeneous regions, it is physically clear that under equilibrium conditions, when the thermodynamic potential of the system is at minimum, they asymptotically approach certain constant values determined by the initial conditions, i.e., by the relative initial quantities of the components and by the chemical reaction constants. These considerations suggest that the precipitates of stoichiometric composition grow in this case, but the corresponding coefficients ν_j are additional internal parameters (for stoichiometric compounds) that must be determined. In one sense, nonstoichiometric compounds can be considered as a solid solution of limited solubility. Thus, all the equations and their solutions, which determine the above precipitate size distribution functions, can be used in this case. The important difference is the existence of additional equations for determining ν_i in the set of algebraic equations, which relate the internal parameters governing the distribution functions $c^s_{l,\infty}$, P^s_l, and J^s to the external parameters K^s_∞, and Q_l.

Let us introduce the parameters $\Delta v^s_j = z^s_j$, which characterize the variation of the stoichiometric coefficients in the homogeneous region of the components of the sth phase $v^s_j = v^s_j(z^s_j)$. If the sth phase has components with finite homogeneous regions, this generally means that various equilibrium chemical reactions can occur at the interphase boundaries in these regions. Under equilibrium conditions, the minimum of the thermodynamic potential is realized,

$$\delta\Phi^s(z^s_1) = \delta\Phi^s(z^s_2) = \cdots = \delta\Phi^s(z^s_n) = 0. \tag{4.388}$$

Hence, for infinitesimally close values we can write

$$\delta\Phi^s(z^s_j + \Delta) - \delta\Phi^s(z^s_j) = \Delta\frac{\partial\delta\Phi^s}{\partial z^s_j} = 0. \tag{4.389}$$

Thus, under quasiequilibrium conditions at the phase boundaries we have

$$\delta\Phi^s = 0, \tag{4.390}$$

$$\frac{\partial\delta\Phi^s}{\partial z^s_j} = \frac{\partial\delta\Phi^s}{\partial\nu^s_j}\frac{\partial\nu^s_j}{\partial z^s_j} = 0. \tag{4.391}$$

The chemical potential per molecule of the sth phase is $\mu^s - \overline{\mu}^s N^s$. The average chemical potential $\overline{\mu}$ (per atom of the compound) generally depends on the normalized nonstoichiometric coefficients. However, because of the presence of a nonstoichiometric compound this dependence is weak in the homogeneous region. Therefore, the chemical potential per atom is assumed to be constant in the homogeneous region, and it sharply increases outside this region to a good approximation.

Let us write down the complete set of equations for the internal parameters of the system ($c^s_{l,\infty}$, ν^s_j, J^s). This set consists of the equations, being consequences of the conservation laws, and of Eqs. (4.390) and (4.391). It is convenient to write it separately for the stoichiometric and nonstoichiometric components, after eliminating, using Eq. (4.391) for a dilute solution, the jth components in Eq. (4.390), and taking into account that $\partial \nu^s_j / \partial z^s_j = 1$ for nonstoichiometric components with subscript j and $\partial \nu^s_i / \partial z^s_i = 0$ for stoichiometric components with subscript i. Equation (4.391) holds identically for these components:

$$\sum_{s,j} P^s_j J^s = Q_l - c^s_{l,\infty}, \tag{4.392}$$

$$\prod_j (\bar{c}_j)^{P^s_j} = \exp\left\{ \frac{1}{k_B T} \sum_j P^s_j \left(\bar{\mu}^s - \psi_j\right) \right\} = A^s_\infty, \tag{4.393}$$

$$A^s_\infty \prod_j (\bar{c}_{j,\infty})^{P^s_j} = (K^s_\infty)^{1/N^s}, \tag{4.394}$$

$$\bar{c}_j = c^s_{j,\infty} = \bar{c}^s_{j,\infty} = \exp\left\{ \frac{\bar{\mu}^s - \psi_j}{k_B T} \right\}. \tag{4.395}$$

The physically meaningful roots of these equations must satisfy the requirement $J^s > 0$, $Q_l > \bar{c}^s_{j,\infty}$ and, in addition, the roots P^s_l must be located in the homogeneous region.

If, under the given conditions, several phases can achieve the concentration $\bar{c}_j = c^s_{j,\infty}$, then $c^{s_0}_{j,\infty} = c_j$ will have the lowest value. It must be assumed for the remaining coexistent phases that $P^s_j = P^s_{j,\min}$ or $P^s_j = P^s_{j,\max}$ depending on whether $\bar{c}^s_{j,\infty} < c^{s_0}_{j,\infty}$ or $\bar{c}^s_{j,\infty} > c^{s_0}_{j,\infty}$. Consequently, Eqs. (4.392)–(4.395) should also be rewritten. The P^s_j ratios should also be taken into account,

$$\sum_j P^s_j = 1, \qquad \frac{P^s_j}{P^s_{j'}} = \frac{\nu^s_j}{\nu^s_{j'}} = \text{const}, \tag{4.396}$$

in all the phases and for all the components, except $P^{s_0}_j$ for which there exists one equation in Eq. (4.395),

$$\bar{c}_j = c^s_{j,\infty} = \bar{c}^s_{j,\infty}. \tag{4.397}$$

If the conditions are such that

$$\cdots \le \bar{c}^{s_{m-2}}_{j,\infty} \le \bar{c}^{s_{m-1}}_{j,\infty} \le \bar{c}_j \le \bar{c}^{s_m}_{j,\infty} \le \bar{c}^{s_{m+1}}_{j,\infty} \le \cdots \tag{4.398}$$

is valid for all possible changes in the homogeneous region, then P^s_j for all the coexistent phases will have the value $P^s_{j,\min}$ on the left-hand side of $\bar{c}_j$ and $P^s_{j,\min}$ on the right-hand side of it. If there are overlapping homogeneous regions, when $\bar{c}^{s_1}_{j,\infty} = \bar{c}^{s_2}_{j,\infty} = \cdots$, and the lowest value of $\bar{c}_j$ is located inside this region, then the P^s_j values for the overlapping phases will lie inside the homogeneous region and will be related as

$$P^{s_1}_j - P^{s_1}_{j,\min} = P^{s_2}_j - P^{s_2}_{j,\min} = \cdots \tag{4.399}$$

Accordingly, in all the remaining coexistent phases $P_j^s = P_{j,\min}^s$ or $P_j^s = P_{j,\max}^s$. The situation discussed above can be generalized to include several components with a finite homogeneous region.

If the concentration $\overline{c}_j$ for the nonstoichiometric component is an external parameter and coincides with one of the parameters $\overline{c}_{j,\infty}$, we must assume that $P_j^s = P_{j,\min}^s$ or $P_j^s = P_{j,\max}^s$, depending on whether $\overline{c}_j$ approaches $\overline{c}_{j,\infty}$ from below or from above. The same procedure is used to prove the unique physical solution for the given conditions and to determine the region of coexistence of the phases.

In the above analysis, we have assumed that precipitates of homogeneous composition are produced. This is correct if the diffusion processes occurring inside the precipitate are not slower by many orders of magnitude than those in the matrix. If, however, the diffusion processes inside the precipitate are slowed down for some reason compared with those in the matrix (e.g., a liquid or gaseous matrix), then precipitates of nonhomogeneous stoichiometric composition are produced inside the homogeneous region. The stoichiometric coefficients are then no longer physically meaningful though the above equations remain valid. This problem is, of course, more complicated and cannot be solved by simply integrating Eq. (4.304) but should incorporate the initial conditions. The quasiequilibrium condition is fulfilled in this case too, thus allowing us to use the laws of mass action at the phase boundaries and the additional conditions of Eqs. (4.390) and (4.391).

4.3.6 Comparison with Experimental Data

An experimental verification of different aspects of the theory outlined has been carried out on different materials of practical importance, such as steels and iron alloys containing carbide, nitride, and other precipitates [10, 37, 52, 68, 87, 101, 172]; ageing alloys containing nickel, aluminium, etc. [10,37,52,87,101]; precipitation-strengthened composite materials [42,204]; internally oxidated alloys [58], and glasses and ionic crystals [105, 288], etc. We shall not attempt now to make a comprehensive survey of the wealth of experimental studies of diffusive decomposition, but will discuss only those studies which are of most interest for comparing theory with experimental data.

In a series of studies [6, 11] Ardell performed careful investigations to obtain statistically representative data on the growth of γ'-precipitates (Ni_3X, where X is Al, Ti, or Si) in nickel alloys which are the basis of various heat-resistant materials for high-temperature applications. Figure 4.8 shows the histograms for γ'-precipitations in the Ni–Al system. The theoretical universal distribution function corresponding to mass transfer through volume diffusion is shown for comparison. It can be seen that the theory fits the experimental data closely. The shape of the histograms resembles the predicted curves: a rather slow growth in the region where the sizes are small, a blocking point $u_0 = 3/2$, and a rapid decline in the region beyond it. The data on the growth kinetics for γ'-precipitates [9] strongly suggest the applicability of the "$t^{1/3}$-law" for the mass transfer attributable to volume diffusion (Figure 4.9). The significance of the experiments in [9, 11] is further increased by the fact that the equilibrium concentration of the dissolved material in the solid solution in nickel was independently assessed by means of magnetic measurements. This made it possible to explore the kinetics of change of the supersaturation in the systems under examination. The supersaturation decreased with time as $t^{-1/3}$, i.e., by the same law as predicted by the theory.

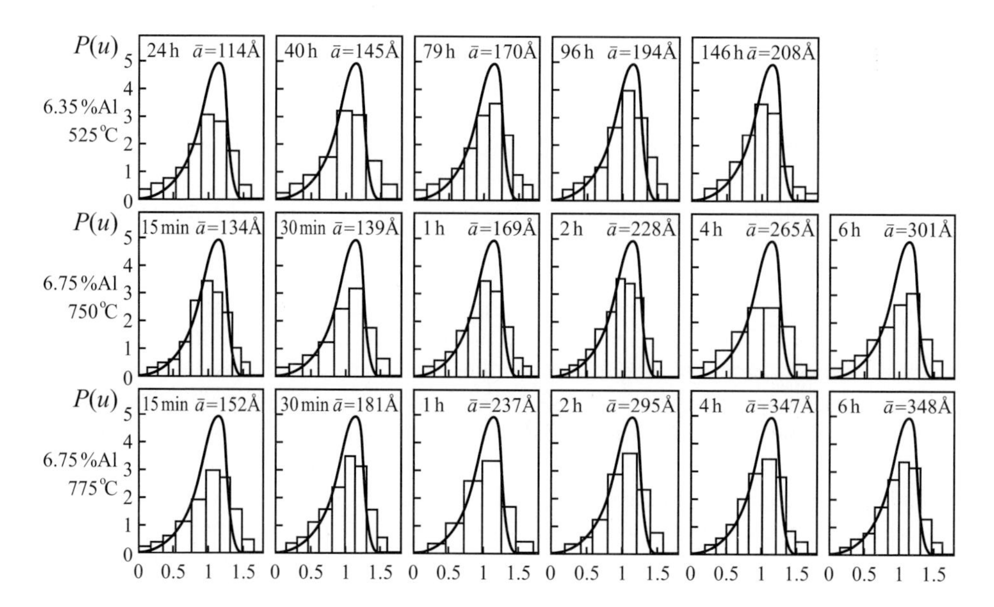

Figure 4.8: Histograms for γ'-precipitations in a Ni–Al system [6] compared with the distribution function $P(u)$.

Figure 4.10 shows the histogram for Ni_3Al precipitates in Ni–22at%Co–13at%Al alloy. The theoretical distribution function computed by the authors [11], using an iterative algorithm to fit the experimental histogram, is also shown in Figure 4.10. The theory and the experiment apparently agree closely with each other.

The evolution of the universal distribution function from an arbitrary initial distribution is clearly illustrated in [335] by the growth of cementite (Fe_3C) particles in Fe–0.79at%C, alloy spheroidized at 704 °C (Figure 4.11). The histograms, which are diffuse at early time of annealing, progressively approach the theoretical distribution while the blocking points approach $u_0 = 3/2$.

A detailed analysis of the precipitate growth in steels containing 0.25% of vanadium was carried out in [18], where the kinetics of precipitation of vanadium carbide, nitride, and carbonitride were studied in the course of steel annealing at a temperature of 790 °C. Early in the annealing process the growth kinetics obeyed the equation $\overline{R}^2 \sim t$ which corresponds, in accordance with the theory, to a decomposition governed by chemical reactions at the interface. A study [70] of the kinetics of ThO_2 particle growth in TD-nickel, carried out by using a correlation analysis of data, showed that a cubic growth law (the "$t^{1/3}$-law") was fulfilled to a very high confidence probability level ($p = 0.995$). A study [188] performed using statistical methods of testing hypotheses (Pearson's χ^2-test, the Kolmogorov test) showed that the time-asymptotic behavior of the size distribution function for ZrO_2 particles in a dispersed system, Mo–ZrO_2, could be described by the universal function to a confidence level of $p = 0.95$. The least-squares method, applied to the points belonging to a specific mass-transfer mechanism, showed that the growth kinetics obeyed the "$t^{1/3}$-law" with a good accuracy. Diffusive decomposition in nonmetallic systems is illustrated by the histograms (Figure 4.12) for silver

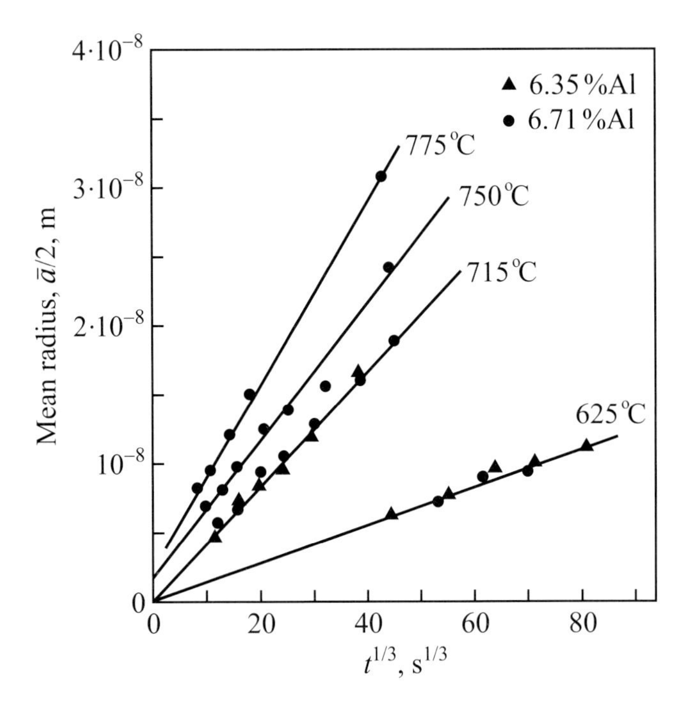

Figure 4.9: Growth kinetics for γ'-precipitations in a Ni–Al system [9].

particles growing in KCl crystals at 700 °C. Comparison of the histograms with the theoretical function for mass transfer by diffusion along the dislocation lines showed good agreement between theory and experiment. The growth of metal precipitates in ionic crystals via migration of the F-centers along the dislocation lines was confirmed by direct observation [48].

Finally, we discuss the results of a numerical computer simulation of the kinetics of the phase segregation in two-component high-temperature heat-hardened alloys. The analysis of these results [201] based on the above concepts of coarsening [153, 155] showed that the diffusive interaction of large and small clusters (larger particles devoring the smaller ones) attributable to the different solubility of differently sized particles was the primary growth mechanism at the later stage of the decomposition, when the overall number of particles tended to decrease. The distribution of large clusters at the later stage was in good agreement with the predicted universal function and the kinetics was in good agreement with the "$t^{1/3}$-law." The experimental data fitted the theory very well both qualitatively and quantitatively.

4.3.7 Conclusions

The above analysis shows that a consistent, comprehensive theory describing coarsening at the later stage of diffusional decomposition has been developed. Diffusive decomposition is of primary importance in creating new materials with specified properties. The problem of the diffusive decomposition of supersaturated multicomponent solid solutions has been formulated rigorously. A method for solving the basic equations defining the decomposition process has been developed. The resulting equations and the method of their solution enable us

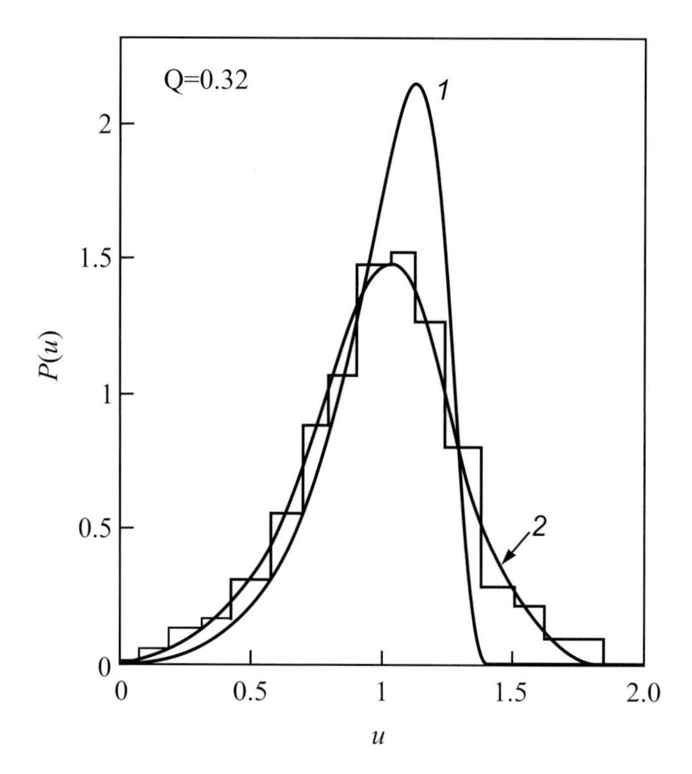

Figure 4.10: Distribution functions, calculated (i) according to Eq. (4.50) and (ii) taking into account the collision integral, in comparison with the histogram for Ni_3Al-precipitations in Ni–22at%Co–13at%Al-alloy [59] (precipitation volume fraction $Q = 0.32$).

to formulate and solve various problems concerning diffusion kinetics in solids. This method is general enough to provide a mathematical description of a variety of physical phenomena such as sintering, swelling, degradation of semiconductors, etc.

The nonlinear kinetics have been shown to determine the peculiar behavior of dispersed systems at a later stage of decomposition, when the system forgets its initial state and enters an asymptotic state, which is stable in the appropriate variables and depends only on the mass-transfer mechanism operating in the system. Physically, the interfacial tension gives rise to a universal (in the appropriate variables) size distribution function, which is independent of the initial distribution, and which is the same for the precipitates of all the existing phases. The phases which asymptotically survive in the process of competitive growth are defined by the laws of mass action and the conservation laws. The boundaries of the coexistence regions of the precipitates of both stoichiometric and nonstoichiometric phases have been determined, thus enabling us to construct phase diagrams of decomposition. The evolution of the mean size of the macrodefects in a dispersed system, at a later stage of the decomposition, obeys the $\overline{R} \sim t^{1/3}$ law for diffusion-limited growth. In general, it depends on the specific mass-transfer mechanism. The kinetics governed by the mass transfer through volume diffusion, which obeys the "$t^{1/3}$-law," is asymptotically stable at the latest stage of decomposition.

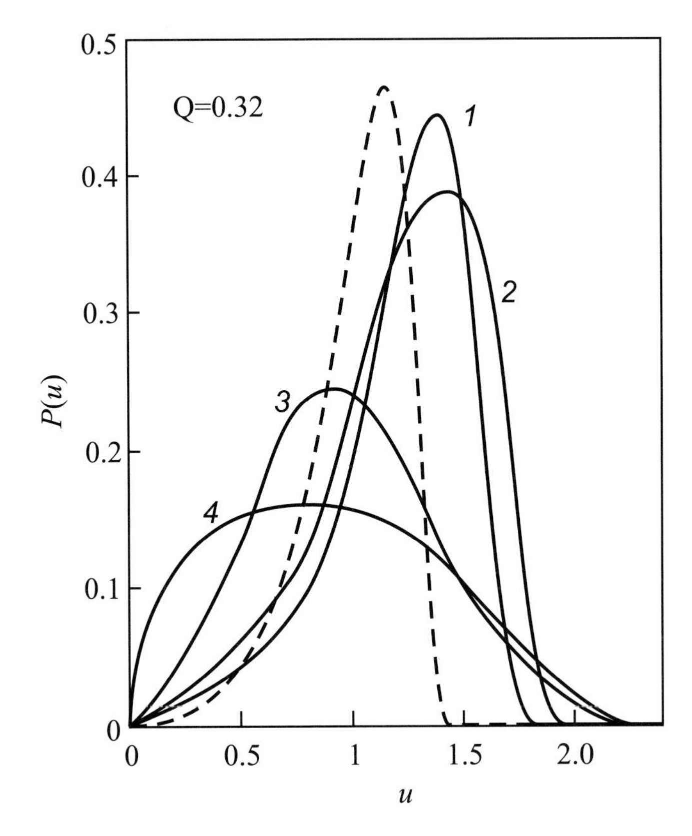

Figure 4.11: Histograms for cementite particles in Fe–0.79at%C steel, spheroidized at 704 °C. The solid curve numbers correspond to different annealing times $t_1 > t_2 > t_3 > t_4$. The dashed curve corresponds to the universal function (Eq. (4.50)).

Numerous experimental studies, on various materials, strongly suggest that the theory describing the late stage of diffusive decomposition fits the experimental data both qualitatively and quantitatively. In the overwhelming majority of cases the growth kinetics follow the theoretical $t^{1/3}$-law, where the growth is governed by volume diffusion, and the distribution function approaches a universal shape. The experimentally observed histograms are broader than the theoretical asymptotic distribution function, a phenomenon which is adequately predicted by the theory when collisions are taken into account and the initial distribution transforms into the universal distribution function.

Investigations of the decomposition kinetics make it possible to determine experimentally the characteristics of multicomponent systems which are important but difficult to measure, such as the partial diffusion coefficients for the components of the precipitates and the coefficients of specific interface energy. The theory allows one to develop an innovative approach to the evaluation of the constants of chemical reactions that may take place inside the solids. This knowledge is of theoretical and practical importance for optimizing the properties of existing materials.

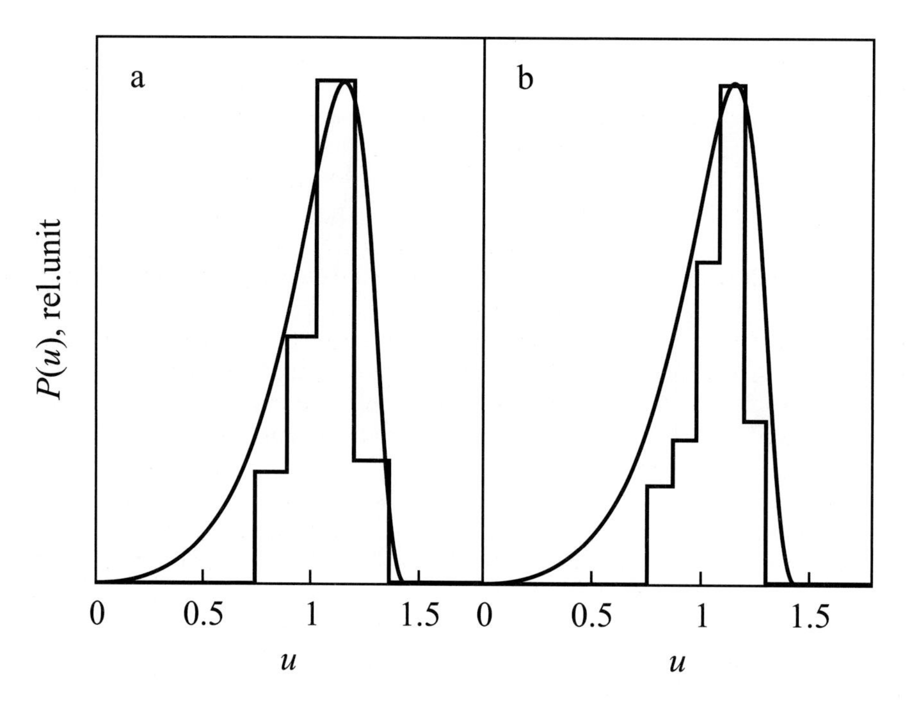

Figure 4.12: Histograms for silver particles, growing in KCl crystal [106] for different decomposition times: (a) for $2R = 330$ Å, (b) $2R = 460$ Å. The universal distribution function for mass transfer by grain boundary diffusion is also shown (see Tables 4.1 and 4.2, $n = 4$).

The theory developed above forms the basis for a quantitative prediction of the evolution of complex multicomponent systems, and hence of their structure-dependent properties associated with the size distribution of macrodefects in the process of diffusive decomposition. The criteria for the stability of dispersed systems, based on this theory, allow us to make practical recommendations concerning the development of advanced multicomponent materials, with improved properties and enhanced size and phase thermal stability. According to the theory, the precipitates grow slowly when the values of specific interface energy σ, solubility c_∞, and the diffusion coefficients D_i, etc., are small. These criteria apply in fact to various heat-resistant materials. Extremely small values of $\sigma \approx 0.2\ \mathrm{J/m^2}$ in nimonic-like alloys, i.e., alloys strengthened by precipitations from the γ'-phase, suggest high stability of these materials at high temperatures.

The principle of low solubility of the dispersed phase was first implemented in the production of tungsten (used in incandescent filaments) and TD–nickel strengthened with thorium oxide [102]. A similar principle was used to create a heat-resistant alloy based on the Fe–Al system, as well as other heat-resistant materials [111] with extremely low solubility of the dispersed phase. The addition of active carbide-producing elements, resulting in the formation of carbide precipitates with low solubility and small diffusive mobility, is a prerequisite for the production of high-quality steels [37, 172]. The same principles have successfully been employed for developing advanced lightweight alloys with enhanced mechanical properties [52].

The problem of diffusive decomposition involves a number of problems that have yet to be resolved. The issues associated with interaction between neighboring macrodefects during decomposition have not been extensively worked out. Recent experimental studies [157] showed that some macrodefects can initiate the growth of others, e.g., in the precipitate–pore systems. Computer simulations of various aspects of decomposition are promising. These activities include the formulation of a model; the construction of phase diagrams of the decomposition of multicomponent systems; the evaluation of the parameters of dispersed systems (D_i, K^s_∞, σ^s), and end with accurate processing of experimental data using statistical methods. The modeling of the stepwise heat treatment techniques, which underly the production technology of numerous modern materials, is a challenging theoretical problem. Some of the advantages of numerical modeling are illustrated in the subsequent chapter (Chapter 5).

The role played by the elastic stresses in the diffusive decomposition process is generally more complicated. The effect of these stresses on the equilibrium geometry of the precipitates may prove important, especially in the presence of anisotropy. For example, cubic-shaped, platelike and acicular precipitates may transform into differently oriented spheroidal or disk-shaped precipitates, as a result of structural transformations during the decomposition. This effect should lead to a considerable modification of the diffusion fields which can affect the characteristics of the decomposition. Two studies, [76, 349], revealed an extremely stable microscopic structure in the Fe–Al alloys which contain fine-grained, coherent, Fe_3Al precipitates with position ordering. These are found even after a long-lasting annealing, such as in the Fe–15at%Al alloy, where the size of the precipitates is smaller than 100 Å while their density is of the order of 10^{23} m^{-3}. Several models based on stress analysis have been proposed to account for such a high stability. This interesting phenomenon will be studied in more detail in Chapter 6.

The morphology, mechanisms, and kinetics of precipitate growth, especially at a later, slower stage [38], require further theoretical and experimental investigation. This is of particular importance because this phenomenon, which is common to many other systems [38], can be useful in enhancing the thermal stability of precipitation-strengthened alloys operating at high temperatures. The effects of exposure to various external factors (hardening, prestraining, high pressures, ultrasound, stepwise heat treatment), which can largely affect the kinetics of the precipitate growth, have not been analyzed thoroughly. A detailed analysis of diffusive decomposition in ionic crystals and heavily disordered materials is another challenge facing the theory. Most of these problems can be attacked within the framework of the above-developed theory.

5 Shapes of Cluster-Size Distributions Evolving in Nucleation and Growth Processes

5.1 Introduction

In the preceding chapters (see also [298, 300, 302, 307]), the basic characteristics of first-order phase transitions, proceeding via the mechanism of nucleation and growth, were established by analytic methods. They include, in particular, the time evolution of the flux and the cluster-size distribution function, characteristic time scales of the different stages of the nucleation–growth process, and estimates of the number of clusters formed in the initial stages of this process and their average sizes. One of the aims of the present chapter consists in the illustration of the analytical results based on the numerical solution of the set of the kinetic equations describing nucleation–growth processes.

In addition to the mentioned characteristics, which have been determined analytically, for a number of applications the evolution of the cluster-size distribution function in the whole course of the nucleation–growth process and its dependence on different constraints has to be known (cf. [22, 114, 178, 350]). An analysis of the literature shows, however, that with respect to possible shapes of cluster-size distributions, at part rather arbitrary, at part even wrong assumptions are made. Some of them will be analyzed in detail in the further discussion. In particular, the knowledge of possible shapes of cluster or bubble-size distributions should be helpful in experimental investigations of the kinetics of phase formation processes (cf., e.g., [100, 343]).

By both mentioned reasons, a detailed analysis of possible shapes of cluster-size distribution functions is believed to be of considerable interest. In the present chapter we concentrate hereby the attention on the following topics:

(i) limits of applicability of statistical cluster-size distributions for an interpretation of experimental results;

(ii) comparison of the steady-state distribution of classical nucleation theory with the distributions evolving in nucleation–growth processes if conservation of the total number of particles is taken into account;

(iii) formulation of conditions for the evolution of monodisperse cluster-size distributions.

Analytical derivations are supplemented and illustrated by the results of numerical calculations.

In order to proceed, the basic kinetic equations describing nucleation–growth process are formulated and specified for different growth mechanisms. It is important to note that the

Kinetics of First-order Phase Transitions. Vitaly V. Slezov

ISBN: 978-3-527-40775-0

method of determination of the kinetic coefficients employed here does not involve the application of the so-called equilibrium distributions of classical nucleation theory, respectively, their modifications and a reference to the principle of detailed balancing [306].

As a model system, precipitation processes in liquid and solid solutions are considered. However, the results are widely independent of specific features of the model system used.

5.2 Analysis of Statistical Approaches: "Equilibrium Distribution" of Classical Nucleation Theory, Fisher's Droplet, and Similar Models

In classical nucleation theory and a variety of its modifications and extensions [28, 29, 71, 93, 306, 358], it is assumed that the evolution of the newly evolving phase proceeds via the formation and growth of clusters in the ambient phase. It is supposed in a commonly good approximation that clusters grow and shrink by aggregation and emission of single particles, only. In particular, such assumptions hold for precipitation processes in solid solutions, where the mobility of dimers, trimers, etc., is much lower as compared with single particles. It has been shown, however, that even for droplet formation in vapor the process is determined in most cases by the considered type of reactions [196]. The following considerations are limited therefore to such kind of reactions, coagulation processes [338, 339] are excluded thus from the analysis.

As was already mentioned in Chapter 2, for the determination of the emission rates of single particles from the clusters and the steady-state nucleation rate, in general, a somewhat artificial model is used introduced originally by L. Szilard (see Figure 2.2). It is assumed that, once a cluster reaches an upper limiting size $n \geq g \gg n_c$, it is instantaneously removed from the system. Moreover, according to Szilard's model, simultaneously to the removal of a g-sized cluster, g single particles are added to the system. In this way, the total number of particles is kept constant.

Starting with a state consisting of single particles only, after some time interval (denoted commonly as time lag in nucleation) a time-independent steady-state cluster-size distribution is established in the system. Assuming that (i) clusters of different sizes can be considered as different components in a multicomponent perfect solution (or a mixture of perfect gases for vapor condensation), (ii) the number of particles aggregated in the clusters is small compared with the total number of solute particles, (iii) conservation of the total number of solute particles is fulfilled, and (iv) the change of the cluster size is possible by emission or aggregation of monomers, only, Frenkel [71] obtained an expression for the stationary cluster-size distribution function $f^{(e)}(n)$ of the form

$$f^{(e)}(n) = f(1) \exp\left\{-\frac{\Delta G(n)}{k_B T}\right\} . \tag{5.1}$$

Here k_B is the Boltzmann constant and T is the absolute temperature. $\Delta G(n)$ denotes the so-called work of formation of a cluster consisting of n monomers. It equals the change of the Gibbs free energy when in a homogeneous initial state at constant values of the external pressure p and temperature T one cluster of size n is formed. $f(n)$ denotes the number of clusters per unit volume consisting of n single particles (atoms, molecules). Going over later on to a continuous description, $f(n)\,dn$ gives the number of clusters in the range of cluster sizes $n, n + dn$. $f(1)$ is the concentration (number per unit volume) of single particles which will be denoted also as c (cf. also [306]).

Expression (5.1) was obtained by Frenkel by minimizing the Gibbs free energy of the heterogeneous system consisting of clusters of different sizes in the otherwise homogeneous ambient phase. Hereby the boundary conditions as formulated above were taken into account. The distribution (5.1) is denoted commonly as equilibrium or constraint equilibrium distribution with respect to cluster sizes [135, 352, 353]. Note, however, that this notation is misleading. The time-independent state in the model system, it was derived for, is not an equilibrium but a nonequilibrium steady state. Therefore, the procedure applied in the derivation of Eq. (5.1) lacks any thermodynamic foundation.

Moreover, one has to take into account that the distribution refers to Szilard's artificial model system which is not realized in nature (except for artificial conditions or by assuming some kind of "Szilard's demon" in analogy with Maxwell's demon (cf., e.g., [117–121])). Therefore, the often found identification of the so-called equilibrium distribution with respect to cluster sizes with real distributions evolving in nucleation–growth processes in thermodynamically unstable systems is, in general, incorrect. However, for thermodynamically stable initial states the distribution can be applied for the description of cluster-size distributions in real systems. In equilibrium states, for any value of n the inequality $\Delta G(n) > 0$ holds. Equation (5.1) represents in such cases Boltzmann-type heterophase fluctuations evolving in thermodynamic equilibrium states (cf. [144]).

In application of classical theory, most of its extensions and generalizations to precipitation processes in solid and liquid solutions, the work of cluster formation is expressed as

$$\Delta G(n) = -n\Delta\mu + \sigma A, \qquad \Delta\mu = \mu(p, T, c) - \mu_\alpha(p, T). \tag{5.2}$$

Here $\Delta\mu$ is the difference of the chemical potential per particle in the ambient and the newly evolving phases (specified by a subscript α) at a pressure p and a temperature T, σ is the specific interfacial energy (or surface tension) and A is the surface area of the cluster. If the surface area is expressed via the number of particles in the cluster according to

$$V_\alpha = \frac{4\pi}{3}R^3 = \omega_s n, \quad A = 4\pi R^2 = 4\pi\left(\frac{3}{4\pi}\omega_s\right)^{2/3} n^{2/3}, \quad \omega_s = \frac{4\pi}{3}a_s^3, \tag{5.3}$$

we may write equivalently

$$\Delta G(n) = -n\Delta\mu + \alpha_2 n^{2/3}, \qquad \alpha_2 = 4\pi\sigma\left(\frac{3\omega_s}{4\pi}\right)^{2/3}. \tag{5.4}$$

Here V_α is the volume and R is the radius of a (spherical) cluster; ω_s denotes the volume per particle and a_s a characteristic size parameter of the particles in the newly evolving phase.

The critical cluster size in nucleation corresponds to a maximum of ΔG; it is determined by

$$n_c^{1/3} = \frac{2\alpha_2}{3\Delta\mu}, \qquad \Delta G_{(c)} = \frac{1}{3}\alpha_2 n_c^{2/3}. \tag{5.5}$$

For thermodynamically unstable initial states, ΔG may be written as

$$\frac{\Delta G}{\Delta G_{(c)}} = 3\left(\frac{n}{n_c}\right)^{2/3} - 2\left(\frac{n}{n_c}\right) . \tag{5.6}$$

The so-called equilibrium distribution function $f^{(e)}(n)$ (cf. Eq. (5.1)) gets in this case the form

$$\left(\frac{f^{(e)}(n)}{f(1)}\right) = \exp\left\{-\frac{\Delta G_{(c)}}{k_B T}\left[3\left(\frac{n}{n_c}\right)^{2/3} - 2\left(\frac{n}{n_c}\right)\right]\right\} . \tag{5.7}$$

It is qualitatively presented in Figure 2.3. The function has a minimum for $n = n_c$ and diverges for large values of (n/n_c).

Based on the methods of equilibrium statistical physics, Fisher [69] developed an alternative derivation of dependences quite similar to Eq. (5.1). Fisher did not employ the second of mentioned Frenkel's assumptions. As a consequence, the expression for the preexponential factor remains undefined in his approach. Fisher introduced, in addition, a term $(k_B T\tau \ln n)$ into the expression for the work of cluster formation resulting thus in

$$\Delta G(n) = -n\Delta\mu + \alpha_2 n^{2/3} + k_B T\tau \ln n . \tag{5.8}$$

The actual value of τ depends on the specific properties of the substance considered. It can vary, according to Fisher's approach, in the range $2.0 < \tau < 2.5$ [69, 125]. In the vicinity of the liquid–gas critical point T_c, the differences between the liquid and the gas vanish, the relations $\Delta\mu \to 0$ and $\alpha_2 \to 0$ hold, and Fisher's model yields

$$\Delta G(n)|_{T\to T_c} = k_B T\tau \ln n , \qquad f_F^{(e)}(n)\Big|_{T\to T_c} \propto n^{-\tau} . \tag{5.9}$$

The occurrence of such dependences in the vicinity of the liquid–gas or percolation critical points was reconfirmed by alternative approaches [31, 61, 74, 199, 319] giving in this way support to Fisher's proposal. Moreover, for thermodynamic equilibrium states Fisher's model is equivalent, again, to Boltzmann-type heterophase fluctuations with a somewhat modified expression for the work of cluster formation as compared with the classical result. However, in thermodynamically unstable states (i.e., below and at a sufficiently large distance from T_c) with respect to Fisher's model the same conclusions of inconsistency have to be drawn as done in the analysis of the so-called equilibrium distribution of classical nucleation theory. Namely, Fisher's model is inappropriate to describe real cluster-size distributions evolving in thermodynamically unstable initial states beyond the liquid–gas critical point.

This conclusion follows from the method applied in the derivation which is based on equilibrium statistical physics. Moreover, it is also evident for physical reasons. For thermodynamically unstable states beyond the critical point, Fisher's model results, again, in qualitatively similar dependences as shown in Figure 2.3. However, the only (time-independent) equilibrium distribution evolving in the course of time consists of one large cluster (the newly evolving bulk phase in the ambient phase) surrounded eventually by a distribution of small clusters (monomers, dimers, etc.).

The same conclusions can be drawn with respect to any other similar expressions resulting from different approaches in the determination of the work of cluster formation, ΔG (cf. [135]). Such distributions may be of use in order to determine the emission coefficients from the expressions for the coefficients of aggregation by applying the principle of detailed balancing to an artificial model state (with all the problems involved in such a procedure (cf. Chapter 2 and [243, 293, 302, 306])). However, the application of these expressions to the description of real cluster-size distributions formed in nucleation–growth processes is, in general, incorrect.

5.3 Thermodynamic Approach: On the Possibility of Evolution of Monodisperse Cluster-Size Distributions

If we consider the process of precipitation in solutions at constant values of the external pressure, p, and temperature, T, the characteristic thermodynamic potential is the Gibbs free energy, G. Provided a spontaneous evolution of an ensemble of clusters into a monodisperse cluster-size distribution would be possible, such a state should correspond then to a minimum of G for a given number, N, of clusters in the system (see, e.g., [133]). To check whether such minimum exists we may assume from the very beginning that all clusters are of the same size. We have to determine then only the value of the size of the clusters at which the Gibbs free energy reaches eventually a minimum.

If N clusters of the same size are formed in the system, the change of the Gibbs free energy, due to the formation of such a monodisperse distribution, may be expressed as [220, 248, 331]

$$\Delta G = (p - p_\alpha)NV_\alpha + \sum_{j=1}^{k} \left[\mu_{j\alpha}(p_\alpha, T, \{x_\alpha\}) - \mu_j(p, T, \{x\})\right] N n_{j\alpha}$$

$$+ \sum_{j=1}^{k} n_{j0} \left[\mu_j(p, T, \{x\}) - \mu_{j0}(p, T, \{x_0\})\right] + N\sigma A. \tag{5.10}$$

Here the general case is considered that the clusters of the newly evolving phase may be composed of particles of all of the different $j = 1, 2, \ldots, k$ components present in the ambient phase.

$\mu_{j\alpha}$ and μ_{j0} denote the chemical potentials of the k different cluster components in the homogeneous initial state and $n_{j\alpha}$ is the number of particles of the component j ($j = 1, 2, \ldots, k$) in a cluster while n_{j0} refers to the number of particles of component j in the homogeneous initial state; μ_j denotes the actual value of the chemical potential in the ambient phase; $\{x_\alpha\}$ is the set of independent molar fractions in the cluster phase, while $\{x\}$ and $\{x_0\}$ refer to the actual and initial values of these quantities in the ambient phase. The second sum on the right-hand side of Eq. (5.10) reflects changes of the state of the system resulting from processes of cluster formation and growth. Such changes are an essential prerequisite for the evolution of monodisperse distributions. If such variations do not occur, no thermodynamic factors exist leading to an inhibition of the growth process.

The necessary equilibrium conditions may be obtained from Eq. (5.10) in the form

$$\mathrm{d}\Delta G = N\left\{\left(p - p_\alpha + \sigma\frac{\partial A}{\partial V_\alpha}\right)\mathrm{d}V_\alpha + \sum_j \left(\mu_{j\alpha} - \mu_j\right)\mathrm{d}n_{j\alpha}\right\} = 0. \tag{5.11}$$

Hereby it is assumed that the number of clusters, N, is kept constant. Since $n_{j\alpha}$ and V_α have to be considered as independent variables, the necessary equilibrium conditions for the considered monodisperse distribution are given by

$$p - p_\alpha + \sigma\frac{\partial A}{\partial V_\alpha} = 0 \qquad \text{or} \qquad p_\alpha - p = \frac{2\sigma}{R}, \tag{5.12}$$

$$\mu_{j\alpha}\left(p_\alpha, T, \{x_\alpha\}\right) = \mu_j\left(p, T, \{x\}\right), \qquad j = 1, 2, \ldots, k. \tag{5.13}$$

However, if more than one cluster is present in the system ($N > 1$), the considered state is an unstable equilibrium state of saddle-point type [220, 225, 229, 331]. Small fluctuations will always lead to processes destroying such monodisperse states. The number of clusters decreases in time while their average size increases. This way, a process takes place in the system denoted commonly as coarsening or Ostwald ripening (see Chapter 4 and [155, 230, 289]). There exists one and only one equilibrium state consisting of one large cluster in the otherwise homogeneous ambient phase (cf. also [32, 230]).

To verify this statement, we write down the Gibbs free energy as a function of all independent variables. These are the volume, V_α, the numbers of particles, $n_{j\alpha}$, of the different components in a cluster, and the total number of clusters, N. The values of the parameters, except N, are given by Eqs. (5.12) and (5.13). They depend on the total number of clusters, N, present in the system (via the molar fractions $\{x\}$ of different components in the ambient phase). For the states, obeying the necessary thermodynamic equilibrium conditions, we thus have

$$\Delta G = \Delta G\left[V_\alpha(N), n_{1\alpha}(N), \ldots, n_{k\alpha}(N), N\right]. \tag{5.14}$$

By taking the derivative of Eq. (5.14) with respect to N we obtain

$$\left.\frac{\mathrm{d}\Delta G}{\mathrm{d}N}\right|_{(\mathrm{eq})} = \left.\frac{\partial\Delta G}{\partial V_\alpha}\right|_{(\mathrm{eq})}\frac{\mathrm{d}V_\alpha}{\mathrm{d}N} + \sum_{j=1}^{k}\left.\frac{\partial\Delta G}{\partial n_{j\alpha}}\right|_{(\mathrm{eq})}\frac{\mathrm{d}n_{j\alpha}}{\mathrm{d}N} + \left.\frac{\partial\Delta G}{\partial N}\right|_{(\mathrm{eq})}. \tag{5.15}$$

Once the necessary thermodynamic equilibrium conditions are fulfilled, the partial derivatives

$$\left.\frac{\partial\Delta G}{\partial V_\alpha}\right|_{(\mathrm{eq})} = \left.\frac{\partial\Delta G}{\partial n_{j\alpha}}\right|_{(\mathrm{eq})} = 0 \tag{5.16}$$

are equal to zero (cf. Eq. (5.11)). Taking the partial derivative of Eq. (5.10) with respect to N, the subsequent substitution of the necessary equilibrium conditions (5.12) and (5.13) yields

$$\left.\frac{\mathrm{d}\Delta G}{\mathrm{d}N}\right|_{(\mathrm{eq})} = \frac{1}{3}\sigma A. \tag{5.17}$$

Comparing different monodisperse cluster-size distributions obeying the necessary thermodynamic equilibrium conditions, the value of the thermodynamic potential decreases with a decreasing number of clusters in the system. Consequently, processes of evolution will occur in the system resulting in the decrease of the number of clusters until only one cluster remains in the system (cf., e.g., [32,230]). This conclusion may be verified in a more general approach also by carrying out a stability analysis of the considered state of the system (cf. [225]).

This stage of evolution, characterized by a decrease of the number of clusters in the system and an increase of their average size, is found quite generally in very different systems. The thermodynamic background is always the same; it consists in the decrease of the surface contributions to the characteristic thermodynamic potential at nearly constant bulk terms. It is the basic origin of coarsening discussed in detail in Chapter 4. Of course, the situation may occur that, though processes of coarsening are thermodynamically favorable, they may not take place due to the inhibition of the kinetics of possible processes of evolution to the more stable states (kinetic stabilization). In particular, this may be the case if the surface tension (or the specific interfacial energy) has relatively low values. Such effects may be of importance in the vicinity of the critical point (see also the analysis performed in subsequent sections). Such possibility is excluded here so far from the consideration.

Reiss and Kegel [209] and Gross [92] (in another context) suggested recently that mixing contributions may also act as a stabilizing factor. However, the determination of the mixing contributions is, in general, a highly complicated problem. For this reason, its possible effect on the shape of the cluster-size distribution function is not considered here in detail. With respect to precipitation in solid and liquid solutions as well as condensation of gases [242], such contributions do not change, in general, the conclusions as outlined above.

As already mentioned, from a thermodynamic point of view the behavior as analyzed above is connected with a minimization of the surface contributions to the Gibbs free energy for an approximately constant value of the amount of the newly evolving bulk phase. Deviations from such behavior occur only when additional factors exist inhibiting the growth. Such factors led to additional terms in the work of cluster formation increasing more rapidly than linear with the volume of a cluster [220, 229, 331]. Indeed, let us assume that the matrix–cluster interaction (e.g., elastic or electric fields) leads to additional terms $N\Delta G^{(\varepsilon)}(V_\alpha)$ in ΔG. The modified expression for ΔG, denoted as $\Delta G^{(\mathrm{mod})}$, then gets the form

$$\Delta G^{(\mathrm{mod})} = \Delta G + N\Delta G^{(\varepsilon)}(V_\alpha), \tag{5.18}$$

where ΔG is given by Eq. (5.10), again. The modified necessary equilibrium conditions read, now,

$$p - p_\alpha + \sigma\frac{\partial A}{\partial V_\alpha} + \frac{\partial \Delta G^{(\varepsilon)}(V_\alpha)}{\partial V_\alpha} = 0, \tag{5.19}$$

$$\mu_{j\alpha}(p_\alpha, T, \{x_\alpha\}) = \mu_j(p, T, \{x\}), \qquad j = 1, 2, \ldots, k. \tag{5.20}$$

Moreover, instead of Eq. (5.17) we obtain

$$\left.\frac{\mathrm{d}\Delta G^{(\mathrm{mod})}}{\mathrm{d}N}\right|_{(\mathrm{eq})} = \frac{1}{3}\sigma A + \left[\Delta G^{(\varepsilon)}(V_\alpha) - V_\alpha\frac{\partial \Delta G^{(\varepsilon)}(V_\alpha)}{\partial V_\alpha}\right]. \tag{5.21}$$

Provided $\Delta G^{(\varepsilon)}(V_\alpha)$ behaves as $\Delta G^{(\varepsilon)}(V_\alpha) = \tilde{\kappa} V_\alpha^\beta$, then a minimum of $\Delta G^{(\mathrm{mod})}$ exists for finite values of N but only when $\beta > 1$ holds.

Ideas of this kind have been successfully applied to the theoretical interpretation of the kinetics of coarsening in highly viscous glass-forming melts in the vicinity of the temperature of vitrification T_g [94, 229]. In such cases, elastic stresses may lead indeed to such additional terms as discussed above. A particular example in this respect will be analyzed in short; a more detailed analysis is given in Chapter 6.

The mechanism of stabilization as discussed here is connected with cluster–matrix interactions. There exist additional mechanisms connected with cluster–cluster interactions as analyzed, e.g., for the case of elastic strains in solid solutions by Kawasaki and Enomoto [122]. Such possibility is also not considered here. Note that in the analysis no restriction was made concerning the properties of the cluster phase. The results are valid thus both for liquid and solid clusters as well as for bubbles (cf. also [227, 228]).

5.4 Dynamical Approach

5.4.1 Basic Kinetic Equations: General Expression

In the considered, now, kinetic approach, the state of the system is characterized by a distribution function with respect to cluster sizes $f(n,t)$. The distribution function $f(n,t)$ represents the number (or number density in a continuous description) of clusters of the newly evolving phase per unit volume containing n monomeric building units. These building units may consist of atoms, molecules, or even complex aggregates with a definite stoichiometric composition (cf. [243, 293]).

As mentioned, the growth, respectively, decay of the clusters proceeds via emission or aggregation of single particles. These processes are considered as independent. The probability of the respective elementary process to proceed in a time interval Δt may be written, therefore, as $w\Delta t$. The respective probabilities that at such a time interval two elementary processes of the same type occur are given thus by $(w\Delta t)^2$. Going over to the limit $\Delta t \to 0$, such higher order terms may be neglected. With above assumptions, the change of the distribution function $f(n,t)$ is governed by the following set of kinetic equations (see also Section 2.2):

$$\frac{\partial f(n,t)}{\partial t} = w^{(+)}_{n-1,n} f(n-1,t) + w^{(-)}_{n+1,n} f(n+1,t) - w^{(+)}_{n,n+1} f(n,t) - w^{(-)}_{n,n-1} f(n,t) \,. \tag{5.22}$$

Here $w^{(+)}_{n-1,n}$ is the probability per unit time that to a cluster of size $(n-1)$ a single particle is added. The coefficients $w^{(-)}_{n,n-1}$ specify similarly emission processes of single particles. The first subscript refers always to the initial state, and the second one to the final state of the process.

The above equation can be written in a more compact form by introducing fluxes, $J(n)$, in cluster-size space. With the notations

$$J(n-1,t) = w^{(+)}_{n-1,n} f(n-1,t) - w^{(-)}_{n,n-1} f(n,t), \tag{5.23}$$

$$J(n,t) = w^{(+)}_{n,n+1} f(n,t) - w^{(-)}_{n+1,n} f(n+1,t), \tag{5.24}$$

we have

$$\frac{\partial f(n,t)}{\partial t} = -\{J(n,t) - J(n-1,t)\}. \tag{5.25}$$

As shown in Chapter 2, this general scheme can be extended easily to cases when more than one parameter is required for an appropriate description of cluster formation. However, in the present analysis we will restrict ourselves in the kinetic description to cases when the state of the cluster is characterized by only one parameter n (for generalizations see, e.g., [28, 29, 296, 306]).

5.4.2 Determination of the Coefficients of Emission

For an application of the general equations as outlined in the preceding section, the kinetic coefficients $w^{(+)}$ and $w^{(-)}$ have to be determined. As a first step, we express the emission coefficients $w^{(-)}$ through the coefficients of aggregation $w^{(+)}$.

The required relation between the kinetic coefficients can be found without reference to the so-called equilibrium distributions with respect to cluster sizes and Szilard's model. As shown in Chapter 2 (see also [293, 306]), quite generally the following relation is fulfilled:

$$\frac{w^{(+)}_{n-1,n}}{w^{(-)}_{n,n-1}} = \exp\left\{\frac{\mu(p,T,c) - \mu^{(n)}_\alpha(p_\alpha,T)}{k_B T}\right\}. \tag{5.26}$$

Here $\mu(p,T,c)$ is the chemical potential of the segregating particles in the solution (at given values of pressure, p, temperature, T, and concentration, c) while $\mu^{(n)}_\alpha$ refers to the respective values in the cluster including interfacial and eventually other possible finite size effects (for the thermodynamic parameters p_α and T specifying the state of the bulk cluster phase).

Equation (5.26) is also valid for nucleation in multicomponent systems, nonisothermal nucleation etc. In limiting cases, like nucleation in one-component systems at given values of pressure and temperature, this equation may be rewritten in an alternative way. For such purposes, we introduce the difference of the Gibbs free energy $\Delta G(n)$, again. As already mentioned, $\Delta G(n)$ is equal to the change of the Gibbs free energy, when, at constant values of external pressure and temperature, a cluster consisting of n single particles is formed. We have

$$\Delta G(n) = G^{(\text{cluster})}(n) - n\mu(p,T,c), \tag{5.27}$$

where $G^{(\text{cluster})}$ is the contribution of the cluster to the thermodynamic potential of the system including interfacial and other possible additional terms. With this equation, we may write

$$\Delta G(n) - \Delta G(n-1) = G^{(\text{cluster})}(n) - G^{(\text{cluster})}(n-1) - \mu(p,T,c). \tag{5.28}$$

Moreover, by a Taylor expansion of $G^{(\text{cluster})}(n-1)$ we get

$$\begin{aligned} G^{(\text{cluster})}(n-1) &= G^{(\text{cluster})}(n) - \frac{\partial G^{(\text{cluster})}}{\partial n} \\ &= G^{(\text{cluster})}(n) - \mu^{(n)}_\alpha(p_\alpha,T). \end{aligned} \tag{5.29}$$

Here the definition of the chemical potential as $\mu = (\partial G/\partial n)$ was taken into account.

A substitution into Eq. (5.28) yields

$$\Delta G(n) - \Delta G(n-1) = \mu_\alpha^{(n)}(p_\alpha, T) - \mu(p, T, c), \tag{5.30}$$

resulting in (cf. Eq. (5.26))

$$\frac{w_{n-1,n}^{(+)}}{w_{n,n-1}^{(-)}} = \exp\left\{-\frac{[\Delta G(n) - \Delta G(n-1)]}{k_B T}\right\}. \tag{5.31}$$

Note that this derivation is valid generally independent of any particular proposals with respect to the detailed form of the expression for the work of cluster formation.

Moreover, by introducing an auxiliary function $f^{(*)}(n)$ as

$$f^{(*)}(n) = \exp\left\{-\frac{\Delta G(n)}{k_B T}\right\} \tag{5.32}$$

we may rewrite Eq. (5.31) in the form

$$\frac{w_{n-1,n}^{(+)}}{w_{n,n-1}^{(-)}} = \frac{f^{(*)}(n)}{f^{(*)}(n-1)}. \tag{5.33}$$

In the approach outlined, $f^*(n)$ does not have, in general, the meaning of a distribution function but is, as noted, some auxiliary mathematical quantity. In this way, in our approach the relation between the coefficients of aggregation and emission of single particles may be expressed by functions of the form as given by Eq. (5.1), again, but without assigning the meaning of a cluster-size distribution to them. This way, as an additional advantage the problem of determination of the preexponential factor in the so-called equilibrium distributions does not occur in our method (cf. Chapter 2 and [306, 352, 353]).

In true thermodynamic equilibrium states, detailed balancing holds and the conditions $J(n) = 0$ have to be fulfilled. With Eqs. (5.24) and (5.31) we then have

$$f^{(\mathrm{eq})}(n) \exp\left\{\frac{\Delta G(n)}{k_B T}\right\} = f^{(\mathrm{eq})}(n+1) \exp\left\{\frac{\Delta G(n+1)}{k_B T}\right\}, \tag{5.34}$$

where $f^{(\mathrm{eq})}(n)$ is a real cluster-size distribution evolving in the course of time in thermodynamic equilibrium states. Since Eqs. (5.34) have to be fulfilled for any value of n, the relation

$$f^{(\mathrm{eq})}(n) = A \exp\left\{-\frac{\Delta G(n)}{k_B T}\right\} \tag{5.35}$$

is obtained as the solution. For thermodynamic equilibrium states, the set of kinetic equations leads, therefore, to statistical cluster distributions or stationary cluster-size distributions of Boltzmann-type heterophase fluctuations (cf. Eq. (5.1)) as discussed in detail in Section 5.2.

5.4.3 Determination of the Coefficients of Aggregation

Considering n as a continuous variable, we get by a Taylor expansion of Eq. (5.25)

$$\frac{\partial f(n,t)}{\partial t} = -\frac{\partial J(n,t)}{\partial n}. \tag{5.36}$$

On the other hand, the expression for the flux $J(n)$ in cluster-size space, Eq. (5.24), may be rewritten as

$$\begin{aligned} J(n,t) &= w^{(+)}_{n,n+1} f^{(*)}(n) \left\{ \frac{f(n,t)}{f^{(*)}(n)} - \frac{f(n+1,t)}{f^{(*)}(n+1)} \right\} \\ &= -w^{(+)}_{n,n+1} f^{(*)}(n) \left\{ \frac{\partial}{\partial n} \left[\frac{f(n,t)}{f^{(*)}(n)} \right] \right\}. \end{aligned} \tag{5.37}$$

A substitution into Eq. (5.36) yields

$$\frac{\partial f(n,t)}{\partial t} = \frac{\partial}{\partial n} \left\{ w^{(+)}_{n,n+1} f^{(*)}(n) \left[\frac{\partial}{\partial n} \left(\frac{f(n,t)}{f^{(*)}(n)} \right) \right] \right\} \tag{5.38}$$

or

$$\frac{\partial f(n,t)}{\partial t} = \frac{\partial}{\partial n} \left\{ w^{(+)}_{n,n+1} \left[\frac{1}{k_B T} \frac{\partial \Delta G(n)}{\partial n} f(n,t) + \frac{\partial f(n,t)}{\partial n} \right] \right\}. \tag{5.39}$$

Equation (5.39) has the same structure as the relation describing the macroscopic deterministic flow in three-dimensional space as well as diffusion processes of particles characterized by a volume concentration c, i.e., [151]

$$\frac{\partial c}{\partial t} = -\nabla \left\{ [c(\mathbf{r},t)\mathbf{v}] - D \nabla c(\mathbf{r},t) \right\}. \tag{5.40}$$

Here $\mathbf{v}$ is the macroscopic (hydrodynamic) velocity of the particles while D is the diffusion coefficient, connected with the diffusive motion of the considered component in real space. It follows that the deterministic (macroscopic) velocity of motion in cluster-size space $v(n,t)$ is given by

$$v(n,t) = -w^{(+)}_{n,n+1} \left\{ \frac{1}{k_B T} \frac{\partial \Delta G(n)}{\partial n} \right\}, \tag{5.41}$$

while the diffusion coefficient in cluster-size space equals $w^{(+)}_{n,n+1}$.

5.4.4 Description of Growth Processes of Clusters

For diffusion-limited growth, the density of fluxes through the surface of a cluster with the radius R is given by (e.g., [94])

$$j_R = -D \left(\frac{\partial c}{\partial r} \right) \bigg|_{r=R} \tag{5.42}$$

or

$$j_R = -D\left(\frac{c - c_R}{R}\right) . \tag{5.43}$$

For the rate of change of the cluster radius with time one thus obtains

$$\frac{\mathrm{d}R}{\mathrm{d}t} = -\frac{1}{c_\alpha} j_R \tag{5.44}$$

or

$$\frac{\mathrm{d}R}{\mathrm{d}t} = \frac{D}{c_\alpha}\left(\frac{c - c_R}{R}\right) = \frac{Dc}{c_\alpha R}\left(1 - \frac{c_R}{c}\right) . \tag{5.45}$$

Here $c_\alpha = (1/\omega_s)$ is the concentration of the segregating particles in the newly evolving phase (cf. Eq. (5.3)).

Often it is assumed, in addition, that in the immediate vicinity of the clusters in the matrix a local equilibrium concentration is established. Thus c_R may be set equal to the equilibrium concentration $c_{\text{eq}}^{(R)}$ of the segregating particles in the vicinity of a cluster of size R. Its value is given by (cf. [94])

$$c_{\text{eq}}^{(R)} = c_{\text{eq}}^{(\infty)}\left[\exp\left(\frac{2\sigma\omega_s}{k_B T}\frac{1}{R}\right)\right] = c_{\text{eq}}^{(\infty)}\left[\exp\left(\frac{2\alpha_2}{3k_B T n^{1/3}}\right)\right] , \tag{5.46}$$

where $c_{\text{eq}}^{(\infty)}$ is the equilibrium concentration of the segregating particles in the ambient phase for an equilibrium coexistence of both phases at a planar interface.

The concentration c in the undisturbed matrix corresponds, on the other hand, to a critical cluster size R_c, determined by

$$c = c_{\text{eq}}^{(\infty)}\left[\exp\left(\frac{2\sigma\omega_s}{k_B T}\frac{1}{R_c}\right)\right] . \tag{5.47}$$

A substitution of Eqs. (5.46) and (5.47) into Eq. (5.45) and a subsequent Taylor expansion of the exponential functions yields

$$\frac{\mathrm{d}R}{\mathrm{d}t} = \frac{2\sigma Dc}{c_\alpha^2 k_B T}\left[\frac{1}{R}\left(\frac{1}{R_c} - \frac{1}{R}\right)\right] . \tag{5.48}$$

Equation (5.48) is the basic relation for the description of diffusion-limited precipitation processes. It can be generalized in order to describe other mechanisms of growth as well. The result of such an extension can be written in the general form (see [289]) as

$$\frac{\mathrm{d}R}{\mathrm{d}t} = \frac{2\sigma D_\gamma c}{c_\alpha^2 k_B T}\left[\frac{a_\gamma^{(\gamma-1)}}{R^\gamma}\left(\frac{1}{R_c} - \frac{1}{R}\right)\right] , \tag{5.49}$$

where a_γ is a length parameter reflecting specific properties of the considered growth mechanism. Different growth mechanisms are described by this equation for different values of γ

($\gamma = 0$: ballistic or interface kinetic limited growth; $\gamma = 1$: diffusion-limited growth; $\gamma = 2$: diffusion along grain boundaries; $\gamma = 3$: diffusion in a dislocation network [289]). Of course, for each particular case, different values of the diffusion coefficient D_γ have to be chosen.

In general, both processes of transport of the segregating particles to the cluster as well as the rate of incorporation may be of importance for the rate of cluster growth. Assuming bulk diffusion as the transport mechanism, the flux in the immediate vicinity of the cluster surface is given by

$$j_R = -D\left(\frac{c - c_R}{R}\right), \tag{5.50}$$

while the flux through the interface is determined by

$$j_R = -D^{(*)}\left(\frac{c_R - c_{\text{eq}}^{(R)}}{a_m}\right). \tag{5.51}$$

In Eq. (5.51) the parameter a_m is identified with an interatomic or average jump distance of the considered solution. It is defined, similar to Eq. (5.3), via the average volume ω_m occupied by a particle in the ambient phase, i.e., as

$$\omega_m = \frac{4\pi}{3}a_m^3. \tag{5.52}$$

The quantity D in Eq. (5.50) is the diffusion coefficient for bulk diffusion, while $D^{(*)}$ in Eq. (5.51) is the respective measure of the mobility for processes of incorporation of particles to the cluster.

Assuming steady-state conditions, we have

$$D\left(\frac{c - c_R}{R}\right) = D^{(*)}\left(\frac{c_R - c_{\text{eq}}^{(R)}}{a_m}\right). \tag{5.53}$$

The concentration in the immediate vicinity of the cluster is then given by

$$c_R = c\left\{\frac{1 + \left[\left(\frac{D^{(*)}}{D}\right)\left(\frac{R}{a_m}\right)\right]\left(\frac{c_{\text{eq}}^{(R)}}{c}\right)}{1 + \left[\left(\frac{D^{(*)}}{D}\right)\left(\frac{R}{a_m}\right)\right]}\right\}. \tag{5.54}$$

With Eqs. (5.44), (5.51), and (5.54) we get similarly as in the derivation of Eq. (5.48)

$$\frac{\mathrm{d}R}{\mathrm{d}t} = \frac{2\sigma D^{(*)}c}{c_\alpha^2 k_B T a_m}\left\{\frac{1}{1 + \left[\left(\frac{D^{(*)}}{D}\right)\left(\frac{R}{a_m}\right)\right]}\left(\frac{1}{R_c} - \frac{1}{R}\right)\right\}. \tag{5.55}$$

This equation contains both limiting cases of kinetic or ballistic growth (prevailing, in general, in the initial stages of the nucleation–growth process), diffusion-limited growth (determining the process at the later stages) as well as the transient behavior.

5.4.5 Application to the Description of Nucleation

Equation (5.49) can be reformulated in terms of the number of particles n in the cluster. Such a reformulation is required in order to apply Eq. (5.41) for a determination of the kinetic coefficients $w_{n,n+1}^{(+)}$. With Eqs. (5.3) and (5.5) we obtain

$$v(n,t) = \frac{\mathrm{d}n}{\mathrm{d}t} = 4\pi D_\gamma c\left(\frac{\Delta\mu}{k_B T}\right)\left[\frac{a_\gamma^{(\gamma-1)}}{R^{(\gamma-2)}}\left(1-\frac{R_c}{R}\right)\right] \tag{5.56}$$

or

$$\frac{\mathrm{d}n}{\mathrm{d}t} = 4\pi D_\gamma c\left(\frac{\Delta\mu}{k_B T}\right)\left\{\frac{a_\gamma^{(\gamma-1)}}{\left[\left(\frac{3\omega_s}{4\pi}\right)n\right]^{(\gamma-2)/3}}\left[1-\left(\frac{n_c}{n}\right)^{1/3}\right]\right\}. \tag{5.57}$$

According to Eqs. (5.4) and (5.41) we get further

$$w_{n,n+1}^{(+)} = -\frac{v(n,t)}{\frac{1}{k_B T}\frac{\partial\Delta G(n)}{\partial n}} = 4\pi D_\gamma c\left\{\frac{a_\gamma^{(\gamma-1)}}{\left[\left(\frac{3\omega_s}{4\pi}\right)n\right]^{(\gamma-2)/3}}\right\}. \tag{5.58}$$

In particular, for bulk diffusion-limited growth ($\gamma = 1$), we find

$$w_{n,n+1}^{(+)} = 4\pi D c\left(\frac{3\omega_s}{4\pi}\right)^{1/3} n^{1/3} \tag{5.59}$$

while for kinetically limited growth ($\gamma = 0$),

$$w_{n,n+1}^{(+)} = 4\pi D^{(*)} c a_m\left(\frac{\omega_s}{\omega_m}\right)^{2/3} n^{2/3} \tag{5.60}$$

holds. Proceeding in the same way with Eq. (5.55), we get

$$w_{n,n+1}^{(+)} = 4\pi D^{(*)} c\left(\frac{3\omega_s}{4\pi}\right)^{1/3} n^{1/3}\left\{\frac{\left(\frac{a_s}{a_m}\right)n^{1/3}}{1+\left[\left(\frac{D^{(*)}}{D}\right)\left(\frac{a_s}{a_m}\right)\right]n^{1/3}}\right\}. \tag{5.61}$$

In this way, the determination of the kinetic coefficients is completed.

5.4.6 Basic Kinetic Equations for Different Important Growth Mechanisms

According to Eqs. (5.22) and (5.31), we may rewrite the basic kinetic equations describing nucleation and growth as

$$\frac{\partial f(n,t)}{\partial t} = w^{(+)}_{n-1,n}\left\{f(n-1,t) - f(n,t)\exp\left[\frac{\Delta G(n) - \Delta G(n-1)}{k_B T}\right]\right\}$$
$$+\, w^{(+)}_{n,n+1}\left\{-f(n,t) + f(n+1,t)\exp\left[\frac{\Delta G(n+1) - \Delta G(n)}{k_B T}\right]\right\}. \tag{5.62}$$

For diffusion-limited growth, the coefficients of aggregation are given by Eq. (5.59). In most applications, the diffusion coefficient of the segregating particles is independent of cluster size. For example, for perfect solutions the diffusion coefficients depend mainly on pressure and temperature. For real solutions, a dependence on the concentrations or molar fractions of the different components may occur as well [21, 304]. Generally, in both cases the diffusion coefficients are determined by the thermodynamic properties of the ambient phase and do not depend on the size of the clusters.

In such cases, a new time scale may be introduced via

$$\mathrm{d}t' = 4\pi D c^{(\infty)}_{\text{eq}}\left(\frac{3\omega_s}{4\pi}\right)^{1/3} \mathrm{d}t. \tag{5.63}$$

Equation (5.62) then gets the form

$$\frac{\partial f(n,t')}{\partial t'} = \left(\frac{c}{c^{(\infty)}_{\text{eq}}}\right)\left\{(n-1)^{1/3} f(n-1,t')\right. \tag{5.64}$$
$$\left. +\, n^{1/3} f(n+1,t')\exp\left[\frac{\Delta G(n+1) - \Delta G(n)}{k_B T}\right]\right\}$$
$$-\left(\frac{c}{c^{(\infty)}_{\text{eq}}}\right)\left\{(n-1)^{1/3}\exp\left[\frac{\Delta G(n) - \Delta G(n-1)}{k_B T}\right] + n^{1/3}\right\} f(n,t').$$

Similarly, we get for kinetically limited growth with

$$\mathrm{d}t'' = 4\pi D^{(*)} c^{(\infty)}_{\text{eq}} a_m \left(\frac{\omega_s}{\omega_m}\right)^{2/3} \mathrm{d}t \tag{5.65}$$

the relation

$$\frac{\partial f(n,t'')}{\partial t''} = \left(\frac{c}{c_{\text{eq}}^{(\infty)}}\right)\Bigg\{(n-1)^{2/3} f(n-1,t'') + n^{2/3} f(n+1,t'') \exp\left[\frac{\Delta G(n+1)-\Delta G(n)}{k_B T}\right]\Bigg\} - \left(\frac{c}{c_{\text{eq}}^{(\infty)}}\right)\Bigg\{(n-1)^{2/3} \exp\left[\frac{\Delta G(n)-\Delta G(n-1)}{k_B T}\right] + n^{2/3}\Bigg\} f(n,t'') . \tag{5.66}$$

The general expression for $w^{(+)}$ (cf. Eq. (5.61)) can be written as

$$w_{n,n+1}^{(+)} = 4\pi D c \left(\frac{3\omega_s}{4\pi}\right)^{1/3} n^{1/3} \left\{\frac{\left[\left(\frac{D^{(*)}}{D}\right)\left(\frac{a_s}{a_m}\right)\right] n^{1/3}}{1+\left[\left(\frac{D^{(*)}}{D}\right)\left(\frac{a_s}{a_m}\right)\right] n^{1/3}}\right\} . \tag{5.67}$$

With the notation

$$g(n) = \left\{\frac{\left[\left(\frac{D^{(*)}}{D}\right)\left(\frac{a_s}{a_m}\right)\right] n^{1/3}}{1+\left[\left(\frac{D^{(*)}}{D}\right)\left(\frac{a_s}{a_m}\right)\right] n^{1/3}}\right\} \tag{5.68}$$

the basic kinetic equation for the description of the time evolution of the cluster-size distribution function gets the form

$$\frac{\partial f(n,t')}{\partial t'} = \left(\frac{c(t')}{c_{\text{eq}}^{(\infty)}}\right)\Bigg\{(n-1)^{1/3} g(n-1) f(n-1,t') + n^{1/3} g(n) f(n+1,t') \exp\left[\frac{\Delta G(n+1)-\Delta G(n)}{k_B T}\right]\Bigg\} - \left(\frac{c(t')}{c_{\text{eq}}^{(\infty)}}\right)\Bigg\{(n-1)^{1/3} g(n-1) \exp\left[\frac{\Delta G(n)-\Delta G(n-1)}{k_B T}\right] + n^{1/3} g(n)\Bigg\} f(n,t') . \tag{5.69}$$

The possibility of introduction of reduced time scales in the limiting cases of diffusion and kinetic or ballistic growth implies that possible shapes of the cluster-size distributions evolving in the course of time are independent of the value of the coefficient of diffusion. Its value determines only the time scale of the process. In the general case, as expressed by Eq. (5.69), the values of the diffusion coefficients retain some minor influence via the ratio $(D^{(*)}/D)$.

Note, however, that the above conclusion was derived under the reasonable assumption that the value of D is independent of cluster size n. As mentioned, this is indeed frequently the case. However, if in the course of growth of a cluster the state of the matrix in the vicinity of the cluster changes qualitatively (e.g., by evolution of elastic fields, segregation of some component inhibiting the flux of the others to the cluster) or the state of the cluster changes in dependence on its size, then effectively a cluster size dependence of the diffusion coefficient occurs. In such cases, qualitative changes of the distribution functions have to be expected (cf. [22]) as compared with the dependences described by Eqs. (5.64), (5.66), and (5.69).

In this way, the basic kinetic equations, describing nucleation and growth for the considered growth equations, are finally established. The specific properties of the system under consideration enter the description only via the respective expression for ΔG. Several cases of particular interest will be analyzed in the subsequent section.

5.5 Numerical Solution of the Kinetic Equations

In the subsequent analysis, we would like to follow the time development of the cluster-size distribution function for two cases: (i) The concentration of single particles per unit volume (the supersaturation in the system) is kept constant, i.e., the equation

$$f(1,t) = c = \text{constant} \tag{5.70}$$

holds. (ii) The conservation of the total number of particles (single particles and particles aggregated in clusters)

$$\sum_{n=1}^{\infty} n f(n,t) = f(1,0) = c_0 = \text{constant} \tag{5.71}$$

is taken into account. In both cases, we assume that in the initial state the segregating particles are distributed in the ambient phase in the form of monomers, only, i.e., the relations

$$f(1,0) = c_0, \qquad f(n,0) = 0 \qquad \text{for} \qquad n \geq 2 \tag{5.72}$$

are fulfilled.

5.5.1 Precipitation in a Perfect Solution

As a first example, we consider the case of precipitation in a perfect solution. $\Delta G(n)$ is chosen hereby in the classical form as expressed by Eq. (5.2) or (5.4). Moreover, for a perfect solution the difference in the chemical potentials can be written as (e.g., [94, 298])

$$\Delta\mu = k_B T \ln\left(\frac{c}{c_{\text{eq}}^{(\infty)}}\right). \tag{5.73}$$

In the derivation of Eq. (5.73), the condition for equilibrium coexistence of both phases at a planar interface,

$$\mu(p,T,c_{\text{eq}}^{(\infty)}) = \mu_\alpha(p,T), \tag{5.74}$$

was employed.

A substitution of these expressions into Eq. (5.69) yields

$$\begin{aligned}\frac{\partial f(n,t')}{\partial t'} &= \left(\frac{c(t')}{c_{\text{eq}}^{(\infty)}}\right)(n-1)^{1/3} g(n-1) f(n-1,t') \\ &+ n^{1/3} \exp\left(\frac{2\alpha_2}{3k_B T (n+1)^{1/3}}\right) g(n) f(n+1,t') \\ &- n^{1/3}\left[g(n)\left(\frac{c(t')}{c_{\text{eq}}^{(\infty)}}\right)\right. \\ &\left. + \left(\frac{n-1}{n}\right)^{1/3} g(n-1)\exp\left(\frac{2\alpha_2}{3k_B T n^{1/3}}\right)\right] f(n,t').\end{aligned} \tag{5.75}$$

In Eq. (5.75), the approximations

$$\Delta G(n+1) - \Delta G(n) \cong -\Delta\mu + \frac{2\alpha_2}{3(n+1)^{1/3}}, \tag{5.76}$$

$$\Delta G(n) - \Delta G(n-1) \cong -\Delta\mu + \frac{2\alpha_2}{3n^{1/3}} \tag{5.77}$$

were used. This was done in order to get some insight into the structure of the different terms on the right-hand side of Eq. (5.75). In the numerical implementation, latter approximation is always omitted.

In Figures 5.1–5.6 results of numerical solutions of Eq. (5.75) are shown (without applying the approximations (5.76) and (5.77)). It is assumed that in the initial state the segregating phase is distributed in the ambient phase in form of single particles only (cf. Eq. (5.72)). The following values of the parameters are used:

$$\frac{2\sigma}{c_\alpha k_B T} = 0.8\ \text{nm}, \qquad c_{\text{eq}}^{(\infty)} = 3.74 \times 10^{-3} c_\alpha, \tag{5.78}$$

$$c_\alpha = 2.3 \times 10^{28}\ \text{m}^{-3}, \qquad D^{(*)} = D, \qquad a_s = a_m. \tag{5.79}$$

Figure 5.1 shows the shape of the cluster-size distribution function for different moments of time provided the supersaturation in the system is artificially kept constant. In the course of time a steady-state distribution is approached, which is, at least qualitatively, well-approximated by the analytical expressions derived earlier in Chapter 3 (see also [298,300,302,307]).

Figure 5.2 shows similarly the evolution of the cluster-size distribution function, however, this time for the case that conservation of the total number of segregating particles is taken into account. As seen, after some initial period of time, where the behavior resembles the previous case, a second maximum for large cluster sizes develops. This maximum is formed by clusters of sizes being of the same order of magnitude or larger than the current value of the critical cluster size (cf. also [178, 350]). By arrows, in Figure 5.2 the actual value of the critical cluster size is indicated. It is determined according to Eqs. (5.5) and (5.73) by

$$n_c^{1/3} = \frac{2\alpha_2}{3k_B T \ln\left(\frac{c}{c_{\text{eq}}^{(\infty)}}\right)}. \tag{5.80}$$

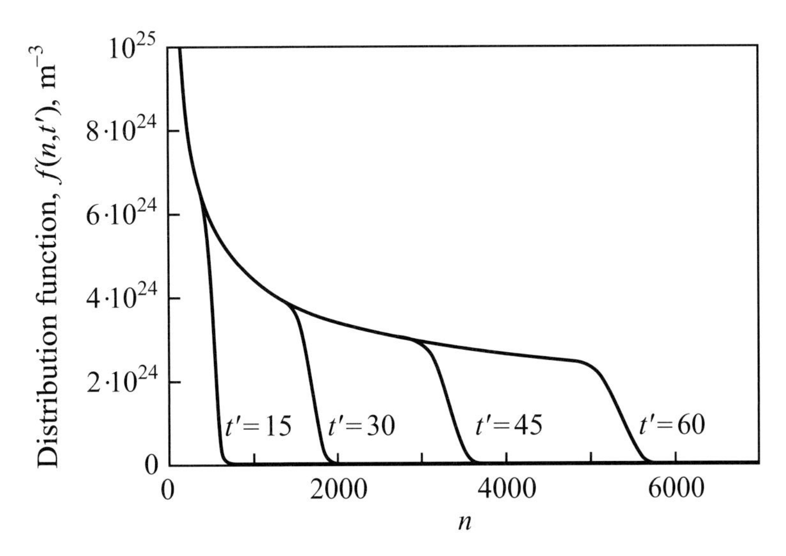

Figure 5.1: Different stages in the time evolution of the steady-state cluster-size distribution, as obtained by a numerical solution of the kinetic equations (5.75), if constancy of the supersaturation (concentration of monomeric building units) in the system is sustained by some appropriate mechanism (cf. Eq. (5.70)). The initial supersaturation was set equal to $(c/c_{\rm eq}^{(\infty)}) = 10$. The following values of the parameters are used here and throughout: Specific interfacial energy $\sigma = 0.08\,{\rm Jm}^{-2}$, particle concentration in the newly evolving phase $c_\alpha = 2.3 \times 10^{28}\,{\rm m}^{-3}$, equilibrium solubility at a temperature $T = 731$ K equals $c_{\rm eq}^{(\infty)} = 8.6 \times 10^{25}\,{\rm m}^{-3}$, radius of the segregating particles $R_1 = 2.2 \times 10^{-10}$ m, $D^{(*)} = D$, $a_s = a_m$. Note, however, that only a combination of these parameters is required for the solution of the kinetic equations.

As shown in Figure 5.3, for $n < n_{\rm max}$ the distribution function with respect to cluster sizes can be well approximated by dependences of the form

$$f(n) \propto n^{-\tau_{\rm eff}}. \tag{5.81}$$

For a given value of the initial supersaturation, the parameters $\tau_{\rm eff}$ as well as $n_{\rm max}$ vary hereby in dependence on time.

In Figure 5.4, the parameters $\tau_{\rm eff}$ and $n_{\rm max}$ are shown for different values of the initial supersaturation in dependence on time. For the considered range of initial supersaturations, values of $\tau_{\rm eff}$ in the range

$$2 \le \tau_{\rm eff} \le 6 \tag{5.82}$$

are observed.

Moreover, at certain moments of time discontinuities in the $\tau_{\rm eff}$ vs t' and $n_{\rm max}$ vs t' dependences occur. This effect is connected with the formation of the second maximum in the cluster-size distributions (cf. Figure 5.2). Once such a maximum appears, dependences of the form as given by Eq. (5.81) can be applied only to the part of the distribution corresponding to relatively small cluster sizes. A qualitatively similar behavior was found also earlier for the limiting cases of diffusion and kinetic limited growth [22, 164, 240].

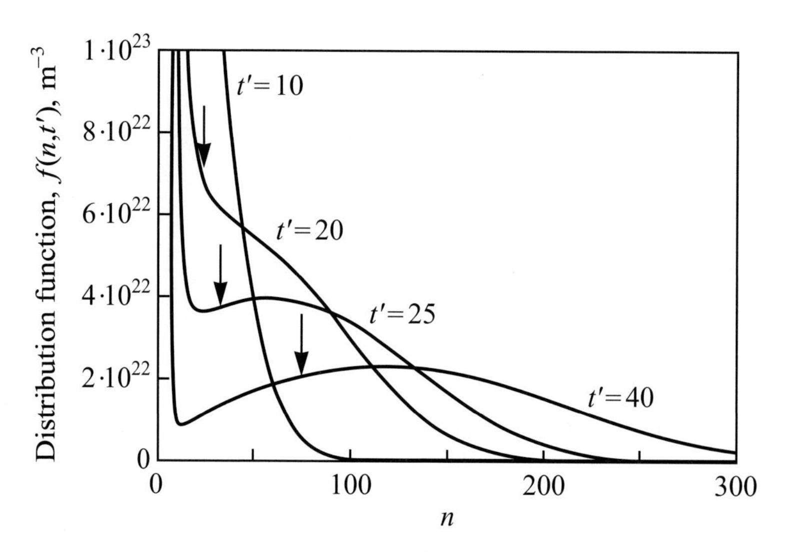

Figure 5.2: Different stages in the evolution of the cluster-size distribution $f(n,t)$ if the condition of conservation of the total number of monomers (Eq. (5.71)) is taken into account. In the first stages of the process monotonically decreasing distributions develop continuously and transform further into a bimodal distribution. By arrows the actual value of the critical cluster size is indicated.

The motion of the maximum to higher values of the cluster size is accompanied by a steepening of the distribution at smaller sizes and an increase of the critical cluster size. After the first stages, characterized by dominating nucleation accompanied by an widely independent growth of the already formed supercritical clusters, a third relatively slow stage of the transformation begins. The further evolution, denoted as coarsening or Ostwald ripening, is determined by dependences described in detail in the preceding Chapter 4 and established theoretically first by the author in cooperation with Lifshitz [155].

For a verification of the statement concerning the general course of the phase transition, in Figure 5.5, typical characteristics of a system undergoing a precipitation process are shown. The three different stages of dominating nucleation, dominating independent growth, and coarsening are clearly distinguishable.

In Figure 5.6 the distribution function

$$\varphi(u,t') = \frac{1}{N(t')} f(R,t') R_c, \qquad u = \frac{R}{R_c} \tag{5.83}$$

in reduced variables is shown. An integration of this equation over all possible values of u in the range $(0 < u < \infty)$ gives

$$\int \varphi(u,t')\,\mathrm{d}u = \frac{1}{N(t')} \int f(R,t)\,\mathrm{d}R = 1, \tag{5.84}$$

which confirms that $\varphi(u,t')$ is normalized to one.

It is seen that in the course of the evolution a time-independent cluster-size distribution in reduced variables is established. The approach of a time-independent distribution in reduced coordinates is an universal feature of the late stages of coarsening. It remains true even if

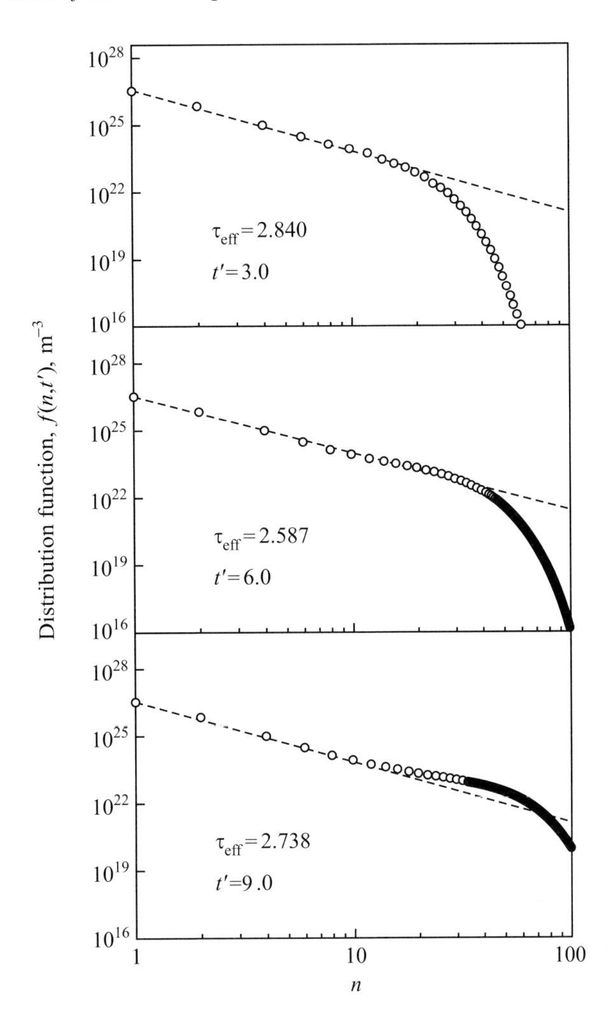

Figure 5.3: Log–log dependence of the size distribution function $f(n,t)$ vs n for different moments of reduced time ($t' = 0.9,\ 1.9,\ 3.9$). For cluster sizes less than a certain value $n_{\max}$ the distributions obtained numerically (circles) can be well fitted by dependences of the form $f(n) \propto n^{-\tau_{\text{eff}}}$ with different values of τ_{eff} (dotted lines). The respective values of τ_{eff} are given in the figures. The initial supersaturation was chosen to be equal to $(c/c_{\text{eq}}^{(\infty)}) = 10$.

additional thermodynamic and kinetic factors, affecting the coarsening process, are taken into consideration (see Chapter 4 and [107, 161, 171, 184, 237, 290, 294]). Such effects may lead to additional terms in the kinetic equations. In such cases, the shape of this distribution may differ from the form as shown on the figure. For example, it was shown that external noise [161] or spatial inhomogeneities [184] may result into a broadening of the distribution function (cf. also Chapter 6). Another example, where a qualitative change of the asymptotic distribution occurs, is discussed below.

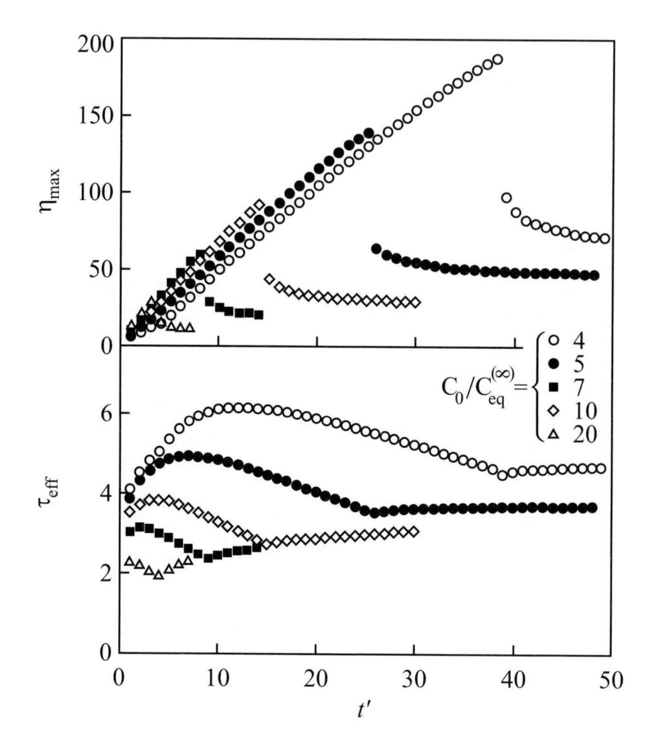

Figure 5.4: Values of the parameters $n_{\max}$ and τ_{eff} as functions of time (in reduced units) for different values of the initial supersaturation $[(c/c_{\text{eq}}^{(\infty)}) = 4,\ 5, \dots,\ 20]$.

5.5.2 Effect of Nonlinear Inhibition of Cluster Growth on the Shape of the Cluster-Size Distributions

Interactions between growing cluster and ambient phase may result in the evolution of monodisperse cluster-size distributions. As a precondition for such behavior, the additional term in the work of formation of critical clusters must grow more rapidly than linear with the volume of the cluster (cf. [94, 220, 229]; Section 5.3). Particular examples, where such effects may occur, are precipitation processes in highly viscous glass-forming melts, polymers or porous materials (see [94, 184, 232, 233, 237, 290] and Chapter 6). In the simplest case, we may write

$$\Delta G(n) = -n\Delta\mu + \alpha_2 n^{2/3} + \kappa n^{\beta}. \tag{5.85}$$

The term κn^{β} describes here the influence of additional inhibiting factors evolving in the course of the transformation on cluster growth (e.g., elastic strains).

For $\beta > 1$, we may reformulate Eq. (5.85) as

$$\Delta G(n) = -n\Delta\mu^{\text{eff}} + \alpha_2 n^{2/3}, \qquad \Delta\mu^{\text{eff}} = \Delta\mu - \kappa n^{(\beta-1)}. \tag{5.86}$$

For the considered conditions, the effective driving force of the transformation $\Delta\mu^{\text{eff}}$ decreases monotonically with increasing cluster size resulting, finally, in a total inhibition of the

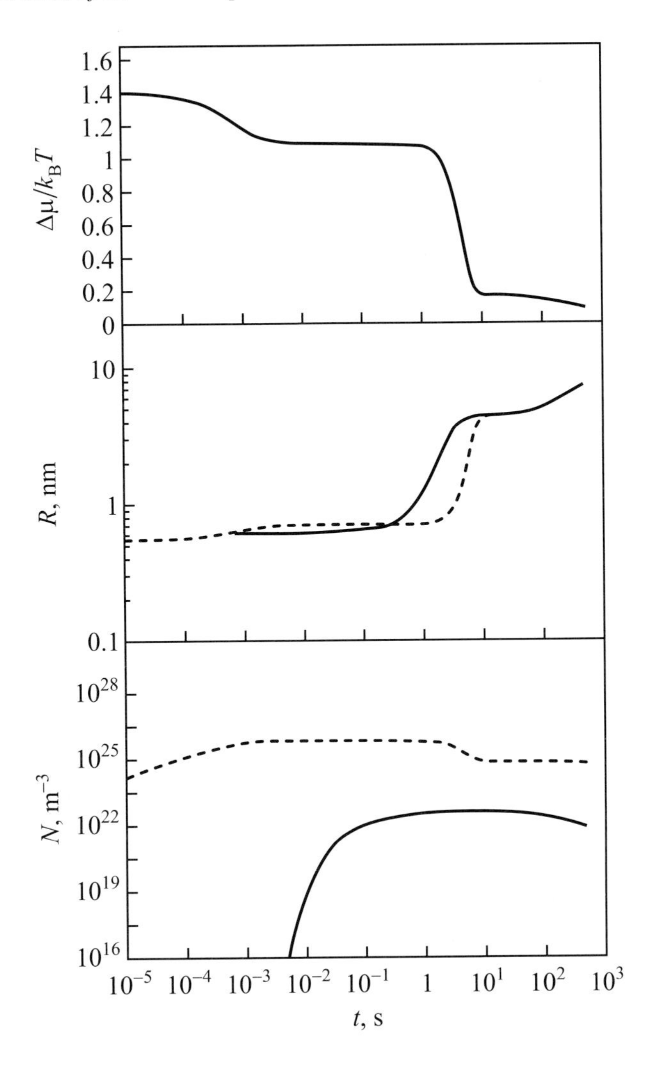

Figure 5.5: Evolution of characteristic properties of a system undergoing a precipitation process. *Top:* Change of the supersaturation as a function of time; *Center:* Change of the average (full curve) and the critical cluster sizes (dotted curve) in the course of the transformation; *Bottom:* Change of the number of clusters in the system. Hereby the dotted curve counts all clusters in the system, while the full curve refers to clusters with a radius $R > 0.6$ nm.

growth. Since this conclusion holds for single clusters, it is also valid for the evolution of ensembles of clusters as pointed out already in Section 5.3.

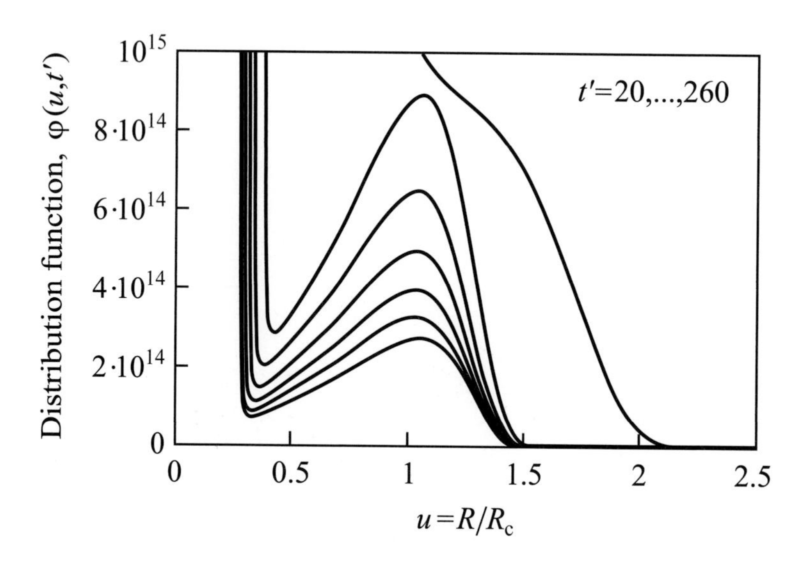

Figure 5.6: Cluster-size distribution function $\varphi(u, t')$ in reduced variables $u = (R/R_c)$ for different moments of time (cf. Eqs. (5.80), (5.83), and (5.84)). In the course of the evolution, a time-independent shape develops as predicted first by Lifshitz and Slezov [155] (see Chapter 4).

If $\Delta\mu$ is expressed, again, via Eq. (5.73), the kinetic equations (5.69) take the form

$$\frac{\partial f(n,t')}{\partial t'} = \left(\frac{c(t')}{c_{\mathrm{eq}}^{(\infty)}}\right)(n-1)^{1/3} g(n-1) f(n-1,t') \tag{5.87}$$

$$+ n^{1/3} g(n) \exp\left(\frac{2\alpha_2}{3k_BT(n+1)^{1/3}}\right) \exp\left(\frac{\kappa\beta(n+1)^{\beta-1}}{k_BT}\right) f(n+1,t')$$

$$- n^{1/3}\left[\left(\frac{c(t')}{c_{\mathrm{eq}}^{(\infty)}}\right) g(n)\right.$$

$$\left. + \left(\frac{n-1}{n}\right)^{1/3} g(n-1) \exp\left(\frac{2\alpha_2}{3k_BTn^{1/3}}\right) \exp\left(\frac{\kappa\beta n^{\beta-1}}{k_BT}\right)\right] f(n,t').$$

For the numerical calculations, shown in Figure 5.7, we set $\beta = 2$ [229] and $\kappa = 0.01k_BT \times \ln[c_0/c_{\mathrm{eq}}^{(\infty)}]$, where c_0 is the concentration of single particles in the homogeneous initial state. As shown in the figure, in the course of time a relatively monodisperse cluster-size distribution develops due to the nonlinear inhibition of cluster growth caused by matrix–cluster interactions.

For other kinds of rheological behavior, i.e., if the matrix acts as a viscoelastic body, a wide spectrum of kinetic laws for the description of the process of Ostwald ripening may be obtained [180, 233]. A particularly interesting example consists in a damped oscillatory approach to the coarsening. Such a kind of behavior was theoretically predicted by Möller

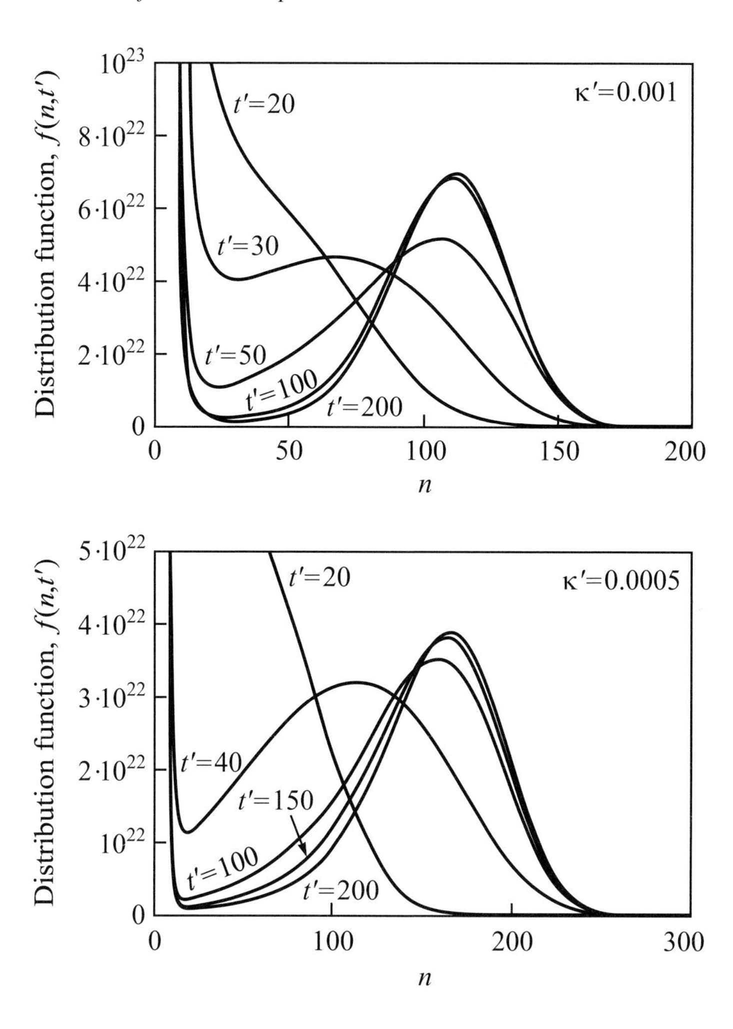

Figure 5.7: Evolution of the cluster-size distribution function for the case of nonlinear inhibition of cluster growth. The evolution is described, again, by the most general expression for the set of kinetic equations (5.87). In the calculations, D is set equal to $D^{(*)}$, $a_m = a_s$, β is chosen equal to 2, while κ is set equal to $\kappa = 0.01 k_B T \ln[c_0/c_{\mathrm{eq}}^{(\infty)}]$. In the course of time, a monodisperse cluster-size distribution is established. Such kind of behavior is always to be expected if the inequality $\beta > 1$ is fulfilled. Note also the peculiarities in the approach to the final distribution.

et al. [104, 183]. As it seems, these results are applicable directly to experimental findings reported by Morosova [181]. A more detailed comparison of experimental and theoretical results is thus highly desirable and given in Chapter 6.

5.5.3 Application of Fisher's Expression for the Work of Cluster Formation

Applying, instead of Eq. (5.4), Fisher's expression (5.8) for the work of cluster formation, Eqs. (5.69) and (5.73) yield, approximately,

$$\begin{aligned}\frac{\partial f(n,t')}{\partial t'} &= \left(\frac{c(t')}{c_{\mathrm{eq}}^{(\infty)}}\right)(n-1)^{1/3} g(n-1) f(n-1,t') \\ &+ n^{1/3}\left(\frac{n+1}{n}\right)^{\tau} \exp\left(\frac{2\alpha_2}{3k_B T (n+1)^{1/3}}\right) g(n) f(n+1,t') \\ &- n^{1/3}\left[\frac{c(t')}{c_{\mathrm{eq}}^{(\infty)}} g(n)\right. \\ &\left. + g(n-1)\left(\frac{n-1}{n}\right)^{1/3}\left(\frac{n}{n-1}\right)^{\tau} \exp\left(\frac{2\alpha_2}{3k_B T n^{1/3}}\right)\right] f(n,t').\end{aligned} \tag{5.88}$$

It is obvious (cf. also [69]) that the exponential terms in Eq. (5.88) will dominate, in general, the behavior compared with the factors $(n/(n+1))^{\tau}$ or $(n/(n-1))^{\tau}$. The evolution of the cluster-size distribution function is, therefore, in general only slightly affected by the additional term $(k_B T \tau \ln n)$. This conclusion can be verified by a numerical solution of the kinetic equations describing nucleation and growth.

The results of the solution of Eq. (5.88) are shown in Figure 5.8. Moreover, in Figure 5.9, the same curves are given in logarithmic coordinates. It can be seen that in the range $n < n_{\max}$ curves of the shape as described by Eq. (5.81) give a good approximation of the results. In Figure 5.10, the values of τ_{eff} and $n_{\max}$ are shown as functions of time for different values of the initial supersaturations. It is verified easily that τ_{eff} varies in the range as indicated in Eq. (5.82), again.

Similarly to calculations carried out with the classical expression for $\Delta G(n)$ (cf. also [240]), in the initial stages of the nucleation–growth processes and in the range $n < n_{\max}$, the distributions can be well approximated by curves of the type

$$f(n) \propto n^{-\tau_{\mathrm{eff}}}, \qquad 2 < \tau_{\mathrm{eff}} < 6, \tag{5.89}$$

again. The values of τ_{eff} and $n_{\max}$ depend hereby on the initial supersaturation and the time the clustering process proceeds. The exponent τ_{eff} is determined thus kinetically, its value is not interrelated with the parameter τ occurring in Fisher's expression for the work of cluster formation (Eq. (5.8)).

The situation becomes, however, qualitatively different in the vicinity of the critical point. As mentioned in Section 5.2, near the critical point the differences between liquid and gas phases disappear. Fisher's statistical droplet model results in this region in a cluster-size distribution function with the shape

$$f(n)|_{T \to T_c} \propto n^{-\tau}, \qquad 2 < \tau < 2.5. \tag{5.90}$$

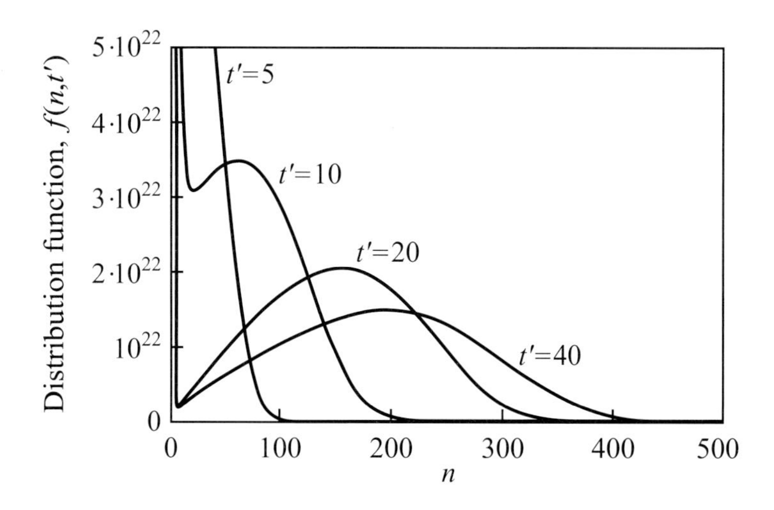

Figure 5.8: Different stages in the evolution of the cluster-size distribution $f(n, t)$ if Fisher's expression for the work of cluster formation is applied in the derivation of the kinetic equations describing nucleation and growth. Again, the condition of conservation of the total number of monomers (Eq. (5.71)) is taken into account. Similarly to Figure 5.2, in the first stages of the process, monotonically decreasing distributions develop continuously and transform further into a bimodal distribution. The initial supersaturation is chosen equal to $(c/c_{\text{eq}}^{(\infty)}) = 10$. The parameter τ was set here equal to 2.5.

This result can be reestablished by the solution of the set of kinetic equations describing nucleation and growth.

With Eq. (5.64) and $\Delta G(n) = k_B T \tau \ln n$ we have in the vicinity of the critical point

$$\begin{aligned}\frac{\partial f(n,t')}{\partial t'} &= (n-1)^{1/3} g(n-1) f(n-1,t') \\ &+ n^{1/3}\left(\frac{n+1}{n}\right)^{\tau} g(n) f(n+1,t') \\ &- \left[n^{1/3} g(n) + (n-1)^{1/3}\left(\frac{n}{n-1}\right)^{\tau} g(n-1)\right] f(n,t').\end{aligned} \tag{5.91}$$

It is easily verified that Eq. (5.90) represents, in this limiting case, the stationary solution of this kinetic equation. As shown in Figure 5.11 the obtained time-dependent solutions tend to this asymptotic solution at large times.

In the immediate vicinity of the critical point, the solution of the set of kinetic equations results, consequently, in cluster-size distributions as predicted by Fisher's statistical droplet model. However, as seen in Figures 5.8–5.10, for thermodynamically unstable states beyond the critical point, a quite different behavior is found in agreement with expectations outlined in Section 5.2 (for a more detailed consideration of the properties of cluster ensembles in the vicinity of the critical point and eventual shortcomings of the applied here method of description in this region see, e.g., [26, 31, 74, 319]).

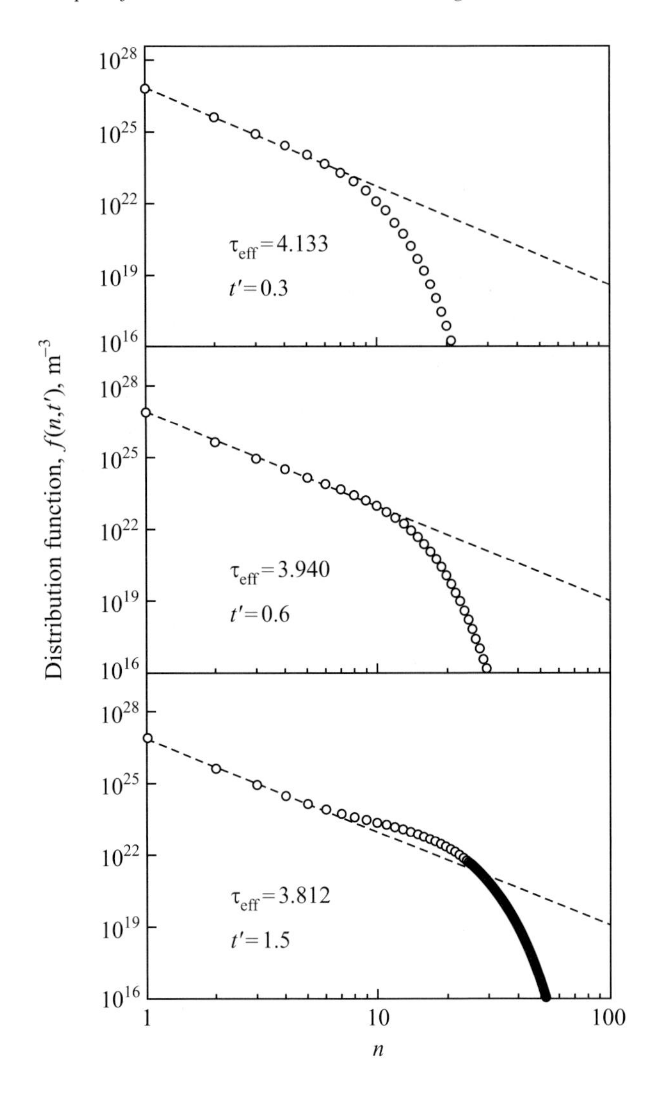

Figure 5.9: Log–log dependence of size distribution function $f(n,t)$ on n at different moments of time for diffusion-limited growth and an initial supersaturation $((c/c_{\rm eq}^{(\infty)}) = 10)$ for the curves shown in Figure 5.8.

5.6 Selected Applications and Conclusions

Clustering processes occur in a variety of scientific and technological applications. As far as the basic premises – existence of a critical cluster size, growth by aggregation or emission of single building units of the newly evolving phase – are fulfilled the methods outlined can be applied with success quite independently of specific features of the system considered. Specific properties of the system under consideration enter the description only via the appropriate

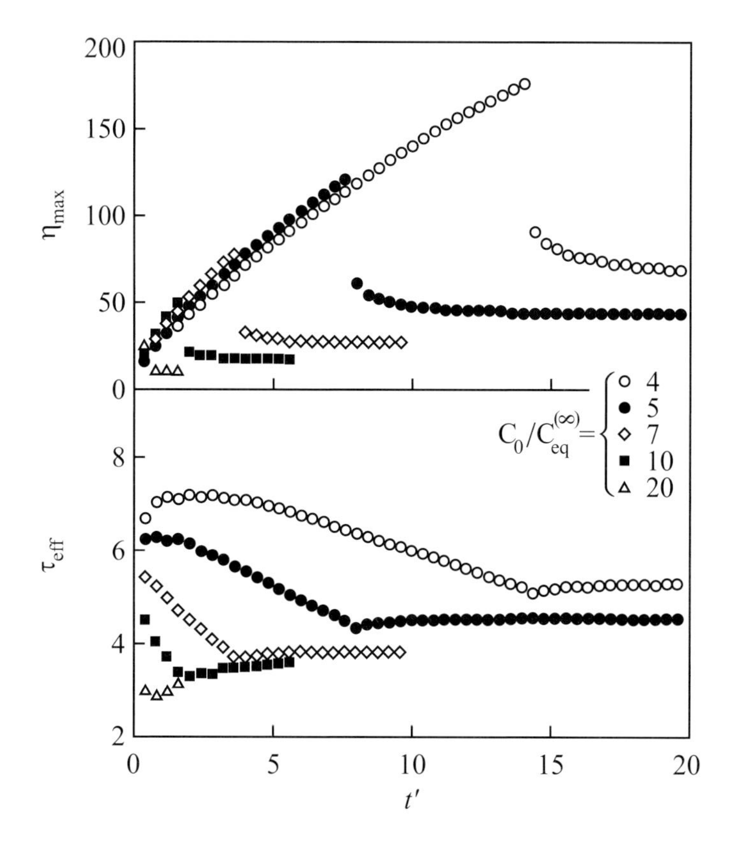

Figure 5.10: Values of the parameters $n_{\max}$ and τ_{eff} as obtained numerically for diffusion-limited growth and different values of the initial supersaturation $[(c/c_{\text{eq}}^{(\infty)}) = 4,\ 5,\ 7,\ 15,\ 20]$ applying Fisher's expression for the work of cluster formation.

choice of the aggregation rates and the work of cluster formation (cf. also [245]). In this way, the results of the analysis may be of significance for very different processes.

In application to such a variety of phase formation processes, the following conclusions can be drawn, among others, from the analysis:

(i) The so-called equilibrium distribution of classical nucleation theory, the statistical droplet model developed by Fisher or similar expressions are, in general, inappropriate means for an interpretation of experimental results on clustering processes. Attempts to connect these distributions with the shapes of distributions observed experimentally (cf., e.g., [316]) or with basic features of the nucleation–growth process (e.g., [53, 327]) are not correct.

(ii) Fisher's statistical droplet model describes experimentally observed cluster-size distributions well in the immediate vicinity of the liquid–gas critical point. Here it leads to

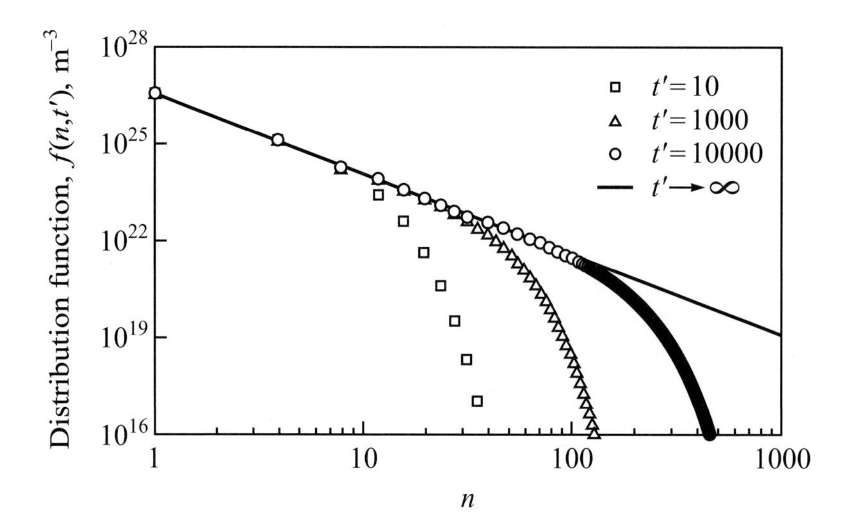

Figure 5.11: Solution of the set of kinetic equations (5.91) for initial states in the immediate vicinity of the liquid–gas critical point. In this case, the curve $f(n) \propto n^{-\tau}$ is approached in the course of time as predicted by Fisher's statistical droplet model. As shown, the same dependence may be obtained also as the stationary solution of the kinetic equations governing nucleation and growth. In the numerical calculations, τ was set equal to 2.5.

dependences of the form $f(n) \propto n^{-\tau}$, where τ is determined by thermodynamic properties of the substance undergoing the transition.

(iii) Beyond the critical point in the region of thermodynamically unstable initial states, the observation of distributions with a shape $f(n) \propto n^{-\tau_{\rm eff}}$ with $2 \leq \tau_{\rm eff} \leq 6$ in a certain range of cluster-size space can be explained in the framework of a dynamic nucleation and growth model as a consequence of clustering in a first-order phase transformation. Hereby $\tau_{\rm eff}$ depends both on the initial supersaturation and the time interval where clustering processes occur.

By applying Fisher's expression for the work of cluster formation, Fisher's statistical distribution is retained in the dynamic approach as a limiting distribution reached for large times for initial states located in the immediate vicinity of the critical point.

In this way, a straightforward explanation of the results of multifragmentation processes in nuclear collisions can be given (cf. [182, 186, 199, 240]). Moreover, one may expect that distributions of such a shape occur also frequently in condensation processes in expanding atomic or molecular gases.

(iv) If the evolution proceeds over longer time intervals in unstable initial states beyond the critical point, conservation of the total number of segregating particles results in the evolution of a maximum at cluster sizes near to the critical one. Such property is found independent of the mechanism of growth of the clusters.

(v) In the majority of applications, the values of the diffusion coefficients affect only the time scale of the processes but not the spectrum of possible shapes of the distribution functions. Different types of distributions may occur only, if by various thermodynamic or kinetic reasons the diffusion coefficients become effectively cluster-size dependent.

(vi) For example, provided kinetic inhibition and direct stabilizing cluster–cluster interactions are of no significance, monodisperse cluster-size distributions may develop spontaneously only, if cluster–matrix interactions lead to additional terms in the expression for the work of cluster formation. Such additional terms have to grow more rapidly than linear with the volume of a cluster. Latter condition (and also the second of the first two mentioned ones) are not fulfilled for bubbles in a bulk liquid. Consequently, the evolution of monodisperse bubble-size distributions is such systems is not possible (cf. in contrast [347,348] and also [354]). Therefore, the conclusion of the authors of [347,348] concerning the possibility of evolution of monodisperse thermodynamically stabilized bubble-size distributions in the bulk of a liquid–gas solution seems to be incorrect.

(vii) The results outlined may be applied as a means to propose realistic shapes of cluster-size distributions for an interpretation of experimental results on scattering measurements (cf. [343]). As shown above, in general, the distributions are not monodisperse. They can be approximated by Gauss curves; however, the average value and the dispersity of these curves increases with time. In the asymptotic stage of the transformation, analytic expressions are available based on the Lifshitz–Slezov theory and its generalizations.

5.7 Discussion

It was shown in the present analysis that statistical approaches in determining cluster-size distributions do not allow one to give, in general, an adequate description of experimental results concerning possible shapes of cluster-size distributions evolving in nucleation–growth processes. By this reason, dynamical approaches have to be applied allowing one to account for the variation of the state of the system in the course of the process.

In the present investigation, only variations of the state of the system are considered connected with the conservation of the total number of particles (free particles and particles aggregated in clusters). The general kinetic equations allow us, of course, also an account of the influence of variations of the external boundary conditions (e.g., pressure, temperature etc.) on the shape of the cluster-size distributions. One important example in this respect consists in the analysis of clustering in freely expanding gases [242]. As discussed in more detail in [250], for precipitation processes in glasses and glass-forming melts – as another important application – relaxation processes of the matrix may affect qualitatively the kinetics of the precipitation processes resulting eventually in the evolution of bimodal cluster-size distributions, where both peaks are found at sufficiently large cluster sizes. In addition, particular properties of the system under consideration may be reflected by an appropriate determination of the work of cluster formation and the aggregation coefficients $w_{n,n+1}^{(+)}$.

The method employed here can also be extended without principal difficulties to cases when more than one parameter is required to determine appropriately the state of the cluster of the newly evolving phase. One example in this respect consists in the process of bubble formation in liquids. Another would be the process of precipitation in a solid solution when one accounts for the possibility that the composition of the clusters is changed in the course of the transformation [3, 249, 253, 254]. Such effects may also result in modifications of the shapes of cluster-size distributions as analyzed above.

6 Coarsening Under the Influence of Elastic Stresses and in Porous Materials

6.1 Introduction

Immediately after the formulation of the theoretical approach to coarsening, reviewed in Chapter 4, the effect of elastic stresses on coarsening was analyzed in detail [154, 155]. Here the case was considered that elastic stresses due to the formation of clusters of a newly evolving phase result in stresses the energy of which increases linearly with the volume of the cluster. The results of the analysis show that in this very frequently realized case, elastic stresses only modify the kinetics of coarsening quantitatively but not qualitatively [154].

However, as it became evident in the experimental investigations of Gutzow and Pascova, elastic stresses may change the kinetics of coarsening qualitatively [94, 200]. The respective results are illustrated in Figure 6.1. The theoretical analysis of the kinetics of coarsening in these systems performed by Schmelzer, Gutzow, and Pascova [220, 223, 224, 229, 232, 233] led the authors to the conclusion that elastic stresses may under certain circumstances result in a qualitative change of the coarsening kinetics. As shown by them, such inhibition of cluster growth and coarsening occurs, when the elastic response of the matrix with respect to cluster growth is accompanied by an increase of the total energy of elastic deformations growing more rapidly than linear with the volume of the clusters. In these cases qualitative modifications of the kinetics of coarsening due to cluster matrix interactions occur (see, e.g., [200]). The basic ideas and main results of this theoretical approach are summarized in Section 6.2.

These analysis have been extended in a variety of common papers to the problem of the kinetics of coarsening in porous materials [236, 237, 290]. The situation was/is here the following: Despite the variety of research done in the theoretical description of coarsening, or Ostwald ripening, as the late stage of first-order phase transformations, most of the attempts developed deal with a restricted problem only, when the matrix where the phase separation process takes place can be considered as a homogeneous body allowing the formation and the growth of clusters at any place with the same probability and the growth to a practically arbitrary size. Spatial correlations occur in these cases via diffusional or elastic interactions of the precipitates [110].

However, in a variety of applications precipitation and coarsening takes place in spatially inhomogeneous systems, where pores may exist and precipitates form and grow inside the pores. In these cases, pores can influence the maximum dimension of the clusters of the newly evolving phase or result, at least, in a significant inhibition of cluster growth once the dimension of the clusters become comparable with the pore sizes. Immediate applications of

Kinetics of First-order Phase Transitions. Vitaly V. Slezov

ISBN: 978-3-527-40775-0

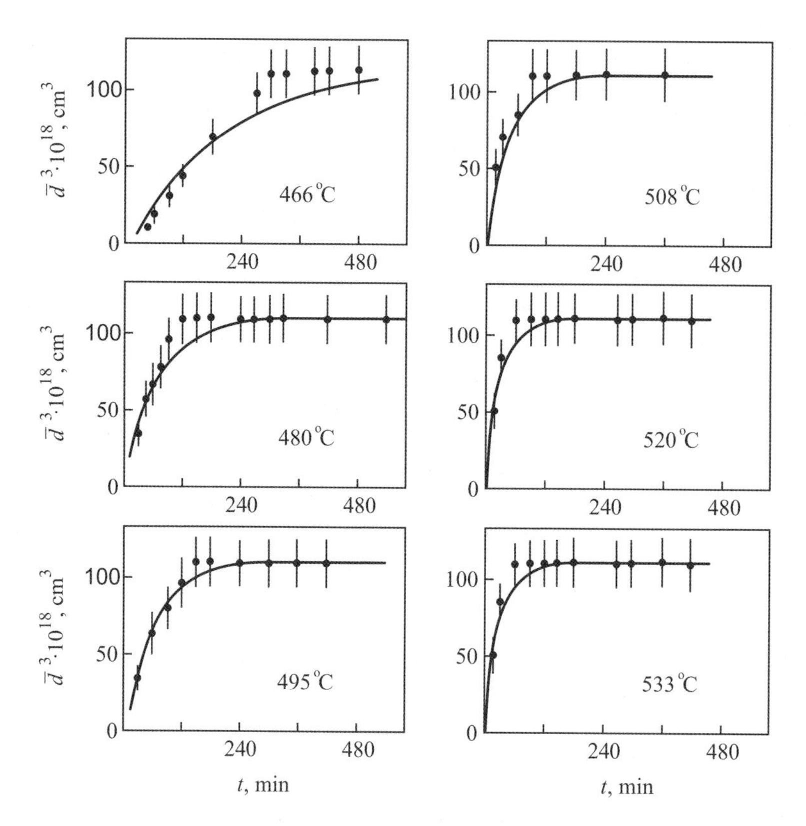

Figure 6.1: Time dependence of the average size of AgCl clusters segregating in a sodium metaborate glass-forming melt. The temperature of the system, at which segregation processes take place, is indicated at each curve. While in the first stage of the process the coarsening kinetics is described by the derived in Chapter 4 asymptotic power laws $\langle R \rangle^3 \propto t$, $N \propto t^{-1}$ [154, 155] a finite stationary value of the average cluster size and a constant number of clusters are established asymptotically in the system [200].

such types of processes are precipitation in porous materials like vycor glasses and zeolithes, they are of relevance also for the understanding of sintering processes (see, e.g., [83]).

The process of Ostwald ripening for the case when clusters are formed and grow only inside the pores of a porous solid matrix was first analyzed in [236, 237, 290] and is described here in Section 6.3. Hereby it is assumed that the solid can be considered as an Hookean elastic body. This assumption implies that after the cluster has reached the size of the pore the evolving elastic strains become sufficiently large as to prevent immediately any further growth of the clusters. As can be seen, the kinetics of coarsening and the type of asymptotic solutions is different as compared with the results obtained first by Lifshitz and Slezov ([153–155] and Chapter 4).

There exists, however, the alternative possibility that the elastic strains evolving in the course of growth of clusters of a new phase in a solid increase only slowly with increasing cluster size but may reach, nevertheless, so large magnitudes with time as to stop the further growth. Coarsening in the system of such weak pores was described first in [290] and is presented in Section 6.4.

In all the cases of coarsening in a porous matrix mentioned so far, the process was considered for the particular situation that the distribution of pores in the solid matrix is monodisperse, i.e., all pores are assumed to be of the same size. This simplifying assumption is, of course, nonrealistic if real porous materials are to be considered. Therefore, in Section 6.5 a generalization of the results outlined is performed for any particular arbitrary size distributions of "hard," nondeformable pores in the solid and in the case when the clusters of the new phase are formed and grow only inside the pores. Finally, in Section 6.6 possible modifications of the results are analyzed when stochastic effects (thermal noise) are taken into account into the description of cluster growth (see also [290]).

Note that the mechanism of the influence of elastic strains on coarsening discussed here is different from the type Kawasaki and Enomoto [122] later dealt with, where the inhibition, respectively, acceleration of coarsening is due to elastic interactions between the growing or dissolving clusters. In the case considered here, the inhibition is due to the cluster–matrix interaction, which may also qualitatively affect the coarsening kinetics as will be shown in the subsequent analysis (see also [220, 232, 233, 236]).

6.2 Cluster Growth and Coarsening Under the Influence of Elastic Stresses Due to Cluster–Matrix Interactions

6.2.1 Models of Elastic Stress in Cluster Growth and Coarsening

The most widely employed model describing the evolution of elastic stresses in phase transformations in solids is directed to the analysis of elastic stresses resulting from a misfit between ambient and newly evolving phases [187]. Suppose, in the bulk of a solid phase, a cluster of a new phase with a different specific volume is formed, the total energy of elastic deformations $\Phi^{(\varepsilon)}$ resulting from the evolution of a cluster of volume V may be written as

$$\Phi^{(\varepsilon)} = \varepsilon V. \tag{6.1}$$

The parameter ε can be expressed via the elastic constants of both phases and an appropriately defined misfit parameter. The change of the Gibbs free energy in cluster formation can then be expressed as

$$\Delta G = -n\Delta\mu + \sigma A + \varepsilon V = -n\left(\Delta\mu - \frac{\varepsilon}{c_\alpha}\right) + \sigma A. \tag{6.2}$$

Here c_α is the particle density of the cluster and n the number of particles in the cluster.

An inspection of Eq. (6.2) shows that the effect of elastic stresses on cluster growth does not depend in this case on cluster size; they lead merely to a constant change of the driving force of cluster growth. Consequently, elastic stresses of the considered type either prevent the formation of a new phase at all or they lead to some redefinition of the parameters in the theory. This is the case analyzed in the first papers on this topic by Lifshitz and the author [154].

The situation may become, however, quite different if elastic stresses in segregation processes are considered [220,223,224,229,232,233]. Let us suppose that one of the components of a binary solution segregates and has a partial diffusion coefficient D considerably larger as compared with the respective parameter of the ambient phase particles (or the second component). In such cases, elastic stresses evolve in segregation resemble the behavior of an elastic spring. For elastic strings, the force is proportional to elongation and the energy of elastic deformation is proportional to elongation squared.

If the initial volume of a cluster, when such type of stresses begin to act, is denoted as V_0, then the elastic stresses in cluster growth caused by such mechanism of evolution of elastic stresses can be written as [220, 224, 229]

$$\Phi^{(\varepsilon)} = \kappa \frac{(V - V_0)^2}{V_0} \theta(V - V_0), \tag{6.3}$$

where

$$\theta(V - V_0) = \begin{cases} 1 & \text{for} \quad V - V_0 > 0 \\ 0 & \text{for} \quad V - V_0 \leq 0 \end{cases}. \tag{6.4}$$

The parameter κ is some combination of elastic constants, again.

The change of the Gibbs free energy in cluster formation can be written in this case as

$$\Delta G = -n\delta\mu + \sigma A + \kappa \frac{(V - V_0)^2}{V_0} \theta(V - V_0), \tag{6.5}$$

or

$$\Delta G = -n \left[\Delta\mu - \frac{\kappa}{c_\alpha} \left(\frac{(V - V_0)^2}{V V_0} \right) \theta(V - V_0) \right] + \sigma A. \tag{6.6}$$

For this mode of evolution of elastic stresses, the effect of stresses increases with increasing cluster size and may, consequently, also qualitatively change the kinetics of cluster growth and coarsening. Therefore, the problem arises to develop a theory of cluster growth and coarsening for such qualitatively different as analyzed in [154] type of stresses, when elastic stresses increase more rapidly than linear with cluster volume.

6.2.2 Theoretical Description of Coarsening at a Nonlinear Increase of the Energy of Elastic Deformations with Cluster Volume: A First Approach

The first approach to the description of coarsening under the influence of elastic stresses – developed by Schmelzer and Gutzow – was based on a thermodynamic analysis of the process of first-order phase transitions [220, 221, 223, 224, 226, 229, 232–234] interpreting coarsening as the evolution along some appropriately defined valleys of the thermodynamic potential describing the system. It results in two differential equations describing the evolution in time of the average cluster size $\langle R \rangle$ and the number of clusters N. The respective equations read for diffusion-limited growth

$$\frac{\mathrm{d}\langle R\rangle}{\mathrm{d}t} = \frac{8Dc}{27c_\alpha^2 k_B T}\frac{1}{\langle R\rangle^2}\left\{\sigma + \frac{3}{4\pi\langle R\rangle^2}\left[\Phi^{(\varepsilon)} - V\frac{\partial\Phi^{(\varepsilon)}}{\partial V}\right]\right\} \times \frac{1}{\Gamma}\left\{1+\Gamma - \frac{\langle R\rangle^2}{2\sigma}\frac{\partial^2\Phi^{(\varepsilon)}}{\partial\langle R\rangle\partial V}\right\}, \tag{6.7}$$

$$\frac{\mathrm{d}\ln N}{\mathrm{d}t} = -\frac{1}{\Gamma}\left\{\left[1 - \frac{\langle R\rangle^2}{2\sigma}\frac{\partial^2\Phi^{(\varepsilon)}}{\partial\langle R\rangle\partial V}\right]\right\}\frac{\mathrm{d}}{\mathrm{d}t}\left[\ln\langle R\rangle\right]. \tag{6.8}$$

Here c is the actual concentration of segregating particles in the ambient phase, D their diffusion coefficient, k_B the Boltzmann constant, and T the absolute temperature. The quantity Γ reflects specific properties of the system under consideration. In general, the relation $\Gamma \leq 1$ holds and the absolute value of this quantity increases with increasing average cluster size. In this limit of large $|\Gamma|$ (and in the absence of stresses), the asymptotic solutions obtained by Lifshitz and Slezov are included in this theoretical approach as a limiting case.

The above theory allows one to describe the whole coarsening process including its initial stages. It allows one to describe in a relatively simple and straightforward way the effect of elastic stresses on coarsening. From the above equations, the following consequences can be drawn:

- If $\Phi^{(\varepsilon)} = 0$ (absence of elastic stresses) or in the case that the energy of elastic deformations increases linearly with the volume of the cluster $\Phi^{(\varepsilon)} = \varepsilon V$, elastic strains do not modify the coarsening process qualitatively. The Lifshitz–Slezov results are obtained asymptotically as special cases.

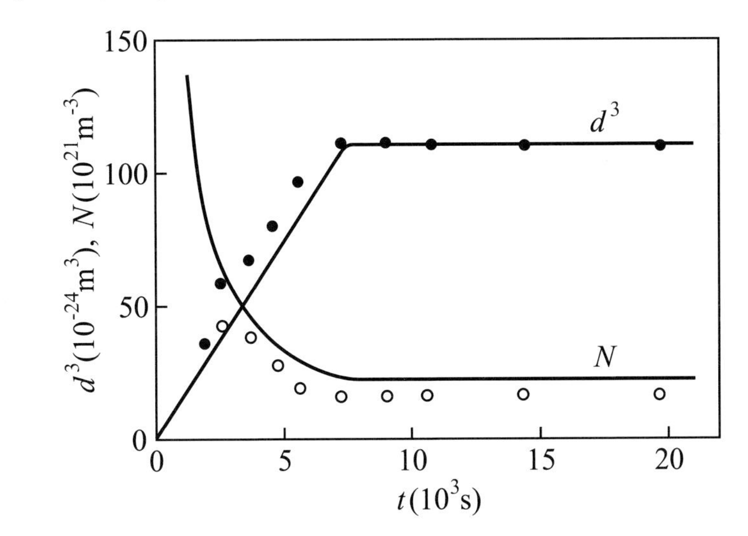

Figure 6.2: Comparison of experimental data (dots for the average cluster size and open circles of numbers of clusters in the system [200]) and theoretical predictions for the process of Ostwald ripening of AgCl clusters in a sodium borate melt (for the details see [233]).

- If elastic stresses result in energies of elastic deformations growing more rapidly than linear with the volume of the clusters, then elastic strains will lead to an inhibition of coarsening.

An example in this respect is shown in Figure 6.2, giving an interpretation of the experimental results shown in Figure 6.1.

As already mentioned, the theory of coarsening in elastic bodies developed by Schmelzer, Gutzow, and Pascova [220,221,223,224,226,229,232–234] has a huge advantage with respect to simplicity and straightforward applicability. However, as a disadvantage, it allows one only to determine the average cluster size and the number of clusters as functions of time and does not allow one to make detailed predictions concerning the evolution of the cluster-size distribution function in time. This problem will be addressed in the following sections (see also [236,237,290]).

6.3 Ostwald Ripening in a System of Nondeformable Pores of Equal Size R_0

6.3.1 Mathematical Formulation of the Problem and General Solution

Analyzing coarsening in porous materials, first the process of competitive growth of an ensemble of clusters in a system of pores of equal size R_0 is considered. The matrix is supposed to be absolutely rigid, i.e., the growth of the clusters is terminated immediately once the cluster size R becomes equal to R_0. It is assumed that for a radius R of a cluster less than R_0 the growth equation has the usual form for diffusion-limited growth, and that the growth is stopped immediately for $R = R_0$. Consequently, the growth equation can be written as

$$\frac{\mathrm{d}R}{\mathrm{d}t} = \frac{R_{c0}^3}{R}\left(\frac{1}{R_c} - \frac{1}{R}\right)\Theta(R_0 - R), \tag{6.9}$$

$$\Theta(R_0 - R) = \begin{cases} 1, & \text{for} \quad R_0 - R > 0, \\ 0, & \text{for} \quad R_0 - R \leq 0, \end{cases} \tag{6.10}$$

$$R_c(t) = \frac{2\sigma}{c_\alpha \Delta\mu(t)}, \qquad t = \frac{2\sigma D c'}{c_\alpha^2 k_B T R_{c0}^3} t_r. \tag{6.11}$$

In Eq. (6.9) a dimensionless time scale is used (see, e.g., [236]); R_{c0} is the critical cluster size at the beginning of the process of competitive growth and R_c is the actual time-dependent critical cluster size, σ is the specific surface energy, c_α the volume density of particles of the segregating phase, while c' denotes the equilibrium density of the segregating component in the matrix. D is the diffusion coefficient in the matrix phase, k_B the Boltzmann constant, T the absolute temperature, t_r the real time in seconds, and $\Delta\mu$ the change of the chemical potential in the precipitation process.

As a second assumption we demand that R_0 is sufficiently large to allow one the establishment of a time-independent distribution with respect to cluster sizes in reduced coordinates before the interactions with the pore walls start to play a dominating role. This assumption guaranties that nucleation processes do not occur any more. Moreover, it supplies a universal initial distribution independent of the prehistory of the process of formation of the cluster ensemble as a starting point for the following considerations.

In the present analysis, we use the original distribution function with respect to cluster sizes in reduced variables $\varphi(u,\tau)$ as introduced in Section 4.1 and proposed first in [153–155]

$$\varphi(u,\tau) = N(\tau)P(u), \tag{6.12}$$

where

$$\tau = 3\ln\frac{R_c(t)}{R_{c0}}, \qquad u = \frac{R}{R_c(t)}, \tag{6.13}$$

$$P(u) = \frac{3^4 eu^2 \exp\left\{-3\Big/\left[2\left(\frac{3}{2}-u\right)\right]\right\}}{2^{5/3}(u+3)^{7/3}\left(\frac{3}{2}-u\right)^{11/3}}, \tag{6.14}$$

and $\varphi(u,\tau)\,\mathrm{d}u$ is the number of clusters per unit volume of the matrix in the interval $u, u+\mathrm{d}u$ at the reduced time τ. The proposed method is, however, not restricted to any special form of the initial distribution function chosen as a starting point.

The function $P(u)$ is depicted in Figure 4.2. It has a relatively sharp maximum and tends to zero for $u = (3/2)$. Moreover, the average cluster size $\langle R(t)\rangle$ coincides with the critical cluster radius $R_c(t)$ thus corresponding to $u = 1$. As a consequence, once such a distribution is formed, the interactions with the walls start to influence the coarsening process when the critical cluster size has reached the value $R_c = (2/3)R_0$. At this moment of time, which we denote as t_0, the size distribution function $f(R,t)$ can be written as

$$f(R,t_0) = N(t_0)P\left(\frac{R}{R_c(t_0)}\right)\frac{1}{R_c(t_0)} \tag{6.15}$$

with

$$R_c(t_0) = \frac{2}{3}R_0. \tag{6.16}$$

In Eq. (6.15), $N(t_0)$ is the number of clusters per unit volume of the matrix at the moment $t = t_0$ and $f(R,t)$ the distribution function with respect to cluster sizes in coordinates R and t. It is connected with $\varphi(u,\tau)$ by

$$\varphi(u,\tau) = f(R,t)R_c(t). \tag{6.17}$$

Since nucleation processes do not occur in this late stage of the transformation, the further evolution of the size distribution function $f(R,t)$ is governed by the continuity equation

$$\frac{\partial f(R,t)}{\partial t} + \frac{\partial}{\partial R}\left(f(R,t)\frac{\mathrm{d}R}{\mathrm{d}t}\right) = 0. \tag{6.18}$$

In the interval $R < R_0$ the solution of Eq. (6.18) with the initial condition (6.15) can be expressed as

$$f(R,t) = \frac{N(t_0)}{R_c(t_0)} P\left(\frac{R_1(R,t)}{R_c(t_0)}\right) \frac{\partial R_1}{\partial R}, \tag{6.19}$$

where $R_1(R,t)$ is the characteristics of the growth equation

$$\frac{\mathrm{d}R}{\mathrm{d}t} = \frac{R_{c0}^3}{R}\left(\frac{1}{R_c} - \frac{1}{R}\right). \tag{6.20}$$

Indeed, the solution of the continuity equation (6.18) for any arbitrary initial distribution $f(R,t_1)$ can be found in the following way. Let $R = R(R_1,t)$ be a solution of the growth equation (6.20) containing some integration constant R_1. This equation can be transformed into the so-called characteristics

$$R_1 = R_1(R,t) \tag{6.21}$$

of the partial differential equation (6.18). The solution of Eq. (6.18) is then given by

$$f(R,t) = f(R_1(R,t),t_1)\frac{\partial R_1}{\partial R}. \tag{6.22}$$

The right-hand side of Eq. (6.22) is obtained by substitution of R_1 for R into the initial distribution $f(R,t_1)$ and multiplying with $(\partial R_1/\partial R)$.

In order to prove that Eq. (6.22) is the solution of Eq. (6.18) let us calculate derivatives of Eq. (6.22),

$$\frac{\partial f}{\partial t} = \left(\frac{\partial f}{\partial R_1}\frac{\partial R_1}{\partial t}\right)\frac{\partial R_1}{\partial R} + f(R_1,t_1)\frac{\partial^2 R_1}{\partial t \partial R}, \tag{6.23}$$

$$\begin{aligned} \frac{\partial}{\partial R}(f\dot{R}) &= \frac{\partial}{\partial R}\left(f(R_1,t)\frac{\partial R_1}{\partial R}\dot{R}\right) \\ &= \left(\frac{\partial f}{\partial R_1}\frac{\partial R_1}{\partial R}\right)\frac{\partial R_1}{\partial R}\dot{R} + f(R_1,t)\frac{\partial^2 R_1}{\partial R^2}\dot{R} + f(R_1,t)\frac{\partial R_1}{\partial R}\frac{\partial \dot{R}}{\partial R}. \end{aligned} \tag{6.24}$$

Since R is, on the other hand, some constant of integration, $R_1 = R_1(R,t)$ determines R as a function of t. Therefore, we have

$$\frac{\mathrm{d}R_1}{\mathrm{d}t} = \frac{\partial R_1}{\partial R}\frac{\mathrm{d}R}{\mathrm{d}t} + \frac{\partial R_1}{\partial t}. \tag{6.25}$$

A second derivation with respect to t and R yields

$$\left[\frac{\partial}{\partial t}\left(\frac{\partial R_1}{\partial R}\right)\right]\frac{\mathrm{d}R}{\mathrm{d}t} + \frac{\partial R_1}{\partial R}\frac{\mathrm{d}^2 R}{\mathrm{d}t^2} + \frac{\partial^2 R_1}{\partial t^2} = 0, \tag{6.26}$$

$$\frac{\partial^2 R_1}{\partial R^2}\frac{\mathrm{d}R}{\mathrm{d}t} + \frac{\partial R_1}{\partial R}\frac{\partial}{\partial R}\left(\frac{\mathrm{d}R}{\mathrm{d}t}\right) + \frac{\partial^2 R_1}{\partial R \partial t} = 0. \tag{6.27}$$

Substituting Eqs. (6.23) and (6.24) into the continuity equation (6.24), taking into account the additional relationships (6.26) and (6.27), proves that Eq. (6.22) is, indeed, the solution of Eq. (6.18).

The validity of Eq. (6.22) may be also verified in an alternative way. Let $f(R_1, t_1)\,\mathrm{d}R_1$ be the number of clusters in the interval $R_1, R_1 + \mathrm{d}R_1$ at the moment of time t_1. The growth of the clusters for $t > t_1$ is determined by Eq. (6.21). Replacing thus R_1 according to Eq. (6.21) by R, we obtain at any arbitrary moment of time

$$f(R_1(R,t), t_1)\frac{\partial R_1}{\partial R}\,\mathrm{d}R,$$

and, consequently, immediately Eq. (6.22).

Substituting Eq. (6.9) into Eq. (6.18) further yields

$$f(R,t) = \left(\left(\int_{t_0}^{t} f(R_0 - \epsilon, t)\left.\frac{\mathrm{d}R}{\mathrm{d}t}\right|_{R_0-\epsilon} \delta(R_0 - R)\,\mathrm{d}t\right)\right)_{\epsilon\to 0}, \qquad R \geq R_0. \tag{6.28}$$

For the determination of the evolution of the size distribution function $f(R,t)$ in the time interval $t \geq t_0$ the time dependence of the critical cluster radius has to be estimated. It is determined by the mass-balance equation, which can be transformed into the following expression:

$$\frac{1}{R_c(t)} = \frac{1}{R_c(t_0)} + \frac{4\pi}{3}\frac{c_\alpha}{c_0 - c'}\frac{1}{R_{c0}}\left[\int_0^{R_0} R^3\left[f(R,t_0) - f(R,t)\right]\mathrm{d}R \right.$$
$$\left. - \left(\left(\int_{t_0}^{t} R_0^3 f(R_0 - \epsilon, t')\left.\frac{\mathrm{d}R}{\mathrm{d}t}\right|_{R_0-\epsilon}\mathrm{d}t'\right)\right)_{\epsilon\to 0}\right]. \tag{6.29}$$

Here c_α is the volume concentration of the segregating particles in the newly evolving phase, and c_0 and c' are the initial and equilibrium concentrations of the segregating particles in the matrix, respectively.

The set of Eqs. (6.19)–(6.29) with the initial condition (6.15) has to be solved, now, in a self-consistent way. As a result one obtains a unique description of the evolution of the size distribution function $f(R,t)$ for any (arbitrary) initial distribution and, moreover, for any form of the growth equation, provided the characteristics $R_1(R,t)$ of the growth equation can be found. In general, the function $f(R,t)$ can only be determined by numerical methods. However, qualitative results can be determined analytically based on certain approximations which are discussed in the next section.

6.3.2 Approximations and Numerical Results

To obtain the characteristics of the growth equation (6.20), we linearize this equation in the vicinity of $R = R_c$. This procedure yields

$$\frac{\mathrm{d}R}{\mathrm{d}t} = \frac{R_{c0}^3}{R_c^3}(R - R_c), \qquad R < R_0. \tag{6.30}$$

A comparison of the original growth equation (6.20) with its linearized version (6.30) is shown in Figure 6.3. Deviations are significant only for small values of the cluster radius R. Since most clusters have a size not differing considerably from R_c, the deviations of the growth rate for small cluster sizes do not play a significant role with respect to the evolution of the distribution function. Moreover, in particular for small values of R, it is problematic to obtain with Eq. (6.20) an accurate description of the decay of the clusters, because this equation was based on the steady-state solution of the diffusion equation [289]. The deviations between an improved decay rate and the linear approximation (6.30) are smaller than indicated in Figure 6.3.

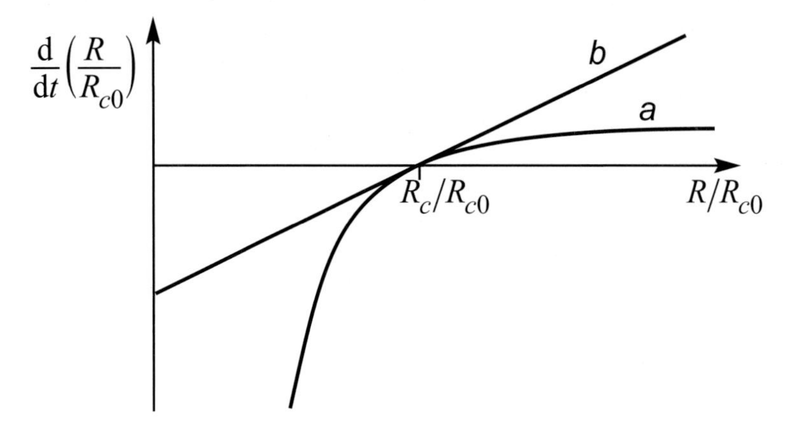

Figure 6.3: Comparison of the original growth equation (6.20) (curve a) with the linearized version (6.30) in the vicinity of R/R_c (curve b).

The characteristics of Eq. (6.30) can be expressed in the following form:

$$R_1(R,t) = [R - R_c(t)] \exp\left(-R_{c0}^3 \int_{t_0}^{t} \frac{\mathrm{d}t'}{R_c^3(t')}\right) + R_c(t_0)$$
$$+ \int_{t_0}^{t} \frac{\mathrm{d}R_c}{\mathrm{d}t'} \exp\left(-R_{c0}^3 \int_{t_0}^{t'} \frac{\mathrm{d}t''}{R_c^3(t'')}\right) \mathrm{d}t', \tag{6.31}$$

or

$$R_1 = R \exp\left(-R_{c0}^3 \int_{t_0}^{t} \frac{\mathrm{d}t'}{R_c^3(t')}\right) + R_{c0}^3 \int_0^t \frac{\mathrm{d}t'}{R_c^2(t')} \exp\left(-R_{c0}^3 \int_{t_0}^{t'} \frac{\mathrm{d}t''}{R_c^3(t'')}\right). \tag{6.32}$$

For the considered time interval $t \geq t_0$, the critical cluster radius is expected to have values in the interval $(2/3)R_0 < R_c(t) < R_0$. Moreover, the inhibition of cluster growth also results in a pronounced inhibition of the rate of growth of the critical cluster size, which is small compared with R_c already in the absence of inhibition effects.

Further taking into account the inequality $R_{c0}^3 \ll R_c^3$, we obtain as a first approximation for $R_1(R,t)$

$$R_1(R,t) = R_c(t_0) + [R - R_c(t)] \exp\left(-R_{c0}^3 \int_{t_0}^{t} \frac{\mathrm{d}t'}{R_c^3(t')}\right). \tag{6.33}$$

With Eq. (6.19) it follows immediately that for large times $f(R,t)$ tends to zero for $R < R_0$. The clusters existing initially in the system either decay or grow up to a size $R = R_0$, thus forming a monodisperse size distribution.

The results of a self-consistent solution of the set of Eqs. (6.19), (6.28), (6.29), and (6.31) or (6.32) are shown in Figure 6.4. As can be seen, the abrupt total inhibition of the growth rate for $R = R_0$ results in the formation of an additional peak in the distribution at R_0 and, consequently, as an intermediate stage, in the formation of a bimodal size distribution. In the course of time the lower peak is shifted to larger R values until both cluster populations unite and a monodisperse distribution is established. Figures 6.5 show the evolution in time of some quantities derived from $f(R,t)$ like the critical cluster size, the average cluster size, the fraction of matter in the peak, and the number of clusters in the system.

As to be expected and is evident from Figure 6.4, starting with some moment of time t_2 the distribution function can be approximated in the interval $R_2 < R < R_0$ by a linear function

$$f(R,t) = \alpha(t)[R - R_2(t)], \qquad t \geq t_2. \tag{6.34}$$

Substituting into the continuity equation and application of Eq. (6.30) results in the following differential equations for α and $R_2(t)$:

$$\frac{\mathrm{d}\alpha}{\mathrm{d}t} + 2\alpha(t)\frac{R_{c0}^3}{R_c^3} = 0, \tag{6.35}$$

$$\frac{\mathrm{d}R_2}{\mathrm{d}t} - \frac{R_{c0}^3}{R_c^3} R_2 = -\frac{R_{c0}^3}{R_c^2}. \tag{6.36}$$

The solutions of these equations are given by

$$\alpha(t) = \alpha(t_2) \exp\left(-2\int_{t_2}^{t} \frac{R_{c0}^3}{R_c^3(t')}\,\mathrm{d}t'\right), \tag{6.37}$$

$$R_2(t) = \exp\left(\int_{t_2}^{t} \frac{R_{c0}^3}{R_c^3(t')}\,\mathrm{d}t'\right) \times \left[R_2(t_2) - \int_{t_2}^{t} \frac{R_{c0}^3}{R_c^3(t'')} \exp\left(-\int_{t_2}^{t''} \frac{R_{c0}^3}{R_c^3(t')}\,\mathrm{d}t'\right) \mathrm{d}t''\right]. \tag{6.38}$$

Since $R_c(t')$ may vary only slightly in the considered stage of the process, we set it equal to some average R_c' and obtain, approximately, the following solutions:

$$R_2(t) = [R_2(t_2) - R_c'] \exp\left(\frac{R_{c0}^3}{R_c'^3}(t - t_2)\right) + R_c', \tag{6.39}$$

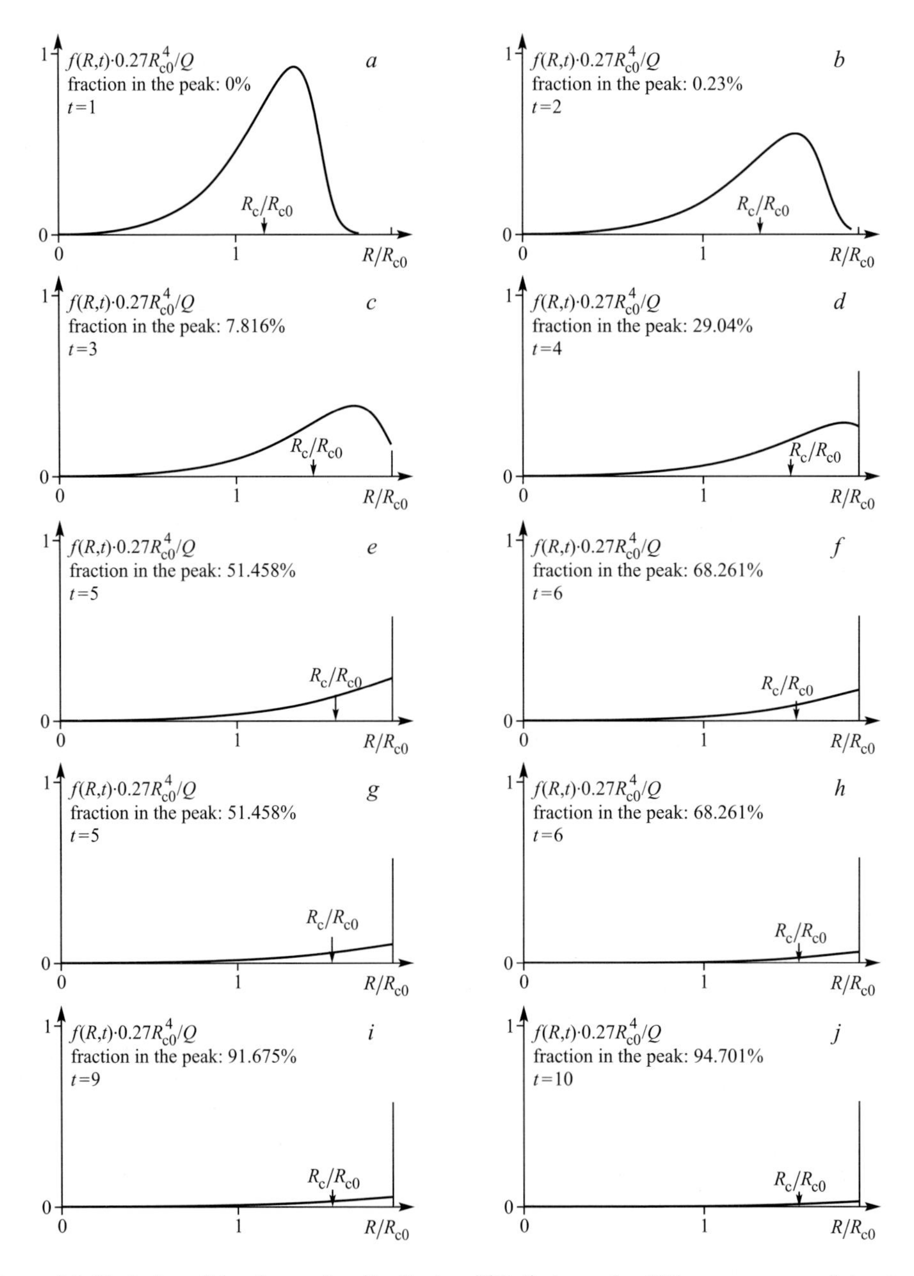

Figure 6.4: Evolution of the cluster-size distribution $f(R,t)$ shown for different moments of time in the time interval when interactions with the pore walls become dominating. The initial distribution is given by Eqs. (6.12) and (6.15). Q is defined as $Q = (c_0 - c')c$.

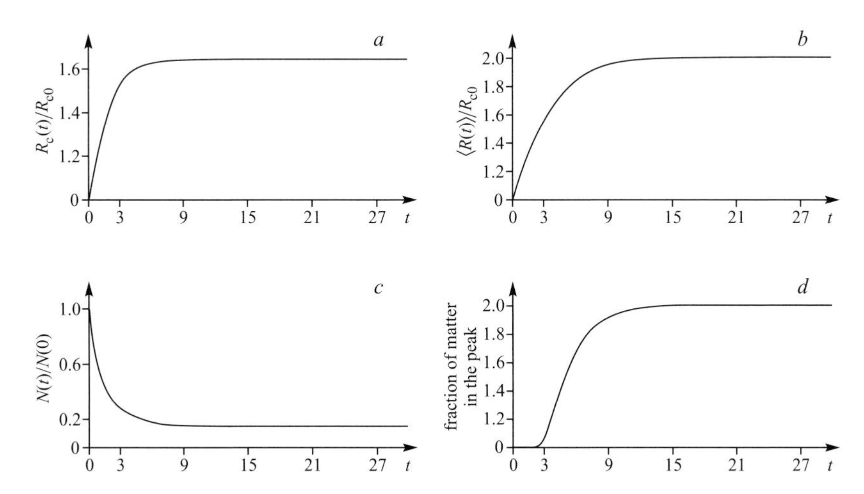

Figure 6.5: Time evolution of some relevant quantities, characterizing the process of coarsening in porous materials: (a) critical cluster size R_c; (b) average cluster size $\langle R\rangle$; (c) number of clusters per unit volume N; (d) fraction of matter in the peak at $R = R_0$.

$$\alpha(t) = \alpha(t_2)\exp\left(-2\frac{R_{c0}^3}{R_c'^3}(t-t_2)\right). \tag{6.40}$$

In this approximation, the limiting value of the critical cluster size for $t \to \infty$ is obtained from Eq. (6.29) as

$$\lim_{t\to\infty}\frac{1}{R_c(t)} = \frac{1}{R_c(t_2)} + \frac{4\pi}{15}\frac{c_\alpha}{c_0 - c'}\frac{\alpha(t)R_0^5}{R_{c0}}\left(1 - \frac{5}{2}\frac{R_0 - R_c'}{R_0}\right). \tag{6.41}$$

Since the average size of the critical cluster R_c' is greater than $2R_0/3$, some constant value greater than $R_c(t_2)$ and less than R_0 is reached asymptotically.

Summarizing the theoretical description of coarsening, or Ostwald ripening, in a system of pores, outlined in the last two sections of the present chapter, note once more that it is not restricted to a particular form of the growth equation. Consequently, it can also be applied to the situation when the inhibiting terms grow continuously over a certain interval of R values. A special case of such a type of behavior has been analyzed already in the papers [221, 232, 233, 236] with particular attention directed to the evolution of the cluster-size distribution in Ostwald ripening [236]. The method applied consists in a modification of the Lifshitz–Slezov theory. However, to some extent the limits of applicability of the method applied in [236] were not finally established. Thus, a verification of the results obtained in [236] with the theoretical approach developed here is desirable. As a further step, the extension of the theory to a system of pores with given size distribution is of interest and will be discussed in Section 6.5.

Not taken into account so far are modifications of the growth equations resulting from the inhomogeneity of the matrix and accompanying disturbances of the diffusion fields of the segregating particles. Such peculiarities will possibly give some quantitative modifications. It is believed, however, that they do not change qualitatively the results outlined here.

6.4 Coarsening in a System of Weak Pores

For a derivation of the equations governing the evolution of an ensemble of precipitate clusters in a system of pores with an arbitrary distribution with respect to pore sizes in a solid matrix as a first precondition two basic equations are required, the growth equation for a single cluster with a radius R in a pore of size R_0 and the distribution function with respect to pore sizes. The velocity of growth of a single cluster in a pore of size R_0 was assumed in Section 6.3 to be of the form

$$\frac{\mathrm{d}R}{\mathrm{d}t} = \frac{R_{c0}^3}{R}\left(\frac{1}{R_c} - \frac{1}{R}\right)\Theta(R_0 - R), \tag{6.42}$$

$$\Theta(R_0 - R) = \begin{cases} 1 & \text{for} \quad R_0 - R > 0 \\ 0 & \text{for} \quad R_0 - R \leq 0. \end{cases} \tag{6.43}$$

This equation implies that the growth of the cluster is terminated at a radius R_0 due to the interactions with the walls of the pore.

In the opposite situation, the elastic strains, starting with some initial value of the precipitate size R_0, begin to inhibit its further growth. We now discuss this situation assuming that the inhibiting effect increases only slowly with an increasing size of the cluster. In fact, in this case Eq. (6.42) has to be modified.

In general, the elastic strains inhibiting the growth are functions of the actual cluster radius R and the initial pore size R_0. We will assume that stress inhibiting effect on cluster growth may be described by a term $\Phi(R, R_0)$ as

$$\frac{\mathrm{d}R}{\mathrm{d}t} = \frac{R_{c0}^3}{R}\left(\frac{1}{R_c} - \frac{1}{R}\right) - \Phi(R, R_0). \tag{6.44}$$

By introducing the reduced variables (see Chapter 4 and [289])

$$u = \frac{R}{R_c(t)}, \qquad x = \frac{R_c(t)}{R_{c0}}, \qquad \tau = \ln\left[\frac{R_c(t)}{R_{c0}}\right], \tag{6.45}$$

Eq. (6.44) is transformed to

$$\frac{\mathrm{d}u^3}{\mathrm{d}\tau} = \gamma(\tau)\left\{u - 1 - \frac{u^2 x^2 \Phi(R, R_0)}{R_{c0}}\right\} - u^3 \tag{6.46}$$

with

$$\gamma(\tau) = \left\{\frac{x^2 \mathrm{d}x}{\mathrm{d}t}\right\}^{-1}. \tag{6.47}$$

The asymptotic stage of Ostwald ripening is reached if the conditions

$$F(u) = \frac{\partial}{\partial u}F(u) = 0, \tag{6.48}$$

$$F(u) = \gamma(\tau)\left\{u - 1 - \frac{u^2 x^2 \Phi(R, R_0)}{R_{c0}}\right\} - u^3 \tag{6.49}$$

are fulfilled. It means that the function $F(u)$ is tangent to the u-axis at some point $u = u_2$ and has a point of intersection with this axis at $u = -u_1$ (see Chapter 4 and [153–155]).

Equations (6.48) and (6.49) yield

$$u = \frac{3}{2} - \frac{x^2 u^2 \Phi}{2R_{c0}} \left\{ \frac{u\Phi_u}{\Phi} - 1 \right\}, \tag{6.50}$$

$$\gamma = \frac{\left\{ \dfrac{3}{2} - \dfrac{x^2 u^2 \Phi}{2R_{c0}} \left[\dfrac{u\Phi_u}{\Phi} - 1 \right] \right\}^3}{\left\{ \dfrac{1}{2} - \dfrac{x^2 u^2 \Phi}{2R_{c0}} \left[\dfrac{u\Phi_u}{\Phi} + 1 \right] \right\}}, \tag{6.51}$$

In these equations the notation $\Phi_u = (\partial\Phi/\partial u)$ is used. By solving both the equations the quantities γ and the particular value of u ($u = u_2$) may be obtained. In the limiting case $\Phi = 0$ the well-known asymptotic values of the Lifshitz–Slezov theory (see Chapter 4)

$$u = \frac{3}{2}, \qquad \gamma = \frac{27}{4} \tag{6.52}$$

result, again.

In the general case $\Phi \neq 0$ these quantities are, of course, modified by Φ and Φ_u, which account for the influence of the pores on the evolution of the cluster ensemble. Provided, as assumed, Φ and Φ_u are smoothly increasing functions of the cluster size, we get in the next approximation, instead of Eqs. (6.50) and (6.51), the following expressions:

$$u = \frac{3}{2} - \frac{9x^2 \Phi}{8R_{c0}} \left[\frac{3\Phi_u}{2\Phi} - 1 \right], \tag{6.53}$$

$$\gamma = \frac{\left\{ \dfrac{3}{2} - \dfrac{9x^2 \Phi}{8R_{c0}} \left[\dfrac{3\Phi_u}{2\Phi} \right] \right\}^3}{\left\{ \dfrac{1}{2} - \dfrac{9x^2 \Phi}{8R_{c0}} \left[\dfrac{3\Phi_u}{2\Phi} + 1 \right] \right\}}. \tag{6.54}$$

Equations (6.53) and (6.54) are obtained by substituting u on the right-hand side of Eqs. (6.50) and (6.51) with its value obtained in first-order approximation ($u = 3/2$). Similarly, the terms Φ and Φ_u have to be understood as

$$\Phi(R, R_0) = \Phi(uR_c, R_0) = \Phi\left(\frac{3R_c}{2}, R_0 \right), \tag{6.55}$$

$$\Phi_u(R, R_0) = R_c \left\{ \frac{\partial \Phi(uR_c, R_0}{\partial(uR_c)} \right\}_{u=\frac{3}{2}}. \tag{6.56}$$

Since Φ is an increasing function of u (the inhibiting effect of the pores increases with an increasing size of the clusters of the evolving phase), Eq. (6.54) may be rewritten in the form (the second term in the numerator of Eq. (6.54) can be neglected)

$$x^2 \frac{\mathrm{d}x}{\mathrm{d}t} = \frac{4}{27} \left\{ 1 - \frac{9x^2 \Phi}{4R_{c0}} \left[\frac{3\Phi_u}{2\Phi} + 1 \right] \right\}. \tag{6.57}$$

Hereby in addition, the definition of $\gamma(\tau)$, as given by Eq. (6.47), was used.

Equation (6.57) represents a differential equation for the determination of the time evolution of the critical cluster radius R_c. Once, having solved Eq. (6.57), the time evolution of the critical cluster radius and, therefore, $\gamma(\tau)$, are known, then the solutions u_1 and u_2 of the equation $F(u) = 0$ can be also determined.

As it was shown already in [236] in the analysis of a special case these solutions u_1 and u_2 determine the shape of the cluster-size distribution function $\varphi(u,\tau)$ in reduced variables (compare Chapter 4 and Eqs. (6.45)). We have (see for the additional details [236])

$$\varphi(u,\tau) = N(\tau)P(u), \tag{6.58}$$

$$P(u) = \frac{3u^2 \exp\left[-\dfrac{3C_1 u}{(u_2-u)u_2}\right] u_2^{3C_2} u_1^{3C_1}}{(u_2-u)^{3C_2+2}\,(u+u_1)^{3C_2+1}} \tag{6.59}$$

with

$$C_1 = \frac{u_2^2}{u_1+u_2}, \qquad C_2 = \frac{u_2(2u_1+u_2)}{(u_1+u_2)^2}, \qquad C_3 = \frac{u_1^2}{(u_1+u_2)^2}. \tag{6.60}$$

In the limiting case $\Phi = 0$ the already known distribution $P(u)$

$$P(u) = \frac{3^4 \mathrm{e} \exp\left\{-\dfrac{3}{[2(3/2-u)]}\right\}}{2^{5/3}(u+3)^{7/3}(3/2-u)^{11/3}} \tag{6.61}$$

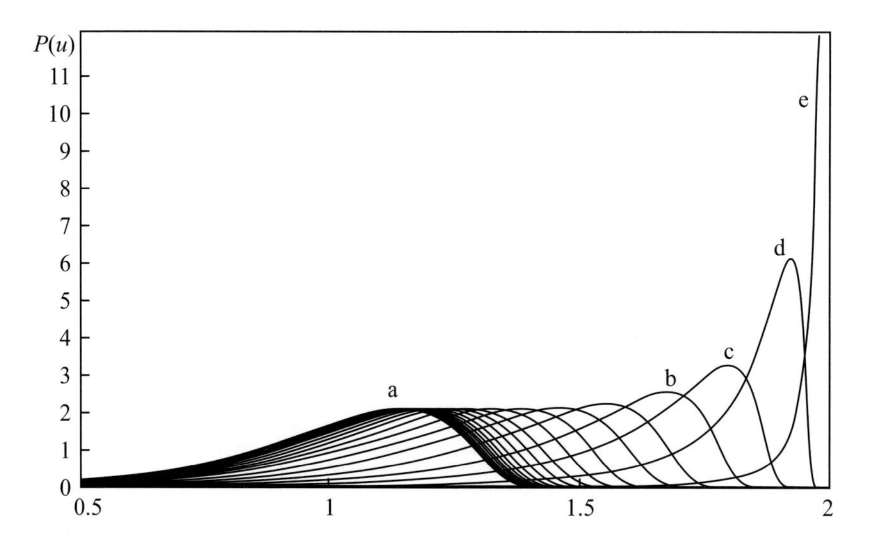

Figure 6.6: Cluster-size distribution functions $P(u)$ for different values of the parameter α corresponding to different moments of time. The numbers refer to the following reduced times: (a) small times, (b) $t' = 6.6 \times 10^5$, (c) $t' = 9.8 \times 10^5$, (d) $t' = 1.47 \times 10^6$, (e) $t' = 2.2 \times 10^6$ (for the details see [236]).

is obtained as a special case (cf. Eqs. (6.12) and (4.50)). However, in general, the parameters u_1 and u_2 are slowly varying functions of time. Consequently, also the distribution function in reduced variables $P(u)$ slowly changes with time.

The outlined method can be applied for any model of cluster growth provided the elastic strains are sufficiently moderately increasing functions of cluster size. For a particular model of cluster growth, developed in [233], the resulting curves for the evolution of the cluster-size distribution function and related quantities are given in Figure 6.6 [236].

6.5 Coarsening in a System of Nondeformable Pores with a Given Pore-Size Distribution

6.5.1 A First Approximation

Whether an arbitrarily chosen cluster with an actual radius R will grow or not depends on the size of the pore it is contained in or more generally on the pore-size distribution, which we denote by $W(R_0, \langle R_0 \rangle)$. The parameter $\langle R_0 \rangle$ has the meaning of the average pore size.

In order to apply the results obtained for ensembles of pores of equal size to the description of, at least, some average characteristics of Ostwald ripening in systems of pores of a given distribution an effective average growth rate may be introduced. This effective growth velocity of a cluster of size R may be obtained by averaging the growth rate with the normalized to unity distribution function $W(R_0, \langle R_0 \rangle)$, i.e.,

$$\left\langle \frac{\mathrm{d}R}{\mathrm{d}t} \right\rangle = \int_0^\infty \mathrm{d}R_0 \left[W(R_0, \langle R_0 \rangle) \frac{\mathrm{d}R}{\mathrm{d}t} \right] \tag{6.62}$$

resulting with Eq. (6.42) in

$$\left\langle \frac{\mathrm{d}R}{\mathrm{d}t} \right\rangle = \frac{R_{c0}^3}{R} \left(\frac{1}{R_c} - \frac{1}{R} \right) \int_0^\infty \mathrm{d}R_0 \left[W(R_0, \langle R_0 \rangle) \Theta(R_0 - R) \right] . \tag{6.63}$$

As mentioned, the normalization condition

$$\int_0^\infty W(R_0, \langle R_0 \rangle) \, \mathrm{d}R_0 = 1 \tag{6.64}$$

has to be fulfilled. It follows from Eq. (6.63) that the effective growth rate $\langle \mathrm{d}R/\mathrm{d}t \rangle$ is a function of the quantities R, R_c, and $\langle R_0 \rangle$, only.

Examples for pore-size distributions which may be of relevance for different applications are the Gaussian distribution

$$W(R_0, \langle R_0 \rangle) = \frac{\exp\left[-\alpha(\langle R_0 \rangle - R_0)^2 \right]}{\int_0^\infty \exp\left[-\alpha(\langle R_0 \rangle - R_0)^2 \right]} \tag{6.65}$$

and a distribution given by

$$W(R_0, \langle R_0 \rangle) = \frac{(n-1)}{\langle R_0 \rangle \left(1 + \dfrac{R_0}{\langle R_0 \rangle}\right)^n} . \tag{6.66}$$

Here, n is some positive number.

The average growth rates are then obtained in the form

$$\left\langle \frac{\mathrm{d}R}{\mathrm{d}t} \right\rangle = \left(\frac{\mathrm{d}R}{\mathrm{d}t} \right) \left\{ \frac{1 + \mathrm{erf}\left[\sqrt[2]{\alpha}(\langle R_0 \rangle - R)\right]}{1 + \mathrm{erf}(\sqrt[2]{\alpha}\langle R_0 \rangle)} \right\} \tag{6.67}$$

for the case given by Eq. (6.65), and

$$\left\langle \frac{\mathrm{d}R}{\mathrm{d}t} \right\rangle = \frac{\left(\dfrac{\mathrm{d}R}{\mathrm{d}t} \right)}{\left(1 + \dfrac{R}{\langle R_0 \rangle}\right)^{n-1}} \tag{6.68}$$

for the distribution described by Eq. (6.66).

The evolution of the cluster-size distribution function $f(R,t)$ and related quantities are governed in this approximation by a continuity equation in cluster-size space of the form (cf. Section 6.3)

$$\frac{\partial f(R,t)}{\partial t} + \frac{\partial}{\partial R} \left(f(R,t) \left\langle \frac{\mathrm{d}R}{\mathrm{d}t} \right\rangle \right) = 0. \tag{6.69}$$

The solution of the problem can be formulated, consequently, in the same way as done in Section 6.3 for the case of monodisperse pore-size distributions where R_0 has to be replaced, now, by $\langle R_0 \rangle$.

Let $f(R, t_1)$ be the cluster-size distribution at some given moment of time t_1 and

$$R_1 = R_1(R,t) \tag{6.70}$$

the characteristics of the growth equation (6.63) (for example, Eq. (6.67) or (6.68)) obeying the condition $R_1(R, t_1) = R$. The cluster-size distribution function in the interval $R < \langle R_0 \rangle$ at any moment of time is then given by

$$f(R,t) = f(R_1(R,t), t_1) \frac{\partial R_1(R,t)}{\partial R} . \tag{6.71}$$

Thus, for a determination of the time evolution of the cluster-size distribution function, similar to those obtained in Sections 6.3 and in [290], only the characteristics of the equation has to be determined.

Suppose, initially a Lifshitz–Slezov distribution is established and the effective growth rate may be expressed in the form as given by Eq. (6.44) with a moderately increasing inhibiting term Φ then also the method developed in Section 6.3 may be used for a first estimation of the time evolution of some characteristic quantities like the number of clusters, the average, and the critical cluster sizes. However, the transition to an average growth rate performed with Eq. (6.62) implies that the details of the evolution of the cluster-size distribution in the system of pores cannot be described adequately by either of both mentioned analytical approaches. Consequently, for a detailed determination of the evolution of the cluster-size distribution in a system of pores of given distribution another method has to be developed which will be discussed in the following subsection.

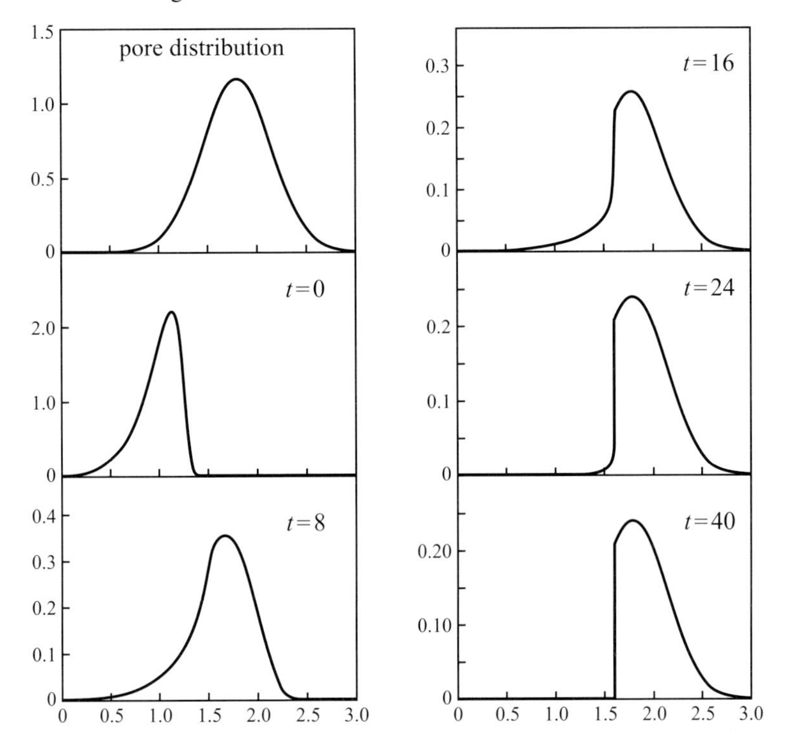

Figure 6.7: Different stages in the time evolution of the cluster-size distribution evolving in a Gaussian-type pore-size distribution (shown first at the top of the figure).

6.5.2 General Approach: Description of the Method

We assume that an ensemble of clusters grows in a system of pores with a size distribution characterized by the normalized to unity distribution function $W(R_0)$. Each pore contains by assumption not more than one cluster, since processes of coarsening inside one pore may be expected to proceed with a much higher rate as compared with the process in the system as a whole.

We further demand that the pores are sufficiently large to allow initially the establishment of a time-independent distribution with respect to cluster sizes in reduced coordinates, as given by the original Lifshitz–Slezov distribution $P(u)$ (Eq. (6.61)). This assumption supplies us with a universal initial distribution independent of the pecularities of nucleation and growth in

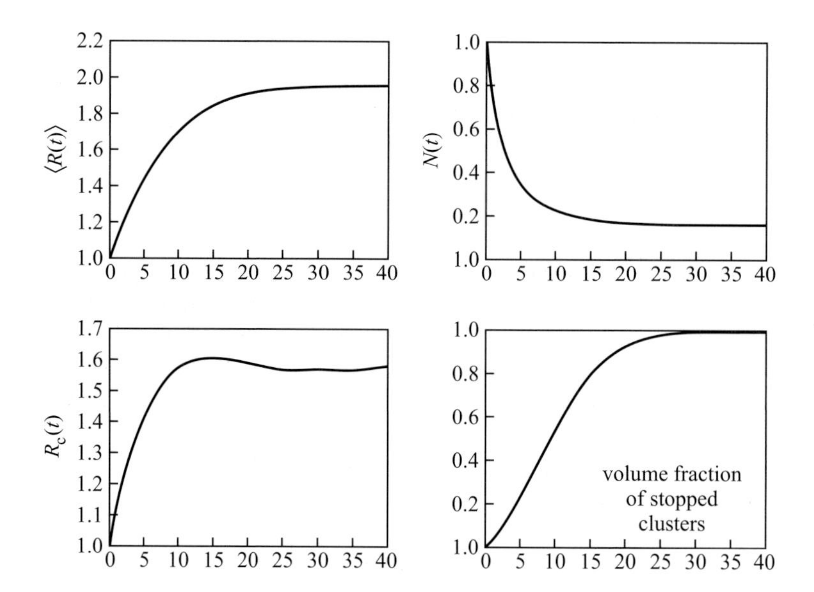

Figure 6.8: Time evolution of the average cluster radius $\langle R \rangle$, the critical cluster radius R_c, the number of clusters $N(t)$ in the system, and the amount of the new phase immobilized in the pores as functions of time for the process shown in Figure 6.7 as obtained from the numerical calculations.

the initial stages of the phase transformation. It means, moreover, that cluster- and pore-size distributions at the beginning of the coarsening process are statistically independent. Note that the described method is applicable, in principle, also for any other initial distribution.

Since we assume that initially an evolution of the cluster-size distribution is not influenced by the interactions with the walls the distribution of clusters and the pore-size distributions are initially well separated in cluster-size space, i.e., the largest cluster in the distribution is less in size than the smallest pore.

Let N be the number of clusters in a unit volume of the porous solid and $\mathrm{d}N$ their share with cluster sizes in the interval $R, R + \mathrm{d}R$ and in pores of radii $R_0, R_0 + \mathrm{d}R_0$, then a distribution function $w(R, R_0, t)$ may be introduced, defined by

$$\mathrm{d}N = w(R, R_0, t)\,\mathrm{d}R\,\mathrm{d}R_0. \tag{6.72}$$

In order to find the cluster-size distribution $f(R, t)$ this equation has to be integrated over all possible pore sizes, i.e.,

$$f(R, t) = \int_0^\infty w(R, R_0, t)\,\mathrm{d}R_0. \tag{6.73}$$

Because of the assumed statistical independence of pore and cluster radii the distribution function $w(R, R_0, t)$ in the initial stage can also be written in the form

$$w(R, R_0, t) = f(R, t) W(R_0). \tag{6.74}$$

The details of the numerical procedure allowing one the determination of the time evolution of the cluster-size distribution are given elsewhere [184]. Here we want to note only that in the numerical calculations no use was made of the continuity equation, which was crucial for guarantying the stability of the numerical procedure.

6.5.3 Results

The results of the numerical determination of the time evolution of the cluster-size distribution function in a given distribution with respect to pore sizes are illustrated in Figures 6.7–6.9. Starting always with an initial distribution (corresponding to the Lifshitz–Slezov distribution) as depicted in Figure 6.7 (for $t = 0$) and assuming a Gaussian type pore-size distribution a number of different stages in the time evolution of the size distribution is depicted (Figure 6.7). Moreover, also the time dependence of some related quantities is shown like for the average cluster radius, the critical cluster radius, the number of clusters in the system and the ratio of the evolving phase immobilized in the pores (Figure 6.8). As expected the cluster-size distribution in the final state reproduces partly the pore-size distribution. The degree of filling hereby depends mainly on the initial density of segregating particles (see Figure 6.9).

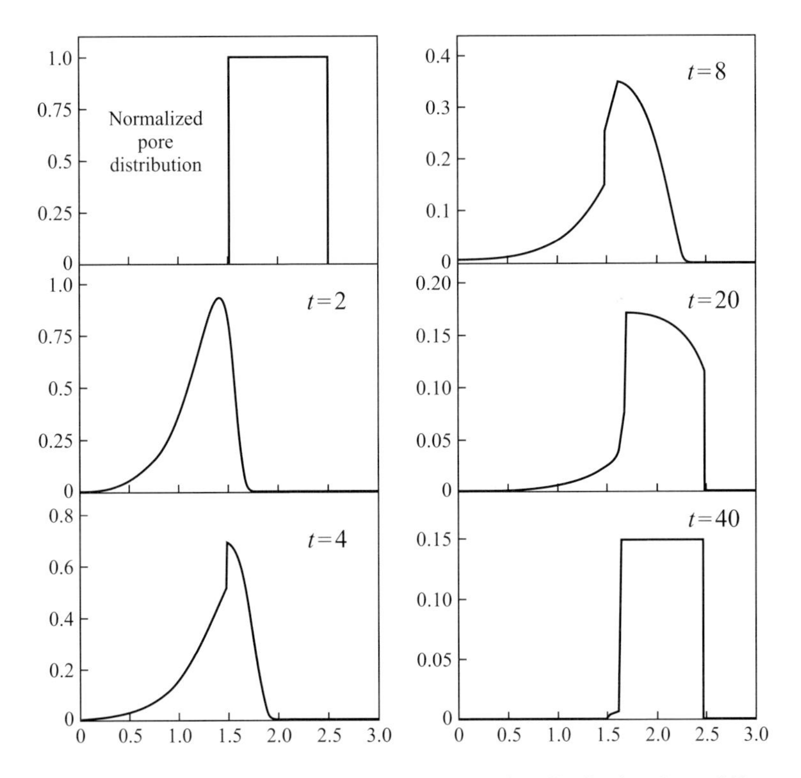

Figure 6.9: Different stages in the evolution of the cluster-size distribution for a different pore-size distribution. As expected, in the final stage the cluster-size distribution models the pore-size distribution (compare also Figure 6.7).

6.6 Influence of Stochastic Effects on Coarsening in Porous Materials

In the preceding discussion the coarsening, or Ostwald ripening process, was analyzed on the basics of deterministic growth equations like Eq. (6.42) or (6.44). By the application of such equations the influence of stochastic effects (like thermal noise) on coarsening cannot be described, in principle.

In [161] it was shown, however, that the incorporation of stochastic effects into the description of Ostwald ripening in homogeneous media results in a broadening of the cluster-size distribution function, at least, for intermediate time scales, and can be considered, thus, as one of the factors which may account for the gap between the theoretical and experimental results concerning the shape of the cluster-size distribution function in Ostwald ripening [107, 171]. Consequently, the same effect is also expected to occur for Ostwald ripening in porous media.

As outlined in detail in [161,293] stochastic effects can be accounted for in the description of coarsening by the numerical solution of the set of basic equations underlying classical nucleation theory, i.e.,

$$\frac{\partial N(j,t)}{\partial t} = J(j-1,t) - J(j,t), \tag{6.75}$$

$$J(j,t) = w^{(+)}(j,t)N(j,t) - w^{(-)}(j+1,t)N(j+1,t). \tag{6.76}$$

Here $N(j,t)$ is the number of clusters in the considered volume consisting at time t of j monomers. The coefficients of attachment $w^{(+)}$ and of detachment $w^{(-)}$ reflect the particular mechanism of growth or decay of the clusters.

By a Taylor expansion of the terms $w^{(+)}(j-1)N(j-1)$ and $w^{(-)}(j+1)N(j+1)$ a Fokker–Planck equation of the form

$$\frac{\partial N(j,t)}{\partial t} = -\frac{\partial}{\partial j}[v(j,t)N(j,t)] + \frac{\partial}{\partial j}\left[D(j,t)\frac{\partial}{\partial j}N(j,t)\right] + \cdots \tag{6.77}$$

is obtained with

$$v(j,t) = w^{(+)}(j,t) - w^{(-)}(j+1,t), \tag{6.78}$$

$$D(j,t) = \frac{w^{(+)}(j,t) + w^{(-)}(j+1,t)}{2}. \tag{6.79}$$

It is seen that while the first term in Eq. (6.77) describes the deterministic flow with the average growth rate v, the second term accounts for stochastic effects on motion in cluster-size space.

In Figures 6.10 results are shown for the time evolution of a cluster ensemble in a system of nondeformable pores of equal size ($j = 5000$) based on the numerical solution of the set of kinetic equations (6.75) and (6.76) with appropriate initial and boundary conditions [161,171]. The distribution curves in Figure 6.10(a) are normalized in such a way that the maximum of all shown curves equals $N = 1$. Qualitatively the results remain the same as discussed so far. However, some broadening of the distribution occurs, again. For the particular case considered this effect is seen most clearly in Figure 6.10(b). Not a delta-function-like distribution is

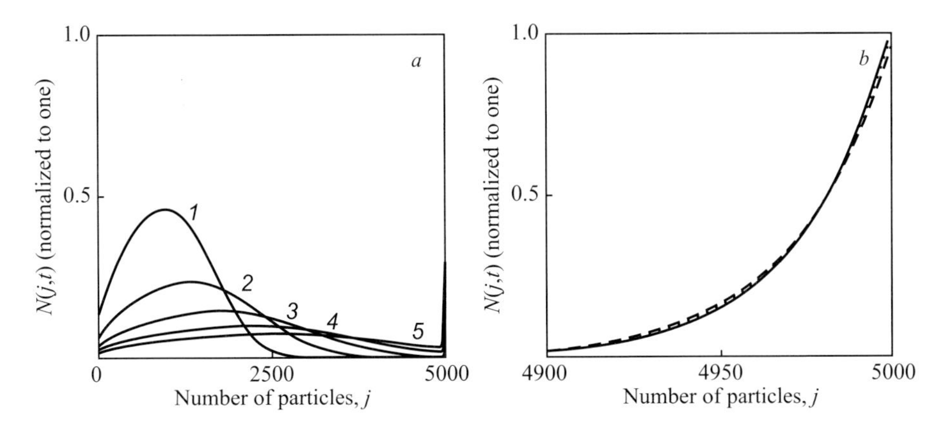

Figure 6.10: (a) Cluster-size distribution function for Ostwald ripening in a porous material containing pores of equal size with radii allowing one to introduce a maximum number of particles $j_{\max} = 5000$ for different moments of time ((1) $t = 992$; (2) $t = 1493$; (3) $t = 1993$; (4) $t = 2493$; (5) $t = 2994$). At $t = 0$ the evolution was started with a monodisperse distribution consisting of monomers, only. (b) Part of the cluster-size distribution for cluster sizes near $j_{\max}$ and relatively late stages of the coarsening process ($t = 11306$, $t = 11794$, and $t = 12282$). As can be seen from the figure, practically the curves coincide. The deviation from a δ-function-like distribution is due to stochastic effects (thermal noise). In both cases (a) and (b), the results were normalized in such a way that the maximum height of all considered curves equals one. The following values are assigned to the parameters employed in the calculations: $c_\alpha = 0.29 \times 10^{29}\,\mathrm{m}^{-3}$; $D = 0.17 \times 10^{-17}\,\mathrm{m}^2\mathrm{s}^{-1}$; $\sigma = 0.08\,\mathrm{Nm}^{-2}$; $c' = 0.86 \times 10^{26}\,\mathrm{m}^{-3}$; $c_\beta(t = 0) = 0.86 \times 10^{28}\,\mathrm{m}^{-3}$; $T = 730$ K.

established finally, but a stationary distribution with a sharp maximum for clusters having the size of the pores and decreasing rapidly with a decreasing number of particles in the clusters.

Analytically, this stationary distribution for $t \to \infty$ can be obtained from Eqs. (6.75) and (6.76) as

$$N(n-1) = \frac{w^{(-)}(n)}{w^{(+)}(n-1)} N(n), \qquad n = j_{\max}, \tag{6.80}$$

$$N(n-k) = \frac{w^{(-)}(n) w^{(-)}(n-1) \cdots w^{(-)}(n-(k-1))}{w^{(+)}(n-1) w^{(+)}(n-2) \cdots w^{(+)}(n-k)} N(n). \tag{6.81}$$

6.7 Discussion

In the present chapter both analytical and numerical methods have been developed allowing one to investigate the kinetics of Ostwald ripening in an ensemble of pores for different pore-size distributions and types of response of the walls of the pores with respect to cluster growth. The notation of pore is used with a generalized meaning either as a real object in materials like vycor glasses or zeolites or as a low-density region of a homogeneous matrix where clusters may be formed and grow preferentially.

In the so far considered examples it was assumed that the diffusional flow of the segregating particles is not influenced by the structure of the porous material. This simplifying assumption can be replaced in a more accurate description including into the analysis an investigation of the influence of the particular structure of the matrix on the diffusional flow of the segregating particles. Qualitatively no changes in the kinetics of coarsening as compared with the outlined here results are expected. Moreover, in all considered cases, the growth is finally stopped by the matrix. In terms of the phenomenological theory of rheology this latter assumption corresponds to the assumption of a Kelvin's like body in modeling the rheological properties of the matrix [72]. Of interest would be, of course, also the consideration of the alternative model, where the properties of the matrix are described by Maxwell's model of viscous flow.

In the last decades, an increased interest in the theoretical description of different aspects of the kinetics of competitive growth of ensembles of clusters (Ostwald ripening), formed as the result of nucleation and growth or spinodal decomposition in first-order phase transition, can be noticed [221, 289]. These investigations include, e.g.: (i) the analysis of the influence of a finite volume fraction of the segregating phase on the coarsening process [12, 40, 65, 103, 170, 197, 324, 340]; (ii) a comparative study of possible reasons for the sometimes experimentally observed deviation of the distribution function with respect to cluster sizes from the form predicted by the original Lifshitz–Slezov–Wagner theory [107, 161, 171]; (iii) a thermodynamic analysis of this process resulting in the estimation of the initial state where Ostwald ripening may start, and in the description of the whole course of Ostwald ripening, including its initial nonasymptotic stage [220, 221, 226, 234]; (iv) the evaluation of the influence of modifications of the growth equations on the ripening kinetics and the form of the distribution function with respect to cluster sizes [55, 86, 278, 285, 289]; (v) the analysis of the influence of elastic strains on the coarsening process, due either to an elastic interaction between the clusters [7, 12, 39, 55, 108, 109, 122, 320] or to external tension or compression [179]; (vi) the description of Ostwald ripening in adiabatically closed systems verifying, in particular, the possibility of a switching from growth limited by diffusion-like processes to growth of the clusters limited by heat conduction [230, 231]; (vii) the analysis of Ostwald ripening for multicomponent systems [284, 289] including the evolution of ensembles of pores [227, 287]; (viii) the description of Ostwald ripening in the isothermal rheocasting process [345] and in nonequilibrium phase transitions [175]. For the respective developments, the reader is referred to the references cited. In the next chapter, we will extend here the analysis of the kinetics of coarsening for the case of a nonconserved amount of the segregating phase (see also [163, 274].

7 Cluster Formation and Growth in Segregation Processes at Given Input Fluxes of Monomers and Under the Influence of Radiation

7.1 Introduction

For a number of technological applications in materials science the understanding of the kinetics of phase transformation processes under varying external and/or internal conditions is of significant importance. One example in this respect consists in the description of the process of formation and growth of ensembles of clusters in segregation processes; if the segregating particles are added homogeneously to the bulk of a solid or liquid solution by a constant or changing in time rate Φ_a.

In application to photography this problem was analyzed both theoretically and experimentally by Leubner [148–150] for diffusion and kinetically limited growth modes of the clusters of the new phase. Leubner argued that for diffusion-limited growth the number of supercritical clusters, N, formed as a result of an interplay of nucleation, growth, and supply of additional monomers, is proportional to Φ_a, while for kinetically limited growth $N\langle R\rangle \approx \mathrm{const}$ should hold [150]. However, Leubner's theory is exclusively based on the consideration of a mass-balance equation interrelating the growth of the supercritical clusters with the input fluxes of monomers of the segregating component, so that an adequate theoretical description of nucleation is lacking. Moreover, a number of important parameters of the theory do not find a self-consistent theoretical determination in Leubner's approach.

Since the problem discussed by Leubner is of general theoretical and practical interest it is revisited here from a more fundamental point of view. Based on the numerical solution of a set of kinetic equations, in the present study, the process of cluster formation and growth at a constant rate of supply of monomers of the segregating component both for diffusion and kinetically limited growth is analyzed. Characteristic quantities like the average ($\langle R\rangle$) and critical (R_c) cluster sizes as well as the time evolution of the cluster-size distribution functions in absolute (R) and reduced variables (R/R_c) are discussed. Particular attention is devoted to the answer of the questions, whether the evolution of the cluster ensemble is governed by power laws, again, and if this is the case, which types of power laws occur, how the number of stable clusters formed in the system as a result of segregation processes depends on the rate of input fluxes of monomers, respectively, which quantities remain constant in the final stage of the segregation processes.

Another topic of huge practical interest is the analysis of cluster formation, cluster growth, and coarsening under the influence of external radiation (see also [1,2]). The respective prob-

Kinetics of First-order Phase Transitions. Vitaly V. Slezov

ISBN: 978-3-527-40775-0

lems are analyzed in the final part of the present chapter. The growth of second phase precipitates from the supersaturated solid solution under irradiation is investigated taking into account a new mechanism of precipitate dissolution. This mechanism is of a purely diffusion origin, i.e., it is based on diffusion outfluxes of point defects produced by irradiation within the precipitates into the host matrix, provided that the interface boundary is transparent for the point defects. The point defect production rate within a precipitate is proportional to its volume while the total diffusion influx of substitutional impurity atoms is proportional to its radius meaning that there exists a maximum size at which the precipitate growth rate equals the rate of its radiation-induced dissolution. This size is shown to be a stable one implying that under irradiation a stationary state can be achieved far away from the thermodynamic equilibrium.

7.2 Coarsening with Input Fluxes of Raw Material

7.2.1 Preliminary Estimates

We consider the process of formation and growth of clusters of a new phase in segregation processes in solid or liquid solutions. In the initial stage, at $t = 0$, the concentration of the segregating particles c is assumed to be equal to the equilibrium solubility $c_{(\mathrm{eq})}$ for a stable coexistence of the evolving phase with the initially existing ambient phase at a planar interface. Starting with such an initial state, particles of the segregating phase are added with a constant rate Φ_a homogeneously to the system. As a result the system is transferred into a metastable state and the spontaneous formation of supercritical clusters becomes possible.

Assuming a perfect solution the relative supersaturation with respect to cluster formation, $(\Delta\mu/k_BT)$ can be expressed as

$$\frac{\Delta\mu}{k_BT} = \ln\left(\frac{c(t)}{c_{(\mathrm{eq})}}\right) \qquad \text{with} \qquad c(t=0) = c_{(\mathrm{eq})}, \tag{7.1}$$

where $\Delta\mu$ is the difference in the chemical potentials referred to one segregating particle in the ambient and newly evolving phases, respectively, k_B is the Boltzmann constant, and T the absolute temperature.

According to classical nucleation theory the probability of formation of stable aggregates of the evolving phase increases with an increasing supersaturation. Consequently, after some period of time an intensive process of nucleation will occur. Under stable clusters hereby such aggregates are understood, which exceed in size the actual critical cluster radius, R_c, given by

$$R_c = \frac{2\sigma}{c_\alpha k_BT \ln\left(\dfrac{c(t)}{c_{(\mathrm{eq})}}\right)}. \tag{7.2}$$

Here c_α is the density of segregating particles in the newly evolving phase, while σ denotes the specific interfacial energy or surface tension.

The further growth of the clusters is described here by the commonly applied growth equation

$$\frac{\mathrm{d}R}{\mathrm{d}t} = \frac{2\sigma D c_{(\mathrm{eq})}}{c_\alpha^2 k_B T} \frac{1}{R} \left(\frac{1}{R} - \frac{1}{R_c} \right) \tag{7.3}$$

for diffusion-limited growth, while for kinetically limited growth the relation

$$\frac{\mathrm{d}R}{\mathrm{d}t} = \frac{2\sigma D c_{(\mathrm{eq})}}{c_\alpha^2 k_B T} \frac{1}{l_0} \left(\frac{1}{R} - \frac{1}{R_c} \right) \tag{7.4}$$

is used. Here D is the diffusion coefficient of the segregating particles in the ambient phase, R is the radius of the cluster, and l_0 a length parameter with a magnitude of the order of molecular dimensions.

Processes of formation and growth of clusters of the newly evolving phase result in a sharp reduction of the supersaturation, so that after some interval of time a steady state may develop, where the formation of new clusters is practically excluded. In this state, the input fluxes of monomers are utilized for the growth of the already existing clusters. Experimental examples for the establishment of such steady states are given by Leubner in the already cited papers [148–150]. In this stage, two limiting mechanisms for the further evolution of the already existing ensemble of clusters can be imagined:

(i) The rate of input of monomers is small, so that the usual dissolution-growth mechanism of Ostwald ripening [94,155] dominates. In this case the behavior of the cluster ensemble is governed by the equations

$$\langle R \rangle^3 \sim t, \qquad N \sim t^{-1} \tag{7.5}$$

for diffusion-limited growth and

$$\langle R \rangle^2 \sim t, \qquad N \sim t^{-3/2} \tag{7.6}$$

for kinetically limited growth.

(ii) The rate of input fluxes of monomers is sufficiently high to allow an independent simultaneous growth of the already formed supercritical clusters. In this case, we have to expect

$$\langle R \rangle^2 \sim t, \qquad N \sim \mathrm{const} \tag{7.7}$$

for diffusion-limited growth and

$$\langle R \rangle \sim t, \qquad N \sim \mathrm{const} \tag{7.8}$$

for kinetically limited growth.

Suppose, a steady state is established in the system, the change of the total amount of the evolving phase has to be equal to the number of monomers added to the system in the same time interval. This condition yields

$$\frac{\mathrm{d}}{\mathrm{d}t} \left[\frac{4\pi}{3 v_\alpha} \langle R \rangle^3 N \right] = \frac{\mathrm{d}N_1}{\mathrm{d}t} = \Phi_a . \tag{7.9}$$

Here v_α is the volume of a monomer in the evolving phase and N_1 is the total number of monomers. Obviously, neither of the mentioned limiting growth mechanisms fulfills the restriction given by Eq. (7.9). It follows that both independent growth at the expense of additionally introduced monomers and growth-dissolution effects have to be taken into account for an understanding of the establishment of the steady state observed experimentally.

If, however, for the considered case the asymptotic behavior is also governed by power laws of the form

$$\langle R\rangle^3 \sim t^\alpha, \qquad N \sim t^\beta \tag{7.10}$$

then according to Eq. (7.9) the additional condition

$$\alpha + \beta = 1 \tag{7.11}$$

has to be fulfilled.

7.2.2 Basic Kinetic Equations

In accordance with the classical theory of nucleation we assume that processes of cluster growth and decay proceed only via addition or evaporation of monomers. The clusters are assumed to be of spherical size and are characterized by the number of monomers i, contained in them, or by a radius R_i. $N(i,t)$ denotes the number of cluster consisting of i monomers.

The time evolution of the cluster-size distribution function $N(i,t)$ is governed under these assumptions by the following set of equations:

$$\begin{aligned}\frac{\partial N(i,t)}{\partial t} &= w^{(+)}(i-1,t)N(i-1,t) + w^{(-)}(i+1,t)N(i+1,t) \\ &\quad - \left[w^{(+)}(i,t) + w^{(-)}(i,t)\right] N(i,t).\end{aligned} \tag{7.12}$$

For relatively large cluster sizes this equation can be transformed by a Taylor expansion into a Fokker–Planck type equation of the form

$$\frac{\partial N(i,t)}{\partial t} = -\frac{\partial}{\partial i}\left[v(i,t)N(i,t)\right] + \frac{\partial^2}{\partial i^2}\left[a(i,t)N(i,t)\right]. \tag{7.13}$$

Hereby the notations

$$v(i,t) = w^{(+)}(i,t) - w^{(-)}(i,t)\,, \tag{7.14}$$

$$a(i,t) = \frac{w^{(+)}(i,t) + w^{(-)}(i,t)}{2} \tag{7.15}$$

are used.

Equations (7.14) and (7.15) allow one to determine the rates of attachment $w^{(+)}$ and dissolution $w^{(-)}$ for different deterministic growth mechanisms. For diffusion-limited growth we have (for the limiting case of a perfect solution)

$$w^{(+)}(i,t) = 4\pi R_i D c_\infty , \tag{7.16}$$

$$w^{(-)}(i,t) = 4\pi R_i D c_{R_i} , \tag{7.17}$$

while for kinetically limited growth

$$w^{(+)}(i,t) = 4\pi R_i^2 D \frac{c_\infty}{l_0} , \tag{7.18}$$

$$w^{(-)}(i,t) = 4\pi R_i^2 D \frac{c_{R_i}}{l_0} \tag{7.19}$$

holds. Hereby c_∞ is the concentration of the segregating particles far away from the growing or dissolving clusters, c_{R_i} specifies the equilibrium concentration of segregating particles in the vicinity of a cluster of size R_i. Its value is given by (e.g., [94])

$$c_{R_i} = c_{(\mathrm{eq})} \exp\left(\frac{2\sigma}{c_\alpha k_B T R_i}\right) . \tag{7.20}$$

Denoting by v_α the volume of a monomeric unit we may write further

$$v_\alpha = \frac{1}{c_\alpha} = \frac{4\pi}{3} R_1^3 . \tag{7.21}$$

Taking into account this relation the radius of a cluster of size R_i may be expressed as

$$R_i = \left(\frac{3v_\alpha}{4\pi}\right)^{1/3} i^{1/3} . \tag{7.22}$$

For the numerical calculations we will further use a dimensionless time scale defined by

$$t' = 4\pi D c_{(\mathrm{eq})} \left(\frac{3v_\alpha}{4\pi}\right)^{1/3} t . \tag{7.23}$$

Moreover, the parameter l_0 we identify in the calculations with R_1.

Substitution of Eqs. (7.16)–(7.23) into the basic equation (7.12) yields

$$\frac{\partial N(i,t')}{\partial t'} = \frac{c_\infty}{c_{(\mathrm{eq})}} (i-1)^{1/3} N(i-1,t') + (i+1)^{1/3} \exp\left(\frac{2\sigma}{c_\alpha k_B T R_{i+1}}\right) N(i+1,t')$$
$$- \left[\frac{c_\infty}{c_{(\mathrm{eq})}} + \exp\left(\frac{2\sigma}{c_\alpha k_B T R_i}\right)\right] i^{1/3} N(i,t') \tag{7.24}$$

for diffusion-limited growth and

$$\frac{\partial N(i,t')}{\partial t'} = \frac{c_\infty}{c_{(\mathrm{eq})}}(i-1)^{2/3}N(i-1,t') + (i+1)^{2/3}\exp\left(\frac{2\sigma}{c_\alpha k_B T R_{i+1}}\right)N(i+1,t')$$
$$- \left[\frac{c_\infty}{c_{(\mathrm{eq})}} + \exp\left(\frac{2\sigma}{c_\alpha k_B T R_i}\right)\right] i^{2/3}N(i,t') \tag{7.25}$$

for kinetically limited growth.

7.2.3 Results of the Numerical Solution of the Kinetic Equations

For the numerical solutions of Eqs. (7.24) and (7.25) Euler's Polygon method is used (cf. [123]), i.e., the change of the number of clusters consisting of i monomers is calculated by

$$N(i,t'+\Delta t') = N(i,t') + \frac{\partial N(i,t')}{\partial t'}\Delta t'. \tag{7.26}$$

The values of the parameters σ, c_α, and $c_{(\mathrm{eq})}$ are taken from Ref. [23].

The method can be made as accurate as desired by decreasing the length of the time step $\Delta t'$ [123]. For the calculations a time step $\Delta t' = 10^{-4}$ has been used. As a check of the reliability of the results obtained a second calculation has been carried out with a time step $\Delta t' = 10^{-5}$. The resulting values for $R_c(t')$, etc., differ only slightly ($< 3\%$) from the values obtained from the calculations with the larger time steps, so that the Polygon method can be considered as sufficiently accurate for our purposes.

In Figures 7.1(a) and (b) the time evolution of the average cluster radius $\langle R\rangle$, the critical cluster radius R_c, and their ratio $\langle R\rangle/R_c$ are shown both for diffusion (a) and kinetically limited growth (b). In Figures 7.2 the change of the relative supersaturation $\ln(c/c_{(\mathrm{eq})})$ is presented starting from an initial value equal to zero (by assumption $c(t=0) = c_{(\mathrm{eq})}$ holds). As

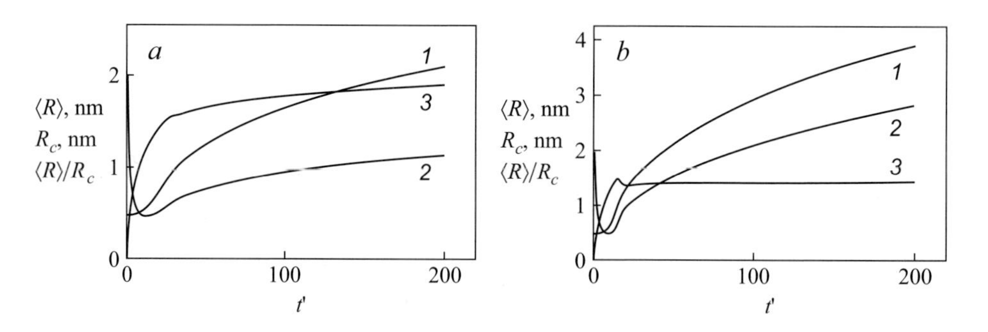

Figure 7.1: Average cluster radius $\langle R\rangle$ *(1)*, critical cluster radius R_c *(2)*, and their ratio $\langle R\rangle/R_c$ *(3)* as functions of time t' (in reduced units) for (a) diffusion-limited growth and (b) for kinetically limited growth. The input flux was chosen to be equal to $\Phi_a = 10^{29}$ monomers/(h m^3) for kinetically limited growth and $\Phi_a = 10 \times 10^{27}$ monomers/(h m^3) for diffusion-limited growth. The values of the parameters are taken from [23].

seen, initially the supersaturation grows monotonically. After some sufficiently large critical value is reached, intensive nucleation occurs. The processes of formation and further growth of the clusters diminish the supersaturation, again.

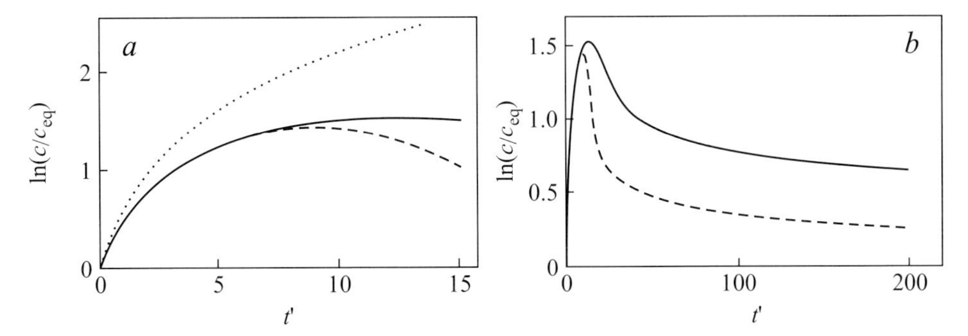

Figure 7.2: Time dependence of the relative supersaturation for diffusion (full curve) and kinetically limited growth (dashed curve). Figure (a) schematically shows the initial stage of the process. Here the dotted line describes the change in the supersaturation for a constant value of the input fluxes when no clusters are formed in the system, while (b) gives an impression of the whole course of the process. For the values of the parameters see the caption to Figure 7.1.

In Figures 7.3 the same processes are illustrated by considering the variation of the number of clusters in the system in time. It is seen that for diffusion-limited growth (full curves) a steady state is reached asymptotically characterized by a practically constant number of clusters. According to Eqs. (7.10) and (7.11) this result implies that the average cluster radius behaves as $\langle R \rangle^3 \propto t'$. Indeed, in agreement with such expectations we obtain from the results of the numerical calculations

$$\alpha = 1.01, \qquad \beta = 0.00 \qquad \text{for diffusion limited growth.} \tag{7.27}$$

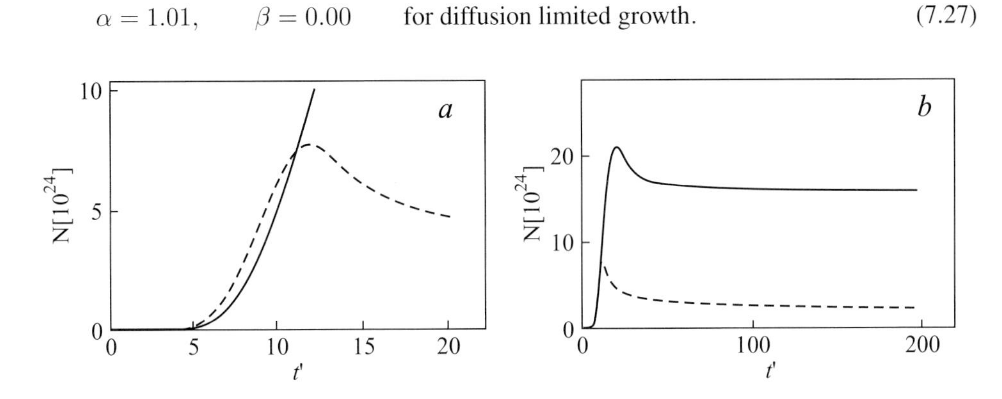

Figure 7.3: Number of cluster per cubic meter as a function of time t' (in reduced units) for diffusion (full curve) and kinetically limited growth (dashed curve). Figure (a) schematically shows the initial stage while (b) illustrates the whole process.

For kinetically limited growth (dashed curves) the rate of decrease of the cluster number becomes slower in the asymptotic region with time; however, it does not tend to zero for this mode of growth. Instead of Eqs. (7.27) we obtain

$$\alpha = 1.24, \qquad \beta = -0.27 \qquad \text{for kinetically limited growth.} \tag{7.28}$$

These differences are also reflected in the variations in time of the shape and the properties of the cluster-size distribution functions $F(R/R_c, t')$ shown in Figure 7.4.

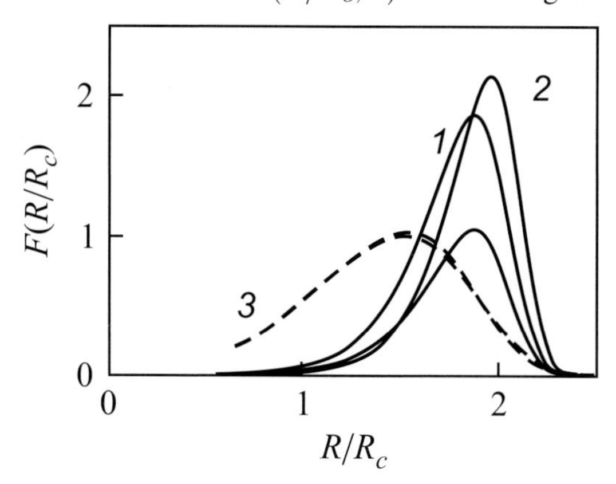

Figure 7.4: Cluster-size distribution function $F(R/R_c, t')$ (cf. Eq. (7.29)) for diffusion (full lines) and kinetically limited growth (dashed lines). The different curves refer to the following values of time (in reduced units): (1) $t' = 100$; (2) $t' = 183$. For kinetically limited growth the distribution curves coincide for both moments of time (3).

The function $F(R/R_c, t')$ is determined in the following way:

$$F\left(\frac{R}{R_c}, t'\right) = \frac{f(R/R_c, t')R_c(t')}{N_{(\text{total})}(t')}. \tag{7.29}$$

Here $f(R/R_c, t')R_c$ is the number of clusters at time t' in the interval $\left[(R/R_c), (R/R_c) + \mathrm{d}(R/R_c)\right]$ and $N_{(\text{total})}(t')$ the total number of clusters in the system at time t'. $F(R/R_c, t')$ fulfills the normalization condition, i.e., it is normalized to unity (cf. [155]). The particular function F introduced with Eq. (7.29) tends, for coarsening in a closed system, to a universal time-independent cluster-size distribution function given by the Lifshitz–Slezov theory of Ostwald ripening (Chapter 4 and [155]). We choose this kind of presentation of the distribution functions to allow one a direct comparison between the results obtained for the boundary conditions considered here compared with the well-known coarsening behavior for large times found for closed systems.

The cluster-size distribution functions in R-space ($f(R, t')$, respectively, $F(R/R_c, t')$) are obtained from $N(i, t')$-dependences, determined by the numerical calculations based on Eqs. (7.24) and (7.25), via the relation

$$N(i, t')\,\mathrm{d}i = N(R_i, t')\,\frac{\mathrm{d}i}{\mathrm{d}R}\,\mathrm{d}R = f(R, t')\,\mathrm{d}R. \tag{7.30}$$

In Figure 7.5, the asymptotic values of the number of clusters are shown as a function of the input fluxes of monomeric building units Φ_a. As is evident, for the considered interval of values of Φ_a, in agreement with Leubner's results, a linear dependence $N(t' \to \infty)$ vs Φ_a is found. For kinetically limited growth such linear dependence of the number of clusters on the value of the input fluxes Φ_a does not exist since N also decreases with time in the asymptotic region (cf. Figures 7.1 and 7.3).

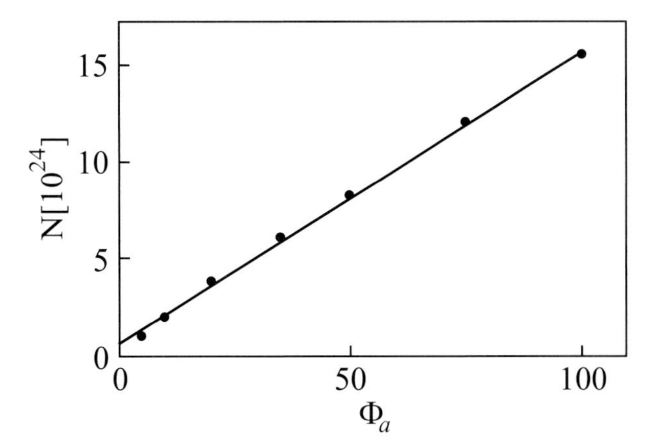

Figure 7.5: Asymptotic value of the number of clusters per cubic meter for diffusion-limited growth as a function of the input flux of monomers Φ_a.

7.2.4 Discussion

As already mentioned above, processes of segregation in a system at a constant rate of supply of monomers can be divided into several distinct stages. Hereby the general scenario of the evolution does not depend on the particular mechanism of growth of the clusters (e.g., kinetically or diffusion-limited growth as considered here).

In the first stage of the process the relative supersaturation increases monotonically with time due to the supply of monomers to the system (Figure 7.2). After the initiation of nucleation, which occurs with perceptible intensity after some critical value of the supersaturation is reached, an ensemble of growing (supercritical) clusters is formed and initially the total number of clusters in the system is increased (Figures 7.3). In the second interval of intensive nucleation, the critical cluster radii are small and almost all supercritical clusters are growing rapidly, while at the same time new supercritical clusters are formed. As a result of both processes the degree of supersaturation (excess of monomers) achieved initially is reduced again (Figure 7.2). The decrease of relative supersaturation leads to an increase of the critical cluster radius (cf. Eq. (7.2)). As a result in a third stage of the process some of the clusters become subcritical and are dissolved, again. Hereby the total number of clusters decreases (see Figures 7.3). Finally, the system approaches a fourth steady state, where the changes in the relative supersaturation or the number of clusters become small and a balance between the input fluxes and the consumption of monomers by the growing clusters is established.

While this general scenario is same for diffusion and kinetically limited modes of growth, nevertheless, important differences are found, in particular, in the late stage of the process. For kinetically limited growth the rate of incorporation of the particles at the interface determines the rate of growth of the aggregates of the newly evolving phase while for diffusion-limited growth the transport of monomers to the interface is the rate-determining step, i.e., the relations

$$w^{(+,-)}(i) \sim i^{1/3} \quad \text{for diffusion-limited growth} \tag{7.31}$$

$$w^{(+,-)}(i) \sim i^{2/3} \quad \text{for kinetically limited growth} \tag{7.32}$$

hold. The differences in the kinetically growth laws result in qualitative differences in the coarsening kinetics.

Indeed, according to Eqs. (7.10), (7.11), (7.27), and (7.28) we have

$$\begin{aligned} &\langle R\rangle \sim t^{1/3} \qquad N \cong \text{const} \qquad \text{for diffusion-limited growth} \\ &\langle R\rangle \sim t^{5/12} \qquad N \sim t^{-1/4} \qquad \text{for kinetically limited growth.} \end{aligned} \tag{7.33}$$

It turns out that the combination $\langle R\rangle N$ is, for kinetically limited growth, a slowly varying quantity ($\langle R\rangle N \sim t^{1/6}$), but not a constant, as supposed by Leubner. According to Eqs. (7.33) it is not $\langle R\rangle N$ but the quantity $\langle R\rangle^{3/5}N$ approaches a constant value for kinetically limited growth

$$\langle R\rangle^{3/5}N \cong \text{const} \quad \text{for kinetically limited growth.} \tag{7.34}$$

On the other hand, Leubner's suggestion concerning nearly constant values of the number of clusters N in the final stage of segregation for diffusion-limited growth and its linear dependence on the input fluxes of monomers Φ_a could be proven to be correct.

Qualitative differences in the kinetics of the evolution of the systems are also clearly reflected in the shapes of the cluster-size distribution functions (Figure 7.4). For diffusion-limited growth, almost all clusters are distributed in the supercritical region $R > R_c$. The whole ensemble of clusters in the system is stable; decay of subcritical clusters does not play an important role and the total number of clusters approaches a constant value. However, the maximum of the cluster-size distribution function still moves to larger values of R/R_c due to the continuous increase of the ratio $\langle R\rangle/R_c$ (see Figure 7.1(a)).

It is seen that the shape of the distribution, the system tends to, resembles the Lifshitz–Slezov distribution (cf. Chapter 4 and [155]). However, the peak is located not at $R/R_c \approx 1$ as for closed systems but at higher values $R/R_c \approx 2$. In the system with kinetically limited cluster growth the distribution function shows for large times a significantly different behavior. The distribution is much broader with a maximum at $R/R_c \sim 1.5$ and practically time independent.

On the other hand, even for large times a nonvanishing part of the cluster ensemble keeps subcritical. This result is connected with a continuous increase of the critical cluster radius (Figure 7.1(b)). The subcritical clusters are dissolved and the total number of clusters decreases, i.e., the ripening is in progress and one cannot speak about an asymptotically approached final number of clusters in the system.

7.3 Void Ripening in the Presence of Bulk Vacancy Sources

7.3.1 Introduction

Practice shows that the exposure of matter to neutron or charged particle irradiation results in an increased porosity of it. This is due to intensive creation of vacancy–interstitial pairs in collision cascades or in individual collisions. The uncompensated source of vacancies, stimulating voidance, can act only if simultaneously (and this is really so) a corresponding outflow of interstitial atoms exists. Indeed, because of higher diffusion of interstitials they are rapidly incorporated into different defects or form new nuclear planes, producing, therefore, swelling of the samples [321, 323]. Without going into further details, we cite, for example, the paper [19], where the aggregation of interstitials into flat clots in copper irradiated by α particles was observed. Expanding the clots will constitute new nuclear planes causing the void swelling of the samples. There exist, apparently, other possibilities of vacancies release.

In this connection it is of interest to investigate possible choices of void ensemble evolution in the presence of bulk sources of vacancies by general analytic methods. Here we study the problem employing the following assumptions: (a) The working interval of temperatures ensures freezing of gas recoils, i.e., the created pores are voids. For uranium, for example, these temperatures are about 350–500 °C [4]. (b) The fluctuation process of nucleation and the growth of vacancy precipitates directly from supersaturated solution are terminated, and the ripening phenomenon starts to play an essential role. In this stage the supersaturation Δ, defined as $\Delta = c - c_\infty$ (c_∞ is the concentration of the saturated solution, c the equilibrium concentration at the void surface, related to the void radius R by the usual relation

$$c = c_\infty + \frac{\alpha}{R}, \qquad \alpha = \left(\frac{\sigma}{k_B T}\right) V c_\infty, \tag{7.35}$$

where σ is the interfacial surface tension, and V the vacancy volume), tends to zero at $t \to 0$, i.e., the voids have time to absorb all vacancies delivered by the sources. (c) The voids are spherical and located far enough apart from each other. (d) The sources of vacancies are described by the function $q(t)$, which is assumed to be monotonic in time. At infinity such functions are well majorized by polynomials. Therefore, not losing in generality, it is possible to consider, that $q(t) \leq q_0 t^{n-1}$, where q_0 is the strength of sources at $t = 1$, n is an arbitrary number (not necessary an integer), and to draw all conclusions for a majorant. Formulating the problem mathematically, we shall adhere to the notations of paper [274], which is the natural continuation of [153].

7.3.2 Basic Equations

Let us first write down the full set of necessary equations describing the kinetics of coarsening under the conditions considered:

(i) The canonical equation of motion in size space (see Eq. (4.24) in Chapter 4) is given by

$$\frac{\mathrm{d}u^3}{\mathrm{d}\tau} = \gamma\,(u-1) - u^3, \qquad \gamma = \frac{3\mathrm{d}t}{\mathrm{d}x^3}. \tag{7.36}$$

Here $u = (\rho/x(t))$ is the reduced radius (in relation to dimensionless critical radius $x(t) = \Delta_0/\Delta$; Δ_0 is the initial supersaturation; t is a dimensionless time ($t = t_1/T$, $T = R_{k0}^3/(\alpha D)$, D is the diffusivity); $R_{k0} = \alpha/\Delta_0$ is the initial critical radius, τ is the dimensionless canonical time ($\tau = \ln x^3$). The times t and τ are related through an unknown function $\gamma(T) = 3\mathrm{d}t/\mathrm{d}x^3$. At $t \to \infty$ one has $\tau \to \infty$.

(ii) The continuity equation in void size space,

$$\frac{\partial f}{\partial t} + \frac{\partial}{\partial \rho}(f v_\rho) = 0, \tag{7.37}$$

where $f(\rho, t)$ is the size-distribution function of voids, normalized per unit volume, so that $\mu = \int_0^\infty f \mathrm{d}\rho$ is the void number in a unit volume.

(iii) The equation of balance of matter, generalizing Eq. (4.21) from Chapter 4, is

$$1 + Q(t) = \frac{\Delta_0}{Q_0 x} + \kappa \int_0^\infty f(\rho, t)\, \rho^3\, \mathrm{d}\rho, \tag{7.38}$$

or

$$1 + Q(t) = \frac{\Delta_0}{Q_0} e^{-\tau/3} + \kappa e^{\tau} \int_{v_0(\tau)}^\infty f_0(v)\, u^3(v, \tau)\, \mathrm{d}v. \tag{7.39}$$

Here

$$Q = \frac{1}{Q_0} \int_0^t q(t)\, \mathrm{d}t \tag{7.40}$$

is the total volume of vacancies in a unit volume at time t, normalized to the total initial volume of vacancies Q_0 in unit volume; f_0 is the initial distribution;

$$\kappa = \frac{4\pi R_{k0}^3}{3Q_0}, \tag{7.41}$$

$v_0(\tau)$ is a solution of the equation $u[v_0(\tau), \tau] = 0$, which is the lower limit of void sizes, not dissolved to a time moment τ.

Equations (7.36) and (7.38) represent a full system. In these equations the unknown function is $\gamma(\tau)$. There are three possibilities of an asymptotical behavior of $\gamma(\tau)$ at $\tau \to \infty$:

$$\gamma(\tau) \to \infty, \tag{7.42}$$

$$\gamma(\tau) \to \mathrm{const} \tag{7.43}$$

with the least value $\gamma_0 = 27/4$, and

$$\gamma(t) \to 0. \tag{7.44}$$

Depending on the value of γ the plot of the rate of growth $\mathrm{d}u^3/\mathrm{d}\tau$ as a function of u can touch the x-axis at the point u_0 (at $\gamma = \gamma_0 = 27/4$) or pass below this axis ($\gamma < \gamma_0$), or to have a zone of positive values ($\gamma > \gamma_0$) (see Figure 4.1).

The analysis made in Section 4.1.2 results in the conclusion that Eqs. (7.36) and (7.38) suppose growth of volume occupied by vacancies (at $t \to \infty$) in the following cases: (i) $\gamma \to \gamma_0$: maximum growth here is $Q(t) \sim t^n$, $n < 1$. (ii) $\gamma \to \mathrm{const}$: it is the case of constant sources, $Q(t) \sim t$. (iii) $\gamma \to \infty$, $Q(t) \sim t^n$, $n > 1$. Let us now consider different special cases in detail.

7.3.3 Damped Sources

Let $q(t) = q_0 t^{n-1}$ with $0 \le n < 1$. It is obvious from the above said that in this case γ should tend to γ_0. Note that for strict equality $\gamma(\tau)_{\tau\to\infty} = \gamma_0$ all points lying to the right of tangency point $u_0 = 3/2$ and moving to the left, cannot "overcome" this point and get "stick" in it. The integral on the right-hand side of the balance equation (7.38), similar to the case $n = 0$, tends to some constant, and the right-hand side itself grows in time as e^τ. At the same time left-hand side of Eq. (7.38) is proportional to $Q(t) \sim t^n \sim e^{n\tau}$, since $x^3 = e^\tau = 3/(\gamma_0 t)$ in case $\gamma = \mathrm{const} = \gamma_0$ (because $\gamma = 3\mathrm{d}t/\mathrm{d}x^3$), and, therefore, the balance equation cannot be satisfied. It means that γ tends to γ_0 from below

$$\gamma_n = \gamma_0 \left[1 - \varepsilon_n^2(\tau)\right], \tag{7.45}$$

where $\varepsilon_n^2(\tau)_{\tau\to\infty} \to 0$ is determined from Eqs. (7.36) and (7.38).

The introduction of $\varepsilon_n^2(\tau)$ is important for understanding of the process kinetics. Practically, similarly to $n = 0$, in a zeroth approximation one has

$$\gamma_n|_{\tau\to\infty} \to \gamma_0, \qquad n < 1. \tag{7.46}$$

Since $\gamma = 3\mathrm{d}t/\mathrm{d}x^3$, then

$$x^3\big|_{t\to\infty} \to \frac{4}{9}t. \tag{7.47}$$

Let us find the form of the distribution function. For this purpose we shall introduce a distribution function $\varphi(u,\tau)$ over relative sizes $u = \rho/x$, connected with $f(\rho,t)$ by the apparent relation

$$\varphi(u,\tau)\,\mathrm{d}u = f(\rho,t)\,\mathrm{d}\rho = xf(\rho,t)\,\mathrm{d}u. \tag{7.48}$$

The continuity equation for $\varphi(u,\tau)$ everywhere, except the neighborhood of the point u_0, at $\tau \gg 1$ takes the form (see Eq. (4.38) in Section 4.1.3)

$$\frac{\partial\varphi}{\partial\tau} - \frac{\partial[\varphi g(u)]}{\partial u} = 0, \tag{7.49}$$

where

$$g(u) = \frac{\left(u - \frac{3}{2}\right)^2(u+3)}{3u^2}. \tag{7.50}$$

The solution of this equation to the left of u_0 can be taken as

$$\varphi = \frac{\chi(\tau + \psi)}{g(u)}, \tag{7.51}$$

where

$$\psi = \int_0^u \frac{du}{g(u)} = \frac{4}{3}\ln(u+3) + \frac{5}{3}\ln\left(\frac{3}{2} - u\right) + \frac{1}{1 - \frac{2u}{3}} - \ln\frac{3^3 e}{2^{5/3}}, \tag{7.52}$$

and $\chi(\tau + \psi)$ is some arbitrary function.

Let us now advert to the equation of balance. In full analogy with the case $n = 0$, points cannot accumulate in the region to the right of u_0, nor directly in the neighborhood of the point u_0, and the cumulative contribution of these regions to the equation of balance is negligible. Therefore, the equation of balance can be used for the determination of an asymptotical behavior of the distribution function in zeroth approximation for $u < u_0$ (distribution function for $u > u_0$ in this approximation is equal to 0). Having substituted the solution (7.51) in the equation of balance (7.39) and, taking into account the above said, we obtain an asymptotic equation for χ

$$\frac{q_0 \gamma_0^n}{3^n n Q_0} e^{n\tau} = \kappa e^{\tau} \int_0^{3/2} \chi(\tau + \psi) \frac{u^3}{g(u)}\, du \tag{7.53}$$

or

$$1 = \kappa^* e^{(1-n)\tau} \int_0^{3/2} \chi(\tau + \psi) \frac{u^3}{g(u)}\, du, \qquad \kappa^* = \frac{\kappa Q_0 n 3^n}{q_0 \gamma_0^n}, \qquad 1 > n > 0. \tag{7.54}$$

In case $n = 0$ one also has to retain in Eq. (7.39) the time-independent term. The solution of Eq. (7.54) will be

$$\chi(\tau + \psi) = A_n e^{-(1-n)(\tau+\psi)}, \tag{7.55}$$

where

$$A_n = \left[\kappa^* \int_0^{3/2} e^{-(1-n)\psi} \frac{u^3}{g(u)}\, du\right]^{-1}. \tag{7.56}$$

The distribution function will take the form

$$\varphi(u, \tau) = A_n e^{-(1-n)(\tau+\psi)} \frac{1}{g(u)}$$
$$= A_n e^{-(1-n)\tau} \left(\frac{3^3 e}{2^{5/3}}\right)^{1-n} \frac{3u^2 \left[-\frac{1-n}{1 - 2u/3}\right]}{(u+3)^{\frac{7-4n}{3}} \left(\frac{3}{2} - u\right)^{\frac{11-5n}{3}}}. \tag{7.57}$$

The number of particles per unit volume, according to normalization in Eq. (7.37), will be equal to

$$\mu(\tau) = \int_0^{3/2} \varphi(u,\tau)\,\mathrm{d}u = \frac{A_n}{1-n} e^{-(1-n)\tau} = \frac{A_n}{1-n}\left(\frac{3}{2}\right)^{2(1-n)} t^{n-1}. \tag{7.58}$$

Let $P(u)\,\mathrm{d}u$ be the probability that the particle size lies between u and $u+\mathrm{d}u$. Then from Eqs. (7.57) and (7.58) it follows

$$\varphi(u,\tau) = \mu(\tau)\,P(u), \tag{7.59}$$

where

$$P(u) = \begin{cases} \dfrac{(1-n)\,3u^2\left(\dfrac{3^3 e}{2^{5/3}}\right)\exp\left[-\dfrac{(1-n)}{1-2u/3}\right]}{(u+3)^{\frac{7-4n}{3}}\,(3/2-u)^{\frac{11-5n}{3}}}, & u < u_0; \\ 0 & u > u_0. \end{cases} \tag{7.60}$$

Figure 7.6 shows the probability $P(u)$ as a function of u for $n = 0$ and $n = 0.9$.

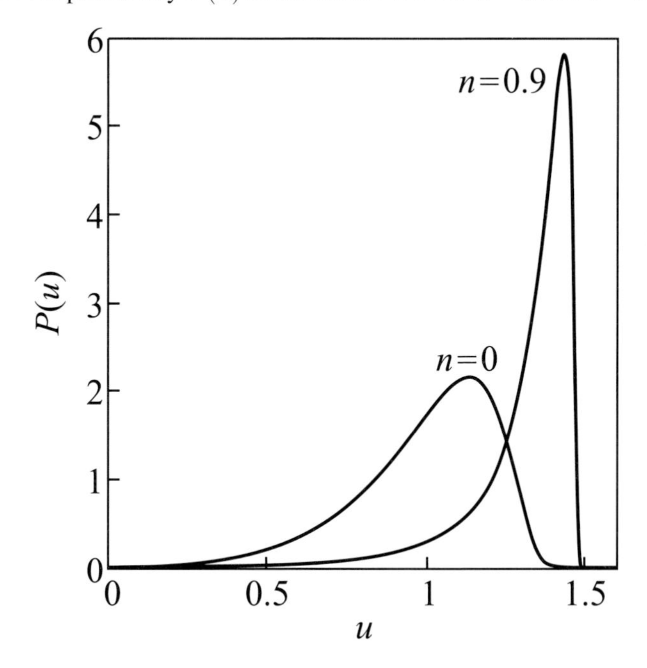

Figure 7.6: The probability $P(u)$ as a function of u for $n = 0$ and $n = 0.9$.

Equations (7.58)–(7.60) completely determine an asymptotical distribution of particles over relative sizes and its time evolution. In order to obtain a distribution function over absolute sizes, we shall also calculate $\overline{u}$.

$$\overline{u} = \frac{\displaystyle\int_0^{3/2} e^{-(1-n)\psi} \frac{u}{g(u)} du}{\displaystyle\int_0^{3/2} e^{-(1-n)\psi} \frac{1}{g(u)} du} = c_n. \tag{7.61}$$

At $n \to 1$ we have $c_n \sim 1.5$, that is $1 \leq \overline{u} < 1.5$. Taking into account that $u = \frac{\rho}{x}$, we find $\overline{u} = c_n = \frac{\overline{\rho}}{x}$, $u = c_n \frac{R}{\overline{R}}$. Besides, from Eq. (7.47) it follows

$$t = \left(\frac{3}{2}\right)^2 x^3 = \frac{9}{4} c_n^{-3} \frac{\overline{R}^3}{R_{k0}^3}. \tag{7.62}$$

Returning to initial dimensional quantities, we have

$$\left.\begin{aligned} f(R,t) &= \mu(t) P\left(c_n \frac{R}{\overline{R}}\right) \frac{c_n}{\overline{R}}, \\ \mu(t) &= \frac{A_n}{1-n} \left(\frac{3}{2}\right)^{2(1-n)} t^{n-1} = \frac{A_n}{1-n} c_n^{3(n-1)} \left(\frac{\overline{R}^3}{R_{k0}^3}\right)^{n-1}, \\ \overline{R}^3 &= \frac{4}{9} c^3 R_{k0}^3 t = \frac{4}{9} c^3 D\alpha t. \end{aligned}\right\}. \tag{7.63}$$

The probability $P(c_n \frac{R}{\overline{R}})$ is given by Eq. (7.60). The supersaturation at the time moment t takes the form

$$\Delta(t) = \frac{\Delta_0}{x(t)} = \left(\frac{3}{2}\right)^{2/3} \Delta_0 \left(\frac{T}{t}\right)^{1/3} = \lambda t^{-1/3}, \tag{7.64}$$

where

$$\lambda = \left(\frac{3}{2}\right)^{2/3} \left(\frac{\alpha^2}{D}\right)^{1/3}. \tag{7.65}$$

Let us find the applicability limits of the obtained formulas. From the above said it is evident that obtained asymptotic expressions are valid under the condition

$$\tau^2 = \left(\ln x^3\right)^2 = 9\left(\ln \frac{\overline{R}}{R_{k0}}\right)^2 \gg 1, \qquad R \gg R_{k0}. \tag{7.66}$$

The time, when the straight void ripening begins, is

$$t \sim \left(\frac{q_0}{M_0 D\alpha}\right)^{\frac{1}{1-n}}. \tag{7.67}$$

This estimation follows from the balance relationship $M_0 R^3 \sim q_0 t^n$ (M_0 is the initial number of voids) and the formula $\overline{R}^3 \sim D\alpha t$.

7.3.4 Undamped Sources

7.3.4.1 Constant Sources

Such kind of sources leads to qualitatively different results. Formally this is connected with the fact that the curve $\mathrm{d}u^3/\mathrm{d}\tau$ (Figure 4.1, case $\gamma > \gamma_0$) to some moment of time necessarily goes into an upper half plane. Thereof, the directions of movement of points are changed: the points lying to the left of u_1 tend to zero, and all the other tend to u_2. The main contribution to the equation of balance, naturally, will give the region close to u_2, and this determines the asymptotic shape of the distribution function taking now a delta-like form with a maximum at u_2.

Curve $\mathrm{d}u^3/\mathrm{d}\tau$ as a function of u, which in the initial stage can totally lie below the axes $0u$, after inclusion of sources begins to move up to the asymptotic position with a velocity defined by the strength q_0 (see Figure 4.1). The number of points which have had time to cross a neighborhood of u_0 and, finally to disappear in the origin of coordinates, depends both on the rate of uprise, and on an initial position of a package of points with respect to u_0. It means that a number of voids N_0, which has no time (to a moment when the curve touches the axes $0u$) to filter out through the locking point and asymptotically seized in an area to the right of u_0, essentially depends on q_0 and $f_0(R)$. Thus, the originating δ-asymptotics of the size distribution of voids is not so universal in a sense of independence from initial conditions as the asymptotics for damped sources which is totally independent from the initial distribution function.

The position of the u_2 point (the value of an averaged relative radius) together with the height of the peak N_0 completely determine the distribution function. To relate u_2 with q_0, we use the equation of balance and equation of motion. In the balance equation (7.39), taking into account the general tendency of points movement to u_2, we have

$$\left. \int\limits_{v_0(\tau)}^{\infty} f_0(v)\, u^3(v,\tau)\, \mathrm{d}v \right|_{\tau\to\infty} \to u_2^3 \int\limits_{u_1}^{\infty} f_0(v)\, \mathrm{d}v = u_2^3 N_0 . \tag{7.68}$$

Since $t = \frac{\gamma_\infty}{3} e^{\tau}$ when $\gamma = \mathrm{const} = \gamma_\infty$, we can divide Eq. (7.39) by common factor e^{τ}, obtaining

$$\frac{\gamma_\infty}{3} q_0 = \kappa u_2^3 N_0 . \tag{7.69}$$

The equation of motion (7.36) gives one more relation between u_2 and γ_∞,

$$\gamma_\infty (u-1) - u^3 = 0 . \tag{7.70}$$

The condition for the roots of this cubic equation to be real is

$$\frac{\gamma^2}{4} - \frac{\gamma^3}{27} < 0 \tag{7.71}$$

(the limiting value is $\gamma_0 = 27/4$), and the roots themselves are

$$u_1 = 2\sqrt{\frac{\gamma_\infty}{3}}\cos\left(\frac{\pi}{3}+\frac{\varphi}{3}\right), \qquad u_2 = 2\sqrt{\frac{\gamma_\infty}{3}}\cos\left(\frac{\pi}{3}-\frac{\varphi}{3}\right),$$

$$u_3 = -2\sqrt{\frac{\gamma_\infty}{3}}\cos\frac{\varphi}{3}. \tag{7.72}$$

Here

$$\varphi = \arccos\left\{\left[2\left(\frac{1}{3}\right)^{3/2}\gamma_\infty^{1/2}\right]^{-1}\right\}. \tag{7.73}$$

Using Eqs.(7.69) and (7.72), we come to the equation for γ_∞,

$$\frac{\gamma_\infty}{3}q_0 = \kappa N_0\left(\frac{\gamma_\infty}{3}\right)^{3/2}\left[2\cos\left(\frac{\pi}{3}-\frac{\varphi}{3}\right)\right]^3, \tag{7.74}$$

which allows us to write down the expression for $u_2(q_0 N_0)$.

At large q_0, which means large γ_∞, we have $\varphi \to \frac{\pi}{2}$. In this case the roots, Eq. (7.72), can be simplified

$$u_1 = 1, \qquad u_2 = \gamma_\infty^{1/2}. \tag{7.75}$$

Together with Eq. (7.69) it allows us to write down u_2 in the form

$$u_2 = \frac{q_0}{3\kappa N_0}. \tag{7.76}$$

The unknown quantity $N_0 = \text{const}$ is connected with the total swelling ΔV through the following relationship:

$$\Delta V = \frac{4}{3}\pi R_2^3 N_0, \quad \text{where} \quad R_2 = u_2 R_{k0} x. \tag{7.77}$$

From this we obtain

$$N_0 = \frac{\Delta V}{\frac{4}{3}\pi\left(u_2 R_{k0} x\right)^3}. \tag{7.78}$$

In the other limiting case ($q_0 \to 0$) $N_0 \to 0$, and the average size is about $\sim u_0$.

7.3.4.2 Regimes of Growth

In this case $\gamma \to \infty$ when $t \to \infty$, and it is already completely not obvious that the point u_2, itself moving with some velocity to the right, will be a point of concentration. Let us introduce a variable $z = u^3$. Then Eq. (7.36) takes the form

$$\frac{\mathrm{d}z}{\mathrm{d}\tau} = \gamma\left(z^{1/3} - 1\right) - z \approx \gamma z^{1/3} - z. \tag{7.79}$$

Let us form an expression $\frac{z}{z_2}$, using Eqs. (7.79) and (7.72) ($z_2 = \gamma_\infty^{5/2}$), and calculate a derivative $\frac{\mathrm{d}}{\mathrm{d}\tau}\left(\frac{z}{z_2}\right)$,

$$\frac{\mathrm{d}}{\mathrm{d}\tau}\left(\frac{z}{z_2}\right) = \left(\frac{z}{z_2}\right)^{1/3} - \left(\frac{z}{z_2}\right) - \frac{z}{z_2}\left(\frac{1}{z_2}\frac{\mathrm{d}z_2}{\mathrm{d}\tau}\right). \tag{7.80}$$

Next, let us transform the term $\frac{1}{z_2}\left(\frac{\mathrm{d}z_2}{\mathrm{d}\tau}\right)$. We have: $\gamma_\infty = \mathrm{const}\cdot t^N$, $0 < N < 1$ (the form of γ_∞ is dictated by $\gamma_\infty \to \infty$, $x^3 = 3\int_0^\infty \frac{\mathrm{d}t}{\gamma_\infty} \to \infty$; here it is pertinently to remind that we deal all the time with majorizing power functions). Besides, it is possible to show that $\frac{\mathrm{d}t}{\mathrm{d}\tau} = \frac{t}{1-N}$. From this it follows

$$\frac{1}{z_2}\frac{\mathrm{d}z_2}{\mathrm{d}\tau} = \frac{3}{2}\frac{N}{1-N}. \tag{7.81}$$

The parameter N is straightforwardly connected to the growth power n (see below). Thus,

$$\frac{\mathrm{d}}{\mathrm{d}\tau}\left(\frac{z}{z_2}\right) = \left(\frac{z}{z_2}\right)^{1/3} - \frac{z}{z_2}\left(1 + \frac{3}{2}\frac{N}{1-N}\right). \tag{7.82}$$

The point of concentration z_{con} has the property that

$$\frac{\mathrm{d}}{\mathrm{d}\tau}\left(\frac{z}{z_2}\right) > 0, \qquad z < z_{\mathrm{con}} \tag{7.83}$$

$$\frac{\mathrm{d}}{\mathrm{d}\tau}\left(\frac{z}{z_2}\right) < 0, \qquad z > z_{\mathrm{con}}. \tag{7.84}$$

It is clear from Eq. (7.82) that such point exists in all cases, which do not contradict the equation of motion, though it does not coincide with z_2. Let us determine z_{con} from the condition $\frac{\mathrm{d}}{\mathrm{d}\tau}\left(\frac{z}{z_2}\right) = 0$,

$$\left.\begin{aligned} \left(\frac{z_{\mathrm{con}}}{z_2}\right)^{1/3} &= \left(\frac{z_{\mathrm{con}}}{z_2}\right)\left(1 + \frac{3}{2}\frac{N}{1-N}\right) \\ z_{\mathrm{con}} &= z_2\left(1 + \frac{3}{2}\frac{N}{1-N}\right)^{-3/2} = Az_2, \quad A < 1 \end{aligned}\right\}, \tag{7.85}$$

i.e., $z_{\mathrm{con}} < z_2$.

Having clarified the existence of z_{con}, let us advert to the equation of balance (Eq. (7.39)). Asymptotically, similar to Eq. (7.69), we have

$$\left.\begin{aligned} &Q = \kappa x^3 z_{\mathrm{con}}(t) N_0 = \kappa x^3 A z_2 N_0, \\ &Q = \frac{1}{Q_0}\int_0^t q(t)\,\mathrm{d}t \quad N_0 = \int_{u_1}^\infty f_0(v)\,\mathrm{d}v \\ &A = \left(1 + \frac{3}{2}\frac{N}{1-N}\right)^{-3/2} \end{aligned}\right\}. \tag{7.86}$$

Substituting here $q = q_0 t^{n-1}$, $z_2 = \gamma^{3/2}$, and $\gamma = \frac{3\mathrm{d}t}{\mathrm{d}x^3}$, we obtain

$$x = \alpha_0 t^{-\frac{2}{3}n+1}, \tag{7.87}$$

where

$$\alpha_0 = \frac{q_0}{Q_0} (\kappa N_0)^{2/3} \frac{A}{1 - \frac{3}{2}N}. \tag{7.88}$$

On summarizing, the following conclusion can be made. Since $x \to \infty$ at $t \to \infty$, Eq. (7.87) is consistent only for $n < 3/2$. With such restriction on n the voids still have time to absorb vacancies delivered by sources, supporting the supersaturation $\Delta \sim 0$, and the matter approaches a destructive threshold, possessing a definite size distribution of voids. In the case $n \geq 3/2$ the supersaturation increases with time, and the swelling process takes an avalanche character, which is impossible to describe without additional model assumptions. So for the existence of δ-asymptotics the fulfillment of the following inequality is needed:

$$1 \leq n < \frac{3}{2}. \tag{7.89}$$

The formula Eq. (7.87) allows us to relate n with N. On the one hand,

$$\gamma = \frac{3\mathrm{d}t}{\mathrm{d}x^3} = \frac{t^{2(n-1)}}{\alpha_0^3 \left(1 - \frac{2}{3}n\right)}, \tag{7.90}$$

on the other, $\gamma = \mathrm{const} \cdot t^N$. Hence $N = 2(n-1)$. For $1 \leq n < 3/2$, as it should be, $0 \leq N < 1$.

Not less important are the restrictions, imposed on the source strength q_0. These restrictions are caused by the requirement that the time necessary for the formation of δ-asymptotics, should be much less than the time of disintegration. The velocity of compression of a package of points is characterized, for example, by the quantity $\frac{\Delta z}{z_2}$ (Δz is a package width). Integrating Eq. (7.82), we obtain

$$\frac{\Delta z}{z_2} \sim \mathrm{const} \cdot \exp\left(-A^{2/3}\tau\right). \tag{7.91}$$

Therefore, the time of compression is $\tau_c \sim A^{2/3}$.

The time of disintegration τ_d by an order of magnitude is

$$\tau_d \sim \beta^{-1} \ln\left[N_0 B\left(q_0\right)\right]. \tag{7.92}$$

This estimate is obtained from a limiting relationship $N_0 z_2 \sim 1$. Besides, according to Eq. (7.90) one has $z_2 = \gamma^{3/2} = B e^{\beta\tau}$, $\beta = \frac{n-1}{1-2n/3}$;

$$B\left(q_0\right) = \left[\alpha_0^3 \left(1 - \frac{2}{3}n\right)\right]^{-3/2} \left[\alpha_0 \frac{9\left(n-1\right)}{2n-3}\right], \qquad \alpha_0 = \alpha_0\left(q_0\right). \tag{7.93}$$

Thus,

$$A^{2/3}\left(n\right) \ll \beta^{-1}\left(n\right) \ln\left[N_0 B\left(q_0\right)\right], \tag{7.94}$$

which, in principle, gives the wanted restriction for q_0.

7.3.5 Conclusions

The results of the present investigation show that the late stage coarsening behavior in segregation processes in solid or liquid solutions with constant rates of input fluxes of monomeric building units can be characterized, again, by power laws (cf. Eqs. (7.33)). The respective exponents depend on the growth laws; they are different from the values obtained for coarsening in closed systems.

With respect to practical applications the results of the present analysis indicate the possibility to generate cluster distributions with definite characteristics (average size, number) by varying the rate of input fluxes of monomers and the time interval the system is exposed to such input fluxes (cf. Eqs. (7.33) and (7.34)).

7.4 Growth and Shrinkage of Precipitates under Irradiation

7.4.1 Introduction

Irradiation is known in some cases to result in dissolution (or shrinkage) of precipitate particles, which may grow in the process of thermal decomposition of supersaturated solid solutions. This effect is usually explained to be due to collision cascades that mix atoms of the precipitate and the matrix near the interface, taking solute atoms away from the inclusion [15,41,73,192]. This dynamic mechanism of precipitate dissolution leads to the existence of a certain maximum size of precipitates that depends on their number density. It is obvious that for this mechanism it is not crucial whether the interface is coherent or not.

Another mechanism describing the evolution of coherent precipitates under irradiation treats the precipitate–matrix interface as a defect sink with finite capacity [46, 332]. In this model annihilation of matrix vacancies and solute interstitials at the interface causes the radiation growth of precipitates. The dissolution of precipitates takes place because the precipitate volume rate of change is proportional to the vacancy concentration. The latter may be low at high precipitate number density. The competition between growth and dissolution provides the possibility of a steady state of existence of the precipitate ensemble, the precipitate dimension depending on their number density.

In the present section we consider a dissolution mechanism of purely diffusional origin based on migration of radiation-induced point defects (PD) out-off precipitate bulk into the matrix [2]. The PD-production rate within a precipitate is proportional to its volume while the total diffusion flux of substitutional solute atoms is proportional to its radius implying that there exists a maximum size at which the precipitate growth rate equals the rate of its radiation-induced dissolution. This size is shown to be a stable one implying that under irradiation a stationary state can be achieved far away from the thermodynamic equilibrium.

7.4.2 Diffusion Mechanism of Radiation-Induced Shrinkage of the Precipitates

Consider the precipitate–matrix interface which is transparent for PD. If a vacancy goes over into the precipitate from the matrix it means that an atom of the precipitate takes the place on its surface and the precipitate volume increases. The reverse motion of vacancies (from the precipitate into the matrix) decreases the precipitate volume in dilute solutions since in this case the matrix atoms are transported inside the precipitate. Under irradiation vacancies are produced both in the matrix and in the precipitate and their fluxes influence the precipitate growth. This effect is taken into account in the present analysis.

Note that the precipitate volume changes at a rate which is different from the rate of variation of the precipitate mass, i.e., the number of atoms in the precipitate, since the volume is determined by the number of lattice sites within the precipitate, both occupied and free. The volume variation rate is proportional to $(j_c + j_v^e)|_R$, where j_c is the volume flux density of substitutional solute atoms (in substitutional solutions under consideration) and j_v^e is the vacancy volume flux density into the matrix at the precipitate surface. The upper index e designates parameters related to the matrix and R is the precipitate radius.

The mass variation rate is proportional to $(j_c + j_i^e)|_R$, where j_i^e is the flux density of interstitial solute atoms at the surface. If the precipitate growth is free from elastic stresses, then the mass variation rate is proportional to the volume variation rate and we may write

$$\frac{\mathrm{d}R}{\mathrm{d}t} = (j_c + j_i^e)|_R\,. \tag{7.95}$$

This expression is valid when the volume deficit due to interstitial fluxes is compensated by the vacancy fluxes which is the case when the precipitate does not contain PD sinks and, hence, vacancy and interstitial fluxes through the surface are equal. We may express it as follows:

$$j_i^i = j_i^e = j_v^i = j_v^e. \tag{7.96}$$

The upper index i designates parameters related to the precipitate.

Assume the fluxes to be positive if they are directed from the matrix into the precipitate. Then sign of j_c is determined by the sign of the supersaturation, while the signs of j_i^i, j_i^e, j_v^i, and j_v^e are negative since the respective fluxes are directed into the matrix. The PD recombination inside precipitates does not change the above equations since vacancies and interstitials recombinate in equal numbers.

Let L^i be the recombination length in the precipitate

$$L^i = \left(\frac{D_v^i D_i^i}{\alpha K^i}\right)^{1/4}, \tag{7.97}$$

where D_v^i and D_i^i are the diffusivities of vacancies and interstitials, respectively, K^i is the production rate of PD in the precipitate, and α is the recombination coefficient. Consider the case of moderate temperatures when $R \ll L^i$ and the thermal PD production rate are still negligible. In this case under stationary conditions all PD generated inside the precipitate come out of it, so that

$$j_i^i = j_v^i = \frac{1}{3} K^i R. \tag{7.98}$$

Using the well-known expression $j_c = D_v^e c_v^e (\bar{c} - c_R)/R$ along with Eqs. (7.96) and (7.98) we obtain from Eq. (7.95)

$$\frac{\mathrm{d}R}{\mathrm{d}t} = \frac{1}{R} D_v^e c_v^e (\bar{c} - c_R) - \frac{1}{3} K^i R, \tag{7.99}$$

where D_v^e and c_v^e are the vacancy diffusivity and concentration in the matrix, respectively, $\bar{c}$ is the mean concentration of substitutional solute atoms,

$$c_R = c_\infty \left(1 + \frac{2\sigma a^3}{k_B T R}\right) \tag{7.100}$$

is the equilibrium concentration at the precipitate surface, c_∞ is the concentration at a planar surface, σ is the surface tension, T is the temperature, and a is the atomic spacing [155].

Equation (7.99) is different from the conventional expression due to the negative term that is proportional to the PD generation rate. It is determined by the PD fluxes from the precipitate into the matrix and is not related to cascade effects. It is the consequence of the transparency of the precipitate–matrix interface to PD diffusion. In reduced variables,

$$r = \frac{R}{R_c}, \qquad \tau = t \frac{\Delta D_v^e c_v^e}{R_c^2} = t \Delta^3 D_v^e c_v^e \left(\frac{k_B T}{2 c_\infty \sigma a^3}\right)^2, \tag{7.101}$$

$$k = \frac{K^i}{\Delta^3 D_v^e c_v^e} \left(\frac{3 c_\infty \sigma a^3}{k_B T}\right)^2,$$

Eq. (7.99) takes form

$$\frac{\mathrm{d}r}{\mathrm{d}t} = \frac{1}{r}\left(1 - \frac{1}{r}\right) - \frac{4k}{27} r^2. \tag{7.102}$$

Here $\Delta = \bar{c} - c_\infty$ is supersaturation and

$$R_c = \frac{2 c_\infty \sigma a^3}{\Delta k_B T} \tag{7.103}$$

is the critical radius.

The dependence of the growth rate given by Eq. (7.102) on r for several values of k is shown in Figure 7.7. In the absence of irradiation ($k = 0$) $\mathrm{d}R/\mathrm{d}t > 0$ for the $R > R_c$ ($r > 1$). Under irradiation ($k > 0$) precipitate grows in the region $R^{(-)} < R < R^{(+)}$ ($r^{(-)} < r < r^{(+)}$) and dissolves for $R < R^{(-)}$ and $R > R^{(+)}$ ($r < r^{(-)}$, $r > r^{(+)}$). So, $R^{(+)}$ is a stable root of the equation $\mathrm{d}R/\mathrm{d}t = 0$ and can be named limiting size of the precipitate. For $k = 1$ the growth interval reduces to the point

$$r^{(-)} = r^{(+)} = r^* = 3/2, \tag{7.104}$$

$$R^{(-)} = R^{(+)} = R^* = \left(\frac{3 D_v^e c_v^e \sigma a^3 c_\infty}{K^i k_B T}\right)^{1/3}. \tag{7.105}$$

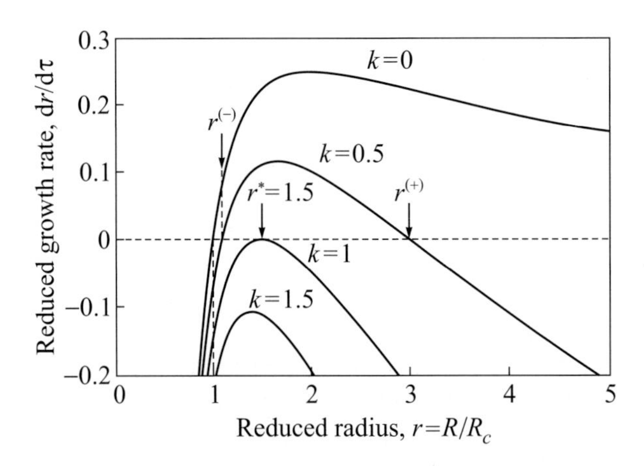

Figure 7.7: Dependence of the growth rate on reduced radius, r, for different values of k.

Here $R^{(-)}$ and $R^{(+)}$ are determined by the equation (see Eq. (7.99))

$$\Delta = \frac{2\sigma a^3 c_\infty}{k_B T R} + \frac{K^i R^2}{3D_v^e c_v^e}. \tag{7.106}$$

Let us neglect the bulk recombination of point defects as compared to losses at sinks like voids and dislocations. If λ^2 is their sink strength, then our assumption means that

$$\lambda L^e \gg 1, \tag{7.107}$$

where L^e is the recombination length in the matrix. In this case

$$c_v^e = \overline{c}_v^e = c_v^T + \frac{K^*}{\lambda^2}, \tag{7.108}$$

where c_v^T is the thermal equilibrium concentration of vacancies in the matrix, $K^* = K^e/D_v^e$, K^e is the production rate of point defects in the matrix.

Neglecting the first term on the right-hand side of Eq. (7.106) we obtain the estimate

$$R^{(+)} \approx \frac{1}{\lambda}\sqrt{\frac{3\Delta}{K^i}\left(K^e + \lambda^2 D_v^e c_v^T\right)}. \tag{7.109}$$

Note that when the radiation-induced vacancy concentration dominates over the thermal equilibrium one, $R^{(+)}$ is almost independent of the radiation dose and is of the order of $\Delta^{1/2}/\lambda \sim$ 10–100 atomic spacings if $\Delta \sim 10^{-2}$–10^{-4}, $\lambda \sim 10^5\mathrm{cm}^{-1}$ [169] (the coherence failure, as a rule, takes place when the precipitate is larger than that). It follows from Eq. (7.109) that $R^{(+)} \ll \lambda^{-1}$. After comparison with Eq. (7.107) we obtain that $R^{(+)} \ll L$, i.e., the recombination of point defects inside the precipitate is really negligible which has been assumed above in the derivation of Eq. (7.98).

Without irradiation precipitate can grow when its size exceeds the critical one determined by the supersaturation. Under irradiation, for their growth it is necessary to fulfill the condition $\Delta > \Delta_{\min}$, where

$$\Delta_{\min} = \left(\frac{K^i}{D_v^e c_v^e}\right)^{1/3} \left(\frac{3\sigma a^3 c_\infty}{k_B T}\right)^{2/3}, \tag{7.110}$$

i.e., under irradiation the phase equilibrium is shifted to the higher supersaturation region. The difference $\Delta - \Delta_{\min}$ may be called an effective supersaturation since when it goes to zero the growth interval reduces to a point (see Eq. (7.105)).

In the above considerations, we have assumed the supersaturation Δ to be the time independent. Actually it decreases while precipitates grow, approaching the $\Delta_{\min}$ value where the precipitate growth stops. At homogeneous nucleation the precipitate radius remains equal to $R^{(+)}$ and precipitate volume density can be determined from the conservation requirement

$$\frac{4\pi}{3} N R^* = \Delta(0) - \Delta_{\min}, \tag{7.111}$$

where $\Delta(0)$ is the initial supersaturation value. Under heterogeneous nucleation N is independent of supersaturation and after replacing R^* by $R^{(+)}(\Delta)$, and $\Delta_{\min}$ by Δ this relationship becomes the equation for defining the stationary supersaturation and the corresponding maximum radius $R^{(+)}(\Delta)$. So, irradiation opens the possibility of existence of a δ-function-like distribution of coherent precipitates with dimensions equal to $\delta(R - R^*)$ in the case of homogeneous nucleation and to $\delta(R - R^{(+)}(\Delta))$ in the case of heterogeneous one.

If the interface absorbs fully PD, the proposed mechanism of diffusion solution does not work since the radiation-induced interstitial atoms and vacancies recombine at the precipitate–matrix interface. This difference in the behavior of coherent and incoherent precipitates can be used for the qualitative verification of the mechanism being proposed. In [346] the irradiation effect on the kinetics of growth of precipitates depositing from the chromium-and-silicium supersaturated solid solution based on copper was investigated experimentally. The alloy being investigated was preaged. So the coherent chromium precipitates and incoherent $CrSi_3$ precipitates were produced. Then the specimens were irradiated with 300 keV Cu^+ ions. The incoherent precipitates continued to grow under irradiation while the coherent ones were dissolved. By simple estimates, the authors of [346] concluded that the collision cascades cannot cause the dissolution and preferred to explain the dissolution by the diffusion mechanism. However, the specific mechanism of such dissolution was not considered.

7.4.3 Effect of the Precipitate Incoherence and the Solute Atom Transition into the Interstitial Sites and Back in the Lattice Sites

The above analysis holds when the following conditions are satisfied: (a) the precipitate–matrix interface is coherent; (b) the vacancy concentration variations in the matrix are insignificant; (c) one can neglect the radiation knock-out of impurity atoms from the matrix lattice sites and their transition into substitutional atoms at sinks; (d) the volume recombination of point defects is negligibly small; (e) the sink strength within precipitates is negligible.

Let us derive the precipitate growth rate without the above assumptions except for the items (d) and (e).

The flux densities j_c and j_i^e from Eq. (7.95) are determined by the set of equations for concentrations of substitution atoms c and interstitial atoms c_i in the matrix. One of them, namely, the condition of the absence of sources of solute atoms in the matrix, $\mathrm{div}\,(j_c + j_i^e) = 0$, can be written assuming spherical symmetry in the form

$$D_v^e\,(c_v^e)^2\,\frac{\mathrm{d}}{\mathrm{d}r}\frac{c}{c_v^e} + D_v^e\frac{\mathrm{d}c_i}{\mathrm{d}r} = \left(\frac{R}{r}\right)^2\frac{\mathrm{d}R}{\mathrm{d}r}. \tag{7.112}$$

The next equation describes the solute atom diffusion in the matrix

$$\frac{1}{r^2}\frac{\mathrm{d}}{\mathrm{d}r}r^2\frac{\mathrm{d}c_i}{dr} - \lambda^2\,(c_i - \overline{c}_i) = \frac{1}{D_i^e}K^e\,(\overline{c} - c)\,. \tag{7.113}$$

The second term on the right-hand side of Eq. (7.113) describes the creation of solute interstitial atoms, the term $\lambda^2\overline{c}_i$ on the left side is the absorption of these atoms at linear sinks. So, the solute interstitial atoms transform in the substitutional positions; thus increasing the supersaturation in the vicinity of the precipitate. The terms $\lambda^2 c_i = K^e\overline{c}/D_i^e$ are introduced for symmetry. Excluding from this system the concentration c we obtain the equation

$$\frac{(c_v^e)^2}{K^*}\frac{\mathrm{d}}{\mathrm{d}r}\left[\frac{1}{c_v^e}\left(\lambda c_i - \frac{1}{r^2}\frac{\mathrm{d}}{\mathrm{d}r}r^2\frac{\mathrm{d}c_i}{dr}\right)\right] + \frac{\mathrm{d}c_i}{\mathrm{d}r} = \frac{1}{D_i^e}\left(\frac{R}{r}\right)^2\frac{\mathrm{d}R}{\mathrm{d}r}. \tag{7.114}$$

Knowing the solution of Eq. (7.114) we can find c from Eq. (7.113). The vacancy concentration c_v^e in the matrix satisfies the equation

$$\frac{1}{r^2}\frac{\mathrm{d}}{\mathrm{d}r}r^2\frac{\mathrm{d}c_v^e}{\mathrm{d}r} - \lambda^2\,(c_v^e - \overline{c}_v^e) = 0 \tag{7.115}$$

from where

$$c_v^e = \overline{c}_v^e + \frac{R}{r}\,(c_v^e(R) - \overline{c}_v^e)\exp\left[-\lambda(r - R)\right]. \tag{7.116}$$

The boundary conditions far from the precipitate for the set of Eqs. (7.112), (7.113), and (7.115) are

$$c|_{r\to\infty} = \overline{c}, \qquad c_i|_{r\to\infty} = \overline{c}_i, \qquad c_v^e|_{r\to\infty} = \overline{c}_v^e. \tag{7.117}$$

At the precipitate boundary $c(R) = c_R$. We obtain the values $c_i(R) \equiv \widetilde{c}_i$, and $c_v^e(R) \equiv \widetilde{c}_v^e$ by considering the point defect transformation at the precipitate–matrix interface.

It is obvious that point defects of the same species (only vacancies or only interstitial atoms) cannot accumulate at the interface because it would lead to an unlimited growth of stresses or to fracture. The absence of cracks implies that the interface absorbs interstitial atoms and vacancies in equal quantities [16,332]. This may be accomplished by some density n of point defect recombinators capturing alternatively interstitial atoms and vacancies at the precipitate–matrix interface.

Let $I_v^i(I_i^i)$ be the trapped-at-the-interface component of vacancy (interstitial) flux directed from the bulk of precipitate into the matrix and $I_v^e(I_{im}^e)$ is the corresponding component of vacancy (self-interstitial atoms) flux directed into the precipitate. We can write

$$I_v^i = \kappa_v^i n(1-W)\widetilde{c}_i, \qquad I_v^e = \kappa_v^e n(1-W)\widetilde{c}_i^e, \tag{7.118}$$

$$I_i^i = \kappa_i^i nW\widetilde{c}_i, \;\; I_{im}^e = \kappa_{im}^e nW\widetilde{c}_{im}^e, \tag{7.119}$$

Here $\widetilde{c}_{im}^e$ is the precipitate self-interstitial atom concentration at the precipitate–matrix interface, W is the probability for a recombinator to absorb an interstitial atom, κ_v^i, κ_v^e, κ_i^i, κ_{im}^e are the kinetic coefficients proportional to capture rates.

The probability W can be written directly via point defect concentrations at the interface using the condition $I_i^i + I_{im}^e = I_v^i + I_v^e$ for the absence of accumulation of point defects of one type (vacancies or interstitials)

$$1 - W = \frac{\kappa_i^i\widetilde{c}_i + \kappa_{im}^e\widetilde{c}_{im}^e}{\kappa_v^i\widetilde{c}_i + \kappa_v^i\widetilde{c}_i + \kappa_i^i\widetilde{c}_i + \kappa_{im}^e\widetilde{c}_{im}^e}. \tag{7.120}$$

To define the values of point defect concentrations at the interface one should solve the equations for transformations at the boundary

$$j_i^i = j_i^e + I_i^i, \tag{7.121}$$

$$j_{im}^e = I_{im}^e, \tag{7.122}$$

$$j_i = \omega\widetilde{c}_v^i - \nu\widetilde{c}_v^e + I_v^i, \tag{7.123}$$

$$j_v^e = -\omega\widetilde{c}_v^i + \nu\widetilde{c}_v^e + I_v^e. \tag{7.124}$$

Here j_{im}^e is the flux density of self-interstitial atoms directed into the precipitate, ω is the probability of the vacancy transition from the precipitate into the matrix, ν is the reverse transition probability. Equation (7.122) is the condition for self-interstitial atom absorption at the precipitate surface rather than in the bulk.

In order to complete the set of equations (7.121) to (7.124) we should express the volume flux density of point defects in terms of corresponding concentrations at the surface. To do this we shall solve the appropriate diffusion problem in the matrix. So, for j_{im}^e we obtain

$$j_{im}^e = D_{im}^e \frac{1+\lambda R}{R}\left(\overline{c}_{im}^e - \widetilde{c}_{im}^e\right). \tag{7.125}$$

Substituting this result into Eq. (7.128) we find

$$\widetilde{c}_{im}^e = \overline{c}_{im}^e \left[1 + \frac{n\kappa_{im}^e WR}{D_{im}^e(1+\lambda R)}\right]^{-1}. \tag{7.126}$$

By analogy we can express $\widetilde{c}_v^i$ and $\widetilde{c}_v^e$ via W using Eqs. (7.123) and (7.124).

Let us assume that the precipitate–matrix interface is coherent. Then the condition $j_v^i = j_v^e$ is fulfilled, from whence we obtain, taking into account Eq. (7.98)

$$\frac{\widetilde{c}_v^e}{\overline{c}_v^e} - 1 = \frac{K^i R^2}{3D_v^e\overline{c}_v^e(1+\lambda R)} < \frac{K^i\lambda^2 R^2}{3K^e}. \tag{7.127}$$

Substituting into this formula the maximum radius $R^{(+)}$ value from Eq. (7.109) we see that c_v^e only weakly depends on r. This statement is also valid when the interface slightly differs from the coherent one, i.e., when it captures only a small fraction of point defects. We proceed to consider this case.

From Eq. (7.114), we obtain

$$\frac{D_i^e}{R}\left(\widetilde{c}_i - \overline{c}_i\right) = \frac{1}{A}\left(j_i^i - \lambda^* bR\frac{\mathrm{d}R}{\mathrm{d}t} - \kappa_i^i n\overline{c}_i\right), \tag{7.128}$$

where

$$\overline{c}_i = \frac{D_v^e}{D_i^e\lambda^2}\overline{c}K^* \tag{7.129}$$

is the mean concentration of solute interstitial atoms far from the precipitate,

$$b = 1 - \frac{\lambda^2}{(\lambda^*)^2}, \tag{7.130}$$

$$(\lambda^*)^2 = \lambda^2\left(2 + \frac{\lambda^2 c_v^T}{K^*}\right)\left(1 + \frac{\lambda^2 c_v^T}{K^*}\right)^{-1}, \tag{7.131}$$

$$A = 1 + \lambda^* R + \kappa_i^i nR/D_i^e. \tag{7.132}$$

It is obvious that

$$\widetilde{c}_i \sim \frac{D_v^e}{D_i^e\lambda^2}\overline{c}K^*. \tag{7.133}$$

It follows from Eq. (7.126) that

$$\widetilde{c}_{im}^e \approx \overline{c}_m^i \approx \frac{D_v^e}{D_i^e\lambda^2}K^*, \tag{7.134}$$

while from Eq. (7.120) we obtain

$$W \approx 1 - \frac{D_v^e}{D_i^e} \approx 1. \tag{7.135}$$

On integrating Eq. (7.112) over r' from r to ∞, we obtain at $r = R$ that

$$D_v^e c_v^e(\overline{c} - c_R) + D_i^e(\overline{c}_i - \widetilde{c}_i) = R\frac{\mathrm{d}R}{\mathrm{d}t}. \tag{7.136}$$

Substituting it into Eq. (7.128), we get

$$\frac{\mathrm{d}R}{\mathrm{d}t} = D_v^e\left(c_v^T + \frac{K^*}{\lambda^2}\right)\left[\frac{A}{R} + \frac{\kappa_v^i n}{D_i^e}\left(1 + \frac{\lambda^2 c_v^T}{K^*}\right)\right]\frac{\Delta - f(R)}{A - b\lambda^* R}, \tag{7.137}$$

where

$$f(R) = \left(B + \frac{\kappa_v^i n}{D_i^e}R\right)^{-1}\left[3K^e K^i\lambda^2 R^2 + \frac{2\sigma a^3 c_\infty}{k_B TR} - \frac{\kappa_i^i nRc_\infty}{D_i^e}\right], \tag{7.138}$$

$$B = A\left(1 + \frac{\lambda^2 c_v^T}{K^*}\right). \tag{7.139}$$

In the absence of irradiation the equation $\Delta = f(R)$ has only one root $R^{(-)} = R_c$ and the expression for $\mathrm{d}R/\mathrm{d}t$ takes the usual form

$$\frac{\mathrm{d}R}{\mathrm{d}t} = D_v^e c_v^T \frac{(\bar{c} - c_R)}{R}. \tag{7.140}$$

Under irradiation besides the critical radius $R^{(-)}$ there appears one more root $R^{(+)} > R^{(-)}$ that is the maximum precipitate size. It follows from the fact that function $f(R)$ is concave and goes to infinity when $R = 0$ or $R \to \infty$. Furthermore, as stated above, the irradiation leads to the shift in the phase equilibrium because the difference $\Delta - \min\left(f(R)/R\right)$ plays the role of the supersaturation. Let us define the criterion of smallness of the parameter n by the inequality

$$\frac{\kappa_i^i n}{\lambda D_i^e} \ll \frac{\Delta^{1/2}}{c_\infty} \tag{7.141}$$

implying that the expression for $\mathrm{d}R/\mathrm{d}t$ is given by Eq. (7.99).

7.4.4 The Case of Incoherent Precipitation

In the limit opposite to that described by inequality (7.141) the precipitate–matrix interface absorbs the majority of point defects coming to it. This may be shown to result in the following: $\tilde{c}_v^e \to c_v^T$, and $\tilde{c}_i, \tilde{c}_{im}^e \to 0$ if $n \to \infty$. Consequently, the second term on the left-hand side of Eq. (7.113) greatly exceeds the right-hand side term that can be safely neglected in this case resulting in the following equation:

$$\frac{c_i(r) - \bar{c}_i}{c_v^e(r) - \bar{c}_v^e} = \frac{\tilde{c}_i - \bar{c}_i}{\tilde{c}_v^e - \bar{c}_v^e}. \tag{7.142}$$

Substituting it into Eq. (7.112) we obtain

$$\frac{\mathrm{d}R}{\mathrm{d}t} = \frac{\bar{c}_v^e}{\tilde{c}_v^e} \frac{1}{RV(r)} \left[D_v^e \bar{c}_v^e \left(\frac{\bar{c}\tilde{c}_v^e}{\bar{c}_v^e} - c_R\right) + D_i^e\left(\bar{c}_i - \tilde{c}_i\right)\right], \tag{7.143}$$

where

$$V(r) = R\left(\bar{c}_v^e\right) \int_r^\infty \frac{\mathrm{d}r'}{(c_v^e)^2 r'^2}. \tag{7.144}$$

Assuming $\tilde{c}_i = 0$, $\tilde{c}_v^e = c_v^T$, $D_{im}^e = D_i^e$ and taking into account that

$$D_{im}^e \bar{c}_{im}^e = D_v^e\left(\bar{c}_v^e - c_v^T\right), \qquad \frac{c_v^T}{\bar{c}_v^e} \ll 1, \tag{7.145}$$

we arrive at

$$\frac{\mathrm{d}R}{\mathrm{d}t} = D_v^e \overline{c}_v^e (\overline{c} - c_R) \frac{1 + \lambda R}{R} \left[1 + 2 \frac{c_v^T}{\overline{c}_v^e (1 + \lambda R)} \ln \frac{\overline{c}_v^e (1 + \lambda R)}{c_v^T} \right]. \tag{7.146}$$

In this expression the term $D_v^e \overline{c}_v^e \overline{c} (1 + \lambda R) / R$ describes the influx of interstitials from the matrix, while $-D_v^e \overline{c}_v^e c_R (1 + \lambda R) / R$ is related to the outflux of substitutional solute atoms to the matrix.

If the irradiation dose rate is low, then $\left(\overline{c}_v^e - c_v^T\right) / c_v^T \ll 1$ and the growth rate expression takes the usual form $\mathrm{d}R/\mathrm{d}t = D_v^e c_v^T (\overline{c} - c_R)/R$ since $V(R) \sim 1$. Note that in this case the above-discussed dissolution mechanism does not work.

7.4.5 Conclusion

We have considered a purely diffusion mechanism of precipitate dissolution under irradiation creating interstitials and vacancies within the precipitate bulk. Diffusion outfluxes of vacancies across the coherent precipitate–matrix interface decrease precipitate volume, while the interstitials outfluxes carry out its material and decrease its mass. This leads to an increase in the critical size and to the appearance of the maximum size that does not depend on the precipitate number density and volume fraction. This mechanism is the most essential one for dilute solid solutions being of first order in vacancy concentration. So one need not to consider effects of second order such as the inverse Kirkendall effect or diffusion of solute atom-point defect complexes (see, e.g., [218]). Note that the interface coherence, which is realized for the case for small enough precipitates, is essential for this mechanism. Usually there exists a threshold size of the coherence failure. If this size exceeds the radiation-induced maximum size then precipitate growth should saturate at the maximum size, and a stationary state should be achieved under irradiation far away from thermodynamic equilibrium. In the opposite case the precipitate coarsening is expected to take place without saturation.

Note that the proposed mechanism of precipitate dissolution can operate simultaneously with a well-known recoil mechanism under the cascade irradiation. However, in the case of low energy irradiation (e.g., electron irradiation) and dilute solid solution the only possible mechanism is the one under consideration in this chapter.

8 Formation of a Newly Evolving Phase with a Given Stoichiometric Composition

8.1 Introduction

A large number of studies, including monographs, have been devoted to the investigation of the kinetics of first-order phase transitions [24, 28, 94, 113, 119, 136, 137, 153, 206, 245, 246, 294, 318, 329, 337, 351]. Nevertheless, a wide spectrum of problems remains unsolved as yet.

As one restriction in the applicability of the theory to experiment, most of the mentioned analyses are devoted to single-component systems. Already homogeneous binary nucleation is more complex as compared to single-component nucleation. In part, these complications arise from uncertainties in the knowledge of the composition of the critical clusters and the related problem of the compositional dependence of the critical cluster surface tension. These limitations in the application of the theory to experiment are even more pronounced in the case of phase formation in multicomponent systems.

Moreover, in the majority of previous studies, attention is directed mainly toward the determination of the so-called steady-state nucleation rate or, in other words, the steady-state flux in cluster-size space. Steady-state conditions can be realized in real systems, however, only for limited periods of time or when the state of the solution is artificially maintained stationary by some appropriate real or supposed mechanism (cf., e.g., [94, 337]).

In most cases of interest, the degree of metastability or the state of the ambient phase changes in the course of the phase transformation because of depletion effects due to cluster formation and growth. Different aspects of the effect of depletion on the course of first-order phase transitions have been studied by various authors (see, e.g., [20, 28, 29, 91, 93, 139, 217, 220, 226, 325, 330, 331, 344, 350]. As it has been shown (cf. [220, 226, 331]), depletion effects affect nucleation quantitatively and determine qualitatively the whole course of first-order phase transformations proceeding by nucleation and growth. As a result of such depletion effects, in particular, only a finite number of clusters develops in the system (cf. Chapters 3 and 5 and [164, 298]).

An extended theoretical analysis of the kinetics of new phase formation of single-component systems via nucleation and growth has been given in Chapter 3. In the present chapter these analyses are extended to multicomponent systems. In order to remedy various problems encountered in the description of phase formation an additional assumption is employed; it is assumed that the new phase has a well defined but arbitrary composition. Using this assumption, the kinetic equations governing nucleation and growth, can be reduced to a relation identical in its form to the respective expression for phase formation in single-component systems [293, 300, 304]. However, as will be shown in the course of the analysis, the effective

Kinetics of First-order Phase Transitions. Vitaly V. Slezov

ISBN: 978-3-527-40775-0

diffusion coefficients and the effective supersaturation have to be expressed as nontrivial combinations of the thermodynamic and kinetic parameters of different components involved in the phase formation process.

In the analyses we assume (cf. Chapters 3 and [298]) that, for the whole course of the process of nucleation, at any time a quasisteady distribution function with respect to the size of the new phase particles $f(n,t)$ is established in the range of cluster sizes $1 \leq n \leq n_c$. Here, as previously, n denotes the number of structural units (or, quasimolecules) in an aggregate (or cluster) of the newly evolving phase, and n_c is the critical size of the aggregate. Remember that the critical cluster is an aggregate of new phase particles in unstable equilibrium with the solid solution. Its size changes with variations in the state of the ambient phase. Similarly, the quasisteady-state distribution with respect to cluster size in the range $n \leq n_c$ is determined by the current state parameters of the solid solution. In other words, for the range of cluster sizes $n \leq n_c$, the time of adjustment t_r of the distribution function $f(n,t)$ and the flux $J(n,t)$ in cluster-size space to the current state parameters of the solid solution is much less when compared to the characteristic times of variation of these parameters. Generally, these conditions are fulfilled.

In line with the classical approach, we assume that the aggregation process of the newly evolving phase proceeds via incorporation and emission of individual structural units exclusively. This way, the smallest aggregate of the newly evolving phase corresponds to $n = 1$. Further, we go over from a discrete description in terms of a set of kinetic equations to a continuous description in terms of the Frenkel–Zeldovich equation. In the domain of cluster sizes $n \geq n_c \gg 1$, we focus on the continuum description which is a reasonable approximation. Moreover, the properties of the aggregates of the newly evolving phase approach the properties of their respective bulk phases. The situation when the aggregates properties may differ from that of the macroscopic phase and depend on their size is considered in Chapter 11.

The kinetic equations are applied to the region of cluster sizes $n \leq n_c$ as well but mainly in order to derive an estimate of the time to establish quasisteady-state conditions and to obtain the boundary conditions at $n = n_c$. Thus, the approximations arising from the application of the kinetic equations to very small clusters are of minor importance for the results of the analysis outlined below.

Once the stage of nonsteady state nucleation is completed (the stage of approach of quasisteady-state conditions in the range of cluster sizes $n \leq n_c$ in cluster-size space), the further evolution can be divided into a stage of dominant quasisteady-state nucleation followed by the stages of independent and competitive growth (Chapters 3 and 4). The analysis of the present chapter is devoted mainly to the description of the stages of quasisteady-state nucleation and independent growth. For these stages, the evolution in time of the basic characteristics of the nucleation–growth process is found in the form of the distribution functions with respect to cluster sizes and the size-dependent flux in cluster-size space. Further, the number of clusters of the new phase and their average size at the end of the independent growth stage are determined as a function of the initial system supersaturation. Finally, estimates for the duration of the stages of quasisteady-state nucleation and independent growth are provided.

The state of the cluster ensemble at the end of the nucleation and independent growth stages simultaneously represents the initial state for the process of competitive growth. The theory of competitive growth or coarsening was presented in detail in Chapter 4. Thus, the present analysis provides a basis for a complete quantitative description of the entire course of first-order phase transitions in multicomponent systems.

Until now it was assumed in our analysis that the different solute components do not interact with each other, i.e., only the solute–matrix interaction was taken into account. This assumption is in line with classical investigations of precipitation processes in solutions where generally the case of a weak (perfect) solution is considered and will be preserved also in Sections 8.1–8.4. In Sections 8.5–8.9 of the present chapter a generalization of the earlier obtained results is presented and such interactions between solute atoms are taken into account in a comprehensive manner (for first attempts in this direction see [21]). It is shown that the general form of the basic equations remains the same but in these equations the concentrations of different components have to be replaced by chemical activities, which are determined both by the concentrations and the type of interactions between different components. A method is developed allowing one an experimental determination of the interaction parameters of the solute components and, thus, of the chemical activities of the solute components. The below outlined theory may also be used to study nucleation processes in other fields of application where interactions between the basic building units of the evolving phase clustering are essential for an understanding of the kinetics of nucleation and growth.

As a next step, a generalization of the kinetic equation commonly used for the description of the evolution of the cluster-size distribution is given. It contains in addition to the regular hydrodynamic term describing the deterministic motion in cluster-size space a term proportional to the first derivative of the cluster-size distribution function. This additional term reflects diffusion-like processes in cluster-size space due to the influence of stochastic effects on the growth kinetics. Moreover, another type of stochasticity is also accounted for connected with a possible touching and merging of clusters in the nucleation–growth process. Such effects are described by a collision integral. The influence of both effects on different stages of the precipitation process is investigated.

8.2 Basic Set of Equations

We consider processes of formation of a new phase with a given stoichiometric composition. The stoichiometric coefficients specifying the composition of the evolving phase are denoted by ν_i and the number of basic structural units in the cluster or aggregate of the newly evolving phase given by n.

The volume V of an aggregate of the newly evolving phase is then given by $V = n\omega_s$ with $\omega_s = \sum_i \nu_i \omega_i$. Here ω_s is the volume of a structural element, ω_i is the volume of the ith component in the ambient solution and in the newly evolving phase (i.e., we do not consider differences in the respective volumes in both phases resulting from elastic stresses).

The basic system of equations, describing the kinetics of nucleation and growth, can then be written in the form (see Chapters 2, 3, and [293, 300, 304])

$$\frac{\partial f(n,t)}{\partial t} = -\frac{\partial J(n,t)}{\partial n}, \quad J(n,t) = -w_{n,n+1}\left(\frac{\partial f(n,t)}{\partial n} + \frac{1}{k_B T}\frac{\partial \Delta\Phi}{\partial n} f(n,t)\right), \tag{8.1}$$

$$f(n,t)\big|_{n\to 0} = \prod_i c_i^{\nu_i}, \qquad f(n,t)\big|_{t=0,n>1} \to 0, \tag{8.2}$$

$$c_{0i} = c_i(t) + \nu_i \int_0^\infty f(n,t)\, n \, \mathrm{d}n. \tag{8.3}$$

The first set of equations (8.1) describes the evolution of the cluster-size distribution function, $f(n,t)$, and the flux in cluster-size space, $J(n,t)$. The distribution function $f(n,t)$ has to obey the boundary and initial conditions as given by Eq. (8.2); T is the temperature. Here it is assumed that the system can be brought suddenly into the respective metastable initial state. Aggregation phenomena and their effects on the phase formation process, which may occur in the course of the transfer of the system into the considered initial state, are excluded from consideration (for the account of such additional transient effects cf., e.g., [140, 244] and references cited therein).

The number of nucleation sites is given by $f(n \to 0, t)$ and is determined by the number of configurations, $\prod_i c_i^{\nu_i}$, which may result in the evolution of the first structural element of the newly evolving phase. Further, conservation of the total number of particles results in the set of balance equations (8.3) for different components forming the newly evolving phase; c_i is the actual concentration and c_{i0} the initial concentration of the ith component in the solution. Note that all concentrations c_i as well as the distribution function $f(n,t)$ refer to the respective numbers per single lattice site.

In order to solve the given system of equations, one has to specify the coefficients of aggregation, $w_{n,n+1}$. As shown in Chapter 2 (see also [293, 300, 304]), these coefficients can be expressed via the macroscopic growth rates $\mathrm{d}n/\mathrm{d}t$ and the derivative $(\partial\Delta\Phi/\partial n)$ as

$$\frac{\mathrm{d}n}{\mathrm{d}t} = -w_{n,n+1}\frac{1}{k_B T}\frac{\partial\Delta\Phi}{\partial n}. \tag{8.4}$$

Here $(\partial\Delta\Phi/\partial n)$ is partial derivative of $\Delta\Phi$ which is the change of the thermodynamic potential if in the solid solution an aggregate of the newly evolving phase with n structural elements is formed. Both the aggregate of the newly evolving phase as well as the solution are considered to be in a state of internal thermal equilibrium while the system as a whole is in a nonequilibrium state. The change of the characteristic thermodynamic potential $\Delta\Phi$, due to the formation of such an aggregate with n structural elements, can then be expressed as

$$\Delta\Phi = n\left(\mu^{(s)} - \sum_i \nu_i\mu_i\right) + 4\pi a_s^2 \sigma n^{2/3}. \tag{8.5}$$

We then obtain immediately

$$\frac{1}{k_B T}\frac{\partial\Delta\Phi}{\partial n} = \frac{1}{k_B T}\left(\mu^{(s)} - \sum_i \nu_i\mu_i\right) + \beta n^{-1/3}, \qquad \beta = \frac{8\pi}{3}\frac{\sigma a_s^2}{k_B T}. \tag{8.6}$$

In the above equations, $\mu^{(s)}$ is the chemical potential per structural element in the new phase, μ_i is the chemical potential of the ith component in the solution, $a_s = (3\omega_s/4\pi)^{1/3}$ is the characteristic size parameter of the structural elements, and σ is the specific interfacial energy of the aggregate of the new phase in the solution. Note that, since the composition of the aggregates and the specific volumes of different components are independent of cluster size, the surface energy σ has to be considered as independent of cluster size as well.

The size of the critical aggregate, n_c, for a cluster, being in unstable equilibrium with the ambient solution, is determined via $(\partial\Delta\Phi/\partial n) = 0$ or via

$$n_c^{1/3} = \frac{\beta}{\sum\limits_i \nu_i\mu_i - \mu^{(s)}}. \tag{8.7}$$

For a weak or perfect solution, where $\mu_i = \psi_i + k_BT\ln c_i$ holds, we get, in particular,

$$r_c = n_c^{1/3} = \frac{\beta}{\ln\left(\frac{\prod\limits_i c_i^{\nu_i}}{k_\infty}\right)}, \qquad k_\infty = \exp\left(\frac{\mu_s - \sum_i \nu_i\psi_i}{k_BT}\right). \tag{8.8}$$

Here with k_∞ the chemical equilibrium constant has been introduced.

For any arbitrary value of n we may write immediately

$$\frac{\Delta\Phi\,(n)}{k_BT} = -\frac{\beta n}{n_c^{1/3}} + \frac{3}{2}\beta n^{2/3}, \qquad \frac{1}{k_BT}\frac{\partial\Delta\Phi}{\partial n} = \beta\left(n^{-1/3} - n_c^{-1/3}\right),$$
$$\frac{1}{2k_BT}\frac{\partial^2\Delta\Phi}{\partial n^2} = -\frac{\beta}{6n^{4/3}}, \tag{8.9}$$

resulting in

$$\frac{\Delta\Phi\,(n_c)}{k_BT} = \frac{\beta}{2}n_c^{2/3}, \qquad \frac{\Delta\Phi\,(8n_c)}{k_BT} = -2\beta n_c^{2/3}, \qquad \delta n_c = \sqrt{\frac{6n_c^{4/3}}{\beta}}, \tag{8.10}$$

$$\frac{1}{2k_BT}\left.\frac{\partial^2\Delta\Phi\,(n)}{\partial n^2}\right|_{n=n_c} = -\frac{1}{\delta n_c^2} = -\frac{\beta}{6}n_c^{-4/3}. \tag{8.11}$$

The quantity δn_c describes the range of the size n values in the vicinity of the critical cluster size, where the relation $\Delta\Phi(n_c) - \Delta\Phi(n) \leq k_BT$ holds.

The expressions for the aggregation coefficients $w_{n,n+1}$ can be found from the analysis of the mode of aggregation of the clusters. For the small-sized clusters, prevailing in the stage of nucleation, the growth is limited kinetically (e.g., Chapter 3 and [304]). In the transient stage, the aggregates of the newly evolving phase are sufficiently large, so that the growth rate $\mathrm{d}n/\mathrm{d}t$ and the aggregation coefficients can be found from the solution of the diffusion equation.

In order to arrive at the respective expressions, first we have to take into account that the partial fluxes j_i of individual components have to fulfill the relations [280, 293, 296, 299, 300, 304]

$$\frac{j_i}{\nu_i} = \frac{j_k}{\nu_k} = \cdots \tag{8.12}$$

In addition, we may write down for any partial flux j_i [284, 304]

$$4\pi R^2 j_i = -w_{n_i,n_i+1}\frac{1}{k_BT}\frac{\partial \Delta\Phi^{(s)}}{\partial n_i} \tag{8.13}$$

$$= \frac{3\alpha_i D_i \widetilde{c}_i n^{2/3}}{a_m^2 k_B T}\left(\frac{\omega_s}{\omega_m}\right)^{2/3}\left(\mu_i\left(\widetilde{c}_i\right) - \mu_i\left(c_{ni}\right)\right),$$

$$\Delta\Phi^{(s)} = \sum_i n_i\left[\mu_i\left(c_{ni}\right) - \mu_i\left(\widetilde{c}_i\right)\right], \qquad \mu^{(s)} = \sum_i \nu_i \mu_i\left(c_{ni}\right). \tag{8.14}$$

Here $\mu^{(s)}$ is, as previously, the chemical potential of a structure element of the new phase, $\mu_i(c_{ni})$ are the chemical potentials of the particles of the ith component in the solution in equilibrium with a cluster of size n, $\mu_i(\widetilde{c}_i)$ are the chemical potentials of the particles of the ith component in the solution in the immediate vicinity of the aggregate, $\{c_{ni}\}$ represents a set of concentrations of the particles of different components which result in an equilibrium with an aggregate of size n, while $\{\widetilde{c}_i\}$ represents the set of concentrations in the vicinity of the surface of the aggregate. We get

$$\frac{\partial \Delta\Phi^{(s)}}{\partial n_i} = -\left(\mu_i\left(\widetilde{c}_i\right) - \mu_i\left(c_{ni}\right)\right), \qquad w_{n_i,n_i+1} = \frac{3\alpha_i D_i}{a_m^2}\widetilde{c}_i\left(\frac{\omega_s}{\omega_m}\right)^{2/3} n^{2/3}. \tag{8.15}$$

Here w_{n_i,n_i+1} is the frequency of incorporation of particles of the component i into the new phase aggregates, α_i is the sticking coefficient ($0 \le \alpha_i \le 1$), D_i is the partial diffusion coefficient of the component i, and a_m is the lattice parameter of the matrix ($\omega_m = 4\pi a_m^3/3$). The radius R and the particle number in the aggregates are related via $n = (4\pi R^3/3\omega_s)$, $\omega_s = 4\pi a_s^3/3$. For the rate of growth of an aggregate of size n, we get from Eq. (8.12) and the equation $\omega_s = \sum_i \nu_i \omega_i$

$$\frac{\mathrm{d}n}{\mathrm{d}t} = \frac{4\pi R^2}{\omega_s}\sum_i \omega_i j_i = 4\pi R^2 \frac{j_i}{\nu_i}. \tag{8.16}$$

As a next step, we insert Eq. (8.13) into Eq. (8.16). Afterward, we may divide the equation at the coefficient in front of the term $[\mu(\widetilde{c}_i) - \mu_i(c_{ni})]$, multiply the resulting equation with ν_i and take the sum over all values of i. As a result, we obtain

$$\frac{\mathrm{d}n}{\mathrm{d}t} = \frac{3D^*}{a_m^2}\left(\frac{\omega_s}{\omega_m}\right)^{3/2} n^{2/3}\sum_i \nu_i\left(\mu_i\left(\widetilde{c}_i\right) - \mu_i\left(c_{ni}\right)\right),$$

$$\frac{1}{D^*} = \sum_i \frac{\nu_i^2}{\alpha_i D_i \widetilde{c}_i}. \tag{8.17}$$

In the limiting case of a weak solution, we get, in particular,

$$\Delta\Phi = -n\ln\left(\frac{\prod_i (c_i)^{\nu_i}}{k_\infty}\right) + 4\pi\sigma a_s^2 n^{2/3}, \qquad k_n = \prod_i (c_{in})^{\nu_i} = k_\infty e^{\beta/(n^{1/3})}, \tag{8.18}$$

$$\frac{\partial\Delta\Phi}{\partial n} = -\ln\left(\frac{\prod_i (c_i)^{\nu_i}}{k_n}\right),$$

resulting in

$$\frac{\mathrm{d}n}{\mathrm{d}t} = -w_{n,n+1}\frac{\partial\Delta\Phi}{\partial n} = \frac{3D^*}{a_m^2}\left(\frac{\omega_s}{\omega_m}\right)^{3/2} n^{2/3}\ln\left(\frac{\prod_i c_i^{\nu_i}}{k_n}\right), \tag{8.19}$$

$$\frac{\mathrm{d}n}{\mathrm{d}t} = \frac{3D^*}{a_m^2}\left(\frac{\omega_s}{\omega_m}\right)^{2/3} n^{2/3}\beta\left(\frac{1}{n_c^{1/3}} - \frac{1}{n^{1/3}}\right), \qquad w_{n,n+1} = \frac{3D^*}{a_m^2}\left(\frac{\omega_s}{\omega_m}\right)^{2/3} n^{2/3}. \tag{8.20}$$

Equations (8.19) and (8.20) describe the flux of particles to the aggregates of the newly evolving phase in the immediate vicinity of the aggregate. The concentrations of different components $\widetilde{c}_i$ in this region are determined by the interplay of losses to aggregation and input fluxes due to the diffusion from the distant environment. For the determination of these concentrations, the respective diffusion problem has to be solved self-consistently (see Chapter 3 and [283]).

If the concentrations of different components in the new phase and the ambient solution differ considerably, then, in order to find the rate of growth of the aggregates of the new phase, one may employ the steady-state solution of the diffusion equation and the effects of the motion of the interface may be neglected (see Chapters 3 and [283]) with an accuracy of the order $c_i/c_i^s \ll 1$, where c_i^s are the concentrations in the solution in equilibrium with a macroscopic aggregate of the newly evolving phase. In such case, we have

$$\frac{\mathrm{d}n}{\mathrm{d}t} = 4\pi R^2\frac{j_i}{v_i} = 3\left(\frac{\omega_s}{\omega_m}\right)^{2/3}\frac{D_i}{a_m^2}\frac{c_i - \widetilde{c}_i}{\nu_i}n^{1/3}. \tag{8.21}$$

From Eqs. (8.13) and (8.21), we arrive at the following expression for the determination of the concentrations $\widetilde{c}_i$ (for the case of a weak solution):

$$\ln\left(\frac{\widetilde{c}_i}{c_{in}}\right) = \frac{1}{\alpha_i n^{1/3}}\frac{c_i - \widetilde{c}}{\widetilde{c}_i}. \tag{8.22}$$

Equations (8.13) and (8.22) and the equilibrium conditions at the interface of an aggregate completely determine the sets of concentrations $\{\widetilde{c}_i\}$ and $\{c_{ni}\}$ for the stage of nucleation. In the limit $\alpha_i n^{1/3} \leq 1$ (which is usually fulfilled in the stage of nucleation), we get $\widetilde{c}_i \cong c_i$ (cf. Eq. (8.22)). In this case, it is more convenient to employ Eq. (8.19) for the determination of the growth rate $\mathrm{d}n/\mathrm{d}t$ with the replacements $\widetilde{c}_i \to c_i$, $\ln(\widetilde{c}_i/c_{in}) \cong \ln(c_i/c_{in})$.

In the transient stage, when the size of aggregates is sufficiently large (i.e., the inequality $\alpha_i n^{1/3} \gg 1$ holds), the relation $\widetilde{c}_i \cong c_{in}$ is fulfilled (cf. Eq. (8.22)). In this case, one has to employ directly Eqs. (8.21) and (8.22) for the determination of $\mathrm{d}n/\mathrm{d}t$.

8.3 The Stage of Nucleation of Clusters of the Newly Evolving Phase

The description of the kinetics of nucleation is significantly simplified after the time interval t_r, when a quasisteady-state flux in a cluster-size space is established in the range $0 \leq n \leq n_c$ (e.g., Chapter 3). Indeed, in this case it is possible to employ a simpler version of the basic equation (8.1) for the determination of the flux in cluster sizes space, $J(n,t)$. The respective relation is valid in the whole stage of steady-state nucleation, $t_r \leq t \leq t_N$. Hereby, the boundary conditions for the flux in the range $0 \leq n \leq n_c$ may be expressed via the boundary conditions for the distribution function $f(n,t)$. Once the flux $J(n,t)$ is known, the distribution function $f(n,t)$ can be found straightforwardly.

An estimate of the time lag, t_r, can be derived in the same way as outlined in Chapter 3. We get

$$t_r = \frac{5}{3}\frac{a_m^2}{D^*}\left(\frac{\omega_m}{\omega_s}\right)^{2/3}\frac{n_c^{2/3}}{\beta}. \tag{8.23}$$

An overview of different alternative attempts to estimate this quantity is given, e.g., in [30, 94, 95]. The results, obtained by different methods, deviate only slightly.

After the completion of the transient stage to steady-state nucleation, the equation for the determination of the flux in cluster-size space may be written in the form (cf. Section 3.4)

$$\frac{\partial J\left(n,t\right)}{\partial t} = w_{n,n+1}\left\{\frac{\partial^2 J(n,t)}{\partial n^2} + \frac{1}{k_B T}\frac{\partial \Delta\Phi}{\partial n}\frac{\partial J(n,t)}{\partial n}\right\} \tag{8.24}$$

with the boundary conditions $J\left(n,t\right)\big|_{n=n_c} = J\left(n_c\right)$. Hereby we chose the moment of time $t=0$ as corresponding to the beginning of the stage of steady-state nucleation, i.e., we make the replacement $t - t_r \to t$.

In the derivation of Eq. (8.24), terms of the order $(\dot{c}/c)J\left(\partial J/\partial t\right)^{-1} \simeq (t_N/t_c) \ll 1$ have been neglected, where t_c is the characteristic time of change of the concentration in the solution. During the time of steady-state nucleation, t_N, the change of the concentration in the solution is insignificant for the case $n_c|_{t=0} \gg 1$. Equation (8.24) is thus valid for any moment of time $t \leq t_N$ or when $J(n,t) > 0$ holds for any value of n. In this stage, the number of supercritical clusters increases.

For $t \geq t_N$, the quantity $J(n_c)$ becomes practically equal to zero and the process of formation of new clusters is terminated. Their number remains then nearly constant at the subsequent transient stage to coarsening for a time $t_f > t_N$. For $t \geq t_f$ the further evolution is governed by processes of competitive growth or coarsening (Chapter 4).

In the considered stage of steady-state nucleation $t < t_N$, it is possible to express $f(n,t)$ via $J(n,t)$ as was done in Section 3.3

$$f(n,t) = \exp\left[-\frac{\Delta\Phi(n)}{k_B T}\right] \int_n^\infty \exp\left(\frac{\Delta\Phi(n')}{k_B T}\right) \frac{J(n',t)}{w_{n',n'+1}} \, dn' . \tag{8.25}$$

With $J(n,t) = J(n_c)$, $0 \le n \le n_c + \delta n_c$ and the boundary condition for $f(n,t)$ at $n \to 0$ we get

$$\prod_i c_i^{\nu_i} = \int_0^\infty \exp\left(\frac{\Delta\Phi(n')}{k_B T}\right) \frac{J(n',t)}{w_{n',n'+1}} \, dn', \qquad \Delta\Phi(0) = 0. \tag{8.26}$$

Since $\Delta\Phi(n)$ has a sharp extremum at $n = n_c$ ($\Delta\Phi(n) = \Delta\Phi(n_c) - k_B T (n - n_c)^2 \delta n_c^{-2} \gg 1$), we get with Eq. (8.11)

$$J(n_c) = \sqrt{\frac{3\beta}{2\pi} \frac{D^*}{a_m^2}} \left(\frac{\omega_s}{\omega_m}\right)^{2/3} \prod_i c_i^{\nu_i} \exp\left[-\frac{\Delta\Phi(n_c)}{k_B T}\right] . \tag{8.27}$$

Equation (8.27) is reduced, evidently, to the respective expression for the steady-state nucleation rate in single-component systems in the limiting case of precipitation of only one component, Eq. (3.79).

For $n < n_c$, in Eq. (8.25) the maximum of $\Delta\Phi$ is located inside the limits of integration. Moreover, $J(n,t) = J(n_c)$ holds and we get

$$f(n)|_{n<n_c} = \prod_i c_i^{\nu_i} \exp\left[-\frac{\Delta\Phi(n)}{k_B T}\right] \frac{1}{2} \left[1 - \mathrm{erf}\left(\frac{n - n_c}{\delta n_c}\right)\right] . \tag{8.28}$$

Here $\mathrm{erf}(x) = -\mathrm{erf}(-x)$ is the error function.

In the limiting case of a saturated system, we have $n_c \to \infty$ and Eq. (8.28) is reduced to the well-known steady-state cluster-size distribution in an equilibrium state, Eq. (3.85).

In the stage of nucleation, we have $J(n,t) > 0$ and for any given value of n this quantity is determined by the respective values at $n' < n$. In the later stages of the process, the situation is different. There we have $J(n,t) < 0$ for $n < n_c$ and $J(n,t) > 0$ for $n > n_c$. It follows that in the stage of nucleation one may determine the functions $J(n,t)$ for different ranges of n values by different methods and take as the boundary conditions the values determined via the solutions on the left-hand side of the respective intervals.

In order to proceed with the analysis, we introduce the dimensionless time $\tau = t/\widetilde{t}$ with $\widetilde{t}^{-1} = D^* a_m^{-2} (\omega_s/\omega_m)^{2/3}$. Further, we note that in the range $1 \le n/n_c \le 8$ the quantity $A = 3\left[(n/n_c)^{1/3} + (n/n_c)^{-1/3} + 1\right]^{-1}$ varies only in between the limits from 1 to $6/7$ (see Chapter 3). Consequently, in this range of cluster sizes we may set $3n^{2/3}\left(n^{-1/3} - n_c^{-1/3}\right) = -(n - n_c)\, n_c^{-2/3}$ and $A \cong -(n - n_c) n_c^{-2/3}$.

The kinetic equation for $J(n, \tau) = J\tilde{t}$ can then be written in this range of cluster sizes as

$$\frac{\partial J}{\partial \tau} = -\frac{\beta}{n_c^{2/3}} (n - n_c) \frac{\partial J}{\partial n} + 3n_c^{2/3} \frac{\partial^2 J}{\partial n^2}, \tag{8.29}$$

$$J|_{n=n_c} = J(n_c) = J_0\tilde{t}, \qquad J(n,\tau)|_{n>n_c,\tau=0} = 0. \tag{8.30}$$

The replacement $n^{2/3} \to n_c^{2/3}$ in the second term on the right-hand side of Eq. (8.29) decreases the diffusion contribution to the flux for $n > n_c$. However, in the considered range this contribution is small (see Chapter 3 and [298]). In the vicinity of $n \simeq n_c$, the replacement represents a quite accurate approximation.

In order to solve Eq. (8.29), we make the ansatz $J(n,\tau) \to J(\psi(x,\tau), t(\tau))$ and determine the functions ψ and $t(\tau)$ via

$$\psi = (n - n_c) \exp(-\delta\tau), \qquad \delta = \beta n_c^{-2/3}, \qquad t(\tau) = \frac{3n_c^{2/3}}{2\delta}\left(1 - e^{-2\delta\tau}\right). \tag{8.31}$$

After this substitution (8.29) takes the form

$$\frac{\partial J}{\partial t} = \frac{\partial^2 J}{\partial \psi^2}, \qquad J|_{n=n_c} = J(n_c), \qquad J(n,\tau)|_{n>n_c,\tau=0} = 0 \tag{8.32}$$

with the solution

$$J = J(n_c)\left[1 - \operatorname{erf}\left(\frac{\psi}{2\sqrt{t}}\right)\right] = J(n_c)\left[1 - \operatorname{erf}\left(\frac{e^{-\delta\tau}(n - n_c)}{2\sqrt{t(\tau)}}\right)\right]. \tag{8.33}$$

It follows that after a time $\tau_p \simeq 1/\delta \simeq \tau_r$ in the range $n_c \le n \le g = 8n_c$ a steady state with the flux $J = J(n_c)$ establishes. Thus, the time of establishment of the steady-state conditions in the range $n \le g = 8n_c$ is of the same order as the time of establishment of the steady state during nucleation in the range $0 \le n \le n_c$.

The distribution function over cluster size in the range $n_c \le n \le g \simeq 8n_c$ can be derived from Eq. (8.25) via a Taylor expansion of $\Delta\Phi(n)$ in the vicinity of n:

$$\begin{aligned} f(n,\tau) &= \frac{J_0}{w_{n,n+1}} \int_n^\infty \exp\left\{-\frac{(\Delta\Phi(n) - \Delta\Phi(n'))}{k_B T}\right\} dn' \\ &= \frac{J_0}{2w_{n,n+1}} \sqrt{\frac{\pi}{b}}\left[1 - \operatorname{erf}\left(\frac{a}{2b}\right)\right] e^{a^2/(4b^2)}, \end{aligned} \tag{8.34}$$

$$a = -\frac{1}{k_B T}\frac{\partial \Delta\Phi}{\partial n} \ge 0, \qquad b = -\frac{1}{2k_B T}\frac{\partial^2 \Delta\Phi}{\partial n^2} > 0, \qquad n \ge n_c. \tag{8.35}$$

The function $f(n,\tau)$, determined via Eq. (8.34), goes over continuously into the expression derived via Eq. (8.28) for $n = n_c$. This way, the boundary conditions for $J(n,\tau) = J(n_c)$ hold after the time interval $\tau \simeq 2\tau_r$ at $g = 8n_c$.

In the range $n \geq g$, it is also possible to further simplify the kinetic equation (8.29) describing the time evolution of $J(n, \tau)$ and obtain an exact analytic solution. However, one has to account for the decrease of the degree of metastability of the system due to the continuous formation of new and growth of the already formed supercritical clusters (depletion effects). In this way, an expression for the time interval of dominating steady-state nucleation may be obtained.

In the range of cluster sizes $n \geq g \simeq 8n_c$, we get for the case of a weak solution

$$\frac{\Delta\Phi\left(n_c\right)}{k_B T} = \frac{\Delta\Phi\left(n_c\left(0\right)\right)}{k_B T} + n_c\left(0\right)\varphi, \qquad \varphi\left(\tau\right) = -\ln\prod_i\left(\frac{c_i\left(\tau\right)}{c_i\left(0\right)}\right)^{\nu_i}. \tag{8.36}$$

Here $c_i(0) = c_{i0}$ is the initial concentration of the particles of the ith component in the solution. These quantities obey the inequality $c_{i0} \geq c_i$.

The boundary conditions at $n = g$ may be written in the form

$$J\left(n_c\right) = J\left(n_c\left(0\right)\right)\exp\left(-n_c\left(0\right)\varphi\left(\tau\right)\right). \tag{8.37}$$

As evident from Eq. (8.37), for $n_c(0) \gg 1$ a small change of the quantity φ results in a significant decrease of the nucleation rate. Processes of nucleation are terminated practically, if the condition $\varphi(\tau_N) = 1/n_c(0)$ is fulfilled.

In the considered range of cluster sizes, the equation for the flux may be formulated most conveniently when the variable $r = n^{1/3}$ is employed (see Chapter 3 and [293, 300]). We get

$$\frac{\partial J}{\partial \tau} = -\frac{\beta}{r_c}\frac{\partial J}{\partial r} + \frac{1}{3r_c^2}\frac{\partial^2 J}{\partial r^2}, \tag{8.38}$$

where the boundary condition is given by Eq. (8.37). In Eq. (8.38) several terms are omitted which are small in comparison with the quantity $(\beta/2r_c)/(\partial J/\partial r)$ (i.e., $(2r^{-3}/3)(\partial J/\partial r)$ and $(\beta/r)(\partial J/\partial r)$, $(\beta > 1)$). Moreover, the substitution $3r^{-2} \to 3r_c^{-2}$ is made (in the considered range the inequality $r > 2r_c$ holds). Such approximation results in a sufficiently accurate description of the spectrum of the viable nuclei, which give the dominating contribution into the law of conservation of particle numbers. However, the mentioned approximations result in some additional broadening of the front of motion of the aggregates in cluster-size space (see Chapters 3 and [293, 300]).

As a next step, we redefine the variable r via $r = n^{1/3} - g^{1/3}$. The solution for J is further expressed as

$$J = \exp\left(\frac{3}{2}\beta r r_c\right)\exp\left(-\frac{3}{4}\beta^2\tau\right)p(r, \tau). \tag{8.39}$$

A substitution of this expression into Eq. (8.38) results in the following equation for the function $p(r, \tau)$:

$$\frac{\partial p}{\partial \tau} = (3r_c^2)^{-1}\frac{\partial p^2}{\partial r^2}. \tag{8.40}$$

The solution of Eq. (8.38) with the boundary conditions (8.37) reads then

$$J = J\left(n_c\left(0\right)\right) e^{3\beta r r_c/2} \frac{r}{2} \left(\frac{\pi}{3r_c}\right)^{-1/2}$$

$$\times \int_0^\tau e^{-n_c(0)\varphi(\tau') - 3\beta^2(\tau-\tau')/4 - 3r^2 r_c^2/(4(\tau-\tau'))} \frac{\mathrm{d}\tau'}{(\tau-\tau')^{3/2}}. \tag{8.41}$$

With the variable $z = r r_c(\frac{4}{3}(\tau - \tau'))^{-1/2}$ we may write down the expression for J in the form

$$J = J\left(n_c\left(0\right)\right) \frac{2}{\sqrt{\pi}} \int_{r r_c (4\tau/3)^{-1/2}}^{\infty} \exp\left[-n_c \varphi\left(\tau - 3r^2 r_c^2 z^{-2}/4\right)\right]$$

$$\times \exp\left[-\left(3\beta r r_c z^{-1}/4 - z\right)^2\right] \mathrm{d}z = J\left(n_c\left(0\right)\right) \frac{2}{\sqrt{\pi}} \exp\left[-n_c \varphi\left(\tau - r r_c \beta^{-1}\right)\right]$$

$$\times \int_{z(\tau'=0)}^{\infty} \exp\left[-\left(3\beta r r_c z^{-1}/4 - z\right)^2\right] \mathrm{d}z. \tag{8.42}$$

In Eq. (8.42), the term $\exp\left[-\left(3\beta r r_c z^{-1}/4 - z\right)^2\right]$ has a sharp maximum at $z = z_0$. In the vicinity of z_0 we may write

$$\left(\frac{3}{4}\frac{\beta r r_c}{z} - z\right)^2 \simeq 4(z - z_0)^2, \qquad z_0 = \sqrt{\left(\frac{3}{4}\beta r r_c\right)}. \tag{8.43}$$

With such approximation, we arrive at

$$J(n,\tau) = J(n_c(0)) \exp[-n_c \varphi(\tau_0(n,\tau))] \frac{2}{\sqrt{\pi}} \int_\xi^\infty \exp(-\xi'^2)\, \mathrm{d}\xi', \tag{8.44}$$

$$\xi = 2\left(z\left(\tau' = 0\right) - z_0\right) = 2\left(\frac{r r_c}{\sqrt{4\tau/3}} - z_0\right), \tag{8.45}$$

$$\tau_0\left(n,\tau\right) = \tau - r r_c \beta^{-1} = \tau - \left(r - r_g\right) \frac{r_c}{\beta},$$

$$\tau_0\left(r_{\max},\tau\right) = 0, \qquad \tau_0\left(g^{1/3},\tau\right) = \tau, \qquad r_{\max}\left(\tau\right) = r_g + \beta r_c^{-1}\tau. \tag{8.46}$$

Here the redefinition of r has been taken into account ($r \to r - r_g$).

The dependence $r_{\max}(t)$ describes the motion of the cluster front in cluster-size space along the characteristics of Eq. (8.44). The diffusiveness of the front is determined by the integral term in Eq. (8.44). This integral is practically equal to unity for $\xi \leq 0$ and equal to zero for $\xi \geq 0$. Approximately, we may thus write

$$J(n,\tau) = J(n_c(0)) \exp\left[-n_c(0)\,\varphi(\tau_0(n,\tau))\right] \theta(r_{\max}(t) - r) \tag{8.47}$$

with

$$r_{\max}(t) = r_g + \frac{\beta\tau}{r_c} = r_g + \ln\left(\prod_i \frac{c_i^{\nu_i}}{k_\infty}\right)\frac{t}{\tilde{\tilde{t}}}\,,$$

$$\theta(x) = 1 \ (\text{for } x > 0)\,, \qquad \theta(x) = 0 \ (\text{for } x < 0)\,. \tag{8.48}$$

Since the broadening of the front of motion of the clusters in cluster-size space is small, we may employ Eq. (8.47) for the application of the laws of conservation of the numbers of particles. These laws can be written as

$$\varphi = -\ln\prod_i \left(\frac{c_i}{c_{i0}}\right)^{\nu_i} = \sum_i \nu_i \ln\left(\frac{c_{i0}}{c_i}\right), \qquad c_{i0} = c_i + \nu_i \int_0^\infty n f(n,\tau)\, \mathrm{d}n, \tag{8.49}$$

$$\dot{c}_i = -\nu_i \int_0^\infty n \frac{\partial f}{\partial \tau}\, \mathrm{d}n = -\nu_i \left(J(\tau)\, g - \int_0^\tau J(\tau_0) \frac{\partial n}{\partial \tau_0}\, \mathrm{d}\tau_0\right). \tag{8.50}$$

By definition of the quantity $\tau_0(n,\tau)$ we may write down (employing Eq. (8.44))

$$n(\tau - \tau_0) = \left[r_g + \ln\left(\prod_i \frac{c_i^{\nu_i}}{K_\infty}\right)(\tau - \tau_0)\right]^3. \tag{8.51}$$

With Eq. (8.51), we may go over in Eq. (8.49) from the variable n to the variable $\tau_0(n,t)$. We get then

$$\int_g^\infty J(\tau_0(n,\tau))\, \mathrm{d}n = \int_g^{n_{\max}} J(\tau_0)\, \mathrm{d}n = \int_{\tau_0(g,\tau)=\tau}^{\tau_0(n_{\max},\tau)=0} J(\tau_0) \frac{\mathrm{d}n}{\mathrm{d}\tau_0}\, \mathrm{d}\tau_0 = -\int_0^\tau J(\tau_0) \frac{\mathrm{d}n}{\mathrm{d}\tau_0}\, \mathrm{d}\tau_0, \tag{8.52}$$

where $\mathrm{d}n/\mathrm{d}\tau = -\mathrm{d}n/\mathrm{d}\tau_0$ is the rate of growth at the moment of time τ of those aggregates of the new phase, which have grown up to the size g at the moment of time τ_0, $n|_{\tau=\tau_0} = g$, $r|_{\tau=\tau_0} = r_g$. This way, we get the following expression for φ:

$$\frac{\mathrm{d}\varphi}{\mathrm{d}\tau} = -\sum_i \frac{\nu_i}{c_i}\frac{\mathrm{d}c_i}{\mathrm{d}\tau} = \left(\sum \frac{\nu_i^2}{c_i}\right)\left(Jg - \int_0^\tau J(\tau_0) \frac{\mathrm{d}n}{\mathrm{d}\tau_0}\, \mathrm{d}\tau_0\right). \tag{8.53}$$

Equation (8.53) has a clear physical meaning (cf. also Chapter 3 and [293, 300]). The decrease of the degree of metastability φ is due both to the formation of new supercritical clusters (first term in Eq. (8.53)) as well as to the growth of already existing aggregates (second term in Eq. (8.53)).

Taking the integral in Eq. (8.53) by parts and employing additionally Eqs. (8.47) and (8.51), we obtain,

$$\frac{\mathrm{d}\varphi}{\mathrm{d}\tau} = \left(\sum \frac{\nu_i^2}{c_i}\right) J_0 n(\tau), \qquad J_0 = J(n_c(0)). \tag{8.54}$$

Here $J(n_c(0))$ is determined by Eq. (8.27) and

$$n(\tau) = n(\tau - \tau_0)|_{\tau_0=0} = (r_g + a\tau)^3 = r_{\max}^3(\tau), \tag{8.55}$$

$$a = \ln\left(\prod_i c_i^{\nu_i}/k_\infty\right), \qquad \tau = \frac{t}{\tilde{t}} = \frac{D^*}{a_m^2}\left(\frac{\omega_s}{\omega_m}\right)^{2/3} t, \qquad r_{\max}(\tau) = r_g + a\tau. \tag{8.56}$$

In Eq. (8.55) a small term of second order in $J_0 \ll 1$ has been neglected (remember, J_0 is by definition the flux per lattice site). Indeed, since we have $\mathrm{d}\varphi/\mathrm{d}\tau_0 \simeq J_0$, we get

$$-n_c \int_0^\tau J_0 e^{-n_c\varphi(\tau_0)} \frac{\mathrm{d}\varphi}{\mathrm{d}\tau_0} n(\tau - \tau_0)\,\mathrm{d}\tau_0 \simeq J_0^2. \tag{8.57}$$

In Eq. (8.55), we may set $c_i \simeq c_{i0}$ since the variations of the concentrations remain small in the stage of nucleation. An integration of this equation then yields (with $\varphi|_{\tau=0} = 0$)

$$\varphi = \left(\sum \frac{\nu_i^2}{c_{i0}}\right) \frac{J_0}{4a}\left[(r_g + a\tau)^4 - r_g^4\right], \quad \varphi(\tau_N) \simeq n_c^{-1}, \quad 0 \le \tau \le \tau_N. \tag{8.58}$$

The time of steady-state nucleation (determined via $\varphi n_c(0) \cong 1$) is then obtained as

$$\tau_N^4 = \left[\left(\sum \frac{\nu_i^2}{c_i}\right) J_0 \frac{a^3}{4}\right]^{-1} \frac{1}{n_c}. \tag{8.59}$$

As evident, τ_N depends weakly on n_c^{-1}. A substitution of the expressions for n_c and a (cf. Eq. (8.56)) yields in the limit $a\tau_N \gg r_g$

$$\tau_N = \frac{t_N}{\tilde{t}} = 4^{1/4}\beta^{-3/4}\left(\sum_i \frac{\nu_i^2}{c_{i0}}\right)^{-1/4} J_0^{-1/4}. \tag{8.60}$$

In order to determine the distribution function $f(r,\tau)$ in the considered range of cluster sizes, $r \ge g^{1/3} = 2r_c$, we first have to find the flux, $J(r,t)$ given by Eq. (8.47). However, in order to get an explicit expression, we first have to find $\varphi(\tau_0(r,\tau))$ from Eq. (8.58). Using $r_g \ll a_\tau$, we have for $n_c\varphi(\tau_0(r,\tau))$

$$n_c\varphi(\tau_0(r,\tau)) = \frac{\varphi(\tau_0(r,\tau))}{\varphi(\tau_N)} \simeq \left(\frac{\tau_0(r,\tau)}{\tau_N}\right)^4 = \left(\frac{r_{\max}(\tau) - r}{r_{\max}(\tau_N) - r_g}\right)^4. \tag{8.61}$$

It follows from Eq. (8.61) that in the considered time interval ($\tau \leq \tau_N$) and cluster size range ($r \leq r_{\max}(\tau_N)$), the quantity $n_c\varphi(\tau_0(r,\tau))$ is much less than unity. For this reason, Eq. (8.47) yields

$$J = J_0\theta\left(r_{\max}(\tau) - r\right), \qquad \theta(x) = 1 \quad (x > 0), \qquad \theta(x) = 0 \quad (x < 0). \tag{8.62}$$

In the range $r > r_g$ and for $n_c \gg 1$, the interfacial effects as well as the influence of the diffusion term in the basic equation can be neglected. Then we get from Eq. (8.34) with $-(k_BT)^{-1}\left(\partial\Delta\Phi/\partial n\right) \gg -0.5k_BT\left(\partial^2\Delta\Phi/\partial n^2\right)$, $(n = 8n_c = r_g^3)$

$$f\left(r,\tau\right)\big|_{r\geq r_g} = J\left(\frac{\mathrm{d}r}{\mathrm{d}t}\right)^{-1} = \frac{J_0}{a}\theta\left(r_{\max}\left(\tau\right) - r\right)\theta\left(r - r_g\right). \tag{8.63}$$

Here $\mathrm{d}r/\mathrm{d}t = a$ holds at $r_g = 2r_c$, as it follows from Eq. (8.17). We also took into account in Eq. (8.63) the relation $f\left(r,\tau\right) = f\left(n,\tau\right)3n^{2/3}$. This way, the distribution function in the stage of nucleation is determined in the range $0 \leq r \leq r_g$ via Eqs. (8.28) and (8.34), while for the range $r_{\max} \geq r > r_g$ this function is given by Eq. (8.63) for $\tau_N > \tau$.

The number of viable clusters per lattice site, formed at $\tau < \tau_N$, is then given by (cf. Eqs. (8.62) and (8.63), $r_{\max} = a\tau$ at $r \gg r_g$)

$$N = \int\limits_0^{\tau} J(g,\tau')\,\mathrm{d}\tau' = \int\limits_{r_g}^{r_{\max}(\tau)} f(n,\tau)\,\mathrm{d}n = J_0\tau. \tag{8.64}$$

The upper limit of N is thus given by

$$N_{\max} = J_0\tau_N = 4^{1/4}\beta^{-3/4}J_0^{3/4}\left(\sum_i \nu_i^2/c_{i0}\right)^{1/4}. \tag{8.65}$$

The largest size of the clusters, which may be formed in the stage of steady-state nucleation, is given by

$$n_{\max}^{1/3} = r_{\max} = a\tau_N = \ln\left(\frac{\prod_i c_{i0}^{\nu_i}}{k_\infty}\right)\tau_N. \tag{8.66}$$

Further, the amount of matter, M, concentrated in the newly evolving phase, is given at $a\tau_N = r_{\max} \gg r_g$ by

$$M = \int\limits_{g^{1/3}}^{r_{\max}} f\left(r,\tau\right)r^3\,\mathrm{d}r = \frac{J_0 r_{\max}^4}{4a} = \frac{1}{n_c}\left(\sum\frac{\nu_i^2}{c_{i0}}\right)^{-1}. \tag{8.67}$$

For the change of the concentrations of different components, we get, consequently,

$$\frac{\Delta c}{c_{i0}} = \frac{c_{i0} - c_i\left(\tau_N\right)}{c_{i0}} = \frac{\nu_{i0}}{c_{i0}}M = \frac{1}{n_c}\frac{\nu_{i0}}{c_{i0}}\left(\sum\frac{\nu_i^2}{c_{i0}}\right)^{-1} < n_c^{-1} \ll 1. \tag{8.68}$$

This way, at $n_c(0) \gg 1$, the stage of steady-state nucleation is terminated at relatively small variations of the concentrations of the components forming the new phase. On physical grounds, this result is a consequence of the exponential decay of the flux in dependence on $n_c\varphi$ (cf. Eq. (8.47)). It follows further that all quantities in the preexponential factors may be set equal to the respective values in the initial state.

8.4 The Transient Stage

After the completion of the stage of intensive nucleation of clusters of the newly evolving phase, a new transient to coarsening stage of the phase transition begins (for $\tau \geq \tau_N$). In order to find the cluster-size distribution function in the transient stage, we have to take into account the following circumstances. First, the initial state for the cluster-size distribution is given by the distribution function formed in nucleation to the moment $\tau = \tau_N$. Second, in the transient stage we may neglect the diffusion term in the basic equation due to the high degree of smoothness of the function $f(n, \tau)$ for $\tau \geq \tau_N$. As will be shown below, in the range $r > r_g = 2r_c$ (most of the matter of the new phase is concentrated in clusters having sizes in this range) one can neglect the effect of surface energies as well. This way, in the transient stage, similar to the late coarsening stage (see Chapter 4) of the process of phase separation, we may write

$$\frac{\partial f}{\partial \tau} + \frac{\partial}{\partial r}\left[\left(\frac{\mathrm{d}r}{\mathrm{d}t}\right) f\left(r, t\right)\right] = 0, \qquad r = n^{1/3}, \tag{8.69}$$

$$f\left(r, \tau\right)\big|_{\tau=\tau_N} = f_H\left(r, \tau_N\right)\theta\left(r_g - r\right) + f_H\left(r, \tau_N\right)\theta\left(r - r_g\right)\theta\left(r_{\max}\left(\tau_N\right) - r\right). \tag{8.70}$$

Here $f_H(r_0, \tau_N)$ is determined by Eqs. (8.28) and (8.34) for $r \leq r_g$ and by Eq. (8.63) for $r \geq r_g$. The growth rate $\mathrm{d}r/\mathrm{d}t$ is given by Eq. (8.19). The solution of Eqs. (8.69) and (8.70) reads $f\left(r, \tau\right) = f\left(r_0, 0\right)\left(\partial r_0/\partial \tau\right)$ or

$$\begin{aligned} f\left(r, \tau\right) = \Big[& f_H\left(r_0, \tau_N\right)\theta\left(r_g - r_0\right) \\ & + f_H\left(r_0, \tau_N\right)\theta\left(r_0 - r_g\right)\theta\left(r_{\max}\left(\tau_N\right) - r_0\right)\Big]\frac{\partial r_0}{\partial r}. \end{aligned} \tag{8.71}$$

Here $r_0 = r_0(r, \tau)$ is the characteristics of Eqs. (8.69) and (8.70) determined by Eq. (8.19). The time $\tau = 0$ corresponds, by definition, to the beginning of the transient stage.

The characteristics is determined from the system of equations, Eq. (8.21), and the conservation law, Eq. (8.3), as

$$\frac{\mathrm{d}r^2}{\mathrm{d}t} = 2\left(\frac{\omega_s}{\omega_m}\right)^{2/3} a_m^{-2} B_{in}, \qquad r\big|_{\tau=0} = r_0, \tag{8.72}$$

$$B_{in} = \frac{D_i}{\nu_i}\left(c_i - c_{in}\right) = B_{jn} = \cdots = B, \qquad \prod \left(c_{in}\right)^{\nu_i} = k_\infty \exp\left(\frac{\beta}{n^{1/3}}\right), \tag{8.73}$$

$$c_{i0} = c_i + \nu_i M, \qquad M = \int_0^\infty f n \mathrm{d}n, \tag{8.74}$$

In the transient stage, for $n > n_c$, we may replace $\beta_{in} \rightarrow \beta_{i\infty}$ or $c_{in} \rightarrow c_{i\infty}$ and finally $k_\infty \exp(\beta n^{-1/3}) \rightarrow k_\infty$.

For $n \leq n_c$, an analytic solution for the characteristics cannot be found in the general case for arbitrary $c_i(\tau)$. However, this range of cluster sizes is not important in the transient stage. The degree of metastability is decreased mainly by the growth of the large clusters, $n \gg n_c(0)$. Small clusters, with $n \leq n_c$, present in the system at the initial moment of time, disappear and give only a small contribution to the supersaturation. The range $n \leq n_c(t)$ will be characterized by the growth of $n_c(t)$ and by the dissolution of clusters with sizes $n_c(\tau) \geq n \gg n_c(0)$.

From Eqs. (8.72), (8.73), and (8.74), we get

$$c_{in} = c_{i0} - \nu_i M - \frac{\nu_i}{D_i A} \frac{\mathrm{d}r^2}{\mathrm{d}t}, \qquad A = \left(\frac{\omega_s}{\omega_m}\right)^{2/3} a_m^{-2}, \tag{8.75}$$

$$\prod_i (c_{i0} - \nu_i M)^{\nu_i} \prod_i \left(1 - \frac{\nu_i}{D_i A\,(c_{i0} - \nu_i M)} \frac{\mathrm{d}r^2}{\mathrm{d}t}\right)^{\nu_i} = k_\infty e^{\beta/r}. \tag{8.76}$$

The main contribution to the characteristic time of the transient stage gives the time interval when $\mathrm{d}r^2/\mathrm{d}t \rightarrow 0$ and $r \gg \beta$, $M \leq M_{\max}$. Here $M_{\max}$ is the maximum amount (at the given conditions) of the newly evolving phase per lattice site. This effect is particularly well expressed for a sufficiently high degree of metastability in the initial state (e.g., for the limit $\prod_i c_{i0}^{\nu_i} \gg k_\infty$ in the case of a weak solution). In this case, the clusters of the newly evolving phase as well as the degree of metastability are sufficiently large, and surface effects may be neglected. The late stage is reached, when the degree of metastability tends to zero. Here surface effects become of importance, again, and determine the asymptotic behavior (Chapter 4).

Taking into account the above comments, we obtain from Eq. (8.76) (employing a Taylor expansion and the condition $\mathrm{d}r^2/\mathrm{d}t \rightarrow 0$) the following sufficiently accurate expression:

$$\frac{\mathrm{d}r^2}{\mathrm{d}t} = \left[1 - \frac{k_\infty}{\prod_i (c_{i0} - \nu_i M)^{\nu_i}}\right] \left[\sum_i \frac{\nu_i^2}{D_i} \frac{1}{A(c_{i0} - \nu_i M)}\right]^{-1}. \tag{8.77}$$

At $M \leq M_{\max}$, Eq. (8.77) can be written in the form

$$\frac{\mathrm{d}r^2}{\mathrm{d}t} = -D_{\text{eff}} A\,(M - M_{\max}), \qquad t_N \leq t \leq t_f \tag{8.78}$$

with

$$D_{\text{eff}} A = \left[\sum_i \frac{\nu_i^2}{D_i A} \frac{1}{(c_{i0} - \nu_i M_{\max})}\right]^{-1} \left[\sum_i \frac{\nu_i^2}{(c_{i0} - \nu_i M_{\max})}\right],$$
$$\prod_i (c_{i0} - \nu_i M_{\max})^{\nu_i} = k_\infty. \tag{8.79}$$

Here t_f is the duration of the transient stage of the process.

Note that the precise expression for $\mathrm{d}r^2/\mathrm{d}t$ for a single-component system is obtained from Eq. (8.77) in the limiting case $r \gg \beta$. A similar limiting result may be derived if one of the components has a diffusion coefficient or a concentration much less as compared with the other components. In these latter cases, the process of phase formation is determined mainly by the behavior of this particular component. In contrast to the single-component case, each growth step remains to be characterized by the addition of one structural element.

The number of particles of the newly evolving phase is given, again, by (cf. Eqs. (8.65) and (8.71))

$$N_{\max} = \int_0^{\infty} f\left(r_0, r\right) \frac{\partial r_0}{\partial r}\,\mathrm{d}r \simeq \int_{r_c}^{r_{\max}(\tau_N)} f\left(r_0, 0\right) \mathrm{d}r_0, \tag{8.80}$$

$$M = \int_0^{\infty} f\left(r_0, 0\right) \frac{\partial r_0}{\partial r} r^3\left(r_0, \tau\right) \mathrm{d}r \simeq \int_{r_c}^{r_{\max}(\tau_N)} f\left(r_0, 0\right) r^3\left(r_0, \tau\right) \mathrm{d}r_0 = N_{\max} r^3(t). \tag{8.81}$$

Here we took into account that, in the transient stage, for the main part of the distribution the inequality $r \gg r_{\max}(\tau_N)$ holds. For this reason, r^3 practically does not depend on r_0.

With Eqs. (8.80) and (8.81), we may reformulate Eq. (8.78) as

$$\frac{\mathrm{d}r^2}{\mathrm{d}t} = -D_{\mathrm{eff}} A N_{\max}\left(r^3 - r_{\max}^3\right), \qquad r\big|_{t=t_N} = r_0, \qquad t_N \le t \le t_f \tag{8.82}$$

with

$$r_{\max} = \left(M_{\max}/N_{\max}\right)^{1/3} = n_{\max}^{1/3}. \tag{8.83}$$

$M_{\max}$ is given by Eqs. (8.80) and (8.81).

In Eq. (8.82), the variables may be separated and we arrive at a solution in the implicit form

$$\int_{y_0}^{y} \frac{y\mathrm{d}y}{1-y^3} = \frac{1}{3}\left[\ln\frac{1-y_0}{1-y} + \frac{1}{2}\ln\frac{y^2+y+1}{y_0^2+y_0+1}\right. \tag{8.84}$$

$$\left. -\sqrt{3}\left(\arctan\frac{2}{\sqrt{3}}(y+1/2) - \arctan\frac{2}{\sqrt{3}}(y_0+1/2)\right)\right] = \frac{t}{t_0},$$

$$y_0 \le y = \frac{r}{r_{\max}} \le 1, \qquad y_0 = \frac{r_0}{r_{\max}} \ll 1, \qquad t_0^{-1} = \frac{D_{\mathrm{eff}}}{2} A N r_{\max}. \tag{8.85}$$

After the substitution of the expression for the quantity A, we have

$$l = a_m N^{-1/3}, \qquad t_0^{-1} = D_{\mathrm{eff}} l^{-2} (\omega_s/\omega_m)^{2/3} M_{\max}^{1/3}. \tag{8.86}$$

The parameter l has the meaning of the average distance between the particles of the new phase. N is determined via Eq. (8.65) and $M_{\max}$ via Eq. (8.81).

Equation (8.84) shows that, in the main part of the spectrum of new phase particles r (for $y < 1$), the term y^3 may be neglected in the denominator. This way, we get

$$y^2 = y_0^2 + 2\frac{t}{t_0}, \qquad t_0 = t_f. \tag{8.87}$$

In the close vicinity of unity the relative size y exponentially goes with time to unity, $y \to 1$. Consequently, in the time interval $t_N \le t \le t_0$ a distribution of new phase particles is formed which represents the initial state for the late stage of the process, called coarsening or Ostwald ripening (Chapter 4). The distribution function is given in the transient state, in dependence on r, by Eqs. (8.71) with a value of r_0 determined by Eq. (8.87). We get

$$f = \frac{J_0 r_0}{ar}\theta(r_{\max}(\tau_N) - r_0)\theta(r_0 - r_g). \tag{8.88}$$

The parameter a is given by Eq. (8.56).

It follows from Eq. (8.87) that the range of cluster sizes Δr, which gives the basic contribution to the new phase in the transient stage of the transformation, is significantly narrower than the range of cluster sizes which is formed in the stage of steady-state nucleation and which serves as the initial distribution in the transient state. Denoting the width of this initial distribution by $r_{\max}(\tau_N) - r_g = \Delta r_0$, we may write

$$\Delta r = \frac{r_0 \Delta r_0}{\sqrt{2r_{\max}^2 t/t_0}} \simeq \frac{r_{\max}(\tau_N)}{r_{\max}(\tau_f)}\Delta r_0 \ll \Delta r_0. \tag{8.89}$$

In other words, the ratio of the widths of the intervals is determined by the ratio $\frac{r_{\max}(\tau_N)}{r_{\max}(\tau_f)}$, where $r_{\max}$ is determined by Eqs. (8.66) and (8.83), respectively, for the both considered cases.

8.5 Kinetic Equations and Thermodynamic Relationships Accounting for Solute–Solute Interactions

The time evolution of an ensemble of clusters in nucleation–growth processes is usually described in terms of the cluster-size distribution function $f(n, t)$. As shown in Section 8.2 (for some additional detail see [293]), the evolution of the distribution $f(n, t)$ of aggregates consisting at time t of n structural elements in the process of formation of a phase with a given stoichiometric composition may be calculated by a Fokker–Planck type equation of the form

$$\frac{\partial f(n,t)}{\partial n} = \frac{\partial}{\partial n}\left\{w_{n,n+1}\left[\frac{\partial f(n,t)}{\partial n} + \frac{f(n,t)}{k_B T}\frac{\partial \Delta\Phi(n)}{\partial n}\right]\right\} + \Upsilon_c(n,t), \qquad n \gg 1. \tag{8.90}$$

As previously, the kinetic coefficient $w_{n,n+1}$ describes here the probability that per unit time a primary building unit is incorporated to an aggregate consisting initially of n such units. Its specific form is determined by the mechanism of growth underlying the temporal evolution of the cluster-size distribution and thermodynamic properties of the considered system.

In addition to earlier considerations, in Eq. (8.90) a "collision integral" term Υ_c is introduced accounting for processes of touching and merging of the aggregates in the course of the precipitation process. In the initial stages of the transformation process, the probability of such collisions is small, at least, if the volume fraction of the solute is sufficiently low. Therefore, as it has been done in the preceding sections, for the consideration of nucleation this term may be neglected. It may have, however, a major impact on the late stages of the transformation accounting, at least at part, for a deviation of the cluster-size distributions observed experimentally from the original Lifshitz–Slezov theoretical predictions (cf. [59, 107, 171]).

In contrast, the second additional term in the kinetic equation is of basic importance in this stage being the only source for processes of formation of supercritical clusters. As will be demonstrated below, its influence on the late stages of the process, however, ceases with time though at intermediate times it may have also an important influence on the shape of the size distribution [161].

To generalize earlier obtained results to precipitation processes in solid solutions both the kinetic coefficients and the boundary conditions have to be reformulated. The respective relations depend on the thermodynamic properties and the type of interaction between the solute particles.

The Helmholtz free energy F and the chemical potentials μ_i of different components in a solid solution, taking into account possible interactions between them, can be calculated easily, if, as it is assumed commonly, only configurational contributions for the determination of the entropy are taken into consideration. Generally we have

$$F = -k_B T \ln Z, \tag{8.91}$$

$$Z = \sum_n \exp\left(-\frac{E_n}{k_B T}\right), \tag{8.92}$$

where k_B is the Boltzmann constant, T the absolute temperature, Z the partition function of the system, and E_n are the different values of the energy of the system in a canonical ensemble.

With the above-mentioned assumption we may write, approximately,

$$Z \cong \left\{\exp\left(-\frac{E_0(S,V)}{k_B T}\right)\Delta\Gamma(T,V)\right\}\left\{\exp\left(-\frac{\Delta E(N,\{n_i\}}{k_B T}\right)\Delta\Gamma(N,\{n_i\})\right\}. \tag{8.93}$$

Here $E_0(S,V)$ is the thermodynamic (most probable value of the) energy of the matrix not containing solute components, while $\Delta E(N,\{n_i\})$ is the correction term accounting for the change of the energy if solute components are introduced into the solid solution.

The statistical weight $\Delta\Gamma$ is determined by the product of the respective quantities for the pure matrix ($\Delta\Gamma(T,V)$) and the configurational part of the solute components ($\Delta\Gamma(N,\{n_i\})$). N is the total number of lattice sites where solute components may be introduced into the matrix, and $\{n_i\}$ describes the set of solute components in the matrix. The thermal contributions to the entropy depend only weakly on the concentration and distribution of solute particles in the matrix, therefore, they can be neglected in the calculation of the chemical potentials.

The energy term $\Delta E(N,\{n_i\})$ may be written in a first approximation [144] as

$$\Delta E(N,\{n_i\}) = \Delta E^{(0)}(N,\{n_i\}) + \frac{1}{2}\sum_{i,k}\beta_{ik}\left(\frac{n_i n_k}{N}\right). \tag{8.94}$$

$E^{(0)}(N, \{n_i\})$ denotes the energy contribution of the solute components taking into account only solute–matrix interactions. The interactions between the solute components themselves are described in a first approximation by the second term in Eq. (8.94).

The configurational statistical weight has the same value independent on whether the interaction of the solute particles with each other is accounted for or not, it depends only on the number of distributions of $\{n_i\}$ solute particles on N lattice sites in the matrix.

Equations (8.91)–(8.94) yield

$$F = F^{(0)} + \frac{1}{2}\sum_{ik} \beta_{ik} \left(\frac{n_i n_k}{N}\right), \tag{8.95}$$

$$\mu_i = \frac{\partial F}{\partial n_i} = \mu_i^{(0)} + \sum_k \beta_{ik} c_k, \tag{8.96}$$

$$c_k = \frac{n_k}{N}, \qquad \mu_i^{(0)} = \psi_i + k_B T \ln c_i . \tag{8.97}$$

Here $F^{(0)}$ and $\mu_i^{(0)}$ are the Helmholtz free energy and the chemical potential of the ith component for the case that the segregating components do not interact with each other.

As a result we get

$$\mu_i = \psi_i + k_B T \ln c_i + \sum_k \beta_{ik} c_k \tag{8.98}$$

or

$$\mu_i = \psi_i + k_B T \ln \left\{ c_i \exp \left(\frac{\sum\limits_k \beta_{ik} c_k}{k_B T} \right) \right\}, \tag{8.99}$$

where ψ_i is the excess enthalpy of the ith solute component in the matrix.

Equation (8.99) indicates that it is reasonable to introduce the notation

$$\varphi_i = c_i \exp \left(\frac{\sum\limits_k \beta_{ik} c_k}{k_B T} \right) \tag{8.100}$$

resulting with Eq. (8.99) in

$$\mu_i = \psi_i + k_B T \ln \varphi_i . \tag{8.101}$$

It it easily verified (cf. Eqs. (8.97) and (8.101)) that the expression for the chemical potential has the same form as for a weak (perfect) solution with the difference that the concentrations c_i have to be replaced by the chemical activities φ_i. Moreover, in the outlined approach, the chemical activities are well-defined quantities expressed through the interaction parameters β_{ik}. A method of experimental determination of these parameters and thus of the activities will be discussed somewhat later.

As a next step we consider the change $\Delta\Phi$ of the Gibbs free energy Φ connected with the formation of an aggregate of a definite stoichiometric composition consisting of n structural elements. We get [293]

$$\Delta\Phi(n) = n\left(\mu_s^{(\infty)} - \sum_i \nu_i\mu_i\right) + 4\pi\sigma\left(\frac{3\omega_s}{4\pi}\right)^{2/3} n^{2/3}. \tag{8.102}$$

Here the following notations are introduced: $\mu_s^{(\infty)}$ is the chemical potential of a structural element of the aggregate; μ_i are the chemical potentials of the solute components in the solid solution; σ is the specific interfacial energy; and ω_s is the volume of one structural element of the evolving phase.

Assuming that both the aggregate and the structural element are of spherical shape we can introduce the radius R of the aggregate and the radius a_s of the structural element as parameters. They are connected by the equations

$$\omega_s = \frac{4\pi}{3}a_s^3, \qquad R = a_s n^{1/3}. \tag{8.103}$$

The first of these equations may also be considered as the definition of the parameter a_s.

A derivation of Eq. (8.102) with respect to n yields

$$\frac{\partial\Delta\Phi(n)}{\partial n} = \left(\mu_s^{(\infty)} - \sum_i \nu_i\mu_i\right) + \frac{8\pi}{3}\sigma\left(\frac{3\omega_s}{4\pi}\right)^{2/3} n^{-1/3}. \tag{8.104}$$

By setting the derivative equal to zero, Eq. (8.104) may be used to determine either the equilibrium concentrations of different solute components (for a given value of n) or the critical size n_c of the aggregate (for given values of the solute concentration in the matrix). The resulting equation reads

$$\sum_i \nu_i\mu_i = \mu_s^{(\infty)} + \frac{8\pi}{3}\sigma\left(\frac{3\omega_s}{4\pi}\right)^{2/3} n_c^{-1/3}. \tag{8.105}$$

With Eq. (8.101) we obtain

$$\sum_i \nu_i\psi_i + k_BT\ln\prod_i \varphi_i^{\nu_i} = \mu_s^{(\infty)} + \frac{8\pi}{3}\sigma\left(\frac{3\omega_s}{4\pi}\right)^{2/3} n_c^{-1/3}. \tag{8.106}$$

If we introduce, in addition, the constant $K_\infty(p,T)$ of the chemical reaction equilibrium for a bulk system ($n_c^{-1/3} \to \infty$) as

$$K_\infty(p,T) = \exp\left(\frac{\mu_s^{(\infty)} - \sum_i \nu_i\psi_i}{k_BT}\right) \tag{8.107}$$

and a similar relation for the respective equilibrium constant for the reaction taking place near an aggregate consisting of n structural elements,

$$K_n(p,T) = K_\infty(p,T) \exp\left\{ \frac{8\pi}{3} \left(\frac{\sigma}{k_B T}\right) \left(\frac{3\omega_s}{4\pi}\right)^{2/3} n^{-1/3} \right\} \tag{8.108}$$

then Eq. (8.106) may be rewritten as

$$\prod_i \varphi_i^{\nu_i} = K_\infty(p,T) \exp\left\{ \frac{8\pi}{3} \left(\frac{\sigma}{k_B T}\right) \left(\frac{3\omega_s}{4\pi}\right)^{2/3} n_c^{-1/3} \right\} = K_{n_c}(p,T). \tag{8.109}$$

It follows that as a special case the equilibrium values of φ_i in the bulk (denoted as $\varphi_i^{(\infty)}$) obey the following relation:

$$\prod_i \left(\varphi_i^{(\infty)}\right)^{\nu_i} = K_\infty(p,T). \tag{8.110}$$

For given values of the quantities φ_i, Eq. (8.105) or (8.106) allow us to estimate the critical number of structural elements in an aggregate as

$$n_c^{1/3} = \frac{2\sigma\omega_s}{a_s k_B T \Delta}, \tag{8.111}$$

where the supersaturation Δ is determined by

$$\Delta = \ln\left\{ \prod_i \left(\frac{\varphi_i}{\varphi_i^{(\infty)}}\right)^{\nu_i} \right\} = \ln\left(\frac{\prod_i \varphi_i^{\nu_i}}{K_\infty(p,T)} \right). \tag{8.112}$$

By using the same notations we may also write

$$\frac{1}{k_B T}\left(\frac{\partial \Delta\Phi(n)}{\partial n}\right) = -\ln\left(\frac{\prod_i \varphi_i^{\nu_i}}{K_n(p,T)} \right) \tag{8.113}$$

or applying Eqs. (8.108) and (8.109)

$$\begin{aligned} \frac{1}{k_B T}\frac{\partial \Delta\Phi(n)}{\partial n} &= -\ln\left(\frac{K_{n_c}(p,T)}{K_n(p,T)}\right) \\ &= -\frac{8\pi}{3}\left(\frac{\sigma}{k_B T}\right)\left(\frac{3\omega_s}{4\pi}\right)^{2/3}\left(\frac{1}{n_c^{1/3}} - \frac{1}{n^{1/3}}\right). \end{aligned} \tag{8.114}$$

These expressions are needed in the subsequent derivations.

8.6 Rate of Change of the Number of Structural Elements of an Aggregate of the New Phase

The flux of particles of the ith component to an aggregate of the new phase consisting of n structural elements may be written, similarly to the case of precipitation of one solute component in the matrix (cf. [293]), in the form of Eq. (8.13)

$$4\pi R^2 j_i = -w_{n_i,n_{i+1}} \frac{1}{k_B T} \left(\frac{\partial \Delta\Phi^{(s)}}{\partial n_i} \right) . \tag{8.115}$$

In this equation $w_{n_i,n_{i+1}}$ denotes the probability per unit time that particles of the ith component are incorporated into an aggregate consisting of n structural elements (and characterized sometimes also by a radius $R(n)$). This quantity may be expressed in the following way (see Chapter 3 and [292]):

$$w_{n_i,n_{i+1}} = \left(\frac{\alpha_i D_i}{a_m^2} \right) \left(\frac{4\pi R^2 a_m \tilde{c}_i}{\omega_m} \right) . \tag{8.116}$$

The parameter α_i, as a sticking coefficient, describes the degree of inhibition of the diffusion process in the immediate vicinity of the aggregate. It has values in the range $0 \leq \alpha_i \leq 1$.

Here D_i denotes the partial diffusion coefficient of the ith component in the immediate vicinity of the aggregate, while a_m is the lattice constant of the matrix. With these notations it becomes evident that the first part of the right-hand side of Eq. (8.116) $(\alpha_i D_i / a_m^2)$ has the meaning of the frequency of jumps of particles of the ith component in the interfacial layer near the aggregate in the direction of the aggregate. Hereby the total number of particles in the surface layer can be written as $(4\pi R^2 a_m / \omega_m)$, where ω_m is the average volume per particle in the matrix. By multiplying this ratio with the concentration (molar fraction) $\tilde{c}_i$ of the ith component in the surface layer we obtain the total number of particles of the ith component in the layer. Consequently, $w_{n_i,n_{i+1}}$ has, indeed, the meaning as specified above.

In addition to the number of jumps, the diffusional fluxes to the surface of the aggregate are determined by the change of the thermodynamic potential $\Delta\Phi^{(s)}$. Let n_i be the number of particles of the ith component in an aggregate of size n. $\Delta\Phi^{(s)}$ is the change of the thermodynamic potential, resulting from the transfer all particles n_i of different components required for the formation of an aggregate of size n from the solution with values of the concentration $\tilde{c}_i$ to a solid state for which an aggregate of size n is in equilibrium with the surrounding solution. The chemical potential of the ith component, being in thermodynamic equilibrium with an aggregate of size n, we denote by $\mu_i(c_{ni})$, by c_{ni} the respective values of the equilibrium concentrations of the components in the matrix are specified. As will be shown later, a determination of these concentrations, although possible, is not necessary for a formulation of the kinetic equations describing nucleation and growth.

To distinguish such a type of change of the characteristic thermodynamic potential from the change due to the formation of an aggregate of size n the superscript (s) in $\Delta\Phi^{(s)}$ is introduced. According to the given definition, $\Delta\Phi^{(s)}$ may be written as

$$\Delta\Phi^{(s)} = \sum_i n_i \left[\mu_i(c_{ni}) - \mu_i(\tilde{c}_i) \right] , \qquad \mu^{(s)} = \sum_i n_i \mu_i(c_{ni}) . \tag{8.117}$$

A derivation of Eq. (8.117) with respect to n_i yields

$$\left(\frac{\partial \Delta \Phi^{(s)}}{\partial n_i}\right) = -\left[\mu_i(\tilde{c}_i) - \mu_i(c_{ni})\right] , \tag{8.118}$$

and after a substitution into Eq. (8.115) we obtain (as in Eq. (8.109))

$$4\pi R^2 j_i = \frac{3\alpha_i D_i \tilde{c}_i}{a_m^2} \left(\frac{\omega_s}{\omega_m}\right)^{2/3} n^{2/3} \left(\frac{\mu_i(\tilde{c}_i) - \mu_i(c_{ni})}{k_B T}\right) . \tag{8.119}$$

In the derivation of Eq. (8.119), in addition, the relations

$$n = \frac{4\pi}{3} \frac{R^3}{\omega_s} , \qquad \omega_m = \frac{4\pi}{3} a_m^3 \tag{8.120}$$

were applied. ω_s is the volume per primary building unit in the segregating phase.

The rate of change of the number n of structural elements may then be written as

$$\frac{dn}{dt} = \frac{4\pi R^2 \sum_i \omega_i j_i}{\omega_s} . \tag{8.121}$$

Here the parameters ω_i denote the volume of a particle of the ith component in an aggregate of the newly evolving phase.

In application of Eq. (8.121) to the description of the process of formation and growth of aggregates of stoichiometric composition it has to be taken into account that the different fluxes are connected by the additional condition

$$\frac{j_1}{\nu_1} = \frac{j_2}{\nu_2} = \cdots = \frac{j_i}{\nu_i} , \tag{8.122}$$

where ν_i denotes the molar fractions of different components in the newly evolving phase. Following the derivation of the aggregate growth rate, dn/dt presented in Section 8.2 (Eqs. (8.16) and (8.17)), we obtain formally the same expression

$$\frac{dn}{dt} = \frac{3n^{2/3}}{a_m^2} \left(\frac{\omega_s}{\omega_m}\right)^{2/3} \frac{1}{\sum_i \left(\frac{\nu_i^2}{\alpha_i D_i \tilde{c}_i}\right)} \sum_i \nu_i \left(\frac{\mu_i(\tilde{c}_i) - \mu_i(c_{ni})}{k_B T}\right) . \tag{8.123}$$

However, the chemical potentials in Eq. (8.123), unlike Eq. (8.17), account already for the solute–solute interactions. Thus introducing the notation

$$\frac{1}{D^*} = \sum_i \frac{\nu_i^2}{\alpha_i D_i \tilde{c}_i} \tag{8.124}$$

and using the following identities (cf. Eqs. (8.101) and (8.109)):

$$\sum_i \nu_i \frac{\mu_i(\tilde{c}_i) - \mu_i(c_{ni})}{k_B T} = \sum_i \nu_i \ln \left(\frac{\varphi_i(\tilde{c}_i)}{\varphi_i(c_{ni})}\right) = \ln \left(\frac{\prod_i [\varphi_i(\tilde{c}_i)]^{\nu_i}}{K_n}\right) \tag{8.125}$$

we finally get

$$\frac{\mathrm{d}n}{\mathrm{d}t} = \frac{3D^* n^{2/3}}{a_m^2} \left(\frac{\omega_s}{\omega_m} \right)^{2/3} \ln \left(\frac{\prod_i [\varphi_i(\tilde{c}_i)]^{\nu_i}}{K_n} \right) . \tag{8.126}$$

For the rate of change of the number of structural elements we may write alternatively an expression similar to Eq. (8.115) as

$$v(n) = \frac{\mathrm{d}n}{\mathrm{d}t} = 4\pi R^2 j = -w_{n,n+1} \frac{1}{k_B T} \left(\frac{\partial \Delta\Phi}{\partial n} \right) , \tag{8.127}$$

where $\Delta\Phi$ is determined by Eq. (8.102). With Eq. (8.113) we may rewrite this relation in the form

$$\frac{\mathrm{d}n}{\mathrm{d}t} = w_{n,n+1} \ln \left(\frac{\prod_i [\varphi_i(\tilde{c}_i)]^{\nu_i}}{K_n} \right) . \tag{8.128}$$

A comparison between Eqs. (8.128) and (8.126) shows that the rate coefficient $w_{n,n+1}$ may be written as

$$w_{n,n+1} = \frac{3D^* n^{2/3}}{a_m^2} \left(\frac{\omega_s}{\omega_m} \right)^{2/3} . \tag{8.129}$$

In this way, the coefficients $w_{n,n+1}$ required for an application of Eq. (8.90) are determined. However, in Eq. (8.128) the activities $\tilde{\varphi}_i = \varphi_i(\tilde{c}_i)$ occur which have to be replaced later on by the known average activities $\varphi_i(c_i)$ of the matrix. This replacement will lead to some further revision of the expression for $w_{n,n+1}$.

Moreover, as is evident from Eq. (8.124) the effective diffusion coefficient D^* is a function of the concentration and depends, therefore, also on the interactions between the solute components. Therefore, as a next step, these dependences have to be specified and expressed through the interaction parameters β_{ik}.

8.7 The Coefficient of Components Mass Transfer

Taking into account the interaction between the solute particles the partial diffusion coefficients D_i depend on the concentration of different components $\{c_j\}$. This dependence is weak for the prefactor D_{i0} in Eq. (8.130)

$$D_i = D_{i0} \exp \left(-\frac{Q_i}{k_B T} \right) , \tag{8.130}$$

but may be of significant importance with respect to the activation energy Q_i of the diffusion process.

This activation energy may be written as a sum of two terms, the first accounting for the contribution only due to solute–matrix interactions ($Q_i(0)$), while the second ($\Delta Q_i(\{c_j\})$) reflects solute–solute interactions, i.e.,

$$Q_i(\{c_j\}) = Q_i(0) + \Delta Q_i(\{c_j\}). \tag{8.131}$$

Note that the solute–solute interaction is of significance only for distances not exceeding several times the respective lattice constants. Moreover, it is evident that the activation energy of the diffusion increases ($\Delta Q_i > 0$) if the solute particles attract each other and decreases ($\Delta Q_i < 0$) if the interaction leads effectively to a repulsion of the dissolved components. If the different solute particles have nearly the same size as the matrix building units (for the case they are occupying vacant lattice nodes) or if they are sufficiently small (occupying interstitial positions) then the solute interaction changes exclusively the potential well for the positions where the solute particles are bound to the lattice without changing the general shape of the energy relief.

The change of the activation energy is given for such cases by the energy of interaction of the solute particles as

$$\Delta Q_i = -\sum_k \beta_{ik} c_k \tag{8.132}$$

resulting in

$$Q_i(\{c_j\}) = Q_i(0) - \sum_k \beta_{ik} c_k. \tag{8.133}$$

The partial diffusion coefficient of the ith component may be then written in the form

$$D_i(\{c_j\}) = D_{i0} \exp\left(-\frac{Q_i(0) - \sum_k \beta_{ik} c_k}{k_B T}\right). \tag{8.134}$$

For an application of the equations describing nucleation and growth of a stoichiometric multicomponent phase not the partial diffusion coefficients themselves but the so-called coefficients of mass transfer $D_i c_i$ have to be known (cf. Eq. (8.124)). Denoting the partial diffusion coefficient in the absence of solute–solute interactions by $D_i^{(0)}$ we obtain from Eq. (8.134)

$$D_i(\{c_j\}) c_i = D_i^{(0)} c_i \exp\left(\frac{\sum_k \beta_{ik} c_k}{k_B T}\right) \tag{8.135}$$

or (cf. Eq. (8.100))

$$D_i(\{c_j\}) c_i = D_i^{(0)} \varphi_i. \tag{8.136}$$

This result allows us to rewrite Eq. (8.124) as

$$\frac{1}{D^*} = \sum_i \frac{\nu_i^2}{\alpha_i D_i^{(0)} \varphi_i}. \tag{8.137}$$

In this way, it turns out that the effective diffusion coefficient is, again, determined by the values of the chemical activities of different components φ_i.

With Eq. (8.136) the relation for the density of fluxes of particles of the ith component

$$\mathbf{j}_i = -\frac{D_i c_i}{k_B T}\mathrm{grad}\mu_i \tag{8.138}$$

may be transformed easily into

$$\mathbf{j}_i = -D_i^{(0)}\mathrm{grad}\varphi_i . \tag{8.139}$$

It turns out that the only modification, one has to introduce into the relations for the description of nucleation and growth for perfect solutions, derived in [293], consists in a replacement $c_i \rightarrow \varphi_i$.

In the case of kinetically limited growth the concentrations (or activities) in the immediate vicinity of the evolving cluster are equal to the average concentrations c_i (average activities φ_i) in the matrix. Taking this into account we obtain from Eqs. (8.124), (8.126), and (8.136) for kinetically limited growth

$$w_{n,n+1} = \frac{3D^* n^{2/3}}{a_m^2}\left(\frac{\omega_s}{\omega_m}\right)^{2/3} , \qquad \frac{1}{D^*} = \sum_i \frac{\nu_i^2}{\alpha_i D_i^{(0)}\varphi_i} . \tag{8.140}$$

By the method outlined in [293], we obtain for the general case

$$w_{n,n+1} = \left(\frac{4\pi R}{\omega_m}\right)\frac{D^* D^{**}}{\left(D^* + D^{**}\frac{a_m}{R}\right)} ,$$
$$\frac{1}{D^*} = \sum_i \frac{\nu_i^2}{\alpha_i D_i^{(0)}\tilde{\varphi}_i} , \qquad \frac{1}{D^{**}} = \sum_i \frac{\nu_i^2}{D_i^{(0)}\tilde{\varphi}_i} . \tag{8.141}$$

In this general case, the chemical activities at the boundary of the aggregate $\tilde{\varphi}_i$ have to be expressed through the average (φ_i) and equilibrium (for a cluster of size n, (φ_{ni})) chemical activities via Eqs. (8.110), (8.122), and (8.139) resulting in [293]

$$\tilde{\varphi}_i = \frac{\varphi_i\left(\dfrac{a_m}{\alpha_i R}\right) + \varphi_{ni}(c_{ni})}{1 + \left(\dfrac{a_m}{\alpha_i R}\right)} . \tag{8.142}$$

For diffusion-limited growth, prevailing for large values of the cluster size R, Eq. (8.141) is reduced to

$$w_{n,n+1} = \frac{4\pi R D^{**}}{\omega_m} = \frac{3D^{**} n^{1/3}}{a_m^2}\left(\frac{\omega_s}{\omega_m}\right)^{1/3} . \tag{8.143}$$

In all considered cases, the rate of deterministic growth of a cluster of size n may be determined from the expressions for $w_{n,n+1}$ via an equation of the form as Eq. (8.128), where, however, $\varphi_i(\tilde{c}_i) = \tilde{\varphi}_i$ has to be replaced by $\varphi_i(c_i)$, i.e.,

$$\frac{\mathrm{d}n}{\mathrm{d}t} = w_{n,n+1} \ln \left(\frac{\prod\limits_i [\varphi_i(c_i)]^{\nu_i}}{K_n} \right) . \tag{8.144}$$

In this way, the determination of the kinetic coefficients $w_{n,n+1}$ and the deterministic growth rates is finally accomplished.

8.8 Steady-State Nucleation Rate

In order to solve the kinetic equation for the evolution of the cluster-size distribution, the boundary conditions for $n \to 0$ and $n \to \infty$ have to be specified. If we express $f(n,t)$ in the form

$$f(n) = \Psi(n) \exp\left(-\frac{\Delta\Phi(n)}{k_B T} \right) \tag{8.145}$$

the physically reasonable boundary condition for large values of n is (compare [293])

$$\lim_{n\to\infty} \Psi(n) = 0, \tag{8.146}$$

while for $n \to 0$ and noninteracting solute particles

$$\lim_{n\to 0} \Psi(n) = \frac{N}{zV} \prod_i c_i^{\nu_i} \tag{8.147}$$

was shown to hold. Here z is the number of lattice sites in the matrix occupied by the particles forming one structural element, N, as mentioned, is the total number of lattice sites in the volume V.

While the boundary condition (8.146) remains unchanged, Eq. (8.147) has to be modified accounting for the change of the number of possible configurations (the change of entropy) due to the interactions between the solute components. Instead of Eq. (8.147) we then get

$$\lim_{n\to 0} \Psi(n) = \frac{N}{zV} \prod_i c_i^{\nu_i} \exp\left(\frac{\Delta S^{(\mathrm{int})}}{k_B} \right) . \tag{8.148}$$

The entropy difference $\Delta S^{(\mathrm{int})}$ can be expressed through the work (with a minus sign) required in a reversible process to transfer different particles constituting a structural element of the new phase into the pretransition state or, equivalently, through the differences in the chemical potentials as

$$T\Delta S^{(\mathrm{int})} = -\sum_i \nu_i \left[\mu_i^{(\mathrm{interface})} - \mu_i \right] . \tag{8.149}$$

In Eq. (8.149), $\mu_i^{(\text{interface})}$ denotes the values of the chemical potentials of different interacting solute components in a group of molecules in the interfacial region capable to be incorporated into the aggregate of the newly evolving phase, while μ_i, as introduced with Eq. (8.98), refers to the respective values in the bulk of the matrix. Denoting further by ε the energy of interactions of the group of molecules in the pretransition state which differs, in general, from the value $\sum_{ik} \nu_i \beta_{ik} c_k$ for a random distribution of the same solute components in the bulk, Eq. (8.149) may be rewritten in the form

$$T\Delta S^{(\text{int})} = -\left(\varepsilon - \sum_{ik} \nu_i \beta_{ik} c_k\right) . \tag{8.150}$$

Substituting Eq. (8.150) into Eq. (8.148) yields

$$\lim_{n\to 0} \Psi(n) = \frac{N}{zV} \prod_i c_i^{\nu_i} \exp\left(-\frac{\varepsilon - \sum_{ik} \nu_i \beta_{ik} c_k}{k_B T}\right) \tag{8.151}$$

and with Eqs. (8.96) and (8.98)

$$\lim_{n\to 0} \Psi(n) = \frac{N}{zV} \exp\left(-\frac{\varepsilon}{k_B T}\right) \prod_i c_i^{\nu_i} \prod_i \left[\exp\left(-\frac{\sum_k \beta_{ik} c_k}{k_B T}\right)\right]^{\nu_i} \tag{8.152}$$

or (cf. Eq. (8.100))

$$\lim_{n\to 0} \Psi(n) = \frac{N}{zV} \exp\left(-\frac{\varepsilon}{k_B T}\right) \prod_i \varphi_i^{\nu_i} \tag{8.153}$$

is obtained.

Once the kinetic coefficients $w_{n,n+1}$ and the boundary conditions are known the Fokker–Planck equation can be solved and the steady-state nucleation rate J can be determined (for details see [293]). Taking into account the interactions of the solute components we get

$$J = \frac{N}{zV} \exp\left(-\frac{\varepsilon}{k_B T}\right) w_{n,n+1}(n_c) \prod_i \varphi_i^{\nu_i} \exp\left(-\frac{\Delta\Phi(n_c)}{k_B T}\right) \times \sqrt{\frac{1}{2\pi k_B T}\left|\left(\frac{\partial^2 \Delta\Phi(n)}{\partial n^2}\right)_{n=n_c}\right|} . \tag{8.154}$$

On substituting the respective values for the kinetic coefficient $w_{n,n+1}$ for $n = n_c$ (denoted by $w_{n,n+1}(n_c)$) and the expressions for the derivatives of the thermodynamic potential with respect to n we obtain

$$J = \left(\frac{N}{zV}\right) \exp\left(-\frac{\varepsilon}{k_B T}\right) \prod_i \varphi_i^{\nu_i} \left(\frac{2D^* a_s}{a_m^2}\right) \left(\frac{\omega_s}{\omega_m}\right)^{2/3} \sqrt{\frac{\sigma}{k_B T}} \times \exp\left\{-\frac{4\pi}{3}\left(\frac{\sigma}{k_B T}\right)\left(\frac{3\omega_s}{4\pi}\right)^{2/3} n_c^{2/3}\right\} \quad \text{– for kinetically limited growth,} \tag{8.155}$$

$$J = \left(\frac{N}{zV}\right) \exp\left(-\frac{\varepsilon}{k_B T}\right) \prod_i \varphi_i^{\nu_i} \left(\frac{2D^{**} a_s}{a_m^2 n_c^{1/3}}\right) \left(\frac{\omega_s}{\omega_m}\right)^{1/3} \sqrt{\frac{\sigma}{k_B T}} \tag{8.156}$$

$$\times \exp\left\{-\frac{4\pi}{3}\left(\frac{\sigma}{k_B T}\right)\left(\frac{3\omega_s}{4\pi}\right)^{2/3} n_c^{2/3}\right\} \quad - \text{ for diffusion-limited growth.}$$

Taking into account Eq. (8.141) for nucleation processes in solid solutions usually Eq. (8.155) should be applied.

The time required for the establishment of the steady-state nucleation rate may be approximated, again, by

$$\Delta t \cong \frac{(\Delta n)^2}{w_{n,n+1}(n_c)} \tag{8.157}$$

while the condition that a constant nucleation rate is found may be written as

$$\frac{(\Delta n)^2}{w_{n,n+1}} \le \frac{\Delta n}{\dot{n}_c}. \tag{8.158}$$

Here Δn characterizes the region in the space of structural elements near the critical cluster size n_c, where the growth of the aggregates proceeds mainly by diffusion-like processes. This interval is given by (cf. [293])

$$\Delta n = \frac{1}{\sqrt{\dfrac{1}{2k_B T}\left|\left(\dfrac{\partial^2 \Delta\Phi(n)}{\partial n^2}\right)_{n=n_c}\right|}} = \left(\frac{3n_c^{2/3}}{2a_s}\right)\sqrt{\frac{k_B T}{\pi\sigma}}\,. \tag{8.159}$$

The time interval Δt in Eq. (8.157) is, however, equal to the mean time of growth of an aggregate from subcritical $(n = n_c - \Delta n)$ to supercritical $(n = n_c + \Delta n)$ size by diffusional motion in cluster-size space. Equation (8.158) implies that during this time interval the critical cluster size does not change appreciably.

The account of solute–solute interactions in the description of precipitation processes allows one in addition to the extension of the region of applicability of the theory also a determination of the interaction parameters for a given solid solution by investigations of the course of precipitation processes in it. In the simplest approach, hereby the quantities β_{ik} may be considered as parameters which have to be determined in such a way as to allow one the best fit of the experimental results.

It is also possible, however, to obtain some additional information by varying the concentrations c_i of different components. Considering the process of precipitation, e.g., of only the ith component one obtains from Eq. (8.98) for a weak solution and an equilibrium coexistence of the pure ith phase with the matrix

$$\psi_i = \mu_i^{(s)} - k_B T \ln c_i^{(\infty)} \tag{8.160}$$

allowing one to determine ψ_i and $\mu_i^{(s)}$ from a set of measurements by varying the temperature. Similarly also additional parameters may be determined by varying the number of components and their concentration in the system.

8.9 Influence of Interaction of the Solute Components on Coarsening Processes

In the case when the inequality, Eq. (8.158), already does not hold, the precipitation process goes over into the late stages which have been studied extensively (cf. Chapter 4 and [155, 341]). In the late stage of precipitation the activities (or concentrations) of different components in the matrix remain nearly constant. Next, although the parameters in the kinetic coefficients, respectively, the deterministic growth equations are complicated functions of the composition, the kinetic equations governing the evolution of the cluster-size distribution function is of the same form as for the case of a perfect solution. Here the same situation is found as for the case of precipitation of only one component in the matrix (see Chapter 4).

Introducing the reduced variables

$$u = \left(\frac{n}{n_c}\right), \qquad \tau = \ln\left(\frac{n_c}{n_{c0}}\right) \tag{8.161}$$

and the cluster-size distribution function $\phi(u, \tau)$ in reduced variables

$$\phi(u, \tau) = f(n, t) n_c \tag{8.162}$$

the Fokker–Planck equation may be written for kinetically limited growth again in the form (for details see Chapter 4 and [295])

$$\frac{\partial \phi}{\partial \tau} + \frac{\partial}{\partial u}\left\{\left[\gamma_0(\tau) u^{1/3}(u^{1/3} - 1) - u\right]\phi\right\} - \left(\frac{\gamma_0(\tau)}{n_c^{2/3}\alpha}\right)\frac{\partial}{\partial u}\left[u^{2/3}\frac{\partial \phi}{\partial u}\right] = \Upsilon_c, \tag{8.163}$$

$$\Upsilon_c = \frac{n_c}{\tau_e}\frac{1}{2}\int_0^u \phi(u - u', \tau)\Omega(u - u', u')\phi(u', \tau)\,\mathrm{d}u' \tag{8.164}$$

$$- \phi(u, \tau)\int_0^\infty \Omega(u, u')\phi(u', \tau)\,\mathrm{d}u', \qquad \tau_e \sim 1 \qquad \Omega(u, u') \sim u + u',$$

$$\gamma_0(\tau) = \frac{\alpha\, n_c^{1/3}}{\tau_0\, \dot{n}_c}, \qquad \alpha = \frac{8\pi}{3}\left(\frac{\sigma}{k_B T}\right)\left(\frac{3\omega_s}{4\pi}\right)^{2/3}, \qquad \frac{1}{\tau_0} = \frac{3D^*}{a_m^2}\left(\frac{\omega_s}{\omega_m}\right)^{2/3}.$$

For the considered here case of kinetically limited growth the critical cluster radius behaves for large times as $R_c \propto t^{1/2}$. It turns out that in intermediate stages of the process (in the initial stages of coarsening) diffusion processes in cluster-size space (stochastic effects, thermal noise) are of significance for the kinetics of coarsening including the shape of the cluster-size distribution function (cf. also [161, 171]). With time such effects become less important. However, processes of touching and merging of aggregates gain in importance for sufficiently large volume fractions of the segregating phase in the transition from nucleation to coarsening and may influence the shape of the cluster-size distribution function significantly (cf. also [59]).

For diffusion-limited growth the collision integral retains the same form as for kinetically limited growth. The Fokker–Planck equation reads, now,

$$\frac{\partial \phi}{\partial \tau} + \frac{\partial}{\partial u}\left\{\left[\gamma_1(\tau)(u^{1/3} - 1) - u\right]\phi\right\} - \left(\frac{\gamma_1(\tau)}{n_c^{2/3}\alpha}\right)\frac{\partial}{\partial u}\left[u^{1/3}\frac{\partial \phi}{\partial u}\right] = \Upsilon_c, \tag{8.165}$$

$$\gamma_1(\tau) = \frac{\alpha}{\tau_1 \dot{n}_c}, \quad \alpha = \frac{8\pi}{3}\left(\frac{\sigma}{k_B T}\right)\left(\frac{3\omega_s}{4\pi}\right)^{2/3}, \quad \frac{1}{\tau_1} = \frac{3D^{**}}{a_m^2}\left(\frac{\omega_s}{\omega_m}\right)^{1/3}. \tag{8.166}$$

The conclusions remain qualitatively the same as for kinetically limited growth.

8.10 Discussion and Conclusion

The present chapter is devoted to the description of the course of the evolution of a phase transformation process encompassing both the quasisteady-state nucleation stage and the transient stage to coarsening in a multicomponent solid solution taking into account solute–solute interactions. A new approach to the theoretical treatment of this problem is proposed and a complete set of equations is formulated describing this process. The expression for the effective diffusion coefficient is derived, which determines the flux of structural units of the new phase through the boundaries of the aggregates of the newly evolving phase. This coefficient can be written as a combination of the partial diffusion coefficients of different components in the solid solution.

All the basic characteristics of the phase transformation process are determined analytically including the following: the distribution function of particles with respect to cluster size, the cluster flux in size space, the maximal number of new phase particles, and estimates of the duration of different stages of the process. Numerical solutions of the basic kinetic equations show an excellent agreement with the results of the theory (see [22, 94, 114, 164, 178, 196, 245, 247, 350] and Figure 8.1).

The degree of dispersity of the system is shown to grow essentially in the initial nucleation stage of the process of phase separation. However, at the transient stage, the width of the distribution with respect to cluster sizes is reduced, but increases again at the later stage, approaching (in reduced variables) a constant value (see Chapter 4). These results allow one to vary the dispersity of the evolving phase by terminating the phase separation at some definite stage of the process.

In addition, the analytic expressions can be utilized for the determination of the interfacial free energy (which can hardly be measured otherwise) by comparing the theoretical results with experimental data. Of course, one has to be sure that the process is dominated, as assumed here, by homogeneous nucleation.

The proposed theory can be applied to a description of phase transitions in liquids as well as for the case of droplet formation with a given stoichiometric composition. Also, in the present chapter the influence of interactions between solute particles in multicomponent solid solutions on processes of formation and growth of phases with a given stoichiometric composition is investigated. The analysis is carried out based on a newly developed general method of treating nucleation and a particular (but rather general) good enough approximation for the description of the interactions of the solute components.

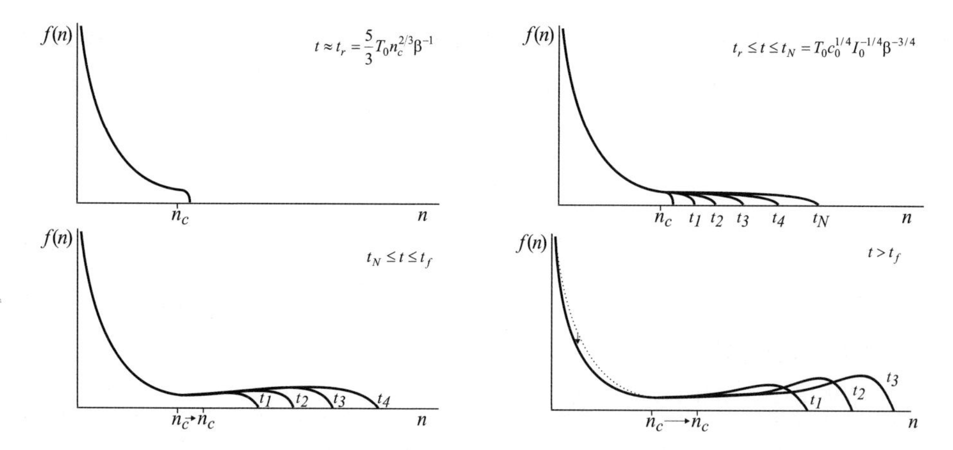

Figure 8.1: Evolution of the cluster-size distribution function, $f(n, \tau)$, for different stages of the process: (a) Establishment of a quasisteady state for the range of cluster sizes $n \leq n_c$ (for $\tau \leq \tau_r$ in the stage of nonsteady state nucleation (left top); (b) Evolution in the stage of quasisteady-state nucleation (right top); (c) Evolution in the transient stage to coarsening (left bottom); (d) Evolution in the late stage of coarsening (right bottom).

It turns out that the basic kinetic equations remain of the same form as for the case of weak (perfect) solutions. However, the kinetic and thermodynamic parameters are complicated functions of the composition of the system and the interaction parameters. In this way, the outlined theory gives the possibility of an adequate quantitative interpretation of experimental results on segregation processes in concentrated multicomponent solutions.

The outlined approach also allows one a straightforward extension to cases, when elastic effects, the influence of radiation on phase formation or other external factors have to be taken into account. Moreover, it gives also the basis for a treatment of the general problem of phase formation under the condition that the monomeric building units of the evolving phase interact with each other in nucleation and growth.

9 Nucleation and Growth of Gas-Filled Bubbles in Liquids

9.1 Introduction

The process of formation of bubbles in a solution supersaturated with a gas is an example of a multicomponent phase transition of large practical importance, for example, it determines major parts of the technology of formation of polymeric foams. In contrast to the cases analyzed here earlier it results in further complications in the theoretical description since the density of the aggregates of the newly evolving phase, the bubbles, is in general not known and has to be determined as well. This property results in the existence of additional degrees of freedom in the description of the cluster properties, which have to be appropriately accounted for.

For the case when the heat conductivity of the solution is sufficiently high, the temperature of a bubble does not differ from that of the solution and the evaporation of the liquid into the bubble can be neglected. The process can then be characterized by the bubble radius and the number of gas atoms in the bubble and a particular consideration of the evolution of the temperature of the bubbles is not required.

As we know from the preceding chapters, processes of phase transitions can be subdivided in a variety of different applications into three stages. Such division holds for the case of bubble formation and growth as well. At the nucleation stage, bubbles with sizes larger than the critical size are formed. In view of a strong dependence of the nucleation rate on the supersaturation level, this stage is characterized by almost time-independent conditions and is terminated by the formation of a spectrum of the finite-size particles (bubbles) of a new phase.

At the second stage, new bubbles are virtually not formed, nucleated earlier supercritical bubbles grow, and the supersaturation decreases substantially. At the third (final) stage, when the supersaturation of the solution almost vanishes, the part of bubbles with subcritical sizes dissolves and supplies the substance (gas) for the growth of supercritical bubbles. In this case, the total amount of gas in the bubbles is conserved, their mean size increases, but the number of bubbles decreases.

The kinetics of multicomponent nucleation was first theoretically treated by Reiss [206] who assumed that, in composition space, the flux of the clusters to the new phase passes the saddle point of the characteristic thermodynamic potential. Later, this idea was elaborated in a number of works (e.g., see [132, 318]). Thus, the problem becomes efficiently one dimensional, because the particle size is completely determined by the position of the saddle point (cf. Chapter 9 and Section 3.8). Such scenario can be considered as the rule, but exceptions from this general rule are possible as well (see, e.g., [130, 152, 328, 356]). Such exceptions correspond to situations when the main flux to the new phase passes not the saddle but some ridge point of the thermodynamic potential. General expectations and detailed anal-

Kinetics of First-order Phase Transitions. Vitaly V. Slezov

ISBN: 978-3-527-40775-0

ysis [168, 219] allow us to conclude that ridge crossing will be the preferred channel of the transformation in the case of significant excess of one of the components in the solution and when the thermodynamic barrier is relatively low.

In [138, 176], the nucleation of gas-filled bubbles in a viscous liquid was studied, and the diffusion regime of their growth was considered. For the case of boiling-up of gas dissolved in low-viscosity liquids in [142] the kinetic equations for homogeneous nucleation were derived and the general regime of bubble growth was discussed. This regime includes both the diffusion regime treated in [138, 176] and free molecular regime, both being considered as limiting cases. The bottleneck of the molecular regime is the boundary kinetics of the addition of gas molecules to the bubble. Subsequent growth of supercritical bubbles was discussed with the account of solvent fugacity in [143].

This chapter is devoted to the study of all three successive stages of the process of the nucleation [313] and growth [314] of gas-filled bubbles in low-viscosity liquids. Note that metal melts can also be considered as low-viscosity liquids within a rather wide temperature range; hence, such an approximation has fairly wide applications. The method employed here makes it possible to determine the main parameters (the rate of bubble growth, characteristic times, and the total volume of supercritical bubbles), and the bubble-size distribution function at all stages of the process.

Nucleation and the growth of supercritical bubbles in high-viscosity liquids were studied in [310], and will not be discussed here in detail.

9.2 Nucleation in a Low-Viscosity Liquid

9.2.1 Reduced Equations Describing the Process of Bubble Nucleation

As a rule, both the liquid and the gas in a bubble in the process of bubble nucleation in a liquid can be considered as being in a local thermodynamic equilibrium, whereas the system as a whole is in a nonequilibrium state. Therefore, to describe the nucleation process, it is convenient to introduce the free energy that determines the corresponding fluctuations under the conditions of thermodynamic equilibrium. Upon the nucleation of a bubble in the metastable medium, its variation $\Delta F(V, N)$ is defined by the expression [308–310, 313, 314]

$$\Delta F(V, N) = V(p^L - p^V) + N(\mu^V - \mu^L) + 4\pi R^2 \sigma. \tag{9.1}$$

Here, V, R, and N are the volume, radius of a bubble, and the number of gas molecules in the bubble, respectively; p^V and p^L are the pressures in the bubble and the liquid after its transition to the metastable state, respectively; $\mu^L = \mu^L\left(p(n^L), T\right)$ and μ^V are the chemical potentials of the gas atom in a liquid and in a bubble, respectively; $p(n^L)$ is the external saturating gas pressure, i.e., the pressure at which the gas in the bubble with a size of R is in equilibrium with the gas dissolved in a liquid with density n^L; and σ is the surface tension of the liquid. Variables describing the bubble are regarded as continuous so that $(1/N) \ll 1$. At the range where this ratio is of the order of unity, the calculation accuracy for the averages will be quite adequate, provided that the range of their variation is rather wide (for the case in question, it is sufficient that the amount of gas in a critical bubble is large, $N_c \gg 1$).

Differentiating ΔF with respect to N at constant volume and taking into account that $N(\partial\mu^V/\partial p^V)|_{T=\text{const}} = N\omega = V$ (ω is the volume per gas molecule in the bubble), we obtain

$$\left.\frac{\partial \Delta F}{\partial N}\right|_{V=\text{const}} = \mu^V - \mu^L, \tag{9.2}$$

and similarly

$$\left.\frac{\partial \Delta F}{\partial V}\right|_{N=\text{const}} = p^L - p^V + \frac{2\sigma}{R}. \tag{9.3}$$

The parameters of the equilibrium bubble (R_c, N_c) are found from the extremum of ΔF. From Eq. (9.3), we obtain

$$p^V = p^L + \frac{2\sigma}{R_c}. \tag{9.4}$$

This expression determines the pressure in a bubble with a size R, which is in mechanical equilibrium with the liquid at pressure p^L. From Eq. (9.2), it follows

$$\mu^V\left(p^V, T\right) = \mu^L\left(p, T\right). \tag{9.5}$$

Since $p = p\left(n^L\right)$ is the external (saturating) pressure of the gas, which is in equilibrium with the gas dissolved in a liquid with density n^L, the chemical potential of the gas in the liquid can be replaced on the right-hand side of Eq. (9.5) by the equal chemical potential of the saturating gas. Assuming that p^V and $p\left(n^L\right)$ are rather small ($\le 10^7$ Pa) and using the equation of state for the ideal gas with constant heat capacity ($\mu = k_B T \ln p + \chi(T)$, k_B is Boltzmann's constant), we obtain from Eq. (9.5)

$$\ln\left[\frac{p^V}{p\left(n^L\right)}\right] = 0, \qquad p^V = p\left(n^L\right). \tag{9.6}$$

Thus, the critical size R_c of the bubble containing N_c gas atoms is determined from Eqs. (9.4) and (9.6)

$$p^V = N_c k_B T\left(\frac{4\pi}{3}R_c^3\right)^{-1} = p(n^L). \tag{9.7}$$

As a rule, the critical size at the nucleation stage is fairly small; then, $p^L \ll p\left(n^L\right) \approx 2\sigma/R_c$ and

$$R_c = \frac{2\sigma}{p^V - p^L} = \frac{2\sigma}{p\left(n^L\right) - p^L} \approx \frac{2\sigma}{p\left(n^L\right)}, \tag{9.8}$$

$$\begin{aligned} p^V &= \frac{N_c k_B T}{V_c} = p\left(n^L\right), \\ N_c &= \left(p^L + \frac{2\sigma}{R_c}\right)\frac{V_c}{k_B T} \approx \frac{2\sigma}{R_c}\frac{V_c}{k_B T} = \frac{8\pi\sigma}{3}\frac{R_c^2}{k_B T}. \end{aligned} \tag{9.9}$$

Hereafter, we assume that for the gas dissolved in a liquid Henry's law is fulfilled, which is applicable within a wide range of parameters

$$n^L(t) = \frac{\delta}{k_B T} p(n^L(t)) = \delta n^G, \tag{9.10}$$

$$n_e^L \equiv n^L(\infty) = \frac{\delta}{k_B T} p^L\left(n^L(\infty)\right) = \delta n_e^G, \tag{9.11}$$

where δ is the solubility parameter, $n^L = n^L(t)$, $n^L(0)$ is the initial gas density, n^G is the saturating gas density equal to the density of the gas in the ambient space (we suppose that the gas in a liquid is distributed homogeneously; it is true in the case when the metastable state is created quickly as compared to the time interval before the start of nucleation [308]) and n_e^G corresponds to the gas equilibrium density n_e^L at the pressure $p^L(n_e^L) \equiv p^L$ in a liquid.

As was shown in [310], depending on the value of the parameters, one can distinguish between two limiting cases: the case of boiling of a high-viscosity liquid where the amount of gas in a bubble is adjusted to its volume and a low-viscosity liquid where the bubble volume is determined by the amount of gas in a bubble. The first case is realized upon the fulfillment of inequality $\tau_g \ll \tau_R$ [310], where $\tau_g = 2lR\left(3\delta D\right)^{-1}$ is the characteristic time of filling the bubble with the gas, $\tau_R = 4\eta\left(3p^L + 8\sigma/R\right)^{-1}$ is the characteristic time of bubble size variations, D is the diffusion coefficient for gas atoms in the liquid, l is the length of an elementary displacement (of the order of the distance between atoms in a liquid), and η is the liquid viscosity.

In the case of a low-viscosity liquid, the inverse inequality is valid

$$\tau_g \gg \tau_R. \tag{9.12}$$

On the adjustment of the bubble size $R = R(N)$ to the amount of gas in the bubble, we get the mechanical equilibrium for the virtually entire spectrum of sizes where the condition

$$p^V = p^L + \frac{2\sigma}{R} \tag{9.13}$$

is fulfilled.

According to Eq. (9.6), $N = 8\pi\sigma R^2(N)/3$ (for N_c, Eq. (9.9) is valid). Substituting the condition (9.4) into Eq. (9.1), we find

$$\Delta F(V,N) = N\left[\mu^V(p^L + \frac{2\sigma}{R}, T) - \mu^L\right] + \frac{4}{3}\pi R^2\sigma. \tag{9.14}$$

After the adjustment, Eq. (9.3) is retained with the account of $\partial p^V/\partial R = -2\sigma/R^2$ and $\partial\mu^V/\partial p^V = \omega$. As a result, we get

$$\begin{aligned}\Delta F(V) = \Delta F(V(N), N) &= N k_B T \ln\frac{p^V}{p(n^L)} + \frac{1}{2}N k_B T \\ &= \frac{N k_B T}{2}\ln\frac{N_c}{N} + \frac{1}{2}N k_B T.\end{aligned} \tag{9.15}$$

The bubble volume $V(N) = 4\pi R^3/3$ is unambiguously determined by the number of gas atoms in the bubble using the equation of state of the gas and Eq. (9.13). Thus, for the ideal gas, we obtain

$$V(N) = Nk_BT\left(p^L + \frac{2\sigma}{R}\right)^{-1}. \tag{9.16}$$

The bubble-size distribution function in a low-viscosity liquid can be represented using Eq. (9.15)

$$f(N,V,t) = \psi(N,t)\,\delta(V - V(N)). \tag{9.17}$$

After integration of the general equation with respect to the adjusting variable V, the reduced distribution function $\psi(N,t)$ satisfies the equation

$$\frac{\partial\psi}{\partial t} = -\frac{\partial J}{\partial N} \tag{9.18}$$

with the boundary conditions

$$\psi(N,t)|_{N=1} = n^L, \qquad \psi(N,0)|_{N>1} = 0, \tag{9.19}$$

where the flux J in the space of gas atom number in the bubble is determined by the following equation (cf. Eqs. (3.5)–(3.7)):

$$J = -D(N)\left(\frac{1}{k_BT}\frac{\partial\Delta F}{\partial N}\psi + \frac{\partial\psi}{\partial N}\right). \tag{9.20}$$

Here, $D(N) = w_{N,N+1}$ is the diffusion coefficient of bubbles in the space of the number of gas particles in the bubble or the probability of absorption of one atom by the bubble from the solution per unit time [308]. The first and the second terms in the brackets of Eq. (9.20) are, respectively, the thermodynamic and the diffusion flow rates in the space of numbers of gas atoms in the bubbles, N. At the initial stage of bubble nucleation with the gas when there are still no diffusion clouds around the bubbles [308], the rate of filling the bubble with the gas has the form

$$\frac{\mathrm{d}N}{\mathrm{d}t} = -D(N)\frac{1}{k_BT}\frac{\partial\Delta F}{\partial N}, \qquad D(N) = \alpha\left(\frac{D}{2l}\right)4\pi R^2 n^L, \tag{9.21}$$

where α is the coefficient accounting for the additional barrier, which can exist for the last jump of gas atoms into the bubble, $0 < \alpha \leq 1$. From Eq. (9.15), we also find

$$\frac{1}{k_BT}\frac{\Delta F}{\partial N} = \frac{1}{2}\ln\frac{N_c}{N}. \tag{9.22}$$

Using Eq. (9.17), we obtain the law of conservation of the total amount of gas in the system [308]

$$n^L(0) - n^L(t) = \left[1 - \int_0^\infty V(N)\psi(N,t)\,\mathrm{d}N\right]^{-1}\int_0^\infty N\psi(N,t)\,\mathrm{d}N. \tag{9.23}$$

Here, $n^L(0) \equiv n_0$ is the initial density of the gas dissolved in the liquid; the expression in square brackets on the right-hand side of Eq. (9.23) accounts for an increase in the total volume of a system due to bubble growth.

Equation (9.18), together with the law of conservation of the total amount of gas in a system (Eq. (9.23)), represents the total set of equations describing the nucleation of bubbles with gas in a low-viscosity liquid.

9.2.2 Time of Establishment of Steady-State Nucleation

At the instantaneous transition of a system into the metastable state, the flow of critically sized nuclei emerges and the quasisteady-state regime is established after a certain time lag t_{lag}. Let us estimate this time.

By the conditions determining the parameters of the critical bubbles, the first derivatives of the thermodynamic potential with respect to N and R are equal to zero. This makes it possible to find the bubble size R_c and the amount of gas in the bubble, N_c. From Eq. (9.5), we obtain $p^V\big|_{R=R_c} = p(n^L(t))$; then

$$R_c = \frac{2\sigma}{p(n^L) - p^L}, \tag{9.24}$$

where p^L is the pressure in a liquid corresponding to the equilibrium density of the dissolved gas $n_e^L = \delta p^L / k_B T$.

For the transition stage, where the metastability of the system decreases substantially but is still rather high, we have

$$R_c = \frac{2\sigma}{p} = \frac{2\sigma\delta}{k_B T n^L} = \frac{2\sigma\delta}{k_B T n^L(0)} \frac{1}{1 - Z(t)} = R_{c0} \frac{1}{1 - Z(t)},$$
$$\frac{n^L}{n^L(0)} = \frac{n^L(0) - n^L(0) + n^L}{n^L(0)} = 1 - Z(t), \tag{9.25}$$

where $Z(t) = \left[n^L(0) - n^L(t)\right] / n^L(0)$ is the relative (per unit volume) number of gas atoms in the bubbles. Taking into account that, as a rule, for subcritical bubbles, the case $p^L \ll 2\sigma/R$ is realized, from the equation of state, we obtain

$$N = \frac{1}{k_B T}\left[p^L V + \frac{8\pi}{3}\sigma R^2\right] \approx \frac{8\pi}{3}\frac{\sigma}{k_B T} R^2, \tag{9.26}$$

$$N_c = \frac{1}{k_B T}\left[p^L V_c + \frac{8\pi}{3}\sigma R_c^2\right] \approx \frac{8\pi}{3}\frac{\sigma}{k_B T} R_c^2, \tag{9.27}$$

$$\frac{N}{N_c} = \frac{R^2}{R_c^2}, \qquad \frac{N}{N_{c0}} = \frac{R^2}{R_{c0}^2}, \tag{9.28}$$

$$R^2 = R_{c0}^2 \frac{N}{N_{c0}}, \qquad N_{c0} = \frac{p\left(n^L(0)\right) V_{c0}}{k_B T}. \tag{9.29}$$

The use of Eq. (9.25) yields

$$\frac{N_c}{N_{c0}} = (1 - Z(t))^{-2}. \tag{9.30}$$

Here, $V_{c0} = 4\pi R_{c0}^3/3$ is the volume of a critical bubble where the pressure is equal to the saturation pressure of the dissolved gas at the initial moment, $p\left(n^L(0)\right)$ and N_{c0} is the amount of gas in the critical bubble at the initial moment. Then, we have

$$n^L(0)Z(t) = \int_0^\infty N\psi(N,t)\,\mathrm{d}N. \tag{9.31}$$

Let us write the expression for the diffusion coefficient $D(N)$ (Eq. (9.21)) as

$$D(N) = \alpha \frac{3D}{2lR_{c0}} \frac{4\pi}{3} R_{c0}^3 n^L(0) \frac{N}{N_{c0}} (1 - Z(t)) = \frac{1}{t_0} N (1 - Z(t)), \tag{9.32}$$

where

$$t_0 = \frac{2lR_{c0}}{3\alpha D\delta}. \tag{9.33}$$

When deriving this relation, we used the Henry law (Eq. (9.10)) and the equality

$$\frac{4\pi}{3} R_{c0}^3 n^L(0) = \frac{p(n^L(0))V_{c0}}{k_B T}\delta = N_{c0}\delta. \tag{9.34}$$

Using Eqs. (9.22), (9.28), and (9.32), we can write Eq. (9.21) in the following form:

$$\frac{\mathrm{d}N}{\mathrm{d}t} = \frac{1}{2t_0} N(1 - Z)\ln\frac{N}{N_c}, \tag{9.35}$$

or, with account of Eq. (9.30), as

$$\frac{\mathrm{d}N}{\mathrm{d}\tau} = N(1 - Z(\tau))\ln\frac{N}{N_c} = N(1 - Z(\tau))\left[\ln\frac{N}{N_c} + 2\ln\frac{1}{1 - Z(\tau)}\right], \tag{9.36}$$

where

$$\tau = \frac{t}{2t_0}. \tag{9.37}$$

With the notation $y = N/N_c$, Eq. (9.18) gets the form

$$\frac{\partial\psi}{\partial\tau} = \frac{\partial}{\partial y} y\left(-\psi\ln y + \frac{2}{N_c}\frac{\partial\psi}{\partial y}\right). \tag{9.38}$$

For the range of subcritical nuclei, $2 \le N < N_c$, it is convenient to introduce the variables

$$x = \ln N, \qquad \widetilde{\psi} = N\psi = e^x\psi. \tag{9.39}$$

Then

$$\ln y = \ln\left(\frac{N}{N_c}\right) = x - x_c, \tag{9.40}$$

and Eq. (9.37) is transformed into

$$\frac{\partial\widetilde{\psi}}{\partial\tau} = b\frac{\partial^2\widetilde{\psi}}{\partial x^2} + a\frac{\partial\widetilde{\psi}}{\partial x}\widetilde{\psi} - d\widetilde{\psi}, \qquad \widetilde{\psi}\Big|_{N=1} = n^L, \tag{9.41}$$

where

$$b = e^{-x}, \qquad a = \frac{1}{2}(x_c - x) - 4e^{-x} > 0, \qquad d = \frac{1}{2}(1 - 2e^{-x}) > 0. \tag{9.42}$$

At variable coefficients a, b, and d, Eq. (9.41) cannot be solved exactly. However, taking into account the constant signs of these coefficients, let us estimate the upper limit of τ_{lag}. Assuming that the coefficients are constant, we get the exact solution of this equation. Using the substitution

$$\widetilde{\psi} = p(x,\tau)\exp\left[-\left(d + \frac{a^2}{4b}\right)\tau\right]\exp\left[-\frac{a}{2b}x\right], \tag{9.43}$$

we obtain

$$\frac{\partial p(x,\tau)}{\partial\tau} = b\frac{\partial^2 p(x,\tau)}{\partial x^2}, \qquad p(x,\tau)|_{x=0} = n^L\exp\left[-\left(d + \frac{a^2}{4b}\right)\tau\right], \tag{9.44}$$

$$N\psi = \widetilde{\psi} = n^L\frac{x}{\sqrt{\pi b}}\int\limits_0^\tau \exp\left(-\frac{x^2}{4b(\tau-\tau')}\right)\exp\left[-\left(d + \frac{a^2}{4b}\right)(\tau-\tau')\right] \tag{9.45}$$
$$\times\frac{\mathrm{d}\tau'}{(\tau-\tau')^{3/2}}.$$

Let us introduce the variable $\xi = x\left(2\sqrt{b(\tau-\tau')}\right)^{-1}$ instead of τ', then we get

$$\widetilde{\psi} = n^L\frac{2}{\sqrt{\pi}}\int\limits_{x(2\sqrt{b\tau})^{-1}}^{\infty}\exp\left(-\frac{\alpha}{\xi} - \xi^2\right)\mathrm{d}\xi, \tag{9.46}$$

where

$$\alpha = \left(d + \frac{a^2}{4b}\right)\frac{x^2}{4b}. \tag{9.47}$$

The integrand in Eq. (9.46) has a sharp maximum at $\xi_{\max} = \alpha^{1/4}$. If this maximum fits the integration domain, $\widetilde{\psi}$ is virtually independent of τ that corresponds to the quasistationary state. Then, we derive the equation for the time of establishment of such state, τ_{lag}, equating (9.47) to the lower integration limit in Eq. (9.46) at $\tau = \tau_{\text{lag}}$

$$\frac{x^2}{4b\tau_{\text{lag}}} = \alpha^{1/2} = \frac{x}{\sqrt{4b}}\sqrt{d + \frac{a^2}{4b}}, \tag{9.48}$$

thus arriving at

$$\tau_{\text{lag}} = \frac{x}{\sqrt{4db + a^2}}. \tag{9.49}$$

The parameters a, b, and d are functions of N and N_c. Let us estimate the maximal time τ_{lag}; to do so, we substitute the minimal values $a_{\min} = 0$, $b_{\min} = e^{-z_c} = N_c^{-1}$, $d = 1/2$, and $x = x_c$ and get

$$\tau_{\text{lag}} \leq \sqrt{N_{c0}} \ln \frac{N_{c0}}{\sqrt{2}}, \tag{9.50}$$

and for the dimensional time

$$t_{\text{lag}} \leq t_0 \sqrt{N_{c0}} \ln \frac{N_{c0}}{\sqrt{2}} = \frac{2}{3\alpha}\frac{lR_c}{D\delta}\sqrt{N_{c0}} \ln \frac{N_{c0}}{\sqrt{2}}. \tag{9.51}$$

Thus, the quasistationary distribution of subcritical bubbles with a distribution function

$$\psi(N,t) = \psi(N, N_c) \tag{9.52}$$

and a flux $J(N,t) = J(N, N_c)$ is established after the time $t > t_{\text{lag}}$.

The number of gas molecules in a critical nucleus, N_c, depends on time due to the depletion of the solution; however (at the stage of intensive nucleation), this dependence, as will be shown below, turned out to be rather weak. Therefore, $\psi(N, N_c)$ and $J(N, N_c)$ can be regarded as slowly varying functions of the parameter N_c.

9.2.3 Quasistationary Distribution of Subcritical Bubbles

As was shown in Refs. [311] and [300], it is convenient to describe the nucleation process after the establishment of the quasistationary state ($t > t_{\text{lag}}$) in terms of the flux $J(N,t)$. For this purpose, let us express the distribution function via $J(N,t)$, using Eq. (9.20)

$$\psi(N,t) = \exp\left(-\frac{\Delta F}{k_B T}\right) \int_N^{N_{\max}} \frac{\mathrm{d}N'}{D(N')} J(N',t) \exp\left(\frac{\Delta F(N')}{k_B T}\right), \tag{9.53}$$

where $N_{\max}$ is the upper bound of the region of nonzero values of flux and distribution function in space N. Since $\exp(\Delta F(N)/k_B T)$ has a sharp maximum in the point $N = N_c$, the integral can be taken using the saddle point approximation and we get

$$\psi(N,t) = \frac{J(N_c)}{D(N_c)} \exp\left(-\frac{\Delta F(N)}{k_B T}\right) \int_N^{N_{\max}} \mathrm{d}N' \exp\left(\frac{\Delta F(N')}{k_B T}\right). \tag{9.54}$$

Expanding the function ΔF into a series in the vicinity of the saddle point, we obtain

$$\frac{\Delta F(N)}{k_B T} = \frac{\Delta F(N_c)}{k_B T} + \frac{1}{k_B T}\left.\frac{\partial \Delta F}{\partial N}\right|_{N_c}(N - N_c) + \frac{1}{2k_B T}\left.\frac{\partial^2 \Delta F}{\partial N^2}\right|_{N_c}(N - N_c)^2, \tag{9.55}$$

$$\frac{1}{k_B T}\frac{\partial \Delta F}{\partial N} = \frac{1}{2}\ln\frac{N_c}{N}, \qquad \frac{1}{(\Delta N)^2} \equiv \frac{1}{2k_B T}\frac{\partial^2 \Delta F}{\partial N^2} = \frac{1}{4N}. \tag{9.56}$$

Taking into account that, at the stage of intensive nucleation, $N_{\max} \geq N_c + \Delta N$ holds and, at homogeneous nucleation, each gas atom is a potential nucleus of a bubble, i.e.,

$$\psi(N,t)|_{N\to 1} = n^L(0), \tag{9.57}$$

we obtain

$$J(N_c) = \frac{n^L(0)}{t_0}\frac{\sqrt{N_c}}{2\sqrt{\pi}}\exp\left(-\frac{N_c}{2}\right), \tag{9.58}$$

$$\psi(N,t) = n^L(0)\exp\left[-\frac{N}{2}\left(\ln\frac{N_c}{N} + 1\right)\right]\frac{1}{2}\left(1 - \mathrm{erf}\frac{N - N_c}{\Delta N_c}\right), \tag{9.59}$$

where

$$\Delta N_c = \left(-\frac{1}{2k_B T}\left.\frac{\partial^2 \Delta F}{\partial N^2}\right|_{N=N_c}\right)^{-1/2} \approx 2\sqrt{N_c}. \tag{9.60}$$

An expression similar to Eq. (9.59) has been first derived in [329]. It demonstrates that the behavior of the distribution function $\psi(N)$ significantly changes at $N \approx N_c$.

9.2.4 Distribution Function of Bubbles in the Range $N_c < N < \widetilde{N}$

Let us calculate the distribution function for bubbles within the range $N_c < N < \widetilde{N}$, where $\widetilde{N}$ is determined from the condition $\Delta F(\widetilde{N}) = 0$. In this range, the diffusion term (second derivative with respect to N in Eq. (9.37)) is still significant, while at $N > \widetilde{N}$, it can be neglected, because the distribution function becomes gently sloping and the growth in size space can be considered as purely hydrodynamic. From Eq. (9.15), we find that $\widetilde{N} = eN_c$. In variables $x = \ln N$ and $\tau = t/2t_0$, the system of equations gets the form [297, 308, 309]

$$\frac{\partial J}{\partial \tau} = -\left(x - x_c + 2e^{-x}\right)\frac{\partial J}{\partial x} + 2e^{-x}\frac{\partial^2 J}{\partial x^2}, \tag{9.61}$$

$$J(x,\tau)|_{x=x_c} = J(N_c(\tau)), \qquad J(x,\tau)|_{\tau=0,\, x>x_c} = 0. \tag{9.62}$$

In order to estimate the time of the establishment of the quasistationary state, one can omit, with a good accuracy (it is true already at $N_c > 10$), the term $2e^{-x} \ll x_c$ on the right-hand side of Eq. (9.62) in the coefficient at the first derivative and substitute e^{-x_c} for e^{-x} in the coefficient at the second derivative. Then

$$\frac{\partial J}{\partial \tau} = -(x - x_c)\frac{\partial J}{\partial x} + 2e^{-x_c}\frac{\partial^2 J}{\partial x^2}, \quad J(x,\tau)|_{x=x_c} = J(N_c(\tau)) = \widetilde{J}(x_c). \quad (9.63)$$

After the substitution $J(x,\tau) = J(\phi(x,\tau), \widetilde{t}(\tau))$, we have

$$\frac{\partial J}{\partial \tau} = \frac{\partial J}{\partial \phi}\frac{\partial \phi}{\partial \tau} + \frac{\partial J}{\partial \widetilde{t}}\frac{\partial \widetilde{t}}{\partial \tau} = -(x - x_c)\frac{\partial J}{\partial \phi}\frac{\partial \phi}{\partial x} + \frac{2}{N_c}\frac{\partial^2 J}{\partial \phi^2}\left(\frac{\partial \phi}{\partial x}\right)^2 + \frac{2}{N_c}\frac{\partial J}{\partial \phi}\frac{\partial^2 \phi}{\partial x^2}. \quad (9.64)$$

The function $\phi(x,\tau)$ is chosen as $\phi(x,\tau) = (x - x_c)\,e^{-\tau}$. Then

$$\frac{\partial \phi}{\partial x} = e^{-\tau}, \qquad \frac{\partial^2 \phi}{\partial x^2} = 0, \qquad \frac{\partial \phi}{\partial \tau} = -(x - x_c)\,e^{-\tau}. \quad (9.65)$$

Selecting the function $\widetilde{t}(\tau)$ such that

$$\frac{\partial \widetilde{t}}{\partial \tau} = \frac{2}{N_c}\left(\frac{\partial \phi}{\partial x}\right)^2 = \frac{2e^{-2\tau}}{N_c}, \qquad \widetilde{t}(\tau) = \frac{1 - e^{-2\tau}}{N_c}, \quad (9.66)$$

we obtain equations

$$\frac{\partial J}{\partial \widetilde{t}} = \frac{\partial^2 J}{\partial \phi^2}, \qquad J(x)|_{x=x_c} = J(N_c), \qquad \phi(x,\tau)|_{x=x_c} = 0, \quad (9.67)$$

which can be solved exactly. As a result, we obtain

$$J = J(N_{c0})\left[1 - \mathrm{erf}\left(\frac{\phi}{2\sqrt{\widetilde{t}}}\right)\right] = J(N_{c0})\left\{1 - \mathrm{erf}\left[p(\tau)(x - x_c)\right]\right\}, \quad (9.68)$$

where

$$p(\tau) = e^{-\tau}\sqrt{N_{c0}/\left(1 - e^{-2\tau}\right)}. \quad (9.69)$$

Hence, we find the relaxation time (induction time) as

$$t_{\mathrm{rel}} \leq t_0 \ln N_{c0}. \quad (9.70)$$

The total relaxation time needed to establish a quasistationary state in the range $N \leq eN_{c0}$ is determined from Eqs. (9.51) and (9.70) as

$$t_{\mathrm{full}} = t_0\sqrt{N_{c0}}\ln\frac{N_{c0}}{\sqrt{2}} + t_0 \ln N_{c0} \approx t_0 \ln N_{c0} \approx t_{\mathrm{lag}}. \quad (9.71)$$

Expanding the integrand in Eq. (9.53) near the lower bound at $t > t_{\text{rel}}$, we obtain the result for the distribution function

$$\begin{aligned}\psi(N,t) &= \frac{J(N_c)}{D(N_c)} \int_N^{N_{\max}} \exp\left\{-\frac{1}{2}\ln\frac{N}{N_c}(N-N') - \frac{1}{4N}(N-N')^2\right\} \mathrm{d}N' \\ &= \frac{J(N_c)}{D(N_c)}\sqrt{\pi N_c}\frac{2}{\sqrt{\pi}} \int_{C(N)}^{A(N)} e^{-\xi^2}\,\mathrm{d}\xi \\ &= \frac{n^L(0)}{2}\sqrt{\frac{N_c}{N}}\left[1-\operatorname{erf}(C(N))\right]\exp\left(\frac{N}{4}\ln^2\frac{N}{N_c} - \frac{N_c}{2}\right),\end{aligned} \tag{9.72}$$

where

$$C(N) = \frac{1}{2}\sqrt{N}\ln\frac{N}{N_c}, \qquad A(N) = \frac{N_{\max}-N}{2\sqrt{N}} + C(N) \gg 1. \tag{9.73}$$

The definitions $D(N_c)$ (Eq. (9.72)) and $J(N_c)$ (Eq. (9.58)) are taken into account in this equation.

Thus, within the size range $2 \le N \le eN_{c0}$, the distribution function of bubbles depends on time only via the critical size $\psi(N,t) = \psi(N, N_c(t))$. Note that, with an accuracy of small terms in the first derivative (of the order of $1/N_c \ll 1$), the derived distribution function is joined at $N = N_c$ with the function represented by Eq. (9.59). In view of quasistationary state, $\partial\psi/\partial\tau = \partial J/\partial N = 0$ holds and

$$J(N,\tau)\big|_{N<\widetilde{N},\tau>\tau_{\text{lag}}} = J(eN_c(\tau)) = J(N_c(\tau)), \tag{9.74}$$

i.e., the flux is almost steady at $t > t_{\text{lag}}$ (Eq. (9.71)) in the range $2 \le N \le eN_{c0}$.

9.2.5 Distribution Function of Bubbles in the Range $N > \widetilde{N}$

Let us now consider the case of large bubbles when $N > \widetilde{N} = eN_c$. We omit the term $2e^{-x}$ in Eq. (9.62) (as will be shown below, it is proportional to the number of gas atoms in the critical bubble at the time of intensive nucleation τ_N) and the second derivative because of the weak dependence $J(N)$ and the smallness of the coefficient in front of this dependence within the studied range $x \ge x_c + 1$ ($N \ge eN_c$). As a result, we obtain the equations

$$\frac{\partial J}{\partial \tau} = -(x - x_c)\frac{\partial J}{\partial x}, \quad J(x)|_{x_c+1} = J(N_c(\tau)), \quad J(N,\tau) = J(N_c(\tau_{\text{ch}})), \tag{9.75}$$

where

$$\tau_{\text{ch}} = \tau - \ln(x - x_c) = \tau - \ln\ln\left(\frac{N}{N_c(\tau)}\right) \tag{9.76}$$

is the characteristics of the differential equation (9.75). Thus, we have

$$N = N_c(\tau)\exp(\exp(\tau - \tau_{\text{ch}})). \tag{9.77}$$

At $\tau_{\text{ch}} = 0$, we obtain that the maximum number of atoms in the bubble is equal to

$$N_{\max}(\tau) = N_c \exp(\exp(\tau)) = N_{c0} \exp(\exp \tau). \tag{9.78}$$

Here, we neglected the small variation in $N_c(\tau)$ in the process of bubble nucleation (the validity of such approximation is shown below, see Eq. (9.104)). This result implies that all bubbles emerge at different times, however, with the identical initial amount of gas, $N_c(\tau) = N_{c0}$.

In the range $N > \widetilde{N}$, the distribution function also changes slowly; hence, the diffusion term in Eq. (9.20) is small. Ignoring the second-order derivative of $\psi(N, t)$ with respect to N and taking into account Eqs. (9.58) and (9.33), we obtain that the bubble-size distribution function within the $eN_c < N \leq N_{\max}$ range has the following form:

$$\psi(N, t) = J(N_c)\left(\frac{\mathrm{d}N}{\mathrm{d}t}\right)^{-1} = n^L(0)\sqrt{\frac{N_c}{\pi}} \exp\left(-\frac{N_c}{2}\right)\frac{1}{N\ln(N/N_c)} \times \frac{1}{2}\left(1 - \operatorname{erf}\frac{N - N_{\max}(t)}{\Delta N_c}\right). \tag{9.79}$$

Here we have introduced (similar to Eq. (9.59)) the error function for the smooth correction of the distribution function at $N > N_{\max}(t)$ (this procedure is proven in Ref. [310]).

To determine the time of intensive nucleation, τ_N, one needs to employ the conservation law for the amount of gas atoms (Eq. (9.23)), which can be conveniently written as

$$n^L(0) - n^L(\tau) = \int_0^{N_{\max}} N\psi(N, \tau)\,\mathrm{d}N, \qquad \psi(N, \tau)|_{N=N_{\max}(\tau)} = 0. \tag{9.80}$$

From Eq. (9.80), we derive the following expression for the variation of the relative amount of gas in the bubbles with time:

$$Z = (n^L(0) - n^L(\tau))/n^L(0), \tag{9.81}$$

$$\frac{\mathrm{d}Z}{\mathrm{d}\tau} = \frac{2t_0}{n^L(0)}\int_0^{\infty}\frac{\partial\psi}{\partial\tau}N'\,\mathrm{d}N' = \frac{2t_0}{n^L(0)}\int_{eN_c(\tau)}^{\infty}\frac{\partial\psi}{\partial\tau}N'\,\mathrm{d}N'. \tag{9.82}$$

It is accounted for in Eq. (9.82) that, at $t > t_{\text{rel}}$, the relations

$$\left.\frac{\partial\psi}{\partial t}\right|_{N\leq eN_c} = 0, \qquad \left.\frac{\partial J}{\partial N}\right|_{N\leq eN_c} = 0 \tag{9.83}$$

hold.

As seen from Eq. (9.58), the determining role in the variation of the flux with time is played by its exponential dependence on $\Delta F(N_c(\tau))/k_B T = N_c(\tau)/2$. Even a small change of this quantity strongly affects the length of the time interval, where the particles of the new phase are intensively formed. Indeed

$$\frac{\Delta F(N_c)}{k_B T} = \frac{N_c}{2} = \frac{1}{2}\frac{pV_c}{k_B T} \approx \frac{1}{2}\frac{4\pi}{3}\frac{(2\sigma)^3}{p^2}$$
$$= \frac{N_{c0}}{2}\left[1 - \frac{n^L(0) - n^L(\tau)}{n^L(0)}\right]^{-2} \approx \frac{N_{c0}}{2}\left(1 + 2Z(\tau)\right). \tag{9.84}$$

When deriving Eq. (9.84), we accounted for the relation

$$V_c = \frac{4\pi}{3} R_c^3 = \frac{4\pi}{3}\left(\frac{2\sigma}{p}\right)^3 \tag{9.85}$$

and the Henry law, Eq. (9.10). From Eqs. (9.58) and (9.84), we obtain that the flux sharply decreases at

$$\frac{n^L(0) - n^L(\tau)}{n^L(0)} = Z(\tau_N) = \frac{1}{N_{c0}} \ll 1, \tag{9.86}$$

where τ_N is the time period of intensive nucleation.

At the same time, we obtain that, at $N_{c0} \gg 1$, the relative change in the amount of gas in the critical nucleus is rather small and almost everywhere (except for the multiplier with exponential dependence on $N_c(\tau)$) one can assume that

$$N_c(\tau) \cong N_{c0} \equiv N_c(\tau)|_{\tau=0} \gg 1. \tag{9.87}$$

Thus, Eq. (9.86) is responsible for the time of intensive nucleation τ_N.

For the dependence of the flux on time at $eN_c(\tau)$ (Eq. (9.74)), we have the same dependence as in the point N_{c0}

$$J(N_c(\tau)) = J(N_{c0}) \exp\left(-N_{c0} Z(\tau)\right), \qquad Z(0) = 0. \tag{9.88}$$

With the account of the characteristics equation (9.76), the general solution as a function of Z can be written as

$$J(N_c(\tau))|_{\tau > \tau_{\text{lag}}} = J(N_{c0}) \exp\left(-N_{c0} Z(\tau_0)\right). \tag{9.89}$$

Expressing the term $\partial\psi/\partial\tau$ via the continuity equation (9.18) and using Eq. (9.89), we obtain the equation for the determination of $Z(\tau)$ as

$$\frac{\mathrm{d}Z}{\mathrm{d}\tau} = \frac{2t_0}{n^L(0)}\left(eN_c J(N_c) + \int\limits_{eN_c(\tau)}^{N_{\max}(\tau)} J(N_{c0}) e^{-N_{c0} Z(\tau_0)}\,\mathrm{d}N'\right). \tag{9.90}$$

Here, we took into account that the flux is equal to zero in the $N \geq N_{\max}$ range.

Let us substitute Eq. (9.89) into (9.90) and going over to the integration variable $\tau_0 = \tau_0\,(N, \tau)$, integrate by part with the account that, according to Eq. (9.76), $\tau_0(\tau, eN_c) = \tau$ and $\tau_0(0, N_{\max}) = 0$. Then we have

$$\frac{\mathrm{d}Z}{\mathrm{d}\tau} = \frac{2t_0}{n^L(0)}\left(eN_c J(N_c) + \int\limits_{\tau}^{0} J(N_c(\tau))\frac{\mathrm{d}N}{\mathrm{d}\tau_0}\,\mathrm{d}\tau_0\right) \tag{9.91}$$

$$= \frac{2t_0}{n^L(0)}\left(J(N_{c0})N_{\max}(\tau) + \int\limits_{0}^{\tau} J(N_{c0})e^{-N_{c0}Z(\tau_0)}\frac{\mathrm{d}Z}{\mathrm{d}\tau_0}N(\tau-\tau_0, N_c)\,\mathrm{d}\tau_0\right).$$

This equation can be solved by the method of successive approximations.

Since the integral term is of second order of smallness with respect to $J(N_{c0}) \ll 1$, in the first approximation, we have

$$\frac{\mathrm{d}Z}{\mathrm{d}\tau} = \frac{2t_0}{n^L(0)}J(N_{c0})N_{\max}(\tau) = A\exp(\exp\tau), \qquad Z(0) = 0, \tag{9.92}$$

where

$$A = \frac{2t_0 J(N_{c0})N_{c0}}{n^L(0)} = \frac{(N_{c0})^{3/2}}{\sqrt{\pi}}\exp\left(-\frac{N_{c0}}{2}\right). \tag{9.93}$$

From Eq. (9.92), we obtain

$$Z(\tau) = A\int\limits_{0}^{\tau}\exp(\exp\tau')\,\mathrm{d}\tau' \approx Ae^{-\tau}\left[\exp(\exp\tau) - e\right] \approx Ae^{-\tau}\left[\exp(\exp\tau)\right]. \tag{9.94}$$

In view of the very fast increase, the integrand gives the main contribution to the integral at the upper limit.

The time period of intensive nucleation can be determined, using Eq. (9.86) and substituting $\tau = \tau_N$ into Eq. (9.94), as

$$N_{c0}Ae^{-\tau_N}\exp\left(\exp\tau_N\right) = 1. \tag{9.95}$$

Taking the logarithm of Eq. (9.95), we find the equation

$$e^{\tau_N} - \tau_N = \ln\frac{1}{AN_{c0}} = \ln\left[\frac{\sqrt{\pi}}{N_{c0}^{5/2}}\exp\left(\frac{N_{c0}}{2}\right)\right], \tag{9.96}$$

which is solved by the method of successive approximations (see Appendix A.1). In the case that the inequality $\ln\frac{1}{AN_{c0}} \gg 1$ is fulfilled, we get

$$e^{\tau_N} \approx \ln\frac{1}{AN_{c0}} + \ln\left(\ln\frac{1}{AN_{c0}} + \ln\left(\ln\frac{1}{AN_{c0}} + \ln(\ldots)\right)\right) \tag{9.97}$$

or, with an accuracy of the terms of the order of $\left(\ln\ln\frac{1}{AN_{c0}}\right)\left(\ln\frac{1}{AN_{c0}}\right)^{-1}$

$$e^{\tau_N} \approx \ln\frac{1}{AN_{c0}} + \ln\ln\frac{1}{AN_{c0}}. \tag{9.98}$$

Equation (9.95) is satisfied with the same accuracy. Retaining principal terms in Eq. (9.94), we obtain

$$e^{\tau_N} = \frac{N_{c0}}{2}\left(1 - \frac{2}{N_{c0}}\ln\frac{(N_{c0})^{5/2}}{\sqrt{\pi}}\right), \qquad \tau_N = \ln\frac{N_{c0}}{2} > 1, \tag{9.99}$$

$$t_N = 2t_0 \ln\frac{N_{c0}}{2} = \frac{4lR_c}{3\alpha D\delta}\ln\frac{N_{c0}}{2}. \tag{9.100}$$

Combining Eqs. (9.94) and (9.95), we find at $\tau_N > 1$ that

$$\frac{Z(\tau)}{Z(\tau_N)} = N_{c0}Z(\tau) = e^{\tau_N - \tau}\exp\left[-\left(e^{\tau_N} - e^{\tau}\right)\right]. \tag{9.101}$$

Differentiating with respect to τ, we obtain

$$N_{c0}\frac{\mathrm{d}Z(\tau)}{\mathrm{d}\tau} = -e^{\tau_N-\tau}\exp\left[-\left(e^{\tau_N}-e^{\tau}\right)\right] + e^{\tau}e^{\tau_N-\tau}\exp\left[-\left(e^{\tau_N}-e^{\tau}\right)\right]. \tag{9.102}$$

Taking into account Eqs. (9.84), (9.86), and (9.102), Eq. (9.76) yields

$$\begin{aligned}\frac{\partial\tau_0}{\partial\tau} &= 1 + \frac{1}{\ln\left(N/N_{c0}(\tau)\right)}\frac{1}{N_c}\frac{\mathrm{d}N_c}{\mathrm{d}\tau} = 1 + \frac{1}{\ln\left(N/N_c(\tau)\right)}2\frac{\mathrm{d}Z}{\mathrm{d}\tau} \\ &= 1 + \frac{1}{\ln\left(N/N_c(\tau)\right)}\frac{2}{N_{c0}}\left(e^{\tau}-1\right)e^{\tau_N-\tau}\exp\left[-e^{\tau_N}\left(1-e^{\tau-\tau_N}\right)\right] \approx 1.\end{aligned} \tag{9.103}$$

It follows from Eq. (9.103) that in the range where the inequality $\exp\left(e^{-\tau_N}\right) \ll 1$ holds true, i.e., almost within the entire range of values

$$0 \le \tau \le \tau_N - e^{-\tau_N} \approx \tau_N, \tag{9.104}$$

we have $\mathrm{d}\tau_0(N,\tau)/\mathrm{d}\tau = 1$, and only in a very narrow range (with a width of $\approx (\ln\tau_N)e^{-\tau_N}$) in the vicinity of τ_N, this value increases up to 2. Hence, at the stage of intensive nucleation, the variations in $N_c(\tau)$ with time can be ignored, as it was the case in Eq. (9.78).

The number of bubbles in unit volume, N_b, is determined by

$$N_b = \int_0^t J(N_c)\,\mathrm{d}t \cong J(N_{c0})t = 2t_0 J(N_{c0})\tau. \tag{9.105}$$

At $\tau = \tau_N$ the number of bubbles formed at the nucleation stage reaches its maximum value

$$N_b^{\max} = 2t_0 J(N_{c0})\tau_N, \tag{9.106}$$

where τ_N is determined by Eq. (9.99).

The same expression can be derived, using the definition of $N_{\max}$ (Eq. (9.78))

$$N_b = \int_{N}^{N_{\max}} \psi(N,t)\,\mathrm{d}N \cong 2t_0 J(N_c) \int_{eN_c}^{N_{\max}} \frac{\mathrm{d}N}{N\ln(N/N_c)}$$

$$\cong 2t_0 J(N_c) \ln\ln \frac{N_{\max}(t)}{N_c} = 2t_0 J(N_{c0})\tau. \tag{9.107}$$

It can be easily seen that this equation yields the expression for the maximum number of bubbles, which is analogous to Eq. (9.106). The passage to the size distribution function $\varphi(R,t)$ is made in accordance with $\varphi(R,t) = \psi(N,t)\,\mathrm{d}N/\mathrm{d}R$, where the dependence of N on R is determined by Eq. (9.16).

9.3 The Intermediate Stage

After the end of nucleation the intermediate stage ($t_N < t < t_f$) starts, when the amount of excess gas in the solution is sufficiently large for the growth of the formed bubbles but too small for intensive nucleation of new bubbles (since the nucleation rate depends on the excess amount of the gas to a very high extent). Therefore, the number of bubbles per unit volume remains virtually constant during this stage. The amount of excess gas at the intermediate stage decreases, tending to the equilibrium value, so Eq. (9.36) can be used to determine the amount of gas in a bubble.

At $N \geq eN_c$, for the main part of the spectra in the N-space one can neglect the diffusion term in the definition of the flux. Then, we obtain a complete set of equations, which includes the continuity equation for the distribution function $\psi(N,t)$ with the initial condition at $\tau = \tau_N$ (which is determined by the distribution function formed at the end of the nucleation stage, Eqs. (9.59) and (9.79)) and the law describing the change in the number of gas atoms

$$\frac{\partial \psi}{\partial \tau} + \frac{\partial J_N}{\partial N} = 0, \qquad J_N = \frac{\mathrm{d}N}{\mathrm{d}\tau}\psi, \qquad \psi(N,\tau)\big|_{\tau=\tau_N} = \psi_0(N,\tau_N). \tag{9.108}$$

The distribution function formed after the completion of the intensive nucleation stage at $N \geq eN_{c0}$ determines the growing bubbles at $\tau > \tau_N$, because bubbles with $N \leq eN_{c0}$ rapidly disappear and make a minor contribution to the law of conservation, owing to small amount of gas. That is the reason why the distribution function $\psi_0(N,\tau_N)$ (which is determined by Eq. (9.79)) can be used with sufficient accuracy.

Let us write the law of conservation with respect to the number of gas atoms in the form of Eq. (9.23) as

$$\left[n^L(0) - n^L(\tau)\right]\left[1 - \widetilde{\varphi}(\tau)\right] = n^L(0) Z(\tau) = \int_0^{N_{\max}} N\psi(N,\tau)\,\mathrm{d}N \approx \int_{eN_{c0}}^{N_{\max}} N\psi(N,\tau)\,\mathrm{d}N = Q, \tag{9.109}$$

where

$$\widetilde{\varphi}(\tau) = \int_0^{\infty} V(N)\psi(N,\tau)\,\mathrm{d}N \tag{9.110}$$

is the relative volume of bubbles and Q is the total number of gas atoms in bubbles, which are contained in a unit volume of the liquid. If $\widetilde{\varphi}$ is not very small, $\widetilde{\varphi} \leq 1$, one should consider the relative increase in the volume of the solution. With Eq. (9.10), we obtain

$$N = \frac{p^V}{k_B T} V = n^V V = \frac{n^L(\tau)}{\delta} V. \tag{9.111}$$

With Eq. (9.111), the law of conservation (Eq. (9.109)) takes the form

$$\left[n^L(0) - n^L(\tau)\right]\left[1 - \widetilde{\varphi}(\tau)\right] = \frac{n^L(\tau)}{\delta}\widetilde{\varphi}(\tau), \tag{9.112}$$

which yields

$$n^L(\tau) = n^L(0)\left[1 + \frac{1}{\delta}\frac{\widetilde{\varphi}(\tau)}{1 - \widetilde{\varphi}(\tau)}\right]^{-1}. \tag{9.113}$$

Thus, we obtain the relative swelling of the liquid $V(\tau)/V(0)$ as

$$\frac{V(\tau)}{V(0)} = \frac{1}{1 - \widetilde{\varphi}(\tau)} = 1 + \frac{\delta n^L(0)}{n^L(\tau)} - \delta \approx 1 + \frac{\delta}{1 - Z(\tau)}. \tag{9.114}$$

When deriving this relationship, we take into account the fact that the solubility coefficient δ is usually much smaller than unity. Further, we will assume $\widetilde{\varphi}(\tau) \ll 1$ and, accordingly, $\delta n^L(0)/n^L(\tau) \ll 1$ (if $\widetilde{\varphi}(\tau) \approx 1$ it is necessary to take into account the increase in the total volume of a solution [310]).

The continuity equation in the space of N (Eq. (9.108)) in the time interval t_f of the existence of the intermediate stage has the following form with allowance for Eq. (9.36):

$$\frac{\partial \psi}{\partial \tau} + (1 - Z(\tau))\frac{\partial}{\partial N}\left(\psi \ln\frac{N}{N_c}\right) = 0, \tag{9.115}$$

and the initial condition is defined by Eq. (9.79) at $\tau = \tau_N$:

$$\psi(N, \tau_N) = \frac{2t_0 J(N_{c0})}{N \ln(N/N_{c0})} \frac{1}{2} \left(1 - \operatorname{erf} \frac{N - N_{\max}(\tau_N)}{\Delta N_c}\right) \theta(N - eN_{c0}). \tag{9.116}$$

Here we have introduced the θ-function to take into account the fact that bubbles with $N \leq eN_{c0}$ rapidly dissolve, making a minor contribution to the balance of gas atoms. The time is calculated referred to the moment τ_N when the stage of nucleation is completed.

As is known, the solution of Eq. (9.115) can be expressed via the characteristics of this equation and the distribution function at the initial time moment as

$$\psi(N, \tau) = \frac{2t_0 J(N_{c0})}{N_0 \ln(N_0/N_{c0})} \frac{1}{2} \left(1 - \operatorname{erf} \frac{N_0 - N_{\max}(\tau_N)}{\Delta N_c}\right) \theta(N_0 - eN_{c0}) \frac{\partial N_0}{\partial N}. \tag{9.117}$$

Here $N_0 = N_0\,(N, \tau)$ is the characteristics of Eq. (9.115), which is determined by Eq. (9.36) with the initial condition $N|_{\tau=0} = N_0$, where N_0 is an arbitrary point of the initial distribution function within the range of values from eN_{c0} to $N_{\max}(\tau_N)$. Let us introduce the variables

$$x = \ln \frac{N}{N_{c0}}, \quad x_0 = \ln \frac{N_0}{N_{c0}}, \quad x_c = \ln \frac{N_c\,(\tau)}{N_{c0}} = \ln \left(\frac{Z_e}{Z_e - Z(\tau)}\right)^2 \tag{9.118}$$

with allowance for the fact that $Z_e \approx 1$ at $n_e^L/n^L(0) \ll 1$ and rewrite Eq. (9.36) as

$$\frac{\mathrm{d}x}{\mathrm{d}\tau} = (1 - Z(\tau))\,(x - x_c)\,, \qquad x|_{\tau=0} = x_0 = \ln \frac{N_0}{N_{c0}}. \tag{9.119}$$

The general solution of Eq. (9.119) has the form

$$x = \beta(\tau)\,(x_0 - f(\tau))\;, \qquad \beta(\tau) = \exp \varphi(\tau), \tag{9.120}$$

$$\varphi(\tau) = \int_0^\tau (1 - Z(\tau'))\,\mathrm{d}\tau', \tag{9.121}$$

$$\begin{aligned} f(\tau) &= \int_0^\tau (1 - Z(\tau'))\,\frac{1}{\beta(\tau')} x_c(\tau')\,\mathrm{d}\tau' \\ &= \int_0^\tau e^{-\varphi(\tau')}\,(1 - Z(\tau')) \ln \left(\frac{Z_e}{Z_e - Z(\tau')}\right)^2 \mathrm{d}\tau'. \end{aligned} \tag{9.122}$$

From these formulas, we derive the relationships

$$\ln \frac{N}{N_{c0}} = \beta(\tau) \left(\ln \frac{N_0}{N_{c0}} - f(\tau)\right), \qquad \ln \frac{N_0}{N_{c0}} = \frac{1}{\beta(\tau)} \ln \frac{N}{N_{c0}} + f(\tau), \tag{9.123}$$

$$N_{\max}(\tau) = N_{c0} \exp\left\{\beta(\tau)\left[\ln\frac{N_{\max}(\tau_N)}{N_{c0}} - f(\tau)\right]\right\}, \tag{9.124}$$

$$\frac{\partial N_0}{\partial N} = \frac{N_{c0}}{N}\frac{1}{\beta(\tau)}\exp\left[\frac{1}{\beta(\tau)}\ln\frac{N_0}{N_{c0}} + f(\tau)\right] = \frac{N_0}{\beta(\tau)N}. \tag{9.125}$$

Using Eq. (9.125), let us write Eq. (9.117) in the form

$$\psi(N,\tau) = \frac{2t_0 J(N_{c0})}{\beta(\tau)N\ln(N_0/N_{c0})}\frac{1}{2}\left[1 - \operatorname{erf}\frac{N_0 - N_{\max}(\tau_N)}{\Delta N_c}\right]\theta(N_0 - N_{\min}(\tau_N)), \tag{9.126}$$

where $N_{\min}(\tau)$ is determined by Eq. (9.123) at $N_0 = eN_{c0}$ and represents the lower bound for the bubbles that give the main contribution to the conservation law for the number of gas atoms (smallest bubbles have been dissolved). The number of bubbles per atom of the liquid N_b remains virtually invariable during the time of the intermediate stage t_f. Indeed, it equals

$$N_b = \int\limits_{N_{\min}(\tau)}^{N_{\max}(\tau)} \frac{2t_0 J(N_{c0})}{N_0\ln(N_0/N_{c0})}\frac{\partial N_0}{\partial N}\,\mathrm{d}N = \int\limits_{eN_{c0}}^{N_{\max}(\tau_N)} \frac{2t_0 J(N_{c0})}{N_0\ln(N_0/N_{c0})}\,\mathrm{d}N_0$$

$$\approx 2t_0 J(N_{c0})\ln\ln\frac{N_{\max}(\tau_N)}{eN_{c0}} = 2t_0 J(N_{c0})\tau_N, \tag{9.127}$$

which coincides with the maximum number of bubbles (Eq. (9.106)) formed at the stage of nucleation.

Note that when the initial lower bound of the spectrum eN_{c0} reaches the lowest value $N_{\min}(\tau) \approx 1$ during the dissolution of small bubbles, the size spectrum will be formed from the initial distribution of bubbles with sizes from eN_{c0} to $N_{\max}(\tau_N)$. The moment τ^* of reaching $N \approx N_{\min}(\tau)$ is determined by Eq. (9.123).

Using the law of conservation for the number of gas atoms, we obtain an equation for the relative number of gas atoms $Z(\tau)$

$$Z(\tau) \equiv 1 - \frac{n^L(\tau)}{n^L(0)} = \frac{1}{n^L(0)}\int\limits_{N_{\min}(\tau)}^{N_{\max}(\tau)} N\psi(N,\tau)\,\mathrm{d}N \tag{9.128}$$

$$= \frac{2t_0 J(N_{c0})}{n^L(0)\beta(\tau)}\int\limits_{N_{\min}(\tau)}^{N_{\max}(\tau)} \frac{\mathrm{d}N}{\ln\left(\frac{N_0(N,\tau)}{N_{c0}}\right)} \approx \frac{2t_0 J(N_{c0})N_{\max}(\tau)}{n^L(0)\beta(\tau)\ln\left(\frac{N_{\max}(\tau)}{N_{c0}}\right)}$$

$$= \frac{2t_0 J(N_{c0})N_{c0}}{n^L(0)\beta(\tau)}\frac{1}{\ln\left(\frac{N_{\max}(\tau)}{N_{c0}}\right)}\frac{N_{\max}(\tau)}{N_{c0}}.$$

When writing Eq. (9.128), we used Eq. (9.124) and replaced the slowly changing logarithm $\ln(N_0(N,\tau)/N_{c0})$ by its value $\ln(N_0(N_{\max}(\tau),\tau)/N_{c0})$ (here we take into account that $N_0(N_{\max}(\tau),\tau)$ is the maximum value at the upper limit of the size spectrum at the nucleation stage, i.e., at the point mainly responsible for the value of the integral) and also considered that $N_{\min}(\tau) \ll N_{\max}(\tau)$. Then Eq. (9.78) yields

$$N_0(N_{\max}(\tau), \tau)/N_{c0} = \exp(\exp(\tau_N)) \tag{9.129}$$

and Eq. (9.128) takes the form

$$Z(\tau) = \frac{1}{N_{c0}} \exp(-e^{\tau_N}) \frac{1}{\beta(\tau)} \exp\left[\beta(\tau)\left(e^{\tau_N} - f(\tau)\right)\right]. \tag{9.130}$$

During the lifetime of the intermediate stage $0 \leq \tau \leq \tau_f$ (remember that the time is taken with reference to the completion of the nucleation stage τ_N), the relative amount of gas in the solution $Z(\tau)$ changes within the limits

$$Z(\tau)|_{\tau=0} = \frac{1}{N_{c0}} \leq Z(\tau) \leq Z(\tau_f) = 1 - \frac{n^*}{n^L(0) - n_e^L} \leq 1 - \frac{n^*}{n^L(0)}. \tag{9.131}$$

The excess amount n^* of gas in the solution at the end of the intermediate stage is determined with an accuracy at which the excess of the substance is regarded as negligible (usually $n^*/n^L(0) \leq 1/e$). The relative decrease in the amount of excess gas in the solution with respect to its initial amount changes within the range $n_e^L/n^L(0) \ll n^*/n^L(0) \leq 1$. When $\tau \to \infty$, n^* tends to the equilibrium amount of gas per unit volume of the liquid n_e^L. The condition (9.131) determines the lifetime of the intermediate regime τ_f.

Using Eqs. (9.118) and (9.121), let us estimate $f(\tau)$ from Eq. (9.122) in the range $0 \leq \tau \leq \tau_f$

$$f(\tau) = \int_0^{\tau} (1 - Z(\tau'))\, e^{-\varphi(\tau')} x_c(\tau')\, d\tau' = \tag{9.132}$$

$$= \int_0^{\tau} (1 - Z(\tau')) \left\{ \exp\left[-\int_0^{\tau'} (1 - Z(\tau''))\, d\tau'' \right] \right\} \ln\left[\frac{1}{(1 - Z(\tau'))^2} \right] d\tau'.$$

Considering that $e^{-\varphi(\tau)} < 1$, we get

$$f(\tau) < \int_0^{\tau} (1 - Z(\tau')) \ln\left[\frac{1}{(1 - Z(\tau'))^2} \right] d\tau' \tag{9.133}$$

$$= -2 \int_0^{\tau} (1 - Z(\tau')) \ln(1 - Z(\tau'))\, d\tau' \leq \frac{2}{e}\tau \leq \frac{2}{e}\tau_f.$$

The expression $(1 - Z(\tau)) \ln(1 - Z(\tau))$ reaches a maximum at $Z(\tau) = 1 - 1/e$; this value lies within the domain of $Z(\tau)$.

Thus, $f(\tau)$ can be neglected at small $\tau_f \ll e^{\tau_N}$. Then, with Eq. (9.120), Eq. (9.130) takes the form

$$Z(\tau) = 1 - \frac{d\varphi}{d\tau} = \frac{1}{N_{c0}} \exp\left[(e^{\varphi} - 1)\, e^{\tau_N} - \varphi \right]. \tag{9.134}$$

For $e^{\tau_N} \gg 1$ the inequality $\varphi \ll 1$ must be fulfilled since $Z(\tau)$ does not exceed unity; therefore, $e^{\varphi} - 1 \approx \varphi$ and Eq. (9.134) can be written as

$$Z(\tau) = \frac{1}{N_{c0}} \exp\left[(e^{\tau_N} - 1)\varphi\right] = 1 - \frac{d\varphi}{d\tau}, \quad \varphi|_{\tau=0} = 0, \quad Z(\tau_f) = 1 - \frac{n^*}{n^L(0)}. \tag{9.135}$$

From these formulas, we find the equation for the determination of φ_f

$$Z(\tau_f) = \frac{1}{N_{c0}} \exp\left[(e^{\tau_N} - 1)\,\varphi_f\right] = 1 - \frac{n^*}{n^L(0)}, \tag{9.136}$$

$$\varphi_f = \varphi|_{\tau_f} = \frac{1}{e^{\tau_N} - 1} \ln\left[N_{c0}\left(1 - \frac{n^*}{n^L(0)}\right)\right], \qquad \frac{n^*}{n^L(0)} \le \frac{1}{e}. \tag{9.137}$$

Solving Eq. (9.135), we determine the behavior of $Z(\tau)$ in the $\tau \le \tau_f$ range as

$$\int_0^{\varphi} \frac{d\varphi'}{1 - (N_{c0})^{-1} \exp\left[(e^{\tau_N} - 1)\,\varphi'\right]} = \tau. \tag{9.138}$$

With the notation

$$\zeta = (N_{c0})^{-1} \exp\left[(e^{\tau_N} - 1)\,\varphi'\right], \tag{9.139}$$

we obtain

$$\int_{1/N_{c0}}^{\zeta_{\max}} \frac{d\zeta}{\zeta\,(1 - \zeta)} = \ln \frac{\zeta}{1 - \zeta}\bigg|_{1/N_{c0}}^{x_{\max}} = (e^{\tau_N} - 1)\,\tau, \tag{9.140}$$

where

$$\zeta_{\max} = (N_{c0})^{-1} \exp\left[(e^{\tau_N} - 1)\,\varphi\right]. \tag{9.141}$$

From this expression, using Eq. (9.135), we obtain

$$\frac{Z(\tau)}{1 - Z(\tau)} = \frac{(N_{c0})^{-1} \exp\left[(e^{\tau_N} - 1)\,\varphi\right]}{1 - (N_{c0})^{-1} \exp\left[(e^{\tau_N} - 1)\,\varphi\right]} = \frac{(N_{c0})^{-1} \exp\left[(e^{\tau_N} - 1)\,\tau\right]}{1 - (N_{c0})^{-1}} \tag{9.142}$$

and

$$Z(\tau) = \frac{\exp\left[(e^{\tau_N} - 1)\,\tau\right]}{N_{c0} + \exp\left[(e^{\tau_N} - 1)\,\tau\right] - 1}. \tag{9.143}$$

Taking into account Eqs. (9.120)–(9.122), one can determine from Eq. (9.124) the maximum number of gas atoms in the bubble

$$N_{\max}(\tau) = N_{c0} \exp\left\{\exp \int_0^{\tau} (1 - Z(\tau'))\, d\tau' \left(\ln \frac{N_{\max}(\tau_N)}{N_{c0}} - f(\tau)\right)\right\} \tag{9.144}$$

and from Eq. (9.28) the radius and number of gas atoms in the critical bubble

$$\frac{N_c}{N_{c0}} = \left(\frac{R_c}{R_{c0}}\right)^2 \approx \left(\frac{n^L(0) - n_e^L}{n^L(\tau) - n_e^L}\right)^2 = \frac{Z_e^2}{(Z_e - Z(\tau))^2}, \tag{9.145}$$

where

$$f(\tau) = \int_0^\tau \exp\left[-\int_0^{\tau'} \left(1 - Z(\tau'')\right) d\tau''\right] (1 - Z(\tau')) \ln\left[\frac{Z_e}{Z_e - Z(\tau')}\right]^2 d\tau'. \tag{9.146}$$

Now let us determine the lifetime of the intermediate regime τ_f using Eqs. (9.143) and (9.136)

$$\exp\left[(e^{\tau_N} - 1)\,\tau_f\right] = \frac{(N_{c0} - 1)\,Z(\tau_f)}{1 - Z(\tau_f)}, \tag{9.147}$$

$$\begin{aligned} \tau_f &= \frac{1}{e^{\tau_N} - 1} \ln\left[(N_{c0} - 1)\left(\frac{n^L(0)}{n^*} - 1\right)\right] \\ &= \frac{1}{e^{\tau_N} - 1}\left[\ln(N_{c0} - 1) + \ln\left(\frac{n^L(0)}{n^*} - 1\right)\right]. \end{aligned} \tag{9.148}$$

From Eqs. (9.120) and (9.135), let us derive the time dependence of the $\beta(\tau)$ coefficient in the expression for the distribution function at the intermediate stage

$$\begin{aligned} \beta(\tau) &= \exp\varphi(\tau) = \left[N_{c0} Z(\tau)\right]^{1/(e^{\tau_N} - 1)} \\ &= e^\tau \exp\left\{-\frac{1}{e^{\tau_N} - 1} \ln\left[1 + \frac{1}{N_{c0}} \exp\left[\tau\left(e^{\tau_N} - 1\right)\right]\right]\right\} \\ &= e^\tau \left[1 - \frac{1}{e^{\tau_N} - 1} \ln\left[1 + \frac{1}{N_{c0}} \exp\left[\tau\left(e^{\tau_N} - 1\right)\right]\right]\right] \approx e^\tau. \end{aligned} \tag{9.149}$$

We derived this relationship employing the inequality $e^{\tau_N} - 1 \gg 1$.

Equation (9.149) sets the lower bound of the lifetime of the intermediate stage. Let us determine the upper bound of this range by equating $\beta(\tau) f(\tau)$ to its maximum value. This is possible, because the $Z(\tau)$ function in Eq. (9.130) monotonically increases within this range. Indeed, using Eqs. (9.120) and (9.149), we obtain

$$\begin{aligned} \beta(\tau) f(\tau) &= \int_0^\tau e^{\varphi(\tau) - \varphi(\tau')} (1 - Z(\tau')) \ln\left(\frac{Z_e}{Z_e - Z(\tau')}\right)^2 d\tau' \\ &< -\frac{2}{e}\int_0^\tau e^{\tau - \tau'}\, d\tau' < \frac{2}{e}\left(e^\tau - 1\right) \\ &\approx \frac{2}{e}\left(e^{\varphi(\tau)} - 1\right) \approx \frac{2}{e}\left(\beta(\tau) - 1\right) \approx \frac{2}{e}\varphi(\tau). \end{aligned} \tag{9.150}$$

In this formula, we replaced $e^{\varphi(\tau)-\varphi(\tau')}$ by a larger quantity $e^{\tau-\tau'}$ and the product

$$\left[(1-Z(\tau'))\ln\left(\frac{Z_e}{Z_e-Z(\tau')}\right)\right]$$

by its maximum value $\frac{1}{e}$. Substituting Eq. (9.150) into Eq. (9.130), we obtain

$$Z(\tau) \approx \frac{1}{N_{c0}}\exp\left[\left(e^{\tau_N}-1-\frac{2}{e}\right)\varphi(\tau)\right]. \tag{9.151}$$

Accordingly, we should replace the term $e^{\tau_N}-1$ in all previous formulas by $e^{\tau_N}-1-2/e$. This means that the upper and lower bounds of the lifetime range of the intermediate stage for $e^{\tau_N} \gg 1$ differ by a small value of the order of $2/\left[e\left(e^{\tau_N}-1\right)\right] \ll 1$.

Using Eqs. (9.133), (9.143), and (9.149), and taking into account Eqs. (9.58) and (9.33), we obtain from Eq. (9.126) the distribution function at the intermediate stage

$$\psi(N,t) = n^L(0)\sqrt{\frac{N_{c0}}{\pi}}\exp\left(-\frac{N_{c0}}{2}\right) \tag{9.152}$$

$$\times\frac{1}{N\left[\ln(N/N_{c0})+f(t)\beta(t)\right]}\frac{1}{2}\left(1-\operatorname{erf}\frac{N-N_{\max}(t)}{\Delta N_c}\right),$$

where

$$f(\tau)\beta(\tau) = 2e^{\tau}\int_0^{\tau}\left\{1+\frac{\exp\left[(e^{\tau_N}-1)\tau'\right]-1}{N_{c0}}\right\}^{-1} \tag{9.153}$$

$$\times\ln\left\{1+\frac{\exp\left[(e^{\tau_N}-1)\tau'\right]-1}{N_{c0}}\right\}\mathrm{d}\tau' \approx 2\tau e^{\tau-1}$$

and τ is defined by Eq. (9.37). Thus, Eqs. (9.127), (9.148), and (9.152) fully describe the kinetics of the intermediate stage ($t_N < t < t_f$).

9.4 The Late Stage

After the termination of the intermediate stage, when the density of the gas dissolved in the liquid or melt becomes close to the equilibrium value $n^L\big|_{t=t_f} \approx n_e^L$, the late stage begins ($t > t_f$). At this stage, the evolution of the system undergoes a fundamental reorganization. The conditions fulfilled at the late stage are

$$\frac{2\sigma}{p^L R_c(t)} \ll 1, \tag{9.154}$$

and the approximation of a low-viscosity liquid (Eq. (9.12)) ($R_c(t) \to \infty$ at $t \to \infty$) turns out to be valid for any viscosity value η. With Eqs. (9.154), we then obtain

$$\frac{l\sigma}{\delta D\eta} \gg \frac{2\sigma}{p^L R_c}. \tag{9.155}$$

The viscosity, according to Eq. (9.155), determines only $R_c(t)$ and, along with it, the time at which the system reaches the late stage.

The kinetics of the separation of gas-supersaturated liquids at initial and intermediate stages significantly differs for viscous and low-viscosity fluids (see [310]). Thus, for the main spectrum of bubbles at the late stage, when conditions (9.154) are fulfilled, Eq. (9.21) yields

$$-\frac{1}{k_B T}\frac{\delta \Delta F}{\delta N} = \ln\frac{p^L + 2\sigma/R_c}{p^L + 2\sigma/R} = \frac{2\sigma}{p^L}\left(\frac{1}{R_c} - \frac{1}{R}\right) \tag{9.156}$$

and

$$\frac{\mathrm{d}N}{\mathrm{d}t} = \alpha\frac{D}{2l}4\pi R^2 \widetilde{n}^L\left(-\frac{1}{k_B T}\frac{\delta \Delta F}{\delta N}\right) = \alpha\frac{D}{2l}4\pi R^2 \widetilde{n}^L \frac{2\sigma}{p^L}\left(\frac{1}{R_c} - \frac{1}{R}\right). \tag{9.157}$$

Here $\widetilde{n}^L$ is the concentration of the gas near the bubble, a value that does not coincide with the volume-average concentration in the general case (see below), and α is determined in Eq. (9.21).

Taking into account that $2\sigma/R \ll p^L$ holds for the main bubble spectrum at the late stage, we obtain

$$N k_B T = \frac{4\pi}{3}R^3\left(p^L + \frac{2\sigma}{R}\right) \approx \frac{4\pi}{3}R^3 p^L, \qquad p^L = p^L(n_e), \tag{9.158}$$

$$\frac{\mathrm{d}N}{\mathrm{d}t} = \frac{p^L}{k_B T}4\pi R^2\frac{\mathrm{d}R}{\mathrm{d}t} \tag{9.159}$$

and

$$\frac{\mathrm{d}R}{\mathrm{d}t} = \alpha\frac{D}{2l}\widetilde{n}^L\frac{2\sigma T}{(p^L)^2}\left(\frac{1}{R_c} - \frac{1}{R}\right). \tag{9.160}$$

In the equations for the distribution function and the initial condition at the late stage, it is convenient to go over from the variable N (number of gas atoms in a bubble) to the adjustable variable R (bubble radius) (note that, according to the terminology of Eq. (9.10), the variables R and N are referred to as thermodynamically stable and unstable, respectively). For this purpose, let us integrate the complete distribution function (9.17) with respect to N resulting in

$$\varphi(V,t) = \int_0^\infty \psi(N,t)\,\delta(V - V(N))\,\mathrm{d}N = \psi(N,t)\left.\frac{\mathrm{d}N}{\mathrm{d}V}\right|_{N=N(V)}. \tag{9.161}$$

The continuity equation

$$\frac{\partial\psi}{\partial t} = -\frac{\partial}{\partial N}\left[D(N)\left(\ln\frac{p^L + 2\sigma/R_c}{p^L + 2\sigma/R}\right)\psi\right] = -\frac{\partial}{\partial N}\frac{\mathrm{d}N}{\mathrm{d}t}\psi \tag{9.162}$$

is transformed into

$$\frac{\partial\varphi}{\partial t} = -\frac{\partial}{\partial R}\left[\frac{\mathrm{d}N}{\mathrm{d}t}\frac{\mathrm{d}R}{\mathrm{d}N}\right]\varphi = -\frac{\partial}{\partial R}\left[\frac{\mathrm{d}R}{\mathrm{d}t}\varphi\right], \tag{9.163}$$

$$\varphi|_{t=t_f} = \psi\left(N\left(R\right)\right) \left.\frac{\mathrm{d}N}{\mathrm{d}R}\right|_{N=N(R)}, \tag{9.164}$$

where $\mathrm{d}N/\mathrm{d}t$ is defined by Eq. (9.159). Deriving this relationship, we employed the fact that

$$\frac{\partial}{\partial N} = \left(\frac{\mathrm{d}R}{\mathrm{d}N}\right)\left(\frac{\partial}{\partial R}\right), \qquad \psi(N,t)\,\mathrm{d}N = \varphi(R,t)\,\mathrm{d}R \tag{9.165}$$

holds (see Eq. (9.161)). Accordingly, the law of conservation (Eq. (9.109)) for the amount of gas in the system takes employing Eq. (9.161) the following form:

$$\begin{aligned} Z(t) &= \frac{1}{n^L(0)} \int_0^\infty \varphi N \frac{\mathrm{d}R}{\mathrm{d}N}\,\mathrm{d}N = \frac{1}{n^L(0)} \int_0^\infty \varphi\left(R\right) N\left(R\right)\,\mathrm{d}R \\ &= \frac{1}{n^L(0)} \frac{4\pi}{3} \frac{p^L(n_e^L)}{k_B T} \int_0^\infty \varphi\left(R,t\right) R^3\,\mathrm{d}R. \end{aligned} \tag{9.166}$$

By virtue of formula (9.158), we arrive then at

$$\frac{\mathrm{d}N}{\mathrm{d}t} = 4\pi R^2 \frac{p^L}{k_B T}\frac{\mathrm{d}R}{\mathrm{d}t} = -4\pi R^2 j_R, \tag{9.167}$$

where j_R is the gas atom flux density at the bubble surface. The normal to the bubble surface is directed outward; therefore, j_R is negative for a flux that is directed from the bulk of the liquid to the bubble.

In limiting cases, depending on the system parameters, the flux j_R can be determined both by the boundary kinetics and by the supply of gas atoms from the ambient medium. To obtain the flux j_R in the general case, let us use the results of Refs. [300] and [329], which show that the entire region of the medium outside the bubbles at $R \gg l$ can be divided into two parts: the near-surface part ($R \leq r \leq R + l$, where the boundary kinetics is important) with a certain density of gas atoms $\widetilde{n}$ and the remaining part ($R + l \leq r \leq \infty$). The parameter $\widetilde{n}$ is determined from the continuity of the flux in the $R + l$ point. At $R \gg l$, one can write

$$j|_{r=R+l-\varepsilon} = j|_{r=R+l+\varepsilon} \approx j|_{r=R-\varepsilon} = j|_{r=R+\varepsilon}, \qquad \varepsilon \to 0. \tag{9.168}$$

Considering the solution of gas atoms as weak, one can write the following expression for the near-surface region:

$$-j|_{r=R+l} \approx -j|_{r=R} = \alpha\frac{D}{2l} 4\pi R^2 \left(\widetilde{n}^L - n_R^L\right), \tag{9.169}$$

where n_R^L is the equilibrium density of gas atoms near the surface of a bubble of radius R. According to Eq. (9.10), we have further

$$n_R^L = \frac{\delta}{k_B T}\left(p^L + \frac{2\sigma}{R}\right), \tag{9.170}$$

and, for a critical-size bubble we have

$$n^L_{R_c} = \frac{\delta}{k_B T}\left(p^L + \frac{2\sigma}{R_c}\right) = \bar{n}^L(t), \tag{9.171}$$

where $\bar{n}^L(t)$ is the average density of gas atoms in the system.

The pressure in the liquid p^L determines the equilibrium value of the gas density n^L_e, and, as follows from Eq. (9.4), $p^L = n^L_e k_B T/\delta$ holds. Substituting this value into Eq. (9.171), we obtain the critical size R_c as

$$R_c = \frac{\delta}{k_B T}\frac{2\sigma}{\bar{n}^L(t) - n^L_e} = \frac{2\sigma}{p^L}\frac{n^L_e}{\bar{n}^L(t) - n^L_e}. \tag{9.172}$$

Apparently, at $t \to \infty$, the critical size $R_c \to \infty$ and the gas density $\bar{n}^L(t)$ tends to the equilibrium value.

Equations (9.172) and (9.154) imply the condition that determines the late stage of evolution

$$\frac{\bar{n}^L(t) - n^L_e}{n^L_e} \ll 1. \tag{9.173}$$

Since the supply of gas atoms from the medium is fairly slow, there is sufficient time for the establishment of a quasisteady-state gas density distribution near the bubble surface (see Chapter 4 and [153, 300, 311]). To find it, let us use the quasisteady-state diffusion equation

$$\frac{\partial n}{\partial t} = D\Delta n = 0, \qquad n|_{r\to R} = \tilde{n}^L, \tag{9.174}$$

$$n|_{r\to\infty} = \bar{n}^L = \frac{\delta}{k_B T}\left(p^L + \frac{2\sigma}{R_c}\right). \tag{9.175}$$

On this basis, we obtain

$$-j|_{r\to R} = 4\pi R^2 \frac{D}{R}\left(\bar{n}^L - \tilde{n}^L\right). \tag{9.176}$$

From the condition of continuity of fluxes (9.169) and (9.174) in the point $r = R$, we determine the gas density near the bubble boundary $\tilde{n}^L$ as

$$\tilde{n}^L = \left(\bar{n}^L - \frac{\alpha R}{2l} n^L_e\right)\left(1 + \frac{\alpha R}{2l}\right)^{-1}. \tag{9.177}$$

For the flux of gas atoms onto the bubble surface, we get

$$-j|_{r\to R} = 4\pi R^2 \frac{D}{R}\left(\frac{\alpha R/2l}{1 + \alpha R/2l}\right)\left(\bar{n}^L - n^L_R\right). \tag{9.178}$$

Apparently, Eq. (9.178) describes both limiting situations: at $\alpha R/2l \ll 1$, we get the "boundary" growth kinetics (the free-molecular regime Eq. (9.10)); at $\alpha R/2l \gg 1$, we have the growth kinetics determined by the diffusion supply of gas atoms to the bubble surface. If α is not very small and $R \gg l$, the case of $\alpha R/2l \gg 1$ is implemented at the late stage. Then, substituting $\bar{n}^L$ and $n^L_{R_c}$ from Eqs. (9.170) and (9.171) into Eq. (9.178) and using Eq. (9.167), we obtain

$$\frac{\mathrm{d}R}{\mathrm{d}t} = D\delta \frac{2\sigma}{p^L}\frac{1}{R}\left(\frac{1}{R_c} - \frac{1}{R}\right). \tag{9.179}$$

Note that the set of equations (9.179), (9.163), and (9.166) fully coincides with the complete set of equations derived in Chapter 4 but one parameter, $\alpha = 2\sigma\omega c_\infty/k_B T$ (where c_∞ is the equilibrium concentration of the admixture near the boundary and ω is the volume per admixture atom in the solution) with the dimension of length is substituted by the parameter $\xi \equiv 2\sigma\delta/p^L$ having the same dimension. Thus, both the method and the results of Chapter 4 are applicable to the solution of this set of equations.

As is shown in the mentioned chapter, this set of equations at $\bar{n}^L(t) - n^L_e \to 0$ starts to "forget" about its initial conditions in time and acquires an increasingly universal nature, asymptotically tending to a self-similar form, which is independent of the initial conditions. The only parameter dependent on the initial distribution function is the time moment at which the bubble distribution becomes fairly well describable by the universal distribution function. In fact, the actual pattern of the time-asymptotic distribution function depends on the mass-transfer mechanism. The process of the transformation of the distribution function from an arbitrary to the asymptotically universal one is thoroughly considered in Chapter 4 for the case where the initial distribution function has an infinitely long tail at $R \to \infty$. In our case, after the intermediate stage, there is a finite tail at $R > R_c$ different from the fluctuational one, and the process of the transformation to the asymptotically universal function occurs sufficiently rapidly.

Let us introduce, just as in Section 4.1.2, the reduced variable $u = R/R_c$ and rewrite Eq. (9.179) in the form

$$\frac{\mathrm{d}u}{\mathrm{d}t} = D\frac{2\sigma}{p^L}\frac{\delta}{R_c^3}\frac{u-1}{u^2} - \frac{u}{R_c}\frac{\mathrm{d}R_c}{\mathrm{d}t}, \tag{9.180}$$

where R_c is determined by Eq. (9.172). Since $\mathrm{d}R_c/\mathrm{d}t > 0$, it is convenient to introduce the new time variable $\tau = \ln\left[R_c^3(t)/R_{c0}^3\right]$ instead of time, t. Then we can rewrite Eq. (9.180) in the canonical form (see Eq. (4.24)) as

$$\frac{\mathrm{d}u^3}{\mathrm{d}\tau} = \gamma\,(u-1) - u^3, \tag{9.181}$$

where $\gamma = 3D\xi\left(\mathrm{d}R_c^3/\mathrm{d}t\right)^{-1}$ is a dimensionless parameter. With $\varphi\,(u,\tau)\,\mathrm{d}u = \varphi\,(R,\tau)\,\mathrm{d}R$, the complete set of equations (9.163) and (9.166) gets the form

$$\frac{\partial\varphi\,(u,\tau)}{\partial\tau} + \frac{\partial}{\partial u}\left[\frac{\partial u}{\partial\tau}\varphi\,(u,\tau)\right] = 0, \tag{9.182}$$

$$\varphi\,(u,\tau)\big|_{\tau=0} = \varphi_0\,(u) = \varphi_0\left(\frac{R}{R_{c0}}\right), \tag{9.183}$$

$$Z(\tau) = \frac{1}{n^L(0)} \frac{4\pi}{3} \frac{p^L(n_e^L) R_{c0}^3}{k_B T} e^{\tau} \int_0^{\infty} \varphi(u, \tau)\, u^3\, du. \qquad (9.184)$$

At the late stage, $n^L(\tau)/n^L(0) \ll 1$; therefore, let us write Eq. (9.184) in a simpler form

$$1 = \kappa e^{\tau} \int_0^{\infty} \varphi(u, \tau)\, u^3\, du, \qquad \kappa = \frac{1}{n^L(0)} \frac{4\pi}{3} \frac{p^L(n_e^L) R_{c0}^3}{k_B T}. \qquad (9.185)$$

As shown in Chapter 4, the resulting set of equations (9.181)–(9.185) has a stable solution at $\tau \to \infty$ if $\gamma|_{\tau\to\infty} \to \gamma_0 = \text{const}$. This condition is fulfilled if $du/d\tau < 0$ for all sizes except for $u = u_0$, for which, in the zeroth approximation,

$$\left.\frac{du}{d\tau}\right|_{u=u_0,\,\gamma=\gamma_0} = 0, \qquad \left.\frac{\partial}{\partial u}\frac{du}{d\tau}\right|_{u=u_0,\,\gamma=\gamma_0} = 0. \qquad (9.186)$$

On this basis, we find

$$u_0 = \frac{3}{2}, \qquad \gamma = \gamma_0 = \frac{27}{4}. \qquad (9.187)$$

In fact, these values are determined by the form of Eq. (9.180), i.e., by the mechanism of mass transfer or even by several simultaneously acting mechanisms. At the same time, $\varphi(u, \tau)$ in our approximation $\tau \to \infty$ tends to the universal function, virtually forgetting about the initial distribution and becoming nullified at $u \geq u_0$ together with all its derivatives with respect to u. This is what determines the only stable (with respect to fluctuations) solution (see Section 4.1.2). Thus, in this approximation and with $\gamma = 27/4$, Eq. (9.181) yields

$$\frac{du}{d\tau} = -g(u) = -\frac{1}{3u^2} 2\left(u - \frac{3}{2}\right)(u+3)\,, \qquad (9.188)$$

and the solution of Eq. (9.183) can be searched out in the form

$$\varphi(u, \tau) = \begin{cases} \chi(\tau + \phi)\dfrac{1}{g(u)}, & u < u_0 = \dfrac{3}{2}, \\ 0, & u \geq u_0 = \dfrac{3}{2}, \end{cases} \qquad (9.189)$$

where $\chi(\tau + \phi)$ is a function to be determined from the conservation law, Eq. (9.184), and

$$\phi(u) = \int_0^u \frac{du}{g(u)} = \frac{4}{3}\ln(u+3) + \frac{5}{3}\ln\left(\frac{3}{2} - u\right) + \frac{1}{1 - 2u/3} - \ln\frac{3^3 e}{2^{5/3}}. \qquad (9.190)$$

Substituting Eq. (9.189) into Eq. (9.184), we find that

$$\chi(\tau + \phi) = A \exp\left[-(\tau + \phi)\right] \qquad (9.191)$$

and obtain

$$\varphi(u,\tau)=\begin{cases} Ae^{-(\tau+\phi(u))}\dfrac{1}{g(u)}, & u<\dfrac{3}{2},\\ 0, & u\geq\dfrac{3}{2},\end{cases} \tag{9.192}$$

where

$$A=\left(\kappa\int_0^{3/2} e^{-\varphi(u)}\frac{u^3}{g(u)}\,\mathrm{d}u\right)^{-1}. \tag{9.193}$$

The condition $\gamma|_{\tau\to\infty}\to\gamma_0$ determines $R_c^3(t)$ from Eq. (9.181). We get

$$R_c^3(t)=\frac{4}{9}\delta D\frac{2\sigma}{p^L}t+R_{c0}^3. \tag{9.194}$$

With this relation, Eq. (9.172) yields

$$\frac{\bar{n}^L(t)-n_e^L}{n_e^L}=\delta\frac{2\sigma}{p^L}\left(\frac{4}{9}\delta D\frac{2\sigma}{p^L}t+R_{c0}^3\right)^{-1/3}. \tag{9.195}$$

The time t is referred here to the end t_f of the intermediate stage.

The number of bubbles per unit volume is diminished with time according to the law

$$N_b(t)=Ae^{-\tau}\int_0^{3/2} e^{-\phi}\frac{\mathrm{d}u}{g(u)}=Ae^{-\tau}\int_0^{3/2} e^{-\phi}d\phi=Ae^{-\tau}=A\left(\frac{R_{c0}}{R_c(t)}\right)^3. \tag{9.196}$$

The bubble-size distribution function for the variable u has the form

$$\varphi(u,\tau)=N_b(t)P(u), \tag{9.197}$$

where

$$P(u)=\begin{cases}\dfrac{3^4 e}{2^{5/3}}\dfrac{u^2\exp\left(-\dfrac{1}{1-2u/3}\right)}{(u+3)^{7/3}(3/2-u)^{11/3}}, & u<\dfrac{3}{2},\\ 0, & u\geq\dfrac{3}{2}.\end{cases} \tag{9.198}$$

Using Eq. (9.181) at $\gamma=27/4$, we obtain

$$\int_0^{3/2} e^{-\phi}(u-1)\frac{\mathrm{d}u}{g(u)}=\int_0^{3/2} e^{-\phi}\left[u(\phi)-1\right]\mathrm{d}\phi$$
$$=\int_0^{3/2}\left(u^3-\frac{du^3}{\mathrm{d}\phi}\right)e^{-\phi}\frac{2\gamma}{u}\,\mathrm{d}\phi=0. \tag{9.199}$$

Hence, $\overline{u}=1,\quad R_c(t)=\overline{R}(t)$; i.e., the average bubble size at $\tau\to\infty$ approaches $R_c(t)$.

To achieve a higher accuracy, one should apply a dependence $\gamma(\tau) = \gamma_0 \left(1 - \varepsilon^2(\tau)\right)$, which was first used in [153–155]. As is shown in those papers and in Section 4.1.2, $\varepsilon^2(\tau) > 0$ and $\tau \to \infty$ at $\varepsilon^2(\tau) \to 0$; also, the behavior of $\varepsilon^2(\tau)$ depends on the form of the tail of $\varphi_0(u)$ at $u \gg u_0$. The distribution function $\varphi(u, \tau)$ at $u > u_0$ is determined by the initial distribution function. At $du/d\tau < 0$, bubbles with $u > u_0$ move in size space in such a way that, passing over the region of the blocking point $\varepsilon^2(\tau)$, they consume the whole gas from the dissolving bubbles, increasing their own dimensions and reducing the total number of bubbles. At the same time, naturally, the total amount of gas in the system (dissolved in the solution and evolving into the bubbles) remains constant. A nonzero $\varepsilon^2(\tau)$ is necessary for bubbles with $u > u_0$ to pass into the $u \leq u_0$ region without sticking in the u_0 point, which would lead to a violation of the conservation law (Eq. (9.184)).

Note that if we consider fluctuations in the nucleation of bubbles, when bubbles formed in the direct vicinity of each other merge, then $\varepsilon^2(\tau) \to \varepsilon^2(\infty) = \text{const}$. These collisions determine the $\varepsilon^2(\infty)$ value and the distribution of bubbles beyond the blocking point [294] (this conclusion has been experimentally confirmed for solid solutions [59]). The $\varepsilon^2(\tau) \to \varepsilon^2(\infty)$ value, which is determined by the decreasing tail of the initial distribution function near the blocking point and beyond it, decreases with time; therefore, the zero approximation asymptotically becomes exact with time. The estimation of the conditions when merging of the colliding bubbles can be neglected is done in Appendix A.2 at the end of this chapter.

Since it is very difficult to determine the time of beginning of the late stage, t_{late}, and, accordingly, the critical size of bubbles in the system after the intermediate stage, one should require good joining of R_c and $R_{\max}$ corresponding to the end of the intermediate and beginning of late stages. Taking into account that in the late stage $R_{\max} = (3/2)R_c$ (see Eq. (9.187)), we obtain the equation for the definition of t_{late} in the form

$$R_{\max}(t_{\text{late}}) = \frac{3}{2} R_c(t_{\text{late}}), \tag{9.200}$$

where $R_{\max}(t)$ and $R_c(t)$ are defined by Eqs. (9.144) and (9.145), respectively. Thus, we prolong a little the intermediate stage until the time t_{late}, when Eq. (9.200) will be fulfilled.

Let us note that such definition does not allow one a smooth connection for both the functions, $R_{\max}(t)$ and $R_c(t)$, simultaneously. Therefore we shall require a smooth connection only for the maximum radius $R_{\max}$; thus the critical radius R_c will have a kink at $t = t_{\text{late}}$ (see Figures 9.3 and 9.5). It is a consequence of the fact that the definitions of $R_{\max}(t)$ and $R_c(t)$ (Eqs. (9.144) and (9.145)) are asymptotical ones, and are not valid in the beginning of the late stage. In this case it is necessary to use a more rigorous analysis describing the transformation of the distribution function which was formed at the end of the intermediate stage, Eq. (9.152), into the asymptotical one, Eq. (9.197), as it has been made in Section 4.2; this topic is beyond the given chapter.

Taking into account that the value of the critical radius, defined by Eq. (9.145), very sharply grows at $t > t_{\text{late}}$, for the definition of the critical radius it is possible to use the simple smoothing procedure

$$R_c(t) = \left(\frac{1}{R_c^{\text{int}}(t)} + \frac{1}{R_c^{\text{late}}(t)} \right)^{-1}, \tag{9.201}$$

where $R_c^{\text{int}}(t)$ is defined by Eq. (9.145), and $R_c^{\text{late}}(t)$ by Eq. (9.194).

9.5 Results of Numerical Computations

Figure 9.1 shows the results of calculation of the time lag (Eq. (9.50)), $\tau_{\rm lag}$, the nucleation time (Eq. (9.99)), τ_N, the time of the intermediate stage (Eq. (9.148)), τ_f, and the time of the beginning of the late stage (Eq. (9.200)), $t_{\rm late}$, in dependence on the initial critical bubble size. One can see that the time lag and the nucleation time increase with increasing critical bubble size, and the time of the intermediate stage and time of the beginning of the late stage decrease with increasing critical bubble size.

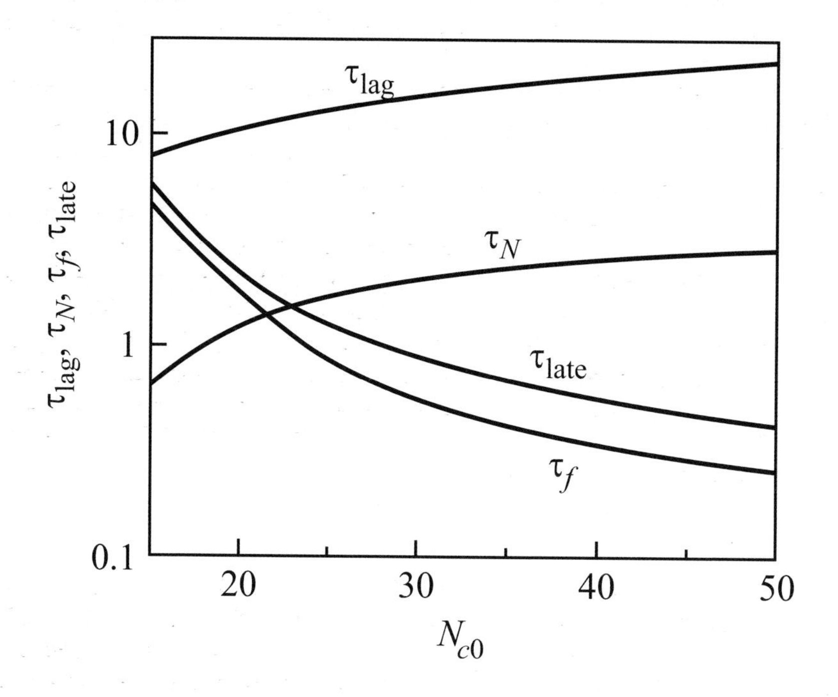

Figure 9.1: Dependences of $\tau_{\rm lag}$, τ_N, τ_f, and $\tau_{\rm late}$ on the initial critical bubble size.

In Figures 9.2 and 9.4 the distribution functions of bubbles are shown for (a) nucleation, (b) intermediate, and (c) late stages for $N_{c0} = 17$ and $N_{c0} = 24$, respectively. In Figures 9.3 and 9.5 the time evolution of critical size $R_c(t)$ (dashed curve (Eqs. (9.145) and (9.194), solid curve (Eq. (9.201)), maximal size $R_{\rm max}(t)$ (Eqs. (9.78), (9.144) and (9.187)), and number of bubbles per unit volume, $N_b(t)$ (Eqs. (9.106) and (9.196)) are shown for $N_{c0} = 17$ and $N_{c0} = 24$, respectively.

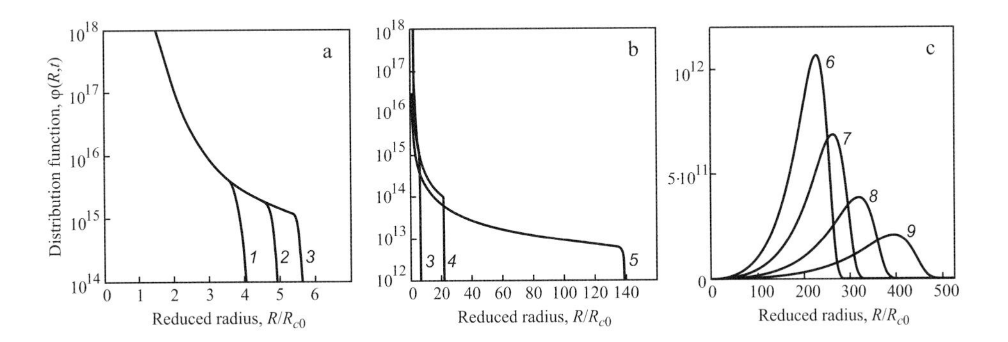

Figure 9.2: Distribution functions of bubbles for $n^L(0) = 3 \times 10^{19}$, $N_{c0} = 17$ at different times: (a) nucleation, (1) $t = 0.2t_N$, (2) $t = 0.75t_N$, (3) $t = t_N$; (b) intermediate stage, (4) $t = t_N + 0.5t_f$, (5) $t = t_N + t_f$; (c) late stage, (6) $t = t_{\text{late}}$, (7) $t = 5t_0$, (8) $t = 6t_0$, (9) $t = 8t_0$.

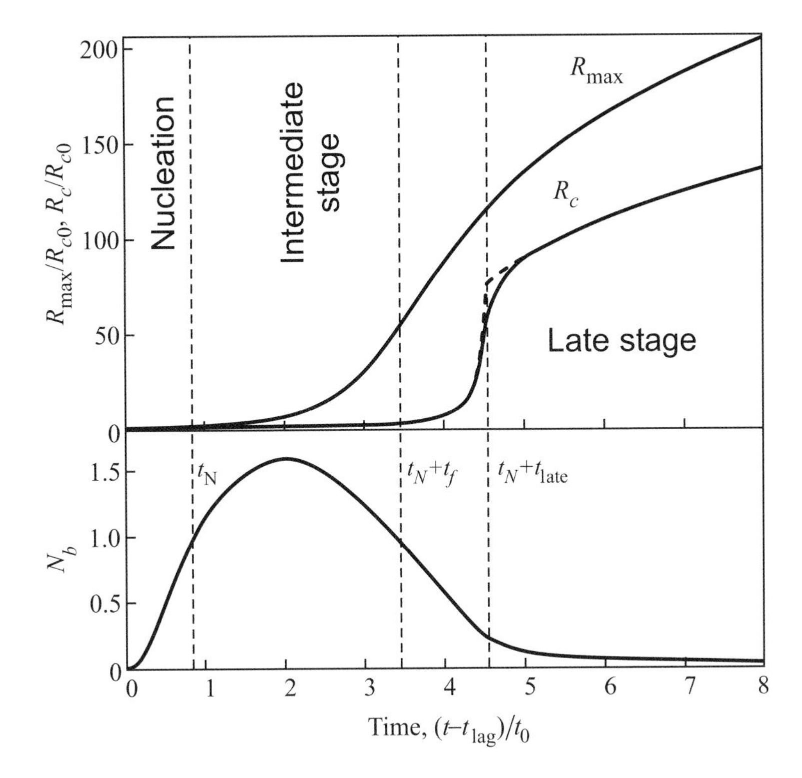

Figure 9.3: Time evolution of $R_{\max}(t)$, $R_c(t)$ (dashed curve (Eqs. (9.145), (9.194)), solid curve (Eq. (9.201)) and $N_b(t)$ for $N_{c0} = 17$.

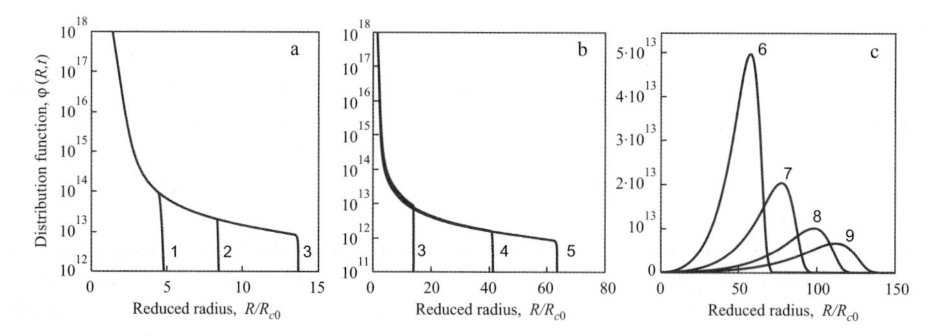

Figure 9.4: Distribution functions of bubbles for $n^L(0) = 3 \times 10^{19}$, $N_{c0} = 24$ at different times: a) nucleation, (1) $t = 0.2t_N$, (2) $t = 0.75t_N$, (3) $t = t_N$; b) intermediate stage, (4) $t = t_N + 0.5t_f$, (5) $t = t_N + t_f$, c) late stage, (6) $t = t_{\text{late}}$, (7) $t = 4t_0$, (8) $t = 6t_0$, (9) $t = 8t_0$.

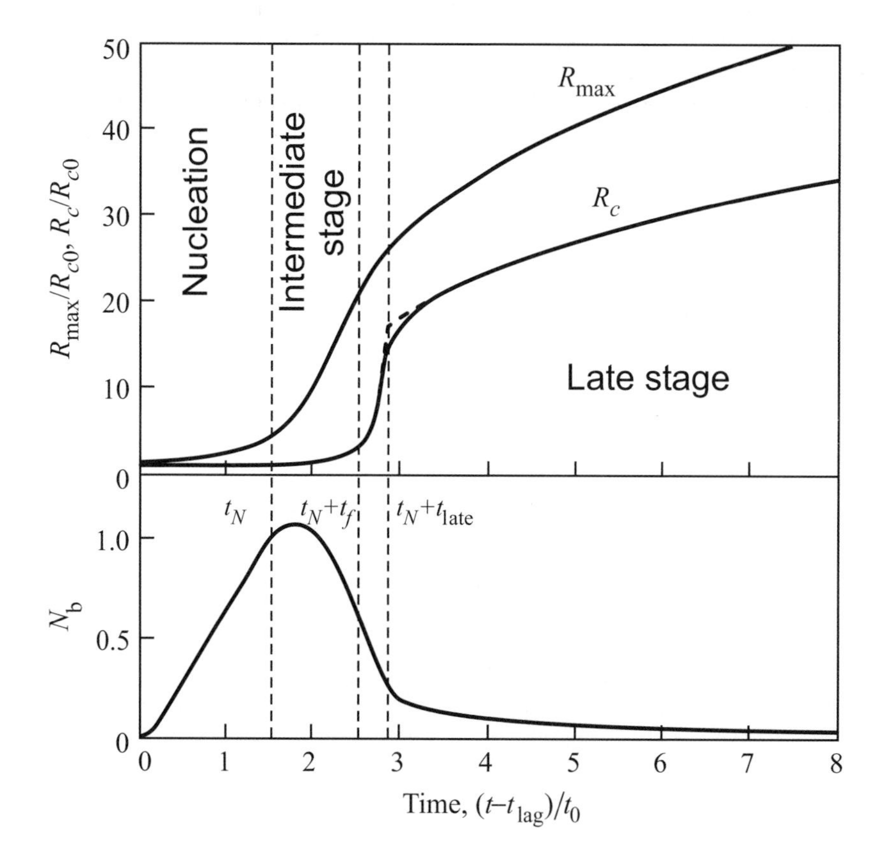

Figure 9.5: Time evolution of $R_{\max}(t)$, $R_c(t)$ (dashed curve (Eqs. (9.145) and (9.194)), solid curve (Eq. (9.201)) and $N_b(t)$ for $N_{c0} = 24$.

9.6 Conclusions

For the case of bubble nucleation in a low-viscosity liquid supersaturated with the gas, we derived expressions for the nucleation time (Eq. (9.99)), bubble-size distribution function (9.135), (9.156), (9.167)), the flux of nuclei in the space of bubble sizes (Eqs. (9.134) and (9.158)), and the maximum number of formed bubbles (Eq. (9.196)) with the assumption of a small volume fraction occupied by bubbles.

The derived size distribution is the initial condition for the next (transient) stage, when still there is a sufficient amount of excess gas, but the number of bubbles does not vary practically remaining at the level reached at the end of the nucleation stage. For such stage the bubble distribution functions (9.126) and (9.136), the flux of nuclei in the size space (Eq. (9.23)), and the maximum number of the formed bubbles (Eq. (9.127)) are obtained. For the late stage, the universal bubble-size distribution function (9.197) is derived; it does not depend on the initial distribution in zero approximation and becomes asymptotically exact in the course of time. The number of bubbles per unit volume (Eq. (9.196)) is also obtained for this stage. Thus, we describe all stages of the process, starting from the initial stage of the nucleation of gas-filled bubbles and ending with coarsening (the late stage).

Our approximation of a small volume fraction occupied by bubbles does not allow us to describe an even later stage, where the volume of bubbles exceeds that of the liquid (foam). This stage requires a separate consideration.

9.A Appendices

9.A.1 Some Mathematical Transformations

To solve the equation

$$\exp(\exp\tau) = Be^{\tau} \tag{9.202}$$

at $B = 1/A + e \geq e$ and $\ln\ln B/\ln B < 1$ (at $B = e$, the solution to this equation is $\tau = 0$), let us use the method of successive approximations

$$e^{\tau} = \ln B + \tau, \quad e^{\tau_0} = 0, \quad e^{\tau_1} = \ln B, \quad e^{\tau_2} = \ln B + \ln\ln B, \tag{9.203}$$

etc., resulting in

$$\begin{aligned} e^{\tau_e} &= \ln B + \ln\left[\ln B\left[1 + \frac{1}{\ln B}\left[\ln\left[\ln B\left[1 + \frac{1}{\ln B}\left[\ln\left[\ln B\left[1 + \cdots\right]\right]\right]\right]\right]\right]\right]\right] \\ &= \ln B + \ln\ln B + \ln\left[1 + \frac{1}{\ln B}\left[\ln\left[\ln B\left[1 + \frac{1}{\ln B}\left[\ln\left[\ln B\left[1 + \cdots\right]\right]\right]\right]\right]\right]\right]. \end{aligned}$$

Expanding this expression for $\ln\ln B/\ln B < 1$, we obtain

$$\begin{aligned} e^{\tau_e} &= \ln B + \ln\ln B + \frac{\ln\ln B}{\ln B} + \frac{\ln\ln B}{(\ln B)^2} + \frac{\ln\ln B}{(\ln B)^3} + \cdots \\ &= \ln B + \ln\ln B \sum_{n=0}^{\infty} \frac{1}{(\ln B)^n} = \ln B + \frac{\ln\ln B}{1 - \dfrac{1}{\ln B}}. \end{aligned} \tag{9.204}$$

Substituting the resulting solution into Eq. (9.202), we have

$$\begin{aligned} \exp e^{\tau_e} &= \exp\left[\ln B + \ln\ln B + \ln\ln B \sum_{n=1}^{\infty} \frac{1}{(\ln B)^n}\right] \\ &= B\ln B \exp\left[\ln\ln B \sum_{n=1}^{\infty} \frac{1}{(\ln B)^n}\right] \approx B\ln B\left[1 + \ln\ln B \sum_{n=1}^{\infty} \frac{1}{(\ln B)^n}\right] \\ &= B\left[\ln B + \ln\ln B \sum_{n=1}^{\infty} \frac{1}{(\ln B)^{n-1}}\right] = B\left[\ln B + \ln\ln B \sum_{n=0}^{\infty} \frac{1}{(\ln B)^n}\right]. \end{aligned} \tag{9.205}$$

9.A.2 Estimation of the Conditions when Merging of Colliding Bubbles can be Neglected

Gas-filled bubbles in a liquid are affected by the buoyancy force; as a result, they float up at a rate depending on their radius. Therefore, collision and merging of bubbles are possible. Let us estimate the conditions when merging can be neglected.

The characteristic time of the diffusion-caused change in the content of gas in a bubble is given by

$$t_{\text{diff}} \approx t_0 = \frac{2}{3\delta\alpha}\frac{lR_c}{D}. \tag{9.206}$$

The change in the content of gas due to merging of bubbles can be approximately described by the equation

$$\left(\frac{\mathrm{d}V}{\mathrm{d}t}\right)_{\text{coll}} = 4\pi R^2 N_b V \Delta v_A. \tag{9.207}$$

Here N_b is the density of bubbles and Δv_A is the spread in the rate of the steady-state motion of bubbles with respect to its average value, which is due to the balance between the Archimedean buoyancy force and the Stokes force of viscous friction

$$\Delta v_A = \frac{2}{9}\frac{\left(R^2 - \overline{R^2}\right) g\rho}{\eta}, \tag{9.208}$$

where g is the gravitational acceleration, ρ is the density of the liquid, and η is its viscosity. Using Eq. (9.207), we can find the characteristic time of the change in the content of gas due to merging of bubbles

$$t_{\text{merg}} = \frac{R}{V N_b \Delta v_A} = \frac{1}{4R^2 N_b \Delta v_A}. \tag{9.209}$$

The effect of merging is not manifested at $\left(\frac{\mathrm{d}V}{\mathrm{d}t}\right)_{\text{coll}} \ll \left(\frac{\mathrm{d}V}{\mathrm{d}t}\right)_{\text{diff}}$, which leads to

$$\frac{1}{4R^2 N_b \Delta v_A} \gg \frac{2}{3\delta\alpha}\frac{lR_c}{D} \tag{9.210}$$

or

$$\frac{3}{2}\alpha\delta\eta D\left[4R_c R^2 N_b l \frac{2}{9} g\rho\left(R^2 - \overline{R^2}\right)\right]^{-1} \gg 1. \tag{9.211}$$

Estimates show that the latter inequality is fulfilled for water if $\overline{R} < 10^{-3}$cm and $N_b > 10^{11}$cm^{-3}, i.e., for the initial and intermediate stages. At the late stage, this condition becomes less strict because of a decrease in N_b and, accordingly, an increase in the interbubble distance, as well as because of a decrease in the bubble size spread, which leads to a decrease in the rate Δv_A of the relative motion of bubbles according to Eq. (9.208).

10 Phase Separation in Solid ^{3}He–^{4}He Mixtures

10.1 Introduction

It has been emphasized frequently in the literature [17, 156, 202] that helium and its isotopic mixtures hold much promise as model systems for studying phase transitions. There was some well-grounded hope that homogeneous nucleation may occur and be studied in detail in dilute liquid ^{3}He–^{4}He mixtures [156]. However, the respective experiments and the comparison of the results with theoretical calculations are encountered by some difficulties. One of them consists in the necessity of realization of the conditions for homogeneous nucleation. Numerous experimental attempts, however, failed to yield unambiguous results; rather, they detected heterogeneous nucleation which may be connected with vortex formation [43, 47].

In the case of solid helium, this problem might be solved provided that high-quality impurity-free samples are available. The quantum character of the diffusion processes in helium ensures fairly high diffusion coefficients, favoring the performance of experiments within reasonable times. There exists an essential difference in the nucleation process in liquid and solid ^{3}He–^{4}He mixtures. In the first case it is impossible to attain a large supersaturation during cooling because of the limited solubility of ^{3}He at $T \to 0$. For solid mixtures there is no such limitation because the equilibrium concentration approaches zero as the temperature tends to zero. In this case, we can achieve a large nucleation rate and the degree of supersaturation can thus be varied over a wide range. This feature makes the realization of homogeneous nucleation much easier.

Considerable attention has been focused recently on phase separation (segregation) in solid ^{3}He–^{4}He mixtures [77, 78, 99, 126, 127, 202, 269, 270]. The phase diagram of solid He isotope mixtures is shown in Figure 10.1. Though it looks symmetrical, small admixtures of ^{3}He to ^{4}He in solid state form hcp lattice unlike bcc lattice for ^{4}He in ^{3}He. Figure 10.1 is taken from [81] where it was drawn on the basis of data of [63] for $P = 33$ bar. The evidence of homogeneous nucleation in ^{3}He–^{4}He solid solutions was obtained for the first time in [202], where experimental results have been successfully compared with the theory presented in [300]. This result permitted an estimation of the most important parameter responsible for nucleation, the surface tension σ at the new-phase cluster boundary, found from the time of pressure relaxation during the phase transition. In [126, 317] the separation of a mixture has been studied by measuring the pressure and nuclear magnetic resonance (NMR) simultaneously. Such measurements can provide additional evidence of homogeneous nucleation in solid ^{3}He–^{4}He mixtures, and permit the estimation of σ, within a single experiment, from two independently measured quantities, the cluster size and the characteristic separation time constant, and thus improve the reliability of this very important parameter.

Kinetics of First-order Phase Transitions. Vitaly V. Slezov

ISBN: 978-3-527-40775-0

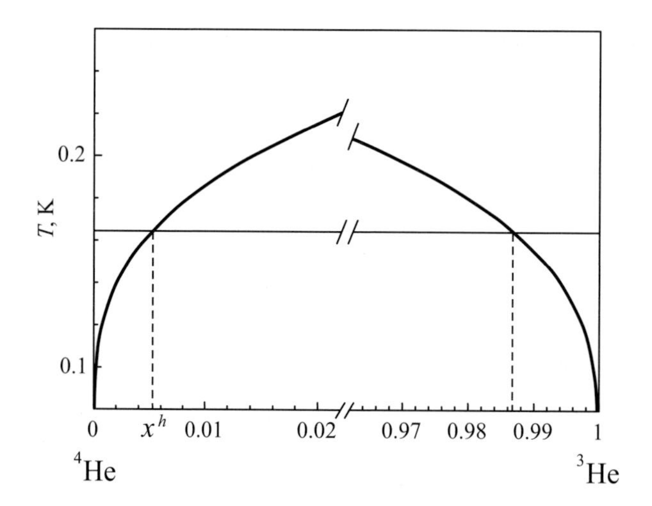

Figure 10.1: Phase diagram of solid He isotope mixtures ^{3}He–^{4}He [81].

Section 10.3 presents NMR measurements in a solid ^{3}He–^{4}He mixture as the temperature was lowered in steps in the course of phase separation [317]. The spin-echo method was used to estimate the diffusion coefficient, size, and cluster concentration in the ^{3}He-enriched phase. The characteristic phase separation time constant of the mixture was found to decrease at lower temperatures. The results convincingly support homogeneous nucleation. From a comparison with theory, the surface tension at the boundary of the phase-separated clusters is found from the cluster concentration, determined by NMR measurements.

Section 10.4 is devoted to study the influence of the degree of supersaturation on the rate of the phase transition in solid solutions of ^{4}He in ^{3}He and a comparison with theoretical results [88]. It describes precision measurements of the pressure during phase separation in solid mixtures of ^{4}He in ^{3}He allowing one to obtain characteristic times of the phase decomposition process. A processing of the measurement results gives additional evidence supporting the view that homogeneous nucleation is realized in ^{3}He–^{4}He solid solutions at significant supercoolings and heterogeneous nucleation at the smallest supercoolings.

The phase separation kinetics at various degrees of supercooling in the solid mixture of ^{4}He in ^{3}He is described in Section 10.5 [202]. The studies were performed in the temperature range of 100–200 mK through precise pressure measurements. The time dependences of the pressure change during phase separation were exponential. At small supercooling the characteristic time constant τ is almost independent of the final temperature T_f and is about 10 h, which is considered to result from heterogeneous nucleation. In a narrow range of T_f-change, τ decreases more than by an order of magnitude. At larger supercoolings, τ is independent of T_f again. This behavior agrees qualitatively with the theoretical consideration of the phase separation kinetics at homogeneous nucleation taking into account the finiteness of the cooling rate. The value of the surface tension has been obtained from comparison of the theoretical results with the experimental one.

We start the presentation of this chapter from the recollection of the basic theoretical relations that will be necessary for interpreting the experimental data (Section 10.2).

10.2 Homogeneous Nucleation in Mixtures: Theory

In Chapter 3 it was explained how homogeneous nucleation in a uniform supersaturated mixture proceeds through the formation of clusters of the new phase at random sites. If the number of particles in a cluster, n, is smaller than a certain critical value n_c, controlled by the competition between the bulk and surface contributions to the thermodynamic potential, such a cluster is unstable and it dissolves. When $n > n_c$ the cluster grows. For a spherical cluster in a dilute and ideal binary mixture n_c is given by

$$n_c = \left(\frac{\beta}{\ln(c_0/c_f)}\right)^3, \tag{10.1}$$

where, in our case, c_0 is the initial ^{3}He concentration in the mixture (the concentration before the supersaturation step), c_f is the equilibrium ^{3}He concentration of the matrix at the cluster boundary at the temperature T_f (after the supersaturation step), and the parameter β is given by

$$\beta = \frac{8\pi}{3}\frac{\sigma a^2}{k_B T_f}, \tag{10.2}$$

where a is the atomic distance, which is determined by $(4\pi a^3)/3 = (V_m/N_A)$, here V_m is the molar volume and N_A is the Avogadro number.

Both nucleation and the subsequent growth of the clusters are dependent on the quantity $J(n)$ characterizing the nucleation rate. $J(n)$ is the flux in the space of cluster sizes n. It is a very sharp exponential function of n, and $J(n_c) \equiv J_0$, which is the flux of the particles in the new phase through the critical point in the space of sizes. It is of fundamental importance in all calculations and can be written as (see Chapter 3 and [300, 337])

$$J_0 = \left(\frac{3\beta}{2\pi}\right)^{1/2} c_0^2 \exp\left[-\frac{\Delta\Phi(n_c)}{k_B T}\right], \tag{10.3}$$

where $\Delta\Phi(n)$ is the change in the thermodynamic potential, when n particles of the initial mixture transform into a cluster. In the approximation considered

$$\Delta\Phi(n) = n\Delta\mu + 4\pi a^2 \sigma n^{2/3}, \tag{10.4}$$

and the difference between the chemical potentials is

$$\Delta\mu = k_B T \ln\frac{c_f}{c_0}. \tag{10.5}$$

Using Eqs. (10.1), (10.2), and (10.5), we obtain

$$J_0 = \left(\frac{3\beta}{2\pi}\right)^{1/2} c_0^2 \exp\left(-\frac{\beta^3}{2\ln^2(c_0/c_f)}\right), \tag{10.6}$$

the flux J_0 is strongly dependent on the degree of supersaturation of the metastable mixture. Assuming $c(T) = \exp(-Q/k_B T))$ (which is quite a good approximation for dilute ^{3}He–^{4}He mixtures) we obtain

$$J_0 = c_0^2 \left(\frac{3\beta}{2\pi} \right)^{1/2} \exp \left[-\frac{\beta_0^3}{2} \left(\frac{k_B T_0}{Q} \right)^2 \frac{1}{x(1-x)^2} \right], \tag{10.7}$$

where Q is the effective heat of separation, $\beta_0 \equiv \beta(T_0)$, and $x = T/T_0$ (T_0 is the phase separation temperature of the mixture).

Although J_0 is finite for all x different from 0 and 1, Eq. (10.7) suggests that for practically any β_0 there is a region of supercooling where J_0, characterizing the nucleation rate, starts changing by orders of magnitude under very slight variation of x. As a result, nucleation is only observable in a narrow range of supersaturations; the process is unobservably slow at low degrees of supersaturation and practically instantaneous when the degree of supersaturation is high. This behavior permits us to introduce the concept of the highest cluster concentration N_m, which corresponds to the end of nucleation at a preassigned temperature and at the initial supersaturation, when $(\Delta c/c_0)\, n_c \approx 1$. The relative cluster concentration per lattice site corresponding to this condition is [300]

$$N_m = (4c_0)^{1/4} \left(\frac{J_0}{\beta} \right)^{3/4}, \tag{10.8}$$

and the nucleation time

$$\tau_N = \left(\frac{4c_0}{\beta^3 J_0} \right)^{1/4} \frac{a^2 c_0}{D}, \tag{10.9}$$

where D is the diffusion coefficient of ^{3}He in the decomposing mixture.

The nucleation stage ends at a certain critical concentration c_c corresponding to the condition [43]

$$(c_0 - c_c)/c_0 \approx n_c^{-1}. \tag{10.10}$$

Then, the stage of the diffusion or independent growth of the nuclei starts. The estimation for the characteristic time of the cluster growth up to the stage of coarsening (Ostwald ripening) is (Chapter 3, [300])

$$\tau_D \approx \frac{a^2}{3D} c_0^{-1/3} N_m^{-2/3} \approx \frac{a^2}{3D} \left(\frac{\beta}{c_0^2} \right) J_0^{-1/2}. \tag{10.11}$$

As we see, N_m is also responsible for the kinetics of the subsequent independent diffusional growth of the clusters. According to Eq. (10.8), N_m is dependent on β and only one unknown parameter σ occurs in the expression for J_0. As soon as we know the cluster concentration, the interfacial tension coefficient can be estimated readily. This will be done in Section 10.3.

One can assume that at the stage of independent diffusional growth of the nucleated clusters, each nucleus of radius r can be conceptually placed into a spherical diffusion zone of mean radius $R = a/N_m^{1/3}$. Then the characteristic time for the diffusional growth will be equal to (see, e.g., [78])

$$\tau_D = \frac{1}{\lambda^2 D}, \tag{10.12}$$

where λ is a solution of the transcendental equation

$$\tan \lambda(R - r) = \lambda R. \tag{10.13}$$

For

$$\lambda(R - r) < 1, \tag{10.14}$$

$\tan \lambda(R - r)$ can be expanded in a series; not going beyond the cubic terms of this expansion, we get

$$\lambda^2 = \frac{3z}{R^2(1 - z)^3}, \tag{10.15}$$

where

$$z = \frac{r}{R} = \left(\frac{c_0 - \bar{c}}{1 - 2\bar{c}}\right)^{1/3} \tag{10.16}$$

and $\bar{c}$ is the mean concentration of the impurity in the matrix. Equation (10.16) was obtained by using the conservation law for impurities in the volume of a sphere of radius R.

Taking into account Eq. (10.15), we can write an expression for τ_D as

$$\tau_D = \frac{R^2}{3D} \frac{(1 - z)^3}{z} = \frac{a^2}{3DN_m^{2/3}} \frac{(1 - z)^3}{z} = \frac{a^2 \beta^{1/2}}{3D\,(4c_0)^{1/6}\, J_0^{1/2}} \frac{(1 - z)^3}{z}, \tag{10.17}$$

and, using J_0 from Eq. (10.3), we find

$$\tau_D = \frac{\alpha \beta^{1/4} a^2}{D c_0^{7/6}} \frac{(1 - z)^3}{z} \exp\left[\frac{\beta^3}{4 \ln^2 (c_0/c_f)}\right], \tag{10.18}$$

where

$$\alpha = \frac{\pi^{1/4}}{3^{5/4} 2^{1/12}} \approx 0.32. \tag{10.19}$$

Thus, for a known value of D, Eq. (10.18) can be used with values obtained for τ_D to determine β and hence the surface tension σ, which enters into it (Section 10.4).

As estimates show that the nucleation time, Eq. (10.9), is often quite short, $\tau_N < 1$ s. It is therefore necessary in many cases to take into account the finiteness of the rate of the temperature decrease. At rather small τ_N, the concentration c_f in Eq. (10.5) becomes strongly dependent on time. When the temperature is lowered, excess admixture concentration in the matrix increases, so does the nucleation rate. The rate of the first process is decreasing with time, while the rate of the other process, which is conditioned by J_0, increases sharply. When these rates become equal at a certain temperature T^*, this corresponds to the minimum of $\Delta\Phi/k_BT$, and T^* is the temperature of intensive nucleation (IN).

For weak ^{3}He–^{4}He mixtures, one can assume, to a quite high accuracy, that the equilibrium concentration is an exponential function of temperature

$$c(T) = e^{-Q/k_BT}. \tag{10.20}$$

In this approximation the equation for $x = T^*/T_0$ found from the condition

$$\frac{\mathrm{d}}{\mathrm{d}t}\left(\frac{\Delta\Phi}{k_BT}\right) = 0, \tag{10.21}$$

is given by

$$\frac{x^{2/3}}{(1-x)^3\,(x-1/3)}\exp\left(-\frac{\beta_0^3}{2}\left(\frac{k_BT_0}{Q}\right)^2\frac{1}{x\,(1-x)^2}\right) = \frac{\tilde{\tau}}{\tau_T}, \tag{10.22}$$

where

$$\beta_0 = \beta(T_0), \qquad \tilde{\tau}^{-1} = \frac{16}{3}c_0^2\sqrt{\frac{3}{2\pi}}\beta_0^{7/2}\left(\frac{k_BT_0}{Q}\right)^4\frac{D_{c_0}}{a^2}, \qquad \tau_T^{-1} = -\frac{1}{T_0}\frac{\partial T}{\partial t}. \tag{10.23}$$

If Eq. (10.22) gives $T^* > T_f$, the IN occurs at $T = T^*$. If $T^* < T_f$ or Eq. (10.22) has no solution, the most IN occurs at T_f. Thus, the T_f (or T^*) values actually dictate the concentration of the new phase nuclei N_m, which, along with the diffusion coefficient, is responsible for the phase separation kinetics as a whole. The investigation of the separation kinetics as a function of T_f is one of the most reliable methods of testing the correspondence between the processes occurring in the samples and the theory of homogeneous nucleation. Section 10.5 describes the corresponding experiments and the obtained results.

10.3 Homogeneous Nucleation in ^{3}He–^{4}He Solid Solutions: Experiment and Comparison with Theory

The present section starts the analysis with a brief summary of bounded diffusion in NMR experiments (Section 10.3.1); this overview is required for interpreting the experimental data. The experimental cell and techniques are described in Section 10.3.2 (see also [317]). The results are presented in Section 10.3.3 together with a discussion, within the framework of the theory of homogeneous nucleation.

10.3.1 Spin Echoes in Restricted Geometry and Cluster Sizes

The key experimental result needed is the determination of the size of the new phase clusters created upon phase formation. It is this parameter that permits us to find the concentration of clusters N_m (calculated in the previous section) from particle conservation. In [317] the size of the droplets was estimated by the pulsed nuclear magnetic resonance (NMR), or spin-echo method.

In segregated dilute mixtures the NMR transverse relaxation is influenced by the small size of the new phase droplets. This dependence is most evident when diffusion is measured by the spin-echo method, where the spin diffusion coefficient is found from the echo signal of the sample placed in a magnetic field with a gradient G after applying two or more resonant RF pulses separated by an interval t^*. With a conventional spin-echo pulse train $90° - t^* - 180°$, the amplitude of the echo signal occurring at time $2t^*$ in a bulk sample can be expressed as

$$E\left(2t^*\right) = \exp\left(-\frac{2}{3}\gamma^2 G^2 D t^{*3}\right), \tag{10.24}$$

where γ is the gyromagnetic ratio. If, however, the sample size d is smaller than the spin diffusion length $\sqrt{Dt^*}$, the dependence $E(t)$ becomes more complicated because the motion is bounded. The description of this effect [190,211] resulted in inconvenient lengthy formulas, which is difficult to use for processing experimental results. However, a simplification is possible through the use of an approximate model.

It is evident that, so far as NMR is concerned, bounded diffusion in a field gradient is equivalent to unbounded diffusion in a triangular field profile, the half-wavelength dimension being d. If the triangular profile is now approximated by a sinusoidal variation then a relatively simple expression for $E(t)$ can be obtained:

$$E\left(2t^*\right) = \exp\left\{-\frac{d^2\gamma^2 G^2}{\pi^2}\tau_c^2\left[\frac{2t^*}{\tau_c} + 4\exp\left(-\frac{t^*}{\tau_c}\right) - \exp\left(-\frac{2t^*}{\tau_c}\right) - 3\right]\right\}, \tag{10.25}$$

where $\tau_c = d^2/\pi^2 D$. Using numerical simulations, in [317] it was proven that Eq. (10.25) gives results in good agreement with the exact expression for a spherical droplet [211]. It is thus evident that through comparison of Eq. (10.25) with the experimental data, one can find both the diffusion coefficient and the cluster size from this expression.

10.3.2 Experimental Details

The experimental cell is shown in Figure 10.2. The cell is supported on a copper cold finger extending down from the experimental plate, which is in good thermal contact with the mixing chamber of a dilution refrigerator. The cell is of modular design in three parts: the cold finger and top third of the cell, the NMR coils in the middle, and the pressure gauge in the bottom third. The copper cold finger is machined to a sharp point to encourage nucleation of the crystal from the liquid mixture at this point. The bottom section of the cell contained a pressure capacitive transducer of the Straty-Adams type.

Thermometry was provided by two Speer carbon resistance thermometers calibrated using a ^{3}He melting curve thermometer and a germanium resistance thermometer (between 0.4 and 10 K). The sample mixture was made using standard volumetric techniques. ^{3}He came from a cylinder and ^{4}He was added to make up the required concentration of $(1.00 \pm 0.01)\%$ from evaporated liquid. The crystal was grown by forcing the mixture into the cell at high pressure using a charcoal filled "bomb" in a ^{4}He transport Dewar (for more details see [317]).

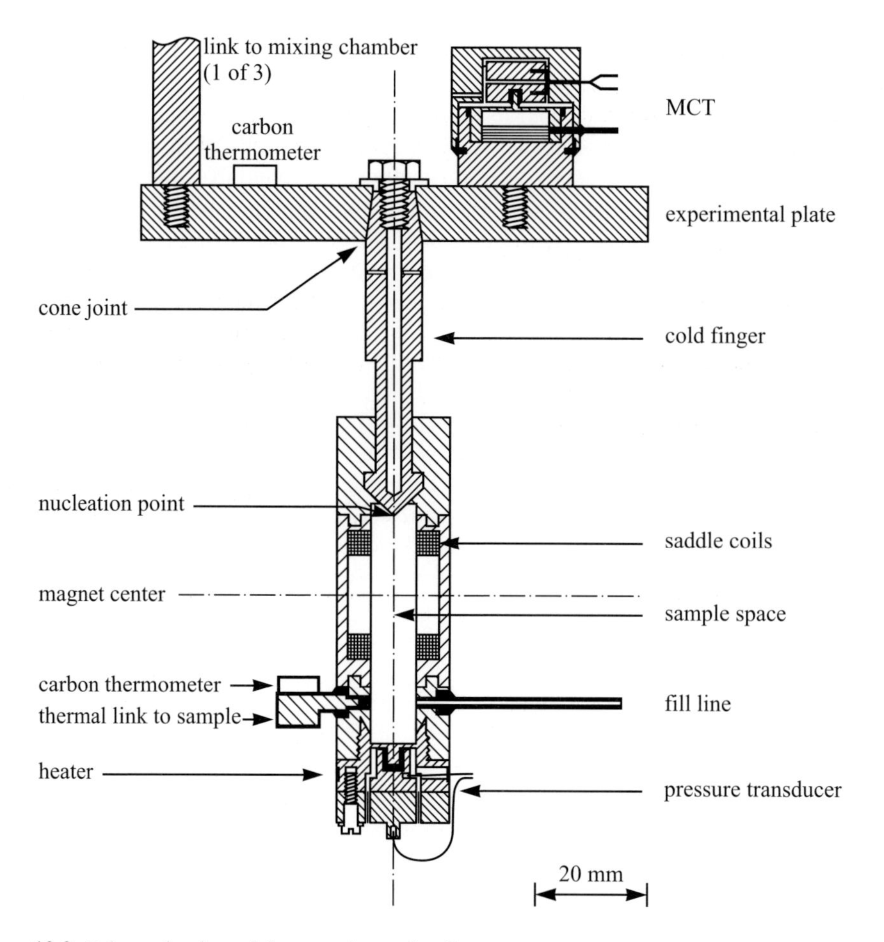

Figure 10.2: Schematic view of the experimental cell.

In order to produce a sample with a minimum of defects the crystals were grown at constant pressure. The cold finger in the cell is cooled below the melting transition until solidification starts, indicated by a dropping pressure on the fill capillary. The nucleation point is then stabilized at this temperature and the other end of the cell is held just above the melting temperature. These conditions are maintained as the solid–liquid interface propagates through the cell.

When the solidification was complete, the fill capillary heaters turned off and the 1-K pot run fully, the refrigerator circulation was stopped and the mixing chamber temperature regulated at 1.2 K for 180 h in order to anneal the crystal and reduce any inhomogeneities in ^{3}He concentration. The sample was then cooled to 500 mK to start the measurements. The pressure $P_0 = 36$ bar was independent of temperature below 0.6 K.

The NMR measurements were all made using a coherent pulse spectrometer. The experimental procedure was as follows. The prepared and annealed sample was smoothly cooled to a temperature close to the separation temperature T_0 ($T_0 = 186$ mK for a 1% mixture).

Then the temperature was lowered in steps and the spin-echo signal, and the pressure also, were recorded. After the onset of phase separation, measurements were repeated at each step until equilibrium was established. In the described experiments [317] the diffusion in the ^{3}He nuclei was measured at the lowest temperatures where nearly all ^{3}He was contained in the new-phase droplets.

10.3.3 Results and Discussion

Figure 10.3 shows the reduced echo amplitude vs interval between the RF pulses. The dependence was measured at different temperatures and magnetic field gradients after a chain of successive coolings of the sample. According to Eq. (10.25), the dependence is universal if the echo signals are raised to the power $(G_0/G)^2$, where G_0 is an arbitrary reference gradient. The data in Figure 10.3 have been scaled in this way. The curve approximates Eq. (10.25), whose parameters were estimated by the least-squares method to give $D = (4.9 \pm 0.3) \times 10^{-8} \mathrm{cm}^2\ \mathrm{s}^{-1}$ and $d = 4.5 \pm 0.5\ \mu\mathrm{m}$. The value of D agrees well with the spin diffusion coefficients for a bulk sample [56, 66, 210], which suggests identical diffusion processes in ^{3}He clusters and in bulk ^{3}He.

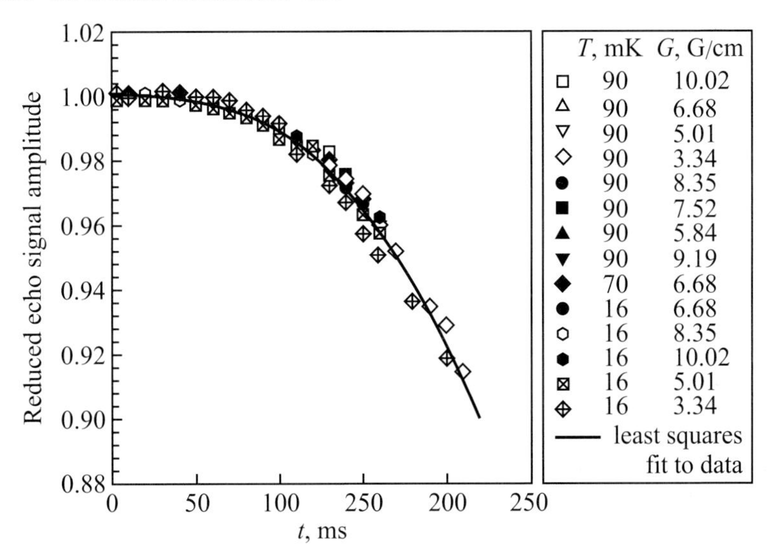

Figure 10.3: Dependence of the reduced echo-signal amplitude on the interval between the pulses at different temperatures and magnetic field gradients. The curve is the least-squares approximation to the results by Eq. (10.25). Data are presented for temperatures between 90 and 16 mK and field gradients between 3.34 and 11.02 G/cm.

The concentration of clusters can be found readily from their sizes. Since d is measured at quite low temperatures assuming that all ^{3}He is contained within the clusters, we obtain from conservation of particles

$$\frac{\pi d^3}{6} \frac{N_A}{V_m} N_m = c_0 \quad \text{and} \quad N_m = 3.2 \times 10^{-24} \frac{V_m}{d^3} c_0, \tag{10.26}$$

where N_A is the Avogadro number. Inserting these parameters, we obtain $N_m = (8.4 \pm 0.8) \times 10^{-15}$. This result can be compared with Eq. (10.8) to estimate σ.

Before estimating σ from Eqs. (10.26) and (10.8), the following comments are appropriate to be made:

(1) As mentioned in Section 10.3.2 the starting mixture for the sample contained 1% ^{3}He. It should also be noted that the initial concentration can differ from that in the crystal because of the isotope fractionation caused both by crystallization and by the desorption-induced increase in pressure [317]. Therefore, the initial ^{3}He concentration in the sample is somewhat uncertain. This uncertainty has, however, no effect on the subsequent steps for which the initial and final concentration (c_i and c_f, respectively) are both determined by the phase diagram of the mixture. Possibilities for the refinement of c_0 are discussed below.

(2) The clusters of the new phase whose sizes were measured in the experiment developed after several successive stepwise coolings (some characteristics of the steps at which the changes in the concentration become measurable are shown in Table 10.1). Basically, new clusters can form at any step. Because of their low concentration at each step ($\sim 10^{-15}$), we can assume that nucleation is an independent process and the experimental N_m values found at low temperatures may be taken as a sum of contributions from all the previous steps.

Table 10.1: Main characteristics of experiments upon step-by-step lowering of the sample temperature.[a]

No.	T_f (mK)	c_f (%)	$\tau \times 10^{-3}$ (s)	$D \times 10^9$ (cm^2/s)	$\sigma^* \times 10^2$(erg/cm^2)	$N_m \times 10^{15}$
1	183	0.91	76.50	0.17	–	–
2	171	0.64	30.24	0.24	1.30	8.8
3	161	0.46	5.40	0.35	1.25	<0.1
4	150	0.31	4.32	0.55	1.24	0.3
5	140	0.20	3.78	0.88	1.24	<0.1
6	130	0.12	3.66	1.40	1.24	<0.1

[a] The final concentration c_f is calculated by Edwards and Balibar's formula [63].

(3) The equations in Section 10.2 refer to the case when the mixture becomes segregated completely after one cooling step. At a multistep cooling, the expression for the nth step can be derived rigorously by solving Eqs. (43) in [300] for different starting conditions, taking into account the previously formed clusters. However, since nucleation takes a shorter time than the diffusion growth and it proceeds independently (see above), the kinetics of the cluster growth may be thought of as invariable, and the refinement of Eq. (10.11) is reduced to substitution of $(c_0 - c_{f_n})^{2/3}/c_{i_n}$ for $c_0^{-1/3}$ (c_{f_n} and c_{i_n} are the final and initial concentrations at the nth step, respectively). In the expressions for β (and J_0) c_0 should also be replaced by c_{i_n}.

Taking into account these comments, we can obtain from Eq. (10.8) the total N_m value corresponding, within the error, to the experimental result 8.4×10^{-15} if we used $\sigma =$

$1.27 \times 10^{-2}\text{erg/cm}^2$. The value of N_m for each step is presented in Table 10.1 (column 7). It is seen that most clusters are formed at steps (2) and (4). In the calculation the number of clusters at step (1) was assumed to be negligible. This is true for $c_0 \leq 1.25\%$. The more prolonged separation time at step (1) supports this assumption.

10.3.4 Conclusion

Nuclear magnetic resonance experiments studying the phase separation in dilute mixtures of ^{3}He in ^{4}He show that the conditions for homogeneous nucleation can be realized in high-quality crystals (due to their growth at constant pressure or to thermocycling in the two-phase region). The main parameter of the theory, namely the interphase surface tension σ of the solid ^{3}He–^{4}He mixture was obtained. Realization of homogeneous nucleation in solid ^{3}He–^{4}He mixtures opens up new possibilities for the comprehensive quantitative correlation with theory.

10.4 Kinetics of Phase Transition in Solid Solutions of ^{4}He in ^{3}He at Different Degrees of Supersaturation

The present section is devoted to a study of the influence of the degree of supersaturation on the rate of the phase transition in solid solutions of ^{4}He in ^{3}He and a comparison with theoretical results [88]. It describes precision measurements of the pressure during phase separation in a solid mixture of ^{4}He in ^{3}He allowing us to obtain characteristic times of the phase decomposition (Section 10.4.1). A processing of the measurement results (Section 10.4.2) gives additional evidence supporting the view that homogeneous nucleation is realized in ^{3}He–^{4}He solid solutions at significant supercoolings and heterogeneous nucleation at the smallest supercoolings.

Two different ways are proposed for comparing the results with a theoretical calculation; taking into account the processes at the boundary of a nucleus of the new phase. Both give roughly similar values of the coefficient of surface tension at the nucleus–matrix boundary, and those values agree with those obtained in other studies. It is conjectured that the bcc–hcp transition has a substantial influence on the kinetics of phase separation at the lowest supersaturations. The results are summarized in Section 10.4.3.

10.4.1 Experimental Results

In the described experiment the time dependence of the pressure in the samples of solid solutions of ^{4}He in ^{3}He was measured after cooling from the region of the stable homogeneous state to various temperatures T_f below the phase separation temperature T_{s0} of the initial solution. The measurement techniques and the experimental setup are described in Section 10.3 and in detail in [78].

An important element of the previous experiments was the development of a technique for obtaining homogeneous samples, making it possible to obtain practically equilibrium solutions. Samples of solid helium in the form of a disk 9 mm in diameter and 1.5 mm in height were located inside a metal chamber cooled by a dilution refrigerator. The samples were grown by the capillary blocking method from a gaseous mixture containing approximately 2% ^{4}He. The bottom of the chamber, which was $\approx$1 mm thick, served as the movable plate of a precision capacitive pressure gauge of the Straty-Adams type. The ^{4}He concentration in the sample was defined according to the value of the change in pressure upon complete separation, ΔP_0, by Mullin's formula [185]

$$\Delta P_0 = 0.4 \frac{V_\mu}{\gamma} c_0(1 - c_0), \tag{10.27}$$

where γ is the compressibility.

Table 10.2: Some characteristics of the samples studied.

Sample no.	c_0 (% ^{4}He)	P_0, (bar)	V_μ (cm^3/mole)	T_{s0} (K)	T^* (K)	$\sigma^* \times 10^2$ (erg/cm^2)	$\bar{\sigma} \times 10^2$ (erg/cm^2)
1	2.25	34.82	23.99	0.199	0.170	2.20	1.4
2	2.43	38.616	23.55	0.201	0.191	1.11	–
3	2.8	31.435	24.42	0.212	0.182	2.21	1.4
4	2.93	35.465	23.89	0.212	0.189	1.84	1.1
5	3.34	32.85	24.20	0.220	–	–	1.2

c_0 is the ^{4}He concentration, P_0 is the pressure, V_μ is the molar volume, T_{s0} is the phase separation temperature of the initial solution, T^* is the temperature of intensive nucleation, and σ^* and $\bar{\sigma}$ are the coefficients of interfacial tension found by different methods.

10.4.2 Discussion

As in the majority of previously studied situations, the time dependence of the pressure in the sample upon a stepwise lowering of the temperature in the phase separation region is described well by an exponential function of the form

$$P_f - P(t) = \Delta P_m \exp\left(-\frac{t}{\tau}\right), \tag{10.28}$$

where $\Delta P_m = P_f - P_i$ is the difference of the equilibrium values of the final pressure P_f and initial pressure P_i, and τ is the characteristic time for establishment of equilibrium at the step under consideration.

The primary data for one of the samples are presented in Figure 10.4, which shows the time dependence of the pressure $\Delta P = P(t) - P_0$ for different values of T_f. All of the curves are described satisfactorily by Eq. (10.28), and the time τ decreases noticeably at large T_f and becomes practically independent of T_f for $T_f \ll 160$ mK. This result is more clearly seen in Figure 10.5, where the values of τ for all the samples are plotted as a function of the relative supercooling $\Delta T/T_s$ ($\Delta T = T_s - T_f$).

At small supercoolings ($\Delta T < 20$ mK) the characteristic times turn out to be extremely long (up to 10^4 s), and they decrease rapidly with increasing ΔT to values of the order of

500–700 s, and at $\Delta T \geq 50$ mK they become practically independent of the degree of supercooling. This behavior agrees qualitatively with what is expected under conditions of homogeneous nucleation. According to Eq. (10.27), the change in pressure is uniquely related to the change in concentration, and the found times τ characterize the progress of the diffusion process in the presence of separation, which is governed by a time $\tau_D \approx R^2/D \approx N_m^{-2/3} D^{-1}$ (see Section 10.2). The concentration of nuclei N_m in the presence of separation grows

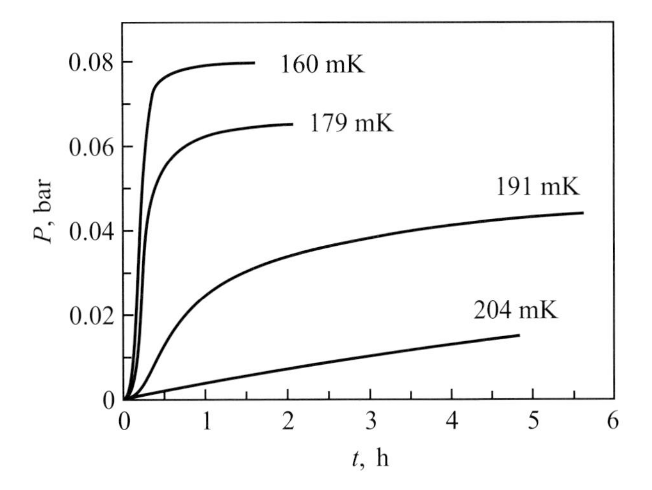

Figure 10.4: Time dependence of the change in pressure in ^{3}He–^{4}He upon cooling from the region of the homogeneous solution to different temperatures T_f, indicated in the figure [88].

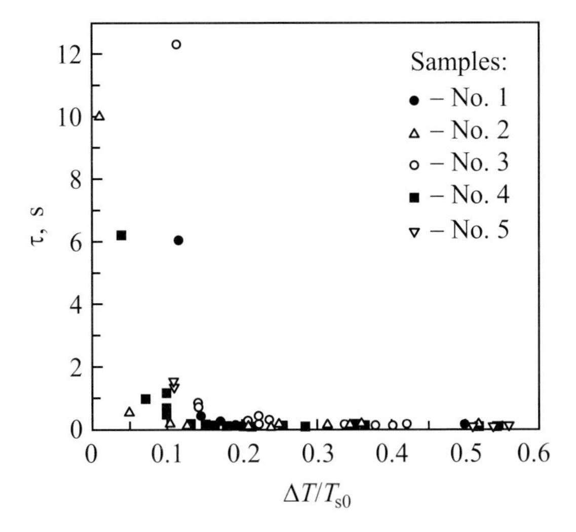

Figure 10.5: Characteristic times τ vs relative supercooling for all the samples studied (indicated in the figure) [88].

sharply with increasing degree of supersaturation (supercooling) (see Eqs. (10.8), (10.9)), and that brings about a corresponding decrease of τ.

The ultimate plateauing of τ can be caused by several reasons:

1. At short diffusion times the impurity atoms at the nucleus–matrix boundary have a resistance to transition characterized by a time τ_s, which is practically independent of T at low temperatures [81, 89].

2. As was shown in [79, 312], at a finite rate of cooling, when it is necessary to take into account the change of the concentration c_f with time, one can introduce the concept of the temperature of intensive nucleation, T^*, at which the decrease of the degree of supersaturation $\varepsilon = c(t) - c_f(t)$ due to the decrease of the matrix concentration $c(t)$ with time owing to the escape of admixture into the new-phase nuclei is comparable to the increase in ε due to the decrease of $c_f(t)$. The value of T^* becomes the determining factor specifying the concentration of nuclei, and decreasing T_f further does not alter the value of N_m, and, hence, of τ for a sufficiently weak temperature dependence of $D(T)$.

3. The minimum times τ found are close to the characteristic times for establishment of a given temperature.

4. With decreasing T^* a fundamental circumstance can come into play: at sufficiently high supersaturations the number of atoms in a nucleus becomes of the order of 1. This fact implies that the macroscopic approach described in Section 10.2 will not be applicable under these conditions, and there is probably a change of nucleation mechanism.

Apparently, the most probable cause of the plateauing of τ with increasing supersaturation is a combination of the first two factors.

The time for establishment of an equilibrium temperature was practically always less than the experimental values of τ. It was shown in [89] that under condition (10.14) the measured characteristic time for the influence of the boundary resistance to be noticeable can be written in the form

$$\tau = \tau_D + \tau_s, \tag{10.29}$$

where τ_D as before is described by Eq. (10.17), and

$$\tau_s = \frac{R}{3K}\frac{1-z^3}{z^2}, \tag{10.30}$$

where K is a constant characterizing the probability of penetration of an impurity through the boundary of a nucleus. If $T_f \leq T^*$, then R ceases to depend on T_f, and then the expression

$$\tilde{\tau} \equiv \tau\frac{z^2}{1-z^3} - \frac{R}{3D}\frac{(1-z)^3 z}{1-z^3} = \frac{R}{3K}, \tag{10.31}$$

obtained by substituting Eqs. (10.15) and (10.30) into Eq. (10.29), will be a constant.

Equation (10.31) can be used to estimate T^* as the temperature starting with which the left-hand side of this formula becomes independent of T_f. The value of R here is figured as the minimum value of R_m corresponding to the temperature T^*. For determining the values of T^* and R_m a plot was constructed in [88] showing the dependence of $\widetilde{\tau}$ vs T_f according to Eq. (10.31) (see Figure 10.6). In the construction the value $K = 2.7 \times 10^{-5}$cm/s from [89] was used, while the values of the diffusion coefficient D obtained in [89] were adjusted to the molar volumes of the samples studied. From the values of $\widetilde{\tau}$ in the plateau region, one can use Eq. (10.30) to estimate the experimental value of R_m corresponding to the concentration of nuclei formed at the temperature T^* (see Table 10.2) and compare it with the calculated value,

$$R_m = \frac{a}{N^{1/3}} \approx \frac{a}{c_0^{7/12}} \exp\left[\frac{\beta^{*3}}{8\ln^2(c_0/c_f^*)}\right], \tag{10.32}$$

where the asterisk denotes values corresponding to the temperature T^*. Such a comparison makes it possible to determine the parameter β and the surface tension σ, which enters into β. The values thus found are listed in Table 10.2 as σ^*. The noticeable scatter of the values of σ^* for different samples is most likely due to an inaccuracy of the purely visual estimate of T^* from the plot in Figure 10.6.

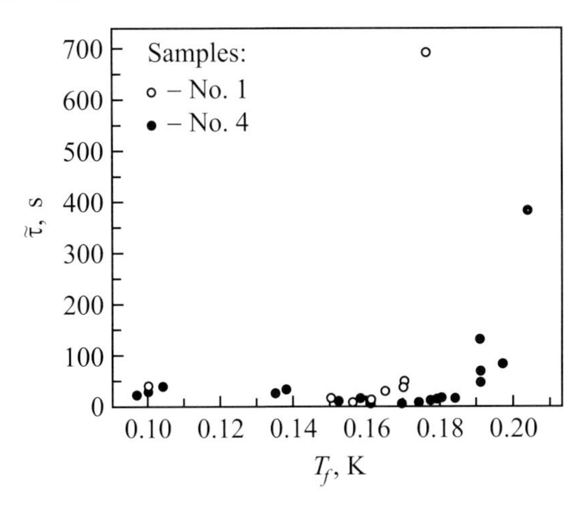

Figure 10.6: Dependence of the reduced time $\widetilde{\tau}$ (see the text) on the final temperature of the ^{3}He–^{4}He samples.

The concentrations of the studied five samples ranged from 2.2 to 3.35% ^{4}He. Note that for all the samples the experimental values of the separation temperature corresponded within the error limits to those calculated according to the formulas of [63]. Some of the characteristics of the samples studied are presented in Table 10.2.

The data obtained at medium supersaturations, i.e., for $T_f > T^*$, can also be used for an estimate of σ. Equation (10.29), written in the form

$$\tau = \frac{R^2}{3D}\frac{(1-z)^3}{z} + \frac{R}{3K}\frac{1-z^3}{z^2}, \tag{10.33}$$

can be considered to be an equation for finding R and can thus be used with the measured values of $\tau(T_f)$ to determine $R(T_f)$ and, with the use of Eq. (10.32), the surface tension σ.

The values of σ thus obtained, averaged for the different T_f, are listed in Table 10.2 as $\bar{\sigma}$. The difference of the mean values of the surface tension ($\sigma^* = 1.8 \times 10^{-2}$ erg/cm^2 and $\bar{\sigma} = 1.3 \times 10^{-2}$ erg/cm^2) estimated by the two methods is about 20%. An averaging of all the data gives a value $\sigma = 1.5 \times 10^{-2}$erg/cm^2, practically the same as the results of previous estimates [81, 89].

Curious features of the time dependence of the pressure difference are observed at the lowest supersaturations (Figure 10.7): the total time for establishment of equilibrium turns out to be very long there, and the $\Delta P(t)$ curves display noticeable irregularities. Since the probability of homogeneous nucleation at such low supersaturations is very small, it can be assumed that heterogeneous nucleation becomes the governing process.

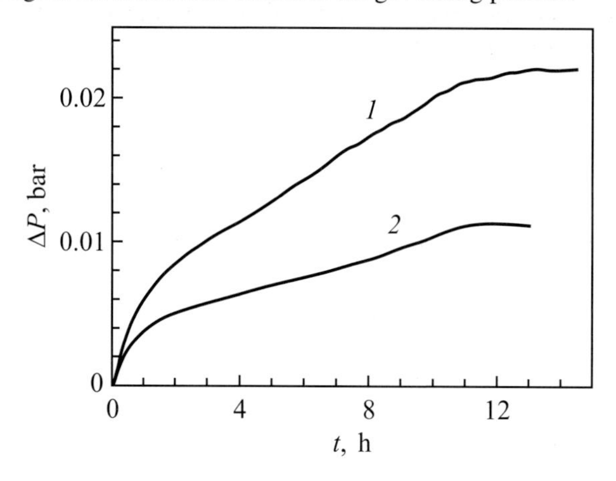

Figure 10.7: Time dependence of the change in pressure at very low degrees of supersaturation for samples no. 1 (curve 2) and no. 3 (curve *1*).

If heterogeneous nucleation occurs on the wall of the cell, then the change in the mean concentration (and, hence, the pressure) should correspond to the solution of the diffusion problem for a sample situated between two planes and is described by the expression (see, e.g., [264])

$$\Delta P = \Delta P_m \left[1 - \frac{1}{a_N} \sum_{n=0}^{N} \frac{1}{(2n+1)^2} \exp\left(-\frac{\pi^2 D t}{d^2} (2n+1)^2 \right) \right], \tag{10.34}$$

where

$$a_N = \sum_{n=0}^{N} \frac{1}{(2n+1)^2}, \tag{10.35}$$

d is the distance between planes and N is the number of terms in the sum.

A processing of the curves in Figure 10.7 with Eq. (10.34) shows that the use of 3 or 4 terms of the sum decreases the rms deviation δ by more than a factor of 3 compared to processing with the use of a single exponential, and increasing N further has little effect on the value of δ. We note that Eq. (10.34) differs from the analogous formula for a spherical geometry (see, e.g., [78]) by a significantly slower decrease of the successive terms of the sum. The relaxation time $\tau = d^2/\pi^2 D$ obtained here is of the order of 4×10^4 s, and when the height of the cell is substituted in τ for d one obtains a value $D \approx 5 \times 10^{-8}\ \mathrm{cm^2/s}$ for the diffusion coefficient, which agrees with the data obtained in [89]. This fact argues strongly in favor of heterogeneous nucleation under these conditions.

The irregularity of the curves in Figure 10.7 may be due to the formation, at low supersaturations, of new-phase nuclei with a bcc structure, which later transforms to the hcp structure. A similar phenomenon has been observed [165] in the crystallization of ^{4}He in the region of the triple point. This possibility is also supported by the fact, registered in X-ray studies of the decomposition of the solutions, that a nonequilibrium bcc structure of the new phase persists to temperatures of about 100 mK [64].

10.4.3 Conclusion

In this section we have reported measurements of the characteristic times for the diffusional decomposition of solid solutions of ^{4}He in ^{3}He. The results of the experiments were compared with theoretical calculations of the parameters of homogeneous nucleation under conditions of a finite resistance at the nucleus–matrix boundary. On the basis of that comparison the value of the coefficient of interphase surface tension was estimated by two methods.

The closeness of the surface-tension results obtained by different methods and their good agreement with values of σ found previously [89] (see also [317]) furnish additional evidence for the realization of conditions for homogeneous nucleation in perfect samples of ^{4}He–^{3}He solid solutions. However, for a final answer to this question it will be necessary to do more careful and systematic studies to minimize if not eliminate the errors existing in the present treatment, viz:

(a) the approximate character of Eqs. (10.29) and (10.31), which are based on an expansion of Eq. (10.13), because of the proximity of $\lambda(R - r)$ to 1;
(b) insufficient accuracy of the estimate of T^*, owing to the small number of experimental points;
(c) possible errors in bringing in the parameters D and K from [89] (it would be desirable to measure them in the same experiment);
(d) possible influence of heterogeneous nucleation and a bcc–hcp transition, especially at low supersaturations.

Experiments with these circumstances taken into account will make it possible to perform a more quantitative comparison with the results of the theory of homogeneous nucleation, extended to the situation with a finite rate of supercooling [312], and to trace the transition from heterogeneous to homogeneous nucleation in ^{4}He–^{3}He solutions.

10.5 Influence of the Degree of Supercooling on the Kinetics of Phase Separation in Solid Mixtures of ^{4}He in ^{3}He

In this section the kinetics of phase separation in the solid mixture of ^{4}He in ^{3}He is studied at various degrees of supercooling in the temperature range of 100–200 mK through precise pressure measurements [202]. The time dependences of the pressure change during phase separation were found to be exponential. At small supercoolings the characteristic time constant τ is almost independent of the final temperature T_f and is about 10 h, which is considered to result from heterogeneous nucleation. In a narrow range of T_f the time τ decreases more than by an order of magnitude. At larger supercoolings, τ is again independent of T_f. This behavior agrees qualitatively with the theoretical considerations of the phase separation kinetics at homogeneous nucleation taking into account the finite value of the cooling rate. The value of interphase surface tension has been obtained from comparison of the theory with the experimental results.

The time dependences of the pressure in the solid ^{3}He–^{4}He mixtures were measured after the samples had been cooled to different temperatures T_f in the phase separation region. The measurement technique and experimental setup are described in [82]. The solid helium samples were located inside a metallic ampoule shaped as a disk 9 mm in the inner diameter and 1.5 mm high. The samples were prepared from a gas mixture ($\sim 2\%$ ^{4}He) through blocking the capillary. Some experimental dependences $P(t)$ measured at different T_f are shown in Figure 10.8.

Note the following features in Figure 10.8:

(i) The dependences taken at $T_f < 170$ mK are well described by the exponential dependences $P_f - P(t) = (P_f - P_0)\exp(-t/\tau)$, where P_f is equilibrium pressure at T_f.
(ii) At a very low supersaturation the dependence $P(t)$ becomes more complicated and its processing through the use of the exponential gives τ about 10 h.
(iii) At $T_f < 150$ mK, τ is practically constant.
(iv) The obtained values of τ decrease sharply as T_f goes below 170 mK.

At intermediate supercoolings the behavior of τ vs $\Delta T/T_0$ (Figure 10.9) agrees qualitatively with what is expected in the context of homogeneous nucleation (see Section 10.2) and the subsequent diffusion growth of the ^{4}He-enriched phase nuclei. The values of σ, cited in the caption to Figure 10.9 as well as in the text and Table 10.3, are corrected as compared with [202] because of different definitions of a in [202] and this chapter. It is seen from Eq. (10.11) that τ_D is dependent on two unknown parameters D and σ. Using the ratio τ_D/τ_N, we can exclude D and estimate σ. At quite low τ_N, we can write

$$c_0 - c_c = -\left.\frac{\mathrm{d}c}{\mathrm{d}t}\right|_{t=t_i} \tau_N, \tag{10.36}$$

where t_i corresponds to the time moment when $T = T_i$ ($T_i = T^*$ when $T^* \geq T_f$ and $T_i = T_f$ when $T^* < T_f$). Using Eq. (10.2), we obtain

$$\tau_N = \frac{c_0}{-(\mathrm{d}c/\mathrm{d}t)} \frac{\ln^3(c_0/c^*)}{\beta^3(T^*)} = \frac{\Delta P_0}{(\mathrm{d}P/\mathrm{d}t)} \frac{\ln^3(c_0/c^*)}{\beta^3(T^*)}, \tag{10.37}$$

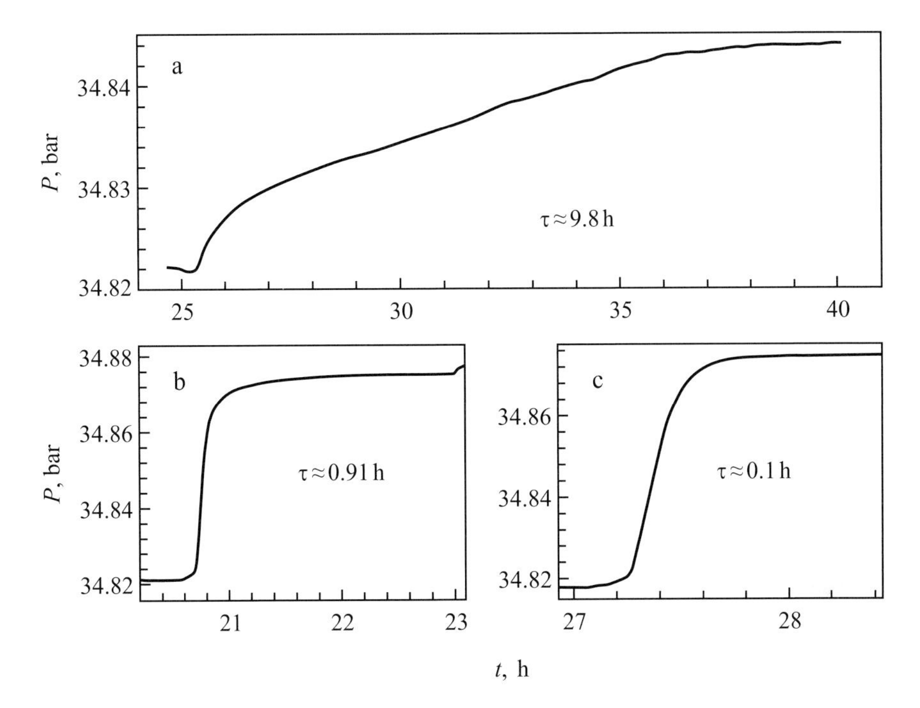

Figure 10.8: The typical kinetics of phase separation at various T_f. (a) 175 mK; (b) 165 mK; (c) 150 mK. The time axis corresponds to the laboratory time and absolute values of time are of no consequence.

where ΔP_0 is the change in the pressure caused by the separation of the mixture with the concentration c_0 into pure components. On finding the ratio τ_D/τ_N from Eqs. (10.9), (10.11) and using Eqs. (10.6) and (10.37), we obtain

$$\tau_D \approx \frac{0.62 c_i \ln^3 (c_0/c_i)}{c_0^{5/4} \beta^{15/8}} \frac{\Delta P_0}{\frac{dP}{dt}\Big|_{t=t_i}} \exp\left(\frac{1}{8} \frac{\beta^3 (T_i)}{\ln^2 (c_0/c_i)}\right), \tag{10.38}$$

where $c_i = c(T_i)$ is the equilibrium concentration at T_i. Using Eq. (10.38), we can find σ from the experimental data for the intermediate supercooling.

The values of σ for different T_f are in good agreement with one another. The average value of $\sigma = 1.43 \times 10^{-2}$ erg/cm^2 is quite close to those taken at the boundary between the solution and a ^{4}He cluster [167].

Of interest is the fact that Eq. (10.38) has no solution at $T_f < 100$ mK. This might be because $T_f < T^*$. Another and more fundamental reason is the following. If the size of the critical nucleus is estimated from the calculated σ, we obtain $n_c < 10$ at $T_f < 160$ mK. Under these circumstances it is problematic to use the macroscopic approach described in Section 10.2.

At small supercoolings, the concentration of the nuclei formed according to Eq. (10.8) is not sufficient to explain the observed values of τ_D, Eq. (10.9). It is natural to expect that the

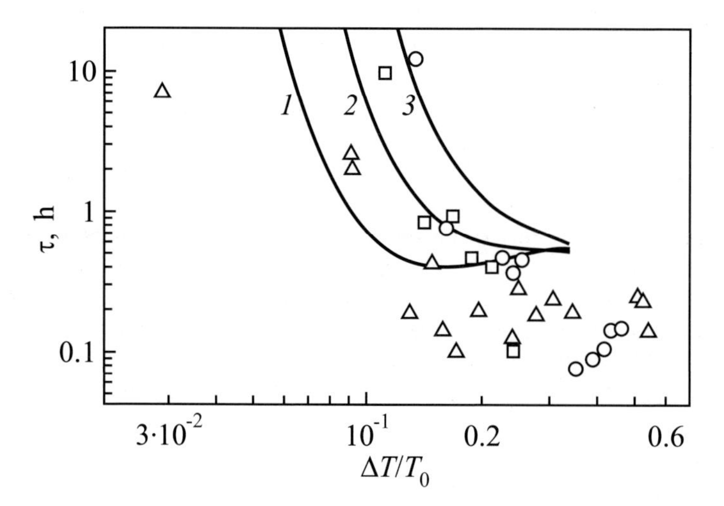

Figure 10.9: Characteristic time constant τ vs $\Delta T/T_0$ for different samples: (○) $c_0 = 2.8\%$ ^{4}He $P_0 = 31.5$ bar; (□) $c_0 = 2.2\%$ ^{4}He, $P_0 = 34.8$ bar; (△) $c_0 = 2.93\%$ ^{4}He $P_0 = 35.5$ bar. Solid lines correspond to the calculation with Eq. (10.38) at $\Delta P_0 = 0.074$ bar, $\mathrm{d}P/\mathrm{d}t|_{t=t_i} = 2 \times 10^{-5}$ bar/s: (1) $\sigma = 1.1 \times 10^{-2}$ erg/cm^2 ; (2) $\sigma = 1.36 \times 10^{-2}$ erg/cm^2; (3) $\sigma = 1.6 \times 10^{-2}$ erg/cm^2.

Table 10.3: Comparison of the results on interfacial surface tension in phase separated ^{3}He–^{4}He mixtures in different experiments.

Coexisting phases	Basic parameters or dependences used to estimate σ	$\sigma \times 10^2$ (erg/cm^2)	Reference
^{3}He-rich cluster – ^{4}He matrix	Cluster concentration	1.27	[317]
^{3}He-rich cluster – ^{4}He matrix	Separation time	1.26	[317]
^{4}He-rich cluster – ^{3}He matrix	Pressure variation	1.5	[88]
^{4}He-rich cluster – ^{3}He matrix	τ vs supersaturation	1.43^a	[202]
^{4}He cluster around vacancy in ^{3}He matrix	Pressure variation during thermocycling	1.48^a	[166]

^{a}These values are corrected as compared with [202] and [166] because the definition of a differs from that in the theoretical consideration [300].

contribution of heterogeneous nucleation becomes appreciable. The experimental nonmonotonic dependence $P(t)$ can be accounted for by the presence of several types of heterogeneous nucleation centers.

The values of τ corresponding to the lowest T_f are dependent on the simultaneous influence of several factors: (i) in this situation $T_f < T^*$; (ii) values of n_c are very small (< 10); (iii) τ_D is close to the time of reaching the stabilization temperature. Finally, these factors lead to the observed τ_D-independence of T_f.

This section has actually revealed a significant effect of the degree of supercooling upon the phase separation kinetics. We observed a transition from heterogeneous nucleation at small supercooling to homogeneous when supercooling is larger. A procedure is proposed to take into account the influence of finiteness of the cooling rate upon the concentration of the new phase nuclei. The surface tension at the nucleus boundary is found from a comparison of the experimental and the calculated data to be $\sigma \approx 1.43 \times 10^{-2}$ erg/cm^2.

10.6 Comparison between Experiments and Conclusions

It is worthwhile to compare the above results between them and with those of other experiments on the kinetics of phase separation in solid ^{3}He–^{4}He mixtures. Experiments have been also performed on dilute ^{4}He–^{3}He mixtures in [166,202], where phase separation leads to the formation of nearly pure ^{4}He inclusions in the nearly pure ^{3}He matrix. The kinetics of phase transitions in solid He was also studied in [317] by pressure measurements. The data obtained are summarized in Table 10.3 along with the results of this chapter.

In [202] the kinetics of phase separation was investigated at different supercoolings of the mixture, $\Delta T = T_i - T_f$, in the two-phase region. At low ΔT the separation time constant is, as expected, high and almost independent of T_f, which is due to heterogeneous nucleation. At high ΔT, τ is small and T_f-independent again, and τ increases sharply with growing T_f only in a narrow region of ΔT. This behavior is consistent with the theory of homogeneous nucleation. The σ data (interfacial tension) are presented in Table 10.3.

Phase separation in dilute ^{4}He–^{3}He mixtures produces different systems [166]. Here temperature cycling of a two-phase crystal leads to the formation of a vacancy cluster in the region of separation, which consists of ordered ^{4}He atoms arranged around a vacancy. In this case the pressure variation in the crystal is determined by the change in the cluster radius, and we can thus estimate σ as a fitting parameter from experimental results (see Table 10.3).

The interphase surface tension coefficients of [202] and [166] are in good agreement (Table 10.3), and they exceed the σ of [317] only by 15–20%. If we take this distinction as real, it could be related to the difference in the cluster density between the ^{4}He or ^{3}He-enriched phases. In experiments [317], for the first time the main parameter of the theory, namely the interphase surface tension of the solid ^{3}He–^{4}He mixture, was obtained through two independent experiments, bounded diffusion measurements and measurements of the separation time constant. Their values obtained in different experiments within the homogeneous nucleation model are in good agreement (Table 10.3). The similarity of the above results supports the validity of the methods used. The calculated σ is about 1.6 times lower than the measured value for segregated liquid mixtures [194]. It is possible that in this case the lower σ values are determined by the small sizes of the new phase clusters and the vacancy clusters.

Phase separation in solid helium isotope mixtures is an example of a first-order transition with a conserved order parameter. This system is attractive because the separation process occurs on an accessible time scale: slower than that in fluids, but faster than that in conventional solids. This is a consequence of the unique nature of the atomic motion in solid helium, where quantum exchange results in a temperature-independent diffusion coefficient, intermediate between that of a solid and a liquid [173]. The time scales in ^{3}He–^{4}He are such that all three stages of the first-order phase transition (nucleation, independent growth and coarsening) may be identified distinctly.

The first observation of all the three stages of homogeneous nucleation and growth in this system was reported in [203]. The experimental approach [202, 317], already described in Section 10.3, involves the simultaneous use of two powerful tools: NMR and high precision pressure measurements, utilized during stepwise cooling through the transition and allowing the observation of phase separation in real time.

As shown in Chapter 4, during the late-stage growth, or coarsening, the droplet size is predicted to increase with a characteristic power law $l(t) \sim t^a$ [173], where the exponent a depends on the "universality class" of the transition. To investigate the coarsening, in [203] the 2%^{3}He–^{4}He crystal was quenched from above the separation down to 100 mK and the size of droplets was followed using NMR. The initial droplet size was 8 μm. The evolution of l with time is shown in Figure 10.10, where $1/l$ is plotted against $1/t^{1/3}$. The asymptotic late-stage behavior is indicated on the left-hand side where the approach to linearity indicates the characteristic exponent $a = 1/3$, in accordance with the Lifshitz–Slezov law for a process

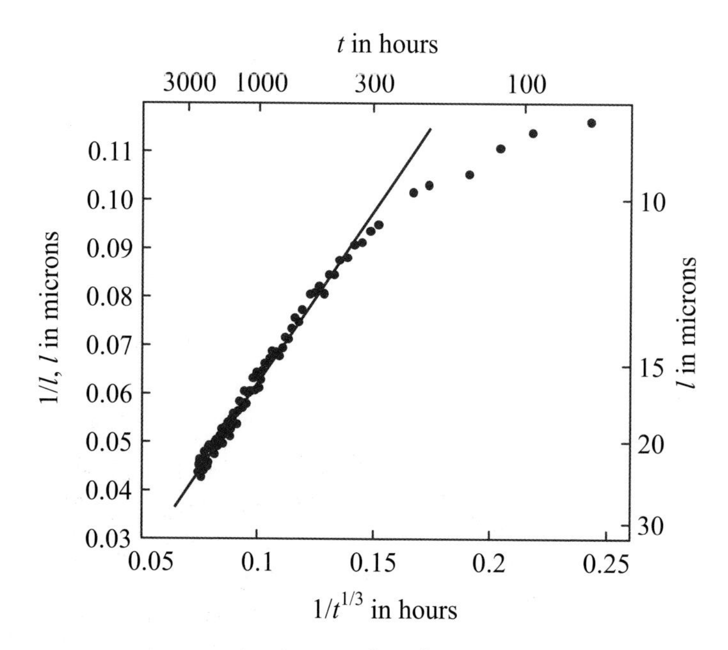

Figure 10.10: Coarsening at phase separation in 2%^{3}He–^{4}He system [203].

with a conserved order parameter (Chapter 4),

$$l(t) = \left(\frac{4}{9} \frac{vc_\infty}{k_B T} \sigma D t \right)^{1/3}, \tag{10.39}$$

where v is the atomic volume and c_∞ is the equilibrium concentration of ^{3}He in the dilute phase. The slope of the asymptote of Figure 10.10 is found in [203] to be equal to $1.09 \times 10^7 \mathrm{s}^{1/3}\mathrm{m}^{-1}$. Thus, the final, coarsening, stage of the phase separation process was also observed in ^{3}He–^{4}He mixtures.

11 Nucleation versus Spinodal Decomposition in Confined Binary Solutions

11.1 Introduction

Nucleation and spinodal decomposition are two major mechanisms how first-order phase transitions may proceed in a variety of systems [14, 33, 44, 45, 71, 93, 256, 272, 273, 337]. Which one of the mentioned mechanisms dominates in the decomposition process is commonly assumed to depend on the degree of instability of the initial state of a phase-separating system. The phase transition is supposed to proceed via nucleation and growth for metastable systems [71, 337], while for thermodynamically unstable systems the mechanism of spinodal decomposition is expected to govern the process [33, 44, 45, 272].

Following the basic ideas anticipated in its basic premises already by Gibbs [85], in nucleation a nucleus of initially small size is supposed to be formed stochastically with state parameters widely similar to the properties of the newly evolving macroscopic phases. In contrast, spinodal decomposition is characterized initially by smooth changes of the state parameters of the system (composition, density, etc.) extended, in general, over large regions in space. These differences in the basic models lead to essentially different theoretical approaches to the description of nucleation–growth and spinodal decomposition processes, respectively. In the simplest formulation, spinodal decomposition is treated as a process of spontaneous growth of a set of long-wavelength fluctuations of the density or composition of the initial state [44, 45]. In such description, the growth increment of these fluctuations is determined in dependence on the wavenumbers of the respective modes as performed for the first time by Cahn and Hilliard [44, 45]. In the decay of initially metastable states via nucleation, the bulk properties of the clusters are supposed to be widely similar to the properties of the respective macroscopic phases [85] and the process of stochastic formation, the further growth and shrinkage of such clusters is analyzed. Briefly speaking, in the initial stages of spinodal decomposition the change of density and/or composition is determined for a more or less fixed size of the new phase regions, while nucleation–growth models draw the attention to a change of the size of the clusters at given values of their intensive state parameters.

Historically, the mentioned different approaches have been developed employing (or reinventing) two different thermodynamic methods of description of thermodynamically heterogeneous systems developed by Gibbs [85] (nucleation) and van der Waals [214, 333] (spinodal decomposition), respectively. The classical Gibbs' theory was and is employed till now as the most frequently used tool basically for the determination of the properties of critical clusters determining the rate of cluster formation in metastable systems and, employing more or less explicitly expressed additional assumptions, to cluster growth and shrinkage processes

Kinetics of First-order Phase Transitions. Vitaly V. Slezov

ISBN: 978-3-527-40775-0

(cf. [249, 253, 254]). Gibbs' classical approach cannot give any predictions about phase formation processes evolving in unstable initial states. In contrast, the van der Waals & Cahn–Hilliard approach allows one to determine the properties of critical clusters for metastable systems and the modes of highest density amplification for phase separation in unstable initial states.

Both Gibbs' and van der Waals' methods of description of heterogeneous systems are hereby conventionally considered as essentially correct and equivalent. However, as shown already long ago by Cahn and Hilliard [44], the predictions concerning the properties of critical clusters in metastable systems derived via the Gibbs and van der Waals methods are in deep contradiction to each other (for more details see [249, 254]). These contradictions are especially significant in the vicinity of the classical spinodal curve separating metastable from unstable initial states. In addition, the above-mentioned comparison of similarities and differences of nucleation and spinodal decomposition processes is somewhat oversimplified. Modern theories of both spinodal decomposition and nucleation exhibit more complicated features in comparison to the models as described briefly above (see, for example, [28, 33, 35, 93, 146, 147, 254, 272]). Moreover, in contrast to the classical picture a smooth transition from metastability to instability has been observed both in computer models of phase-separating systems [28, 33, 35, 93] and in experiment [263].

In the papers [249, 254], it was shown that the contradictions between Gibbs' and van der Waals' methods of description of thermodynamically heterogeneous systems in application to phase formation processes can be resolved by generalizing Gibbs' approach. In this generalization (for details see [258, 259]), Gibbs' idealized cluster model is employed again for the theoretical treatment of density or composition fluctuations; however, the basic equations are generalized allowing one to consider, in contrast to Gibbs' classical approach, the interfacial tension, in general, as a function of the state parameters of both ambient and cluster phases. It was shown that this generalization of Gibbs' approach leads, in addition to a variety of other consequences, to the reconciliation of both the mentioned thermodynamic methods of description of heterogeneous systems. Moreover, the generalized Gibbs method is shown to allow one to also arrive at an understanding of basic features of the kinetics of spinodal decomposition [3].

In the present chapter, we first consider spinodal decomposition in solid solutions based on the classical Cahn–Hilliard–Cook approach [44, 54]. However, in generalization of the mentioned theories, we assume that the system is adiabatically closed (see also [162, 177, 235, 238]). For this reason, in the course of evolution of the phase separation process a feedback of the phase transformation kinetics on the state of the ambient phase occurs with the change of the temperature of the system. As will be shown the existence of such feedback leads to a similar general scenario of the transformation kinetics as compared with cluster formation and growth and to self-similarity of the final stage of evolution.

The main attention here is directed, however, to another problem. It is shown that basic features of spinodal decomposition, on one side, and nucleation, on the other side, and the transition between both mechanisms can be treated within the framework of above-described generalized thermodynamic cluster model [3, 254]. Hereby the clusters, representing the density or composition variations in the system, may change with time both in size and in their intensive state parameters (density and composition, for example). In the first part of the analysis, we consider phase separation processes in dependence on the initial state of the system for the case when changes of the state parameters of the ambient system due to the evolution of the clusters can be neglected as this is the case for cluster formation in an infinite system. As a next step, the effect of changes of the state parameters on cluster evolution is analyzed. Such depletion effects are of importance both for the analysis of phase formation in confined systems [33, 84, 189, 205, 222, 262, 266, 331] and for understanding of the evolution of ensembles of clusters in large (in the limit infinite) systems [229, 230, 233, 234, 331]. The results of the thermodynamic analysis are employed in both cases to exhibit the effect of thermodynamic constraints on the dynamics of phase separation processes. As a model system for the analysis, we again consider phase separation in a binary regular solution, (see also [25, 249, 254]).

The chapter is organized as follows. In Section 11.2, spinodal decomposition in adiabatically closed systems is analyzed. In Section 11.3, basic equations employed for the thermodynamic analysis of phase separation in solutions in terms of the generalized Gibbs approach are summarized. In Section 11.4, these results are applied to the analysis of phase formation in infinite domains in the sense as specified above. In Section 11.5, finite size effects in the kinetics are studied both in application to phase separation in systems of finite size and with respect to the understanding of the evolution in macroscopic systems described in terms of formation and competitive growth of ensembles of clusters [155, 331]. A discussion of the results and possible further developments in Section 11.6 completes the chapter.

11.2 Spinodal Decomposition in Adiabatically Isolated Systems

11.2.1 The Cahn–Hilliard–Cook Equation

Following van der Waals ([333], see also Rowlinson [214], Cahn and Hilliard [44]) the free enthalpy G of a binary inhomogeneous solution can be written in a first approximation in the form

$$G = \int \left[g(c) + \kappa (\nabla c)^2 \right] \, \mathrm{d}V. \tag{11.1}$$

Here c is the volume concentration of one of the components in the solution, again, $g(c)$ is the volume density of the free enthalpy, and κ $(\kappa > 0)$ a coefficient describing the contributions to the thermodynamic potential due to the inhomogeneities in the system.

If the deviations from the initial homogeneous concentration c_0 are relatively small, then a Taylor expansion of Eq. (11.1) results in the following expression for the change of the free enthalpy connected with the evolution of the concentration field $c(\mathbf{r}, t)$:

$$\Delta G = \int \left[\frac{1}{2} \left(\frac{\partial^2 g}{\partial c^2} \right)_{c_0, T} (c - c_0)^2 + \kappa (\nabla c)^2 \right] \, \mathrm{d}V. \tag{11.2}$$

In agreement with the thermodynamic stability conditions (see, e.g., Kubo [133]), a spontaneous growth of the density fluctuations takes place only for $g''(c_0, T) < 0$, since only in this case the amplification of the density profile may be accompanied by a decrease of the free enthalpy of the system.

In the framework of the Cahn–Hilliard–Cook theory, the kinetics of the decomposition process is described by a generalized diffusion equation connecting the variations in the thermodynamic potential G with the kinetics of the decomposition process [44, 54]. It follows from the set of equations

$$\frac{\partial c}{\partial t} + \mathrm{div}\mathbf{j} = 0, \qquad \mathbf{j} = \mathbf{j}_D + \mathbf{j}_A, \tag{11.3}$$

$$\mathbf{j}_D = -M\nabla \frac{\delta G}{\delta c}, \qquad \mathbf{j}_A = -\nabla A(\mathbf{r}, t) \tag{11.4}$$

and has the form

$$\frac{\partial c(\mathbf{r}, t)}{\partial t} = Mg''(c_0, T)\nabla^2 c(\mathbf{r}, t) - 2M\kappa\nabla^4 c(\mathbf{r}, t) + \nabla^2 A(\mathbf{r}, t), \tag{11.5}$$

where $\mathbf{j}_D$ is the vector describing the deterministically determined density of fluxes of particles, $\mathbf{j}_A$ represents the flow connected with the fluctuating scalar field $A(\mathbf{r}, t)$ superimposed on the deterministic flow, and M is a mobility coefficient.

For a solution of this equation, the $c(\mathbf{r}, t)$ and $A(\mathbf{r}, t)$ fields are expressed through Fourier expansions via

$$c(\mathbf{r}, t) = c_0 + \sum_{-\infty}^{\infty} S(\mathbf{k}_n, t) \exp(i\mathbf{k}_n\mathbf{r}), \tag{11.6}$$

$$S(\mathbf{k}_n, t) = \frac{1}{V}\int [c(\mathbf{r}, t) - c_0] \exp(-i\mathbf{k}_n\mathbf{r})\, d\mathbf{r}, \tag{11.7}$$

$$A(\mathbf{r}, t) = \sum_{-\infty}^{\infty} B(\mathbf{k}_n, t) \exp(i\mathbf{k}_n\mathbf{r}), \tag{11.8}$$

$$B(\mathbf{k}_n, t) = \frac{1}{V}\int A(\mathbf{r}, t) \exp(-i\mathbf{k}_n\mathbf{r})\, d\mathbf{r}. \tag{11.9}$$

Here V is the volume of the system under consideration.

Equations (11.3)–(11.9) result in the following differential equation for the spectral function $S(\mathbf{k}_n, t)$:

$$\frac{\partial S(\mathbf{k}, t)}{\partial t} = R(\mathbf{k}, t)S(\mathbf{k}, t) - \kappa^2 B(\mathbf{k}, t), \tag{11.10}$$

where the amplification factor $R(\mathbf{k}, t)$ is determined by

$$R(\mathbf{k}, t) = -M\mathbf{k}^2 \left[g''(c_0, T) + 2\kappa\mathbf{k}^2\right]. \tag{11.11}$$

The subscript n in $\mathbf{k}_n$ is omitted here and subsequently for simplicity of the notations.

Based on Eq. (11.11), in analogy with the critical cluster size in nucleation a critical wavenumber k_c may be introduced. It is defined by the condition that the deterministic amplification rate $R(k_c, t)$ is equal to zero. This condition yields

$$\mathbf{k}_c^2 = -\frac{1}{2\kappa} g''(c_0, T). \tag{11.12}$$

Concentration waves in the Fourier spectrum with wavenumbers $k > k_c$ decay while the modes with $k < k_c$ grow. The value of the wavenumber corresponding to the highest amplification rate is given by

$$k_{\max} = \frac{1}{\sqrt{2}} k_c . \tag{11.13}$$

Moreover, in experimental studies of phase transformation processes not only the spectral function itself but a quantity proportional to the average of the square of the spectral function $\langle S^2 \rangle = \langle S S^* \rangle$ is of relevance. Assuming that in the average the concentration fluctuations are equal to zero and uncorrelated

$$\langle A(t) \rangle = 0, \qquad \langle A(\xi) A(\chi) \rangle = Q(\mathbf{k}) \delta(\xi - \chi) \tag{11.14}$$

we get (see, e.g., [162])

$$\frac{\partial \langle S^2(\mathbf{k}, t) \rangle}{\partial t} = 2R(\mathbf{k}, t) \langle S^2(\mathbf{k}, t) \rangle + \mathbf{k}^4 Q(\mathbf{k}), \tag{11.15}$$

$$Q(\mathbf{k}) = \frac{2 M k_B T}{V} \frac{1}{k^2} . \tag{11.16}$$

Finally, by introducing dimensionless wavenumbers $\widetilde{k}$ and a dimensionless time scale $\widetilde{t}$ via

$$\widetilde{k} = ak, \qquad \widetilde{t} = \frac{4 \kappa M}{a^4} t, \tag{11.17}$$

we obtain with Eq. (11.12)

$$\frac{\partial \langle S^2(\widetilde{k}, \widetilde{t}) \rangle}{\partial \widetilde{t}} = \widetilde{k}^2 \left\{ \left[\widetilde{k}_c^2 - \widetilde{k}^2 \right] \langle S^2(\widetilde{k}, \widetilde{t}) \rangle + \frac{k_B T a^2}{2 \kappa V} \right\} . \tag{11.18}$$

The parameter a is a measure of the intermolecular distance in the solution.

11.2.2 Thermodynamic Aspects

The process of spinodal decomposition is considered conventionally by assuming that the temperature in the system remains constant in the course of the decomposition process. However, both from a principal point of view and with respect to a number of possible technological applications an important question is, how is the picture changed if isothermal conditions are replaced by the constraint of an adiabatic closure of the system.

Under the condition of an adiabatic closure, the latent heat of the transformation results in an increase of the temperature in the system and, consequently, modifications of the decomposition kinetics are expected to occur. This qualitative difference is due to the fact that the critical wavenumber becomes temperature and thus time dependent and decreases with increasing temperature, respectively, increasing degree of advancement of the decomposition process. In this way, nonlinearities enter the description already in the framework of the linear Cahn–Hilliard–Cook theory.

The aim of the present section consists in the analysis of these modifications and the characterization of the whole course of spinodal decomposition in adiabatically isolated systems. Since we are interested mainly in qualitative results, we here restrict ourselves to a description of spinodal decomposition at the level of the Cahn–Hilliard–Cook theory to reveal the specific role of the nonlinearity arising from the chosen boundary conditions. The results obtained in the framework of the Cahn–Hilliard–Cook theory for isothermal systems serve as a reference for the variations caused by the temperature changes in the system. Moreover, the system, where the transformation occurs is assumed to be a binary regular solution as discussed by Becker [25], Cahn and Hilliard [44]. Heat conduction processes inside the system are considered to proceed fast as compared with diffusional fluxes. In this way, at any moment of time an internal thermal equilibrium is established.

The free enthalpy g^* and the potential energy u^* (both quantities referred to one particle) of a regular solution can be expressed as (cf. [25, 44])

$$g^* = \omega^* x(1-x) + k_B T \left[x \ln x + (1-x)\ln(1-x) \right] + \kappa^* (\nabla x)^2, \tag{11.19}$$

$$u^* = \omega^* x(1-x) + \kappa^* (\nabla x)^2. \tag{11.20}$$

The molar fraction x in Eqs. (11.19) and (11.20) can be replaced by the volume concentration c via

$$x = \frac{c}{c_t}, \qquad c_t = \frac{N}{V}. \tag{11.21}$$

Here N is the total number of particles and V the volume of the solution.

The change in the temperature of the adiabatically isolated system due to the evolution of the concentration field $c(\mathbf{r}, t)$ can be expressed as [235]

$$\Delta T = \frac{1}{(N C_p^*)} \int \left\{ \omega (c-c_0)^2 - \kappa \left[\nabla (c - c_0) \right]^2 \right\} \, \mathrm{d}V. \tag{11.22}$$

Here C_p^* is the specific heat per particle of the solution and the notations

$$\omega = \frac{\omega^*}{c_t}, \qquad \kappa = \frac{\kappa^*}{c_t} \tag{11.23}$$

are used.

In terms of the spectral function S the temperature variations may be written as

$$\Delta T = \left(\frac{V\omega}{C_p^* N} \right) \sum_{\mathbf{k}} |S(\mathbf{k}, t)|^2 \left(1 - \frac{1}{2} a^2 \mathbf{k}^2 \right). \tag{11.24}$$

Hereby the relation

$$\kappa = \frac{1}{2} \omega a^2 \tag{11.25}$$

was used in addition, valid for the considered regular solutions [44]. The parameter a in Eq. (11.25) is, as previously, a measure of the intermolecular distance in the solution.

It turns out in this way that the structure function S determines not only the concentration field but also the actual temperature of the system. To obtain results from the above equations, in this way a self-consistent solution both for the concentration fields and the temperature changes has to be developed.

In application of the above-derived equations, one should bear in mind that in the Cahn–Hilliard–Cook approach one deals with a coarse-grained free energy, averaged over volume elements considerably exceeding in size the intermolecular distances (lattice constants). Similarly, the applied macroscopic concept of a wavelength λ looses any meaning if $\lambda = 2\pi/k$ is less than the lattice constant. This condition gives a natural estimate of the upper limit of the number of modes, which have to be taken into account in the numerical calculations carried out to solve the above given equations.

11.2.3 Results of Numerical Calculations

Going over to the dimensionless variables $\widetilde{t} = t(4\kappa M)/a^4$ and $\widetilde{k} = ak$, and using Eq. (11.25) the basic equation (11.18), describing the time dependence of the structure factor, can be written in the form (see also [162])

$$\frac{\partial \left\langle S^2(\widetilde{k},\widetilde{t})\right\rangle}{\partial \widetilde{t}} = \widetilde{k}^2 \left[\left(\widetilde{k}_c^2 - \widetilde{k}^2\right) \left\langle S^2(\widetilde{k},\widetilde{t})\right\rangle + \left(\frac{k_B T}{\omega V}\right)\right], \tag{11.26}$$

$$\widetilde{k}_c^2 = 2 - \Omega_1 T, \qquad T = T_0 + \Delta T, \tag{11.27}$$

$$\Delta T = \Omega_2 \sum_{\mathbf{k}} \left\langle S^2(\widetilde{k},\widetilde{t})\right\rangle \left(1 - \frac{\widetilde{k}}{2}\right), \tag{11.28}$$

where Ω_1 and Ω_2 are two parameters characterizing specific properties of the solution considered.

In the homogeneous thermodynamically unstable initial state, the temperature T equals T_0. Correspondingly, the initial value of the critical wavenumber $\widetilde{k}_c$ is $\widetilde{k}_{c0}$. In the course of the decomposition process the temperature increases and $\widetilde{k}_c$ decreases.

The results of the numerical evaluation of Eqs. (11.26)–(11.28) are shown in Figures 11.1–11.3. In Figure 11.1, the structure factor is shown for different moments of time. In contrast to the picture observed for the isothermal case the maxima of the S^2-curves are shifted to lower values of $\widetilde{k}$ with time. In Figure 11.2 (left curve), the temperature T is shown as a function of time. Clearly three different stages of the transformation may be distinguishably characterized by different types of functional dependences $T = T(\widetilde{t})$. The first stage of spinodal decomposition is characterized by the stochastic generation of the random concentration field [54]. Remember that here we start the process with a homogeneous initial state. In experimental investigations, such a stage may not occur since already in any equilibrium state density fluctuations with well-defined properties are present. Such fluctuations will be retained in the process of transformation of the system into the respective nonequilibrium state (cf. also [250]).

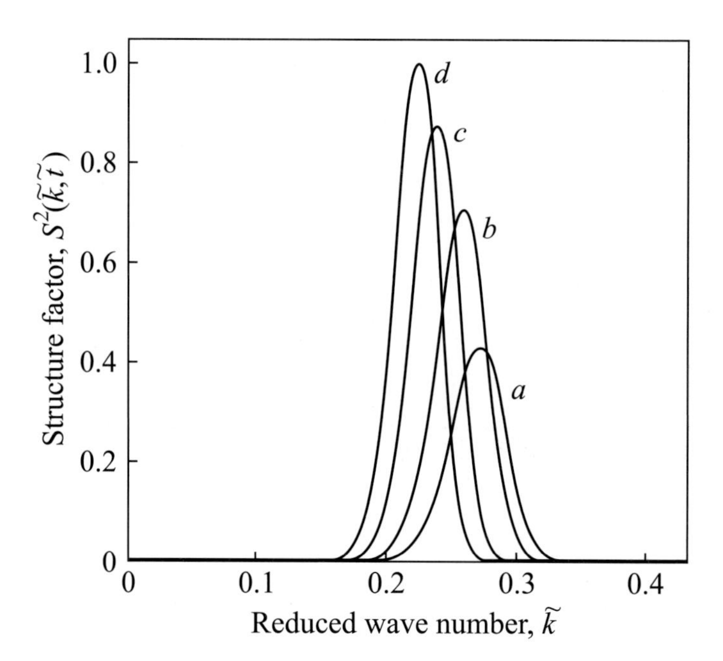

Figure 11.1: Structure factor $S^2(\widetilde{k}, \widetilde{t})$ for different moments of time ($\widetilde{t} = 3000$ (a), 4000 (b), 5000 (c), 7000 (d), and 9000 (d) (in reduced units)). With increase of time a shift of the maximum to lower values of the wavenumber and an increase of the value of the maximum is observed.

This first stage of spinodal decomposition is followed by a stage of independent growth of different modes. Hereby such modes are amplified, in particular, corresponding to wavenumbers $\widetilde{k}_{\max} = \widetilde{k}(0)/\sqrt{2}$. This type of evolution then goes over into a stage of moderate competitive growth of the modes characterized by self-similarity and scaling laws.

The behavior of temperature in dependence on time is also reflected in the $\widetilde{k}_c = \widetilde{k}_c(\widetilde{t})$-curve, shown in logarithmic coordinates in Figure 11.2 (right curve). An analysis of the curve verifies that in the second and third (late) stages of the transformation a scaling behavior for the critical wavenumber of the form

$$\widetilde{k}_c \propto \widetilde{t}^{-\alpha}, \qquad \alpha \sim \begin{cases} \frac{3}{2} & \text{second stage} \\ \\ \frac{1}{4} & \text{third stage} \end{cases} \tag{11.29}$$

is established. This type of functional dependence of the critical wavenumber on time is confirmed both by experimental results [250, 334] and by Monte Carlo simulations (MC) of spinodal decomposition in adiabatically closed systems [177].

The scaling law as expressed by Eq. (11.29) is different from the well-known power law; one has to expect according to the Lifshitz–Slezov theory of coarsening [155]. Such a Lifshitz–Slezov behavior is indeed found for isothermal MC simulations of the same process. It follows that the temperature variations in the system result in a different scaling behavior as compared with the isothermal case, at least, for the considered interval of time.

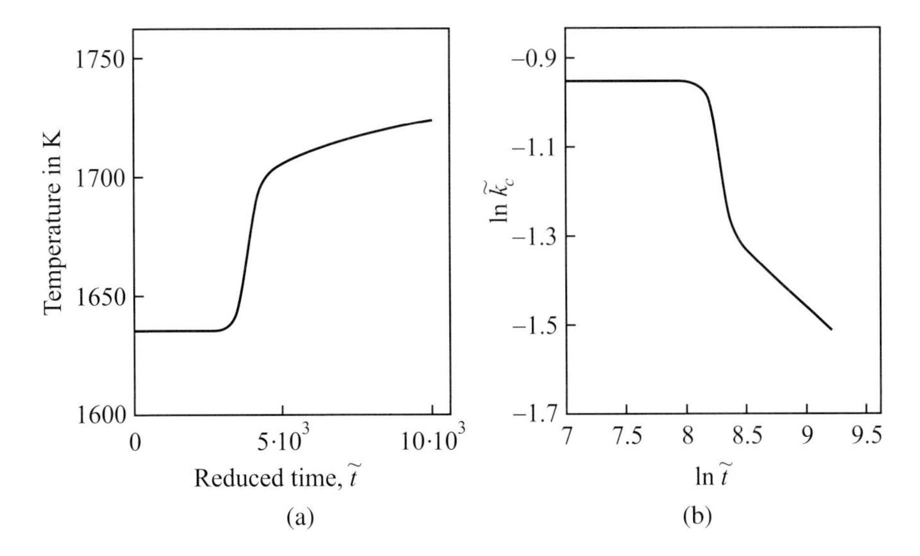

Figure 11.2: (a): Temperature vs time curve for spinodal decomposition in adiabatically closed systems. The curve results from the solution of Eqs. (11.26)–(11.28) (for the details of the computation and the values of the parameters see [162]); (b):The dependence of the critical wavenumber $\widetilde{k}_c$ on time in a double logarithmic plot. In the second and third stages of the decomposition process a linear behavior is found indicating the existence of power laws of the form $\widetilde{k}_c \sim t^{-\alpha}$. As a result of a linear regression a value $\alpha \sim 1/20$ was found for the second stage of the process while for the third stage $\alpha = 0.245 \sim 1/4$ was obtained. Similar dependences were observed recently in experimental investigations of spinodal decomposition [250, 334] in sodium silicate glasses.

The existence of a power law for k_c in the late stage of the transformation, where the temperature changes only slightly, suggests that the structure factor also obeys some kind of self-similarity. Thus, we may propose an equation of the form [103, 161]

$$\langle S^2 \rangle = f(t)g(u), \qquad u = \frac{\widetilde{k}}{\widetilde{k}_c(\widetilde{t})} \tag{11.30}$$

to be valid. Moreover, we demand, that at any moment of time the normalization condition

$$\int_0^\infty g(u)\, du = 1 \tag{11.31}$$

is fulfilled.

In Figure 11.3 the function $g(u)$, as calculated from the numerical solution of the set of Eqs. (11.26)–(11.28), is shown for different moments of time in the third stage of the transformation. It can be easily noticed that, indeed, a universal distribution function $g(u)$ is established, which gives a verification of the proposal expressed by Eqs. (11.30) and (11.31).

The results of the numerical solution of the basic equations (11.26)–(11.28) describing decomposition in adiabatically closed systems lead to the conclusion that in analogy with nu-

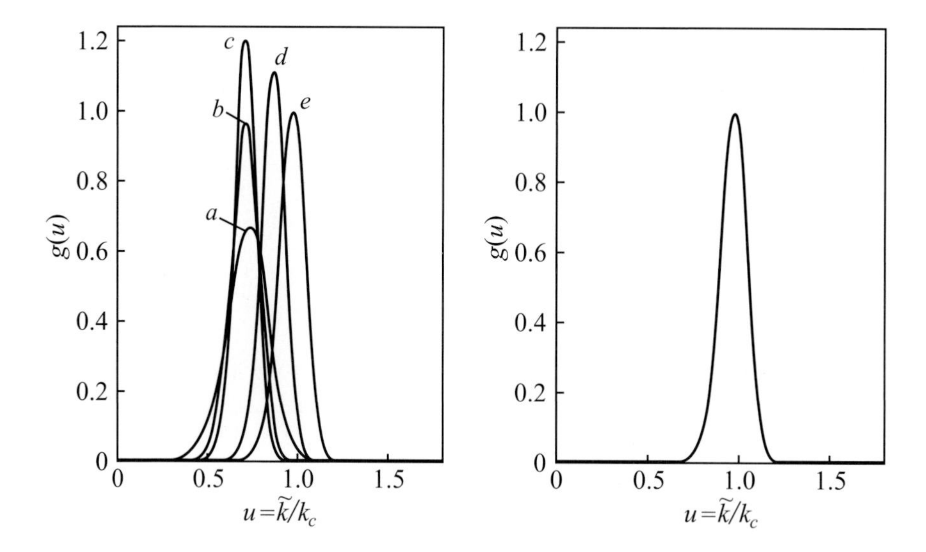

Figure 11.3: Normalized structure factor $g(u)$ in dependence on the reduced wavenumber $\widetilde{k}/\widetilde{k}_c$. The different curves on the left-hand side of the figure correspond to $\widetilde{t} = 1000$ (a), $\widetilde{t} = 2000$ (b), $\widetilde{t} = 3000$ (c), $\widetilde{t} = 4000$ (d), $\widetilde{t} = 5000$ (e). In the course of the evolution, a time-independent distribution is approached as shown on the right-hand side. Here the curves obtained for $\widetilde{t} = 5000$ and for $\widetilde{t} = 9000$ are shown. As easily verified both curves coincide practically.

cleation and growth processes in finite systems [220,230], the whole course of the transformation can be roughly divided into three different stages, a first stage of formation and moderate growth of the initial density inhomogeneities, a second stage of rapid increase of these density differences, connected in adiabatic systems with a sharp increase of temperature, and a third stage of reorganization of the concentration field, characterized by self-similarity and scaling laws. The following section is devoted to a theoretical interpretation of this scaling behavior [238].

11.2.4 Theoretical Interpretation

For the analytical verification of the scaling laws we omit stochastic terms in the Cahn–Hilliard–Cook equation being of significant importance only with respect to the formation of the initial concentration profile. A substitution of Eq. (11.30) into the differential equation (11.26) then yields

$$\frac{\mathrm{d}}{\mathrm{d}t}\ln f(t) - \frac{\widetilde{k}}{\widetilde{k}_c^2}\frac{\mathrm{d}\widetilde{k}_c}{\mathrm{d}\widetilde{t}}\frac{\mathrm{d}}{\mathrm{d}u}\ln g(u) = \widetilde{k}_c^4 u^2(1-u^2). \tag{11.32}$$

Assuming (as it is verified by the numerical calculations) that

$$\frac{\mathrm{d}}{\mathrm{d}t}\ln f(t) \cong 0 \tag{11.33}$$

holds we arrive at

$$\frac{\mathrm{d}}{\mathrm{d}t}\left(\frac{\widetilde{k}_c^{-4}}{4}\right)\frac{\mathrm{d}}{\mathrm{d}u}\ln g(u) = u(1-u^2). \tag{11.34}$$

Since the right-hand side of Eq. (11.34) is only a function of the reduced variable u, the left-hand side of this equation must also be a function of u. Consequently, the relation

$$\frac{\mathrm{d}}{\mathrm{d}t}\left(\frac{\widetilde{k}_c^{-4}}{4}\right) = C_1^{-1} \tag{11.35}$$

has to be fulfilled. The general solution of this equation may be written as

$$\widetilde{k}_c^4 = \frac{C_1}{4(\widetilde{t}+C_1^*)}, \tag{11.36}$$

which for large times results in

$$\widetilde{k}_c^4 \propto \widetilde{t}^{-1}. \tag{11.37}$$

This result immediately confirms the scaling law (11.29) obtained numerically.

Moreover, for the determination of the function $g(u)$ we get the equation

$$\frac{\mathrm{d}}{\mathrm{d}u}\ln g(u) = C_1 u(1-u^2) \tag{11.38}$$

with the general solution

$$g(u) = C_2 \exp\left[C_1\left(\frac{u^2}{2}-\frac{u^4}{4}\right)\right] \tag{11.39}$$

or, equivalently,

$$g(u) = A\exp\left[-B\left(u^2-1\right)^2\right], \tag{11.40}$$

$$A = C_2\exp\left(\frac{C_1}{4}\right), \qquad B = \frac{C_1}{4}. \tag{11.41}$$

This result shows, in agreement with the numerical solution, that, indeed, the maximum of the distribution function $g(u)$ is located at $u = 1$ or $\widetilde{k} = \widetilde{k}_c$. Moreover, the shape of the distribution function is also reproduced in a correct way.

One of the constants A and B in Eq. (11.40) may be determined from the normalization condition (11.31). However, it is easily seen from Eqs. (11.26)–(11.28) or

$$2\widetilde{k}_c\frac{\mathrm{d}\widetilde{k}_c}{\mathrm{d}t} = -\Omega_1\frac{\mathrm{d}T}{\mathrm{d}t} \tag{11.42}$$

that the coefficient C_1 depends on specific properties of the system under consideration (Ω_1 and Ω_2) and is not a universal parameter. In this way, the width of the distribution is well defined but not universal, it depends on specific properties of the system under consideration, in particular, on the rate of change of the critical wavenumber.

11.2.5 Discussion

The theoretical description of spinodal decomposition in the framework of the Cahn–Hilliard–Cook theory comes across two serious difficulties. The first consists in the correct determination of the free energy density for the initial thermodynamically unstable state. The second problem is connected with a correct introduction of nonlinear terms into the original Cahn–Hilliard–Cook equation, to avoid an unlimited amplification of the density fluctuations. Different attempts have been developed to overcome these difficulties (see, e.g., [34, 145]), however, so far a final solution of these problems is missing.

An analysis of phase separation processes, starting from metastable initial states and proceeding via nucleation and growth, leads to the conclusion that the change of the state of the system (change of supersaturation or thermodynamic driving force of the transformation) resulting from the transition determines the whole scenario of the process [94, 220, 230]. In spinodal decomposition the thermodynamic driving force of segregation is given by the second derivative of the free energy with respect to molar fraction or concentration of one of the components. By analogy and in agreement with the thermodynamic stability conditions, one may expect that variations of this quantity should also determine basically the course of the phase transformation in spinodal decomposition.

In the special case, considered here, the variations of the thermodynamic driving force of the transformation can be calculated relatively easily. They are due to, in the applied approach, exclusively temperature changes in the system. This is the only factor taken into account and as already mentioned, in analogy with nucleation and growth processes it determines qualitatively the whole scenario of the transformation. Since, as discussed, the change of the driving force of the transformation is expected to be the major factor, governing the phase transformation, a similar scenario should be always expected to occur in experiments and also in theoretical approaches, when the change of the state of the system (or nonlinearities in spinodal decomposition) are described in a proper way.

Indeed, Binder [29] and Mazenko et al. [174] found theoretically a $(1/4)$-power law behavior based on a field theoretical approach for the description of spinodal decomposition. Moreover, it has been shown by Velasco and Toxvaerd [336] in recent molecular dynamics and Monte Carlo simulations of spinodal decomposition that in binary systems under isothermal conditions in the late stages such $(1/4)$-power laws are also found; giving a additional support of our hypothesis. While in one-component closed systems under adiabatic constraints, as considered here, the change of the thermodynamic driving force of the transformation is due to the increase in temperature, for isothermal conditions in binary systems it is connected with a change in the composition. The existence of such stages in spinodal decomposition has also been predicted theoretically in an alternative way (cf. [195]) and observed experimentally [250, 334].

The exponent in the power law for the critical wavenumber k_c may be, of course, different in dependence on the specific mechanism governing the decomposition process at a certain stage. Note that the equations considered do not include hydrodynamic effects (compare [29, 336]) as well as restrictions on the amplitude of the density variations. In this sense, the considered process describes not the evolution of an already formed two-phase system – as in the Lifshitz–Slezov theory of coarsening – but the process of formation of two distinct phases. It can – and will be, in general, – followed by laws of the type $\langle R \rangle \propto t^{1/2}$ and $\langle R \rangle \propto$

$t^{1/3}$ describing independent and competitive growth of ensembles of clusters, respectively [250, 334]. In the final part of this chapter we will show that – based on the generalized Gibbs approach – a generalized cluster model can be developed taking into account both density (composition) and size variations of the clusters and allowing one to treat both nucleation–growth and spinodal decomposition in a unique way.

11.3 Generalized Cluster Model Approach to the Description of Phase Separation: The Model System

We consider thermodynamic aspects of new phase formation in a binary solid or liquid solution; both in finite domains and infinite space. Since here we are mainly interested in the discussion of the basic principles and consequences of the newly developed generalized Gibbs' approach in application to phase separation, the solution is considered as a regular one representing one of the simplest models of a system consisting of two kinds of interacting molecules. The domain, where the processes of nucleation and/or spinodal decomposition are assumed to proceed, is considered similarly to [205] as a sphere of radius R_0. The limiting situation of an infinite system is thus reached for $R_0 \to \infty$, while finite-size effects take place for finite values of R_0 (see Figure 11.4).

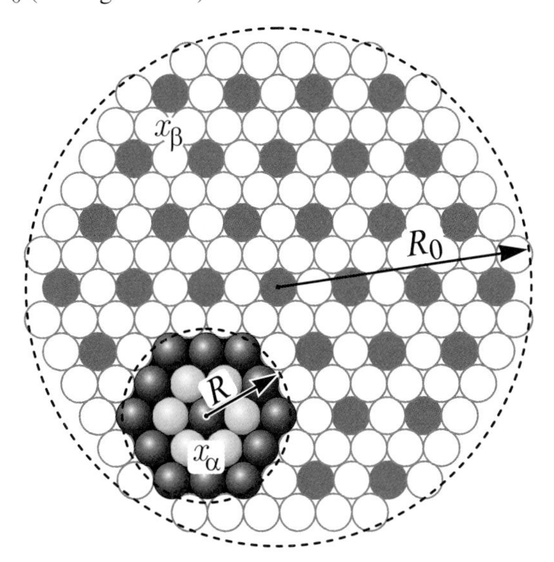

Figure 11.4: Model employed in the analysis: A cluster of size, R, and molar fraction, x_α, is formed in a volume, $V = 4\pi R_0^3/3$, of an initially homogeneous binary solid or liquid solution with a composition given by the molar fraction, x_β. The initial composition of the ambient solution is denoted by x.

Cluster formation in a binary solution results from a redistribution of molecules. Following Gibbs' model approach, we consider a cluster as a spatially homogeneous part of the domain volume with a composition different from the ambient phase. Both size and composition of the cluster may vary in a wide range. As the dividing surface, separating the cluster from the ambient phase in the thermodynamic description underlying the method of analysis, we here always employ the surface of tension [85, 258, 259]. As in an analysis [222, 262] performed in terms of the classical Gibbs approach and investigations of related problems employing different methods [189, 205, 266], the effect of the finite size is taken into account only by the conservation laws for the number of particles of different components in the cluster (specified by the subscript α) and in the ambient phase (specified by β). We may then write

$$\begin{aligned} n_j &= n_{j\alpha} + n_{j\beta} = \text{const}, \qquad j = 1, 2, \\ n &= n_\alpha + n_\beta = \text{const}, \\ n_\alpha &= n_{1\alpha} + n_{2\alpha}, \qquad n_\beta = n_{1\beta} + n_{2\beta}. \end{aligned} \tag{11.43}$$

The molar fractions of the second component in the ambient phase (x_β) and the cluster (x_α) are defined as

$$x_\beta = \frac{n_{2\beta}}{n_{1\beta} + n_{2\beta}}, \qquad x_\alpha = \frac{n_{2\alpha}}{n_{1\alpha} + n_{2\alpha}}. \tag{11.44}$$

The initial state is either a metastable or unstable homogeneous state, characterized by $x_\alpha(0)$ $=x_\beta(0) \equiv x$.

In line with the basic assumptions underlying the model of regular solutions [25] and for simplicity of the notations, the volume per particle, ω, is assumed to be the same for both components and independent on composition ($\omega_\alpha = \omega_\beta \equiv \omega$). Cluster size and particle number in a cluster are then related by the following simple expression:

$$\frac{4\pi}{3} R^3 = n_\alpha \omega. \tag{11.45}$$

Assuming further that a cluster of radius R and composition x_α is formed in a spherical domain of radius R_0 and initial composition x, Eqs. (11.43), (11.44), and (11.45) yield

$$x_\beta = \frac{n x - n_\alpha x_\alpha}{n - n_\alpha} = \frac{x - x_\alpha (R/R_0)^3}{1 - (R/R_0)^3}. \tag{11.46}$$

The change of the Gibbs free energy, ΔG, connected with the formation of one cluster in the initially homogeneous ambient phase in a commonly good approximation can be written as [249, 253, 254]

$$\Delta G = \sigma A + \sum_j n_{j\alpha} (\mu_{j\alpha} - \mu_{j\beta}) + \sum_j n_j (\mu_{j\beta} - \mu_{j0}). \tag{11.47}$$

The first term on the right-hand side of Eq. (11.47) reflects cluster surface effects (σ is the interfacial tension and A is the surface area of the cluster) and the second term reflects cluster bulk contributions to the change of the Gibbs free energy. The third term describes the influence of depletion effects (change of the composition of the ambient phase due to cluster formation) resulting in differences of the chemical potentials per particle in the initial state (μ_{j0}) and the state of the ambient phase once a cluster has been formed ($\mu_{j\beta}$).

For a binary regular solution the chemical potentials of different components in the cluster, $\mu_{j\alpha}$, and ambient solution, $\mu_{j\beta}$, are defined by [25]

$$\mu_{1\alpha/\beta} = \mu^*_{1\alpha/\beta} + k_B T \ln(1 - x_{\alpha/\beta}) + \Omega x^2_{\alpha/\beta},$$
$$\mu_{2\alpha/\beta} = \mu^*_{2\alpha/\beta} + k_B T \ln x_{\alpha/\beta} + \Omega \left(1 - x_{\alpha/\beta}\right)^2, \tag{11.48}$$

where k_B is the Boltzmann constant, T the absolute temperature, and Ω is an interaction parameter describing specific properties of the considered system. The parameter, Ω, can be expressed via the critical temperature, T_c, of the system (cf. also Figure 11.5) as

$$T_c = \frac{\Omega}{2k_B}. \tag{11.49}$$

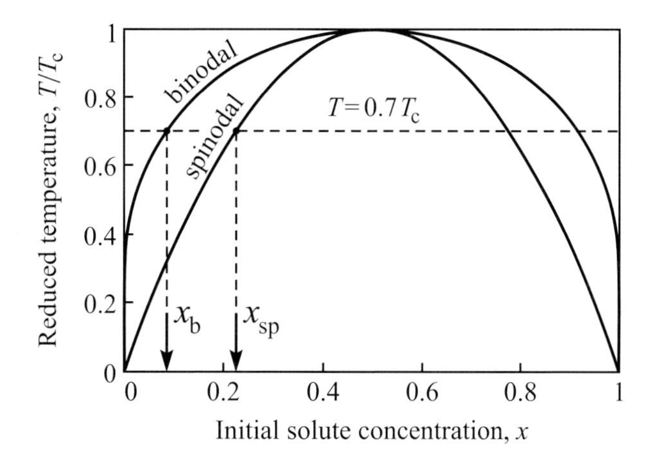

Figure 11.5: Phase diagram of a binary regular solution with binodal and spinodal curves. The spinodal curve separates thermodynamically stable from thermodynamically unstable states of the homogeneous ambient phase. In the present analysis, we assume that the temperature is equal to $T = 0.7T_c$ and can vary the driving force of the phase transformation process by changing the initial composition of the ambient phase, x.

The surface tension between two macroscopic phases with compositions x_α and x_β, respectively, is given, according to Becker ([25], see also [50]) by

$$\sigma = \widetilde{\sigma} \left(x_\alpha - x_\beta\right)^2. \tag{11.50}$$

From Eqs. (11.47), (11.48), (11.49), and (11.50), we have

$$\frac{\Delta G}{k_B T} = \frac{3}{2} n_\sigma^{1/3} n_\alpha^{2/3} \left(x_\alpha - x_\beta\right)^2 + n_\alpha f(x_\beta, x_\alpha) - n f(x_\beta, x), \tag{11.51}$$

where

$$f(x_\beta, x_\alpha) = (1 - x_\alpha)\left\{\ln\frac{1 - x_\alpha}{1 - x_\beta} + 2\frac{T_c}{T}\left(x_\alpha^2 - x_\beta^2\right)\right\}$$
$$+ x_\alpha\left\{\ln\frac{x_\alpha}{x_\beta} + 2\frac{T_c}{T}\left[(1 - x_\alpha)^2 - (1 - x_\beta)^2\right]\right\} \tag{11.52}$$

holds and the scaling parameter, n_σ, for the particle number in the cluster is specified as

$$n_\sigma^{1/3} = \frac{2\widetilde{\sigma}}{k_B T}\left(\frac{4\pi}{3}\right)^{1/3}\omega^{2/3}. \tag{11.53}$$

In addition, we also introduce via Eqs. (11.45) and (11.53) a scaling parameter, R_σ, for the cluster radius as

$$R_\sigma = \left(\frac{3n_\sigma\omega}{4\pi}\right)^{1/3} = \frac{2\widetilde{\sigma}\omega}{k_B T}. \tag{11.54}$$

In further analysis, we will always assume for an illustration of the results that the temperature in the system is equal to $T = 0.7T_c$. The concentration of the solute in the initially homogeneous system is varied in the range from $x = x_b \cong 0.086$ (left branch of the binodal curve) to $x = x_{\text{sp}} \cong 0.226$ (left branch of the spinodal curve) covering metastable initial states and $x_{\text{sp}} < x \leq 0.5$ covering unstable initial states (see Figure 11.5). Since the phase diagram of a regular solution is symmetric [254], we may restrict the analysis to initial states in the considered range with initial concentrations, $x \leq 0.5$. A specification of further parameters like $\widetilde{\sigma}$ and ω is not required, since we compute reduced characteristics, so that such system parameters can enter the description only via the scaling quantities (see also [254]).

11.4 Phase Separation in Infinite Domains

11.4.1 Thermodynamic Analysis

The above-given equations allow us to determine the thermodynamic potential surface as a function of the number of particles, $n_{1\alpha}$ and $n_{2\alpha}$, in the cluster. The results are shown for different values of the initial supersaturation in Figure 11.6 both for metastable ((a) $x = 0.15$, (b) $x = 0.19$, (c) $x = 0.22$) and unstable ((d) $x = 0.3$, (e) $x = 0.4$, (f) $x = 0.5$) initial states. Depletion effects are neglected so far (we consider infinite systems), so we set $R_0 \to \infty$. As far as we are interested mainly in the demonstration of the basic qualitative features, like in Figure 11.6 and similar ones, the numbers are omitted at the axes.

For each of the metastable initial states, the thermodynamic potential surface has, in the vicinity of the critical cluster coordinates, a typical saddle shape. The position of this saddle-point is determined by the set of equations

$$\frac{\partial \Delta G(n_{1\alpha}, n_{2\alpha})}{\partial n_{1\alpha}} = 0\,, \qquad \frac{\partial \Delta G(n_{1\alpha}, n_{2\alpha})}{\partial n_{2\alpha}} = 0, \tag{11.55}$$

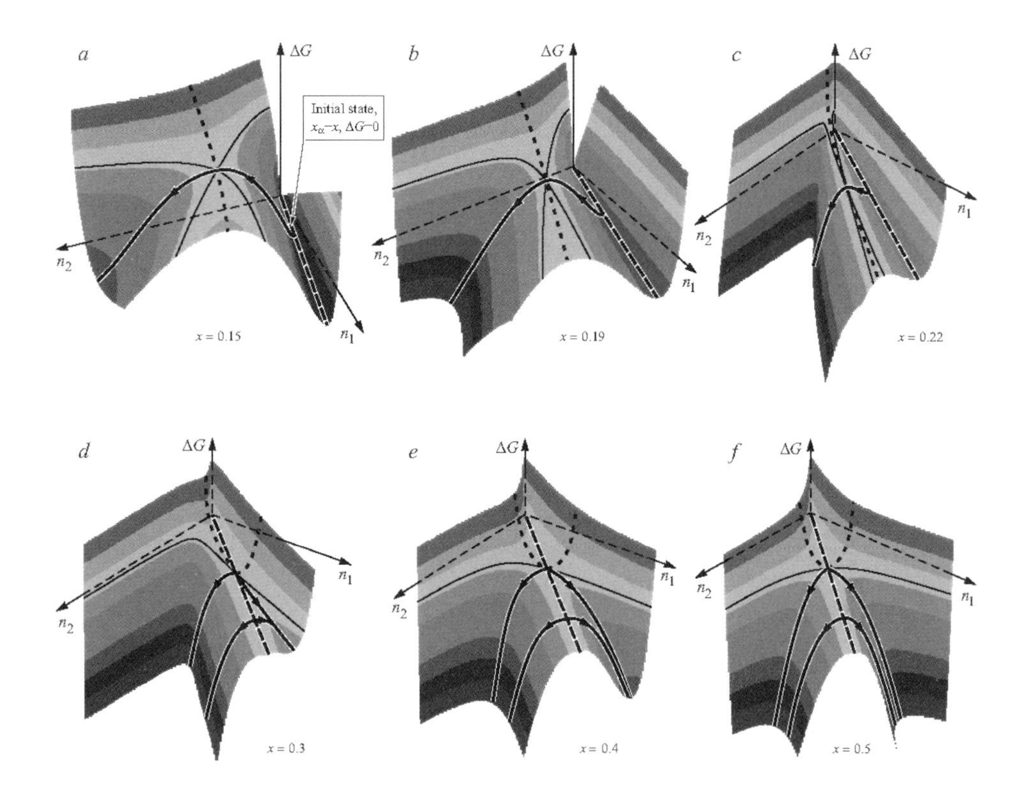

Figure 11.6: Shape of the Gibbs free energy surface for metastable ($x = 0.15$, $x = 0.19$, and $x = 0.22$; (a)–(c)) and unstable initial states ($x = 0.3$, $x = 0.4$, and $x = 0.5$; (d)–(f)). As mentioned, the temperature is chosen equal to $T/T_c = 0.7$ (for further details, see the text).

if we employ Eq. (11.47) for the description of the thermodynamic potential surface. In an alternative approach utilizing Eq. (11.51), we may first determine the size of the cluster for any fixed value of x_α via

$$\frac{\partial \Delta G\,(n_\alpha, x_\alpha)}{\partial n_\alpha} = 0 \qquad \Longrightarrow \qquad n_\alpha^{1/3}(x_\alpha) = -n_\sigma^{1/3}\frac{(x_\alpha - x)^2}{f(x, x_\alpha)}. \tag{11.56}$$

A substitution of the expression for n_α into Eq. (11.51) yields (cf. also Eq. (11.53))

$$\frac{\Delta G(n_\alpha(x_\alpha), x_\alpha)}{k_B T} = \frac{1}{2} n_\sigma \frac{(x_\alpha - x)^6}{f^2(x, x_\alpha)}\,, \qquad n_\sigma = \frac{32\pi}{3}\frac{\widetilde{\sigma}^3\omega^2}{(k_B T)^3}. \tag{11.57}$$

The composition of the critical cluster and the work of critical cluster formation is then obtained by searching for the minimum of $\Delta G(n_\alpha(x_\alpha), x_\alpha)$ with respect to the cluster composition, x_α, [249, 253, 254, 259] as

$$\frac{\mathrm{d}\Delta G\,(n_\alpha(x_\alpha), x_\alpha)}{\mathrm{d}x_\alpha} = 0. \tag{11.58}$$

In order to allow us a better understanding of the shape of the thermodynamic potential surface, contour lines through the saddle are included in the figures by full curves and the curve of steepest increase of the potential surface starting from the critical cluster coordinates by dotted curves. The full curve with arrows describes the most probable trajectory of cluster evolution. It starts at some point along the dashed curve determined by the initial conditions $x_\alpha = x$ and $\Delta G = 0$ (in the initial state, the composition of the cluster is the same as in the ambient phase). Then it passes the saddle point and follows further the trajectory of macroscopic growth with an initial cluster size slightly above the critical size. As discussed in detail in [254], the trajectory of evolution from the initial state to the saddle point can be assumed to coincide, in general, with the path of cluster dissolution starting with initial states slightly below the critical cluster size.

The most probable trajectory of evolution is thus determined for both regions by the macroscopic growth equations. For segregation in solutions, these equations can be written in the form

$$\frac{dn_{1\alpha}}{dt} = -D_1(1-x_\beta)\Theta_0 n_\alpha^\kappa \frac{d\Delta G}{dn_{1\alpha}},$$

$$\frac{dn_{2\alpha}}{dt} = -D_2 x_\beta \Theta_0 n_\alpha^\kappa \frac{d\Delta G}{dn_{2\alpha}}, \tag{11.59}$$

where D_1 and D_2 are the partial diffusion coefficients of different components in the ambient phase. The parameter κ has the value $\kappa = 2/3$ for kinetically limited growth, $\kappa = 1/3$ for diffusion-limited growth, and Θ_0 is a parameter depending only on temperature (we set, as mentioned, the temperature equal to $T = 0.7T_c$). As evident from the above considerations and the structure of Eqs. (11.59), the path of cluster evolution depends on the partial diffusion coefficients of both components of the solution (see for the details [254]); however, qualitatively the picture remains always the same. In Figure 11.6, the trajectories are shown for $D_1(1-x_\beta) = D_2 x_\beta$. In this case, the kinetic prefactors to the partial derivatives of ΔG with respect to $n_{j\alpha}$ in Eqs. (11.59) are the same and the evolution proceeds along the valley of the thermodynamic potential surface, $\Delta G\,(n_{1\alpha}, n_{2\alpha})$, passing the saddle point.

The analysis of Eqs. (11.56) and (11.58) shows [249, 253, 254] that the composition of the critical cluster decreases with increasing supersaturation and approaches the value of the composition of the ambient phase for initial states near to the spinodal curve (cf. Figure 11.7). The work of critical cluster formation decreases monotonously with increasing supersaturation and tends to zero at the spinodal curve (Figure 11.8). Taking into account Eq. (11.52) and the relations

$$\begin{aligned}
\frac{\partial f(x,x_\alpha)}{\partial x_\alpha} &= \ln\left(\frac{x_\alpha}{x}\right) - \ln\left(\frac{1-x_\alpha}{1-x}\right) + 4\left(\frac{T_c}{T}\right)(x - x_\alpha), \\
\frac{\partial^2 f(x,x_\alpha)}{\partial x_\alpha^2} &= \frac{1}{x_\alpha} + \frac{1}{1-x_\alpha} - 4\left(\frac{T_c}{T}\right), \\
\frac{\partial^3 f(x,x_\alpha)}{\partial x_\alpha^3} &= -\frac{1-2x_\alpha}{x_\alpha^2(1-x_\alpha)^2},
\end{aligned} \tag{11.60}$$

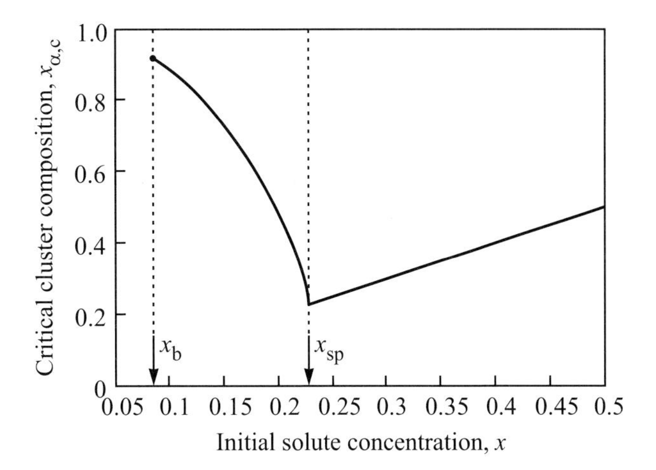

Figure 11.7: Dependence of the composition of the critical cluster, $x_{\alpha,c}$, on the initial solute concentration. For $x > x_{\rm sp}$, the identity $x_{\alpha,c} = x$ always holds [254].

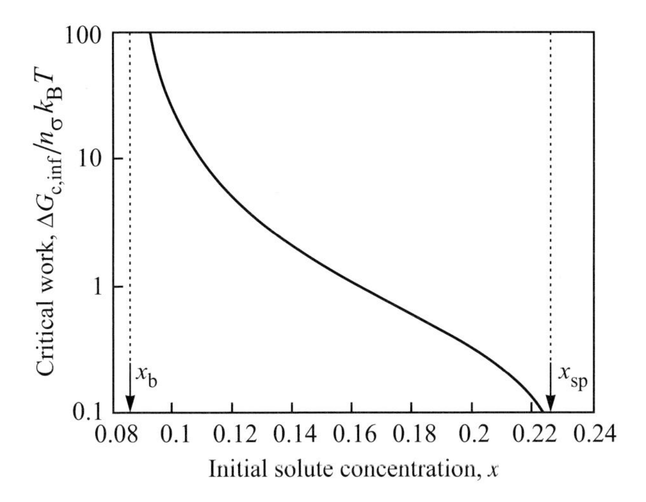

Figure 11.8: Dependence of the minimum work of critical cluster formation, $\Delta G_{c,\rm inf}/(n_\sigma k_B T)$, on the initial solute concentration, x, for infinite domains (specified by the abbreviation, inf) when changes of the state of the ambient solution due to cluster formation can be neglected.

it can be shown that the critical cluster radius, R_c, behaves as

$$\lim_{x \to x_{\rm sp}} R_c \propto \lim_{x \to x_\alpha} \frac{1}{(x_\alpha - x)}. \tag{11.61}$$

In this derivation, the equation $(\partial^2 f(x, x_\alpha)/\partial x_\alpha^2)\big|_{x_\alpha = x_{\rm sp}} = 0$ has been employed [249]. The dependence of the critical cluster size on supersaturation is illustrated in Figure 11.9.

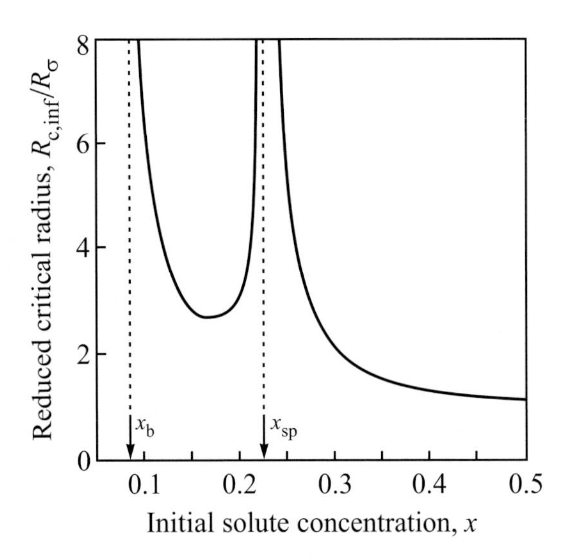

Figure 11.9: Dependence of the critical cluster radius, $R_{c,\inf}/R_\sigma$, on the initial solute concentration, x, for phase formation in an infinite domain at metastable ($x_b < x < x_{sp}$) and unstable ($x > x_{sp}$) initial states of the ambient solution.

The results summarized above are also reflected in Figures 11.6(a)–(c). As evident from the figures, with an increase of the supersaturation (molar fraction in the ambient phase) the nucleation barrier decreases, the location of the saddle point is shifted closer to the line of initial states, $x_\alpha = x$, and at the spinodal, $x = x_{sp}$, the position of the critical cluster tends to the composition of the ambient phase, i.e., the critical cluster is located in this limiting case on the line of initial states. Since, for the initial states corresponding to the spinodal curve, the condition $x_\alpha \to x$ holds, for such states the work of critical cluster formation (determined via the generalized Gibbs approach employed here) tends to zero.

For unstable initial states, $x_{sp} < x \leq 0.5$, the situation is different. Here the critical cluster has always a composition equal to the composition of the ambient phase (cf. Figure 11.7 and [254]). The critical cluster size cannot be expressed here directly by employing Eqs. (11.56) and (11.60), since for $x_\alpha = x$ the relations

$$\left.\frac{\Delta G}{k_B T}\right|_{x_\alpha = x} = \frac{\partial}{\partial x_\alpha}\left(\frac{\Delta G}{k_B T}\right)\bigg|_{x_\alpha = x} = \frac{\partial}{\partial n_\alpha}\left(\frac{\Delta G}{k_B T}\right)\bigg|_{x_\alpha = x} = 0 \tag{11.62}$$

hold independently of the value of n_α in Eq. (11.51). In the definition of the critical cluster size for unstable initial states we have to thus rely on the second-order differential of ΔG with respect to n_α and x_α. As can be proven easily, the second-order differential of ΔG for states with $x_\alpha = x$ is given by $\mathrm{d}^2\Delta G\big|_{x_\alpha = x} = (\partial^2 \Delta G/\partial x_\alpha^2)\big|_{x_\alpha = x}\,\mathrm{d}x_\alpha^2$. The second-order derivative of ΔG with respect to x_α at $x_\alpha = x$ can be expressed as

$$\frac{\partial^2}{\partial x_\alpha^2}\left(\frac{\Delta G}{k_B T}\right)\bigg|_{x_\alpha = x} = 3 n_\sigma^{1/3} n_\alpha^{2/3}\left\{1 - \left(\frac{n_\alpha}{n_{\alpha,c}}\right)^{1/3}\right\}, \tag{11.63}$$

where the notations

$$n_{\alpha,c}^{1/3} = \frac{3n_\sigma^{1/3}}{2K}, \qquad K = -\frac{1}{2}\frac{\partial^2 f(x, x_\alpha)}{\partial x_\alpha^2}\bigg|_{x_\alpha = x} = \frac{1}{2}\left[4\left(\frac{T_c}{T}\right) - \frac{1}{x} - \frac{1}{1-x}\right] \quad (11.64)$$

are employed.

In the range of metastable initial states $x_b < x < x_{\rm sp}$, the critical cluster corresponds to a minimum of $\Delta G(n_\alpha(x_\alpha), x_\alpha)$ with respect to the variations of the state of the cluster, i.e., $(\mathrm{d}^2\Delta G/\mathrm{d}x_\alpha^2) > 0$ holds [259]. For unstable initial states, states along the trajectory $x_\alpha = x$ with $\Delta G = 0$ again correspond to minima with respect to variations of the state parameters of the cluster phase at fixed values of the cluster sizes if the inequality $(n_\alpha/n_{\alpha,c})^{1/3} < 1$ is fulfilled (cf. Eq. (11.63)). However, there exists a cluster size, $n_{\alpha,c}$, where the state along the line $x = x_\alpha$ switches from a minimum to a maximum of ΔG with respect to variations of the cluster composition at fixed values of the cluster sizes. The possible trajectories of evolution for $n_\alpha \geq n_{\alpha,c}$ are shown by full curves with arrows in Figures 11.6(d)–(f).

Moreover, in Figures 11.6(d)–(f), the solid curve, $\Delta G = 0$, divides regions with $\Delta G \geq 0$ and $\Delta G \leq 0$ as compared with the states corresponding to $x_\alpha = x$. In the first region (that is in the region with $\partial^2\Delta G/\partial x_\alpha^2\big|_{x_\alpha=x} > 0$ or $(n_\alpha/n_{\alpha,c}) < 1$) cluster composition changes lead to the growth of the Gibbs free energy, and the cluster is stable in such region. In the second region ($\partial^2\Delta G/\partial x_\alpha^2\big|_{x_\alpha=x} < 0$ or $(n_\alpha/n_{\alpha,c}) > 1$) any compositional change (both increase and decrease of the cluster concentration) results in a decrease of the Gibbs free energy. In such region, the cluster is unstable and the decomposition proceeds via growth of the concentration differences, i.e., according to the basic mechanism commonly assigned to spinodal decomposition. Thus, for $x > x_{\rm sp}$, the system is stable for small clusters and unstable for clusters with a size $n_\alpha > n_{\alpha,c}$. So, changing the size of the clusters with composition equal to the composition of the ambient phase, we arrive at a transition from metastable to unstable states and at $n_\alpha = n_{\alpha,c}$ the minimum transforms into a maximum via a singular point of third order. Recalling the physical meaning of a critical cluster size as the lowest size of a cluster for which a spontaneous further growth in accordance with the thermodynamic evolution laws is possible, $n_{\alpha,c}$ as defined via Eq. (11.64) is obviously an appropriate definition of the critical cluster size for unstable initial states.

In terms of the radius, we may express the critical cluster size in infinite domain as

$$R_{c,\rm inf} = \frac{3R_\sigma}{2K}. \quad (11.65)$$

The parameter K is positive for values of x in the range of unstable initial states and tends to zero at the spinodal curve (cf. [249, 254]); resulting in a divergence of the critical cluster size for unstable initial states near the spinodal curve (cf. Figure 11.9). The work of formation of such critical cluster is, in the range of unstable initial states of the ambient phase, always equal to zero (cf. Eq. (11.51)).

11.4.2 Kinetics versus Thermodynamics in Phase Separation

In discussing the trajectories of evolution in phase separation processes, we assumed here in line with the commonly employed assumption that the evolution to the new phase proceeds via the saddle point of the thermodynamic potential surface. We believe ridge crossing as another possible channel of formation of the new phase [212, 213, 328] to be of importance only in the vicinity of the spinodal curve [254] since otherwise the increase of the potential barrier required for ridge crossing as compared with the evolution via the saddle point overcompensates as a rule the advantages connected with the eventually easier realization of the kinetics of the process. Of course, the trajectories of evolution via the saddle point will depend on the kinetics and, for the model system considered, on the ratio of the partial diffusion coefficients of both components. The different paths of evolution of the critical clusters and their further growth in dependence on the ratio of diffusion coefficients of both components are illustrated in Figures 11.10 and 11.11.

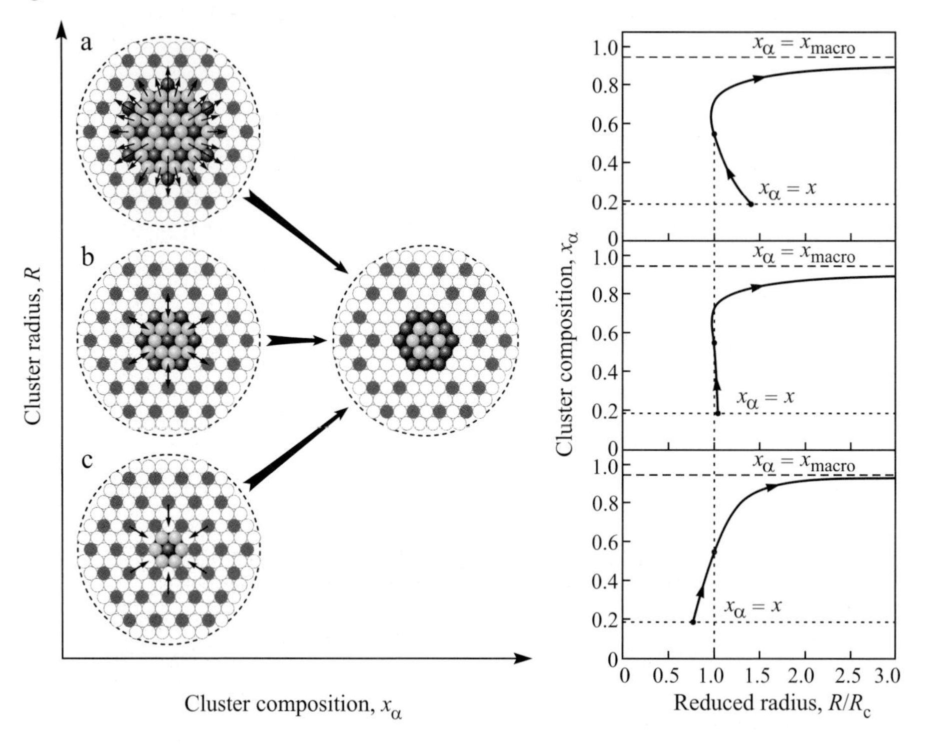

Figure 11.10: Schematic illustration of trajectories of evolution in dependence on the ratio of the partial diffusion coefficients of different components in the solution. On the right-hand side, the change of the composition of the clusters in dependence on reduced cluster sizes, R/R_c, is shown for the three different cases considered: (a) $D_2 \ll D_1$, (b) $D_2 \approx D_1$, and (c) $D_2 \gg D_1$.

For this purpose, we choose a volume of radius R in the center of some spherical domain (see Figure 11.10, left side). This selected volume has initially the same composition as the ambient phase, therefore it is not a cluster yet. If atoms of the second component are incorporated into this volume, the concentration, x_α, of this component in the cluster increases, its size increases, and it becomes a (super)critical nucleus of a new phase. Such scenario is realized (cf. [254]) when the mobility of the atoms of the second component is higher than for

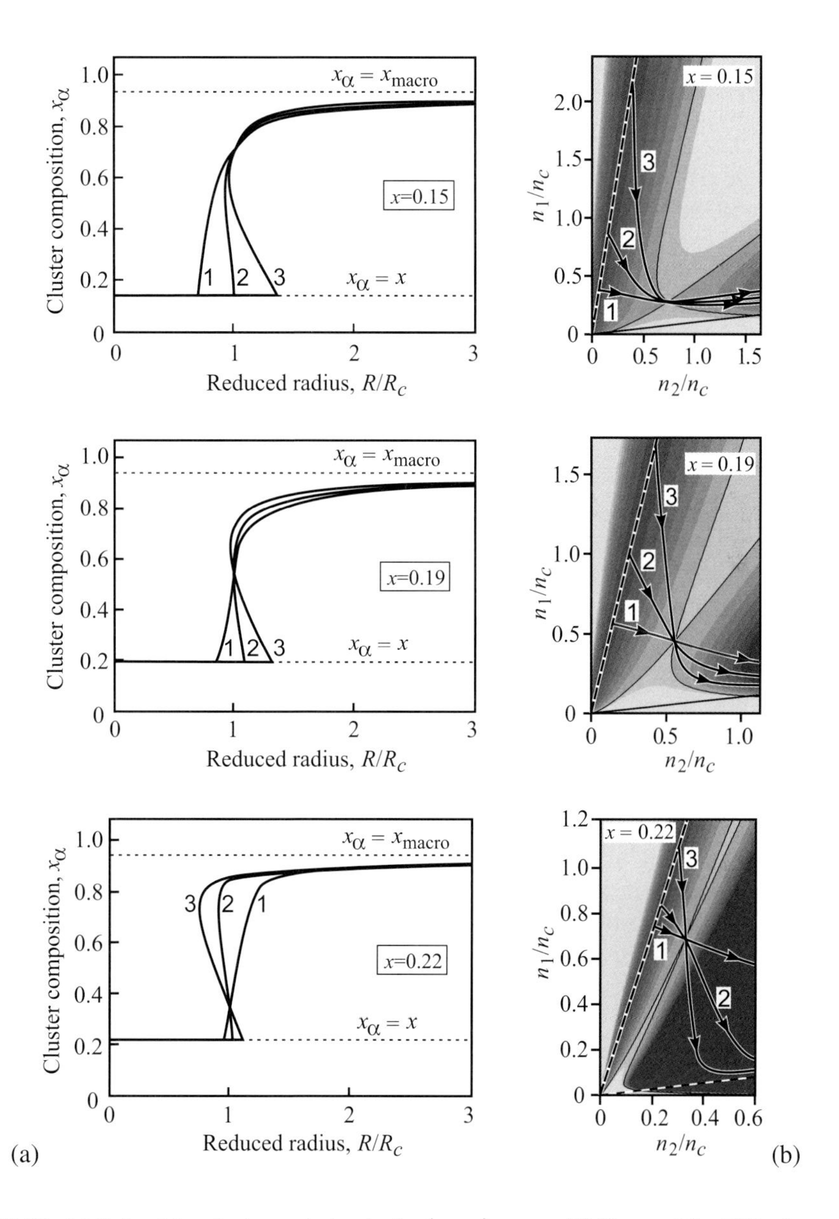

Figure 11.11: (a) Path of the cluster evolution in the (r, x_α)-space. (b) Shape of the Gibbs free energy surface and path of the cluster evolution in the $(n_1/n_c,\ n_2/n_c)$-space. The computations are made for a regular solution with molar fractions $x = 0.15$, $x = 0.19$, and $x = 0.22$ of the second component in the ambient phase for different values of the ratio D_1/D_2: $D_1/D_2 = 0.1$ *(1)*, $D_1/D_2 = 1$ *(2)*, $D_1/D_2 = 10$ *(3)*.

the atoms of the first sort (Figure 11.10(a)). The dependence of concentration of the cluster on its radius is given on the right-hand side.

In the opposite case, when the mobility of atoms of the first kind is higher, the formation of the critical cluster proceeds in such a way that atoms of the first kind leave the region where the cluster will be formed. In such case, the concentration, x_α, of the cluster increases but its size decreases (Figure 11.10(c)). Again, the dependence of cluster composition on cluster size is shown on the right-hand side. Once the critical cluster is formed in such process, its further growth is then again determined by the motion of the second less mobile component. So, here we have the situation that growth processes will proceed with much smaller effective diffusion coefficients as the nucleation process. And, finally, in the case when atoms of the first kind in the cluster are replaced by atoms of the second one, i.e., when the mobilities of both components are nearly equal (or more precisely, if the relation $D_1(1-x) \cong D_2 x$ holds, cf. Eq. (11.59)), the change of the composition of the cluster proceeds at nearly constant size (Figure 11.10(b)). In all these cases, the critical cluster is the same (determined thermodynamically) but the trajectories of evolution differ (see also [254]) due to different ratios of the partial diffusion coefficients of different components involved in the process of formation of the new phase.

11.5 Phase Separation in Finite Domains

11.5.1 Thermodynamic Analysis

It was shown in the preceding analysis that, neglecting depletion effects, the critical cluster size diverges in the vicinity of the spinodal curve (Figure 11.9). Taking into account that phase separation processes in real systems always proceed in systems of finite size, the model of an infinite domain is not appropriate in a variety of cases already by this reason. In the further analysis, the effects of finite domain size on the phase separation processes are studied.

Similar to Figure 11.6, Figures 11.12–11.15 represent the shape of the Gibbs free energy surface $\Delta G(n_{1\alpha}, n_{2\alpha})$ in dependence on the domain size, R_0, for different values of the initial solute concentration (Figure 11.12: $x = 0.15 < x_{\rm sp}$, Figure 11.13: $x = x_{\rm sp} \approx 0.226$, Figure 11.14: $x = 0.3 > x_{\rm sp}$ and Figure 11.15: $x = 0.4 > x_{\rm sp}$). As evident, at a given value of the supersaturation, the degree of instability of the system decreases with the decrease of the domain size. For example, for the case of an initial molar fraction equal to $x = 0.15$ (Figure 11.12), the critical cluster sizes, R_c, and the nucleation barrier, ΔG_c, increase with the reduction of the size of the domain, R_0. The free energy difference, ΔG_f, corresponding to a stable coexistence of a single cluster with radius, R_f, in the ambient phase noticeably grows, the size of this stable cluster, R_f, decreases considerably. The free energy difference, ΔG_f, reaches a value equal to zero at $R_0/R_\sigma = 14.55$. At such system size, initial (homogeneous) and final (heterogeneous) states become equivalent from a thermodynamic point of view. With the further reduction of R_0, ΔG_f becomes positive, and for $R_0/R_\sigma = 12.66$ the relation $\Delta G_f = \Delta G_c$ holds and the transition to a two-phase system becomes impossible due to finite-size effects.

For the considered supersaturation, in the range $R_0/R_\sigma > 14.55$, the initial state of the finite system is metastable. The final two-phase state is characterized by smaller values of

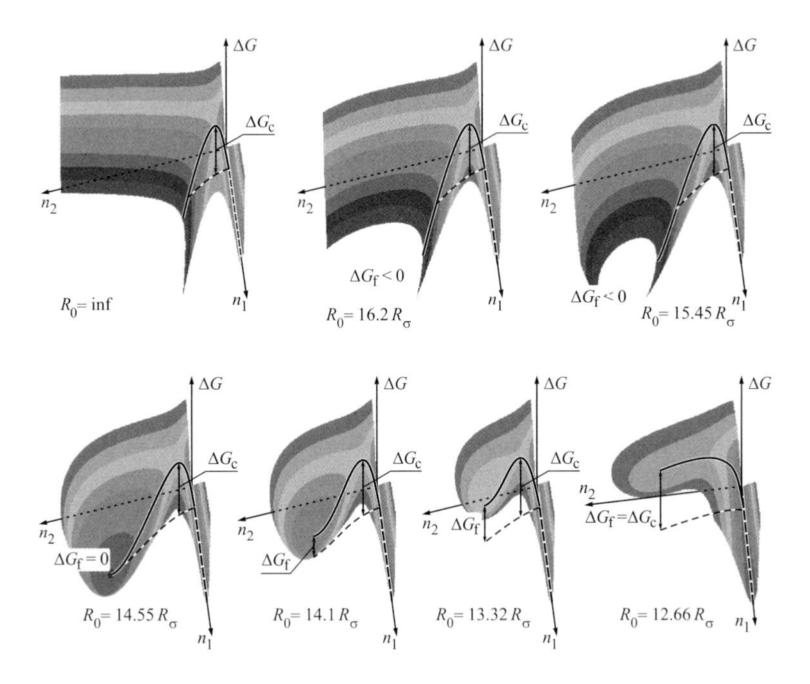

Figure 11.12: Shape of the Gibbs free energy surface for $x = 0.15$ and for different values of the domain size, R_0.

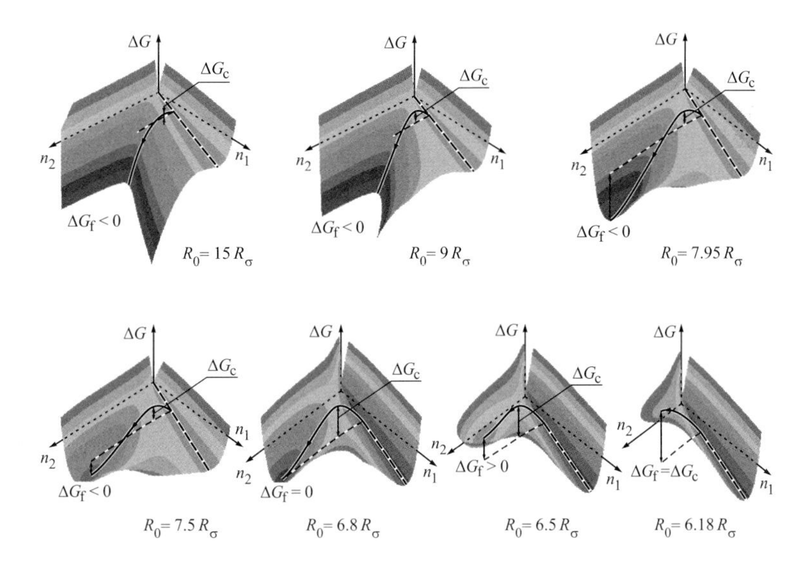

Figure 11.13: Shape of the Gibbs free energy surface for $x = x_{\rm sp}$ and for different values of the domain size, R_0.

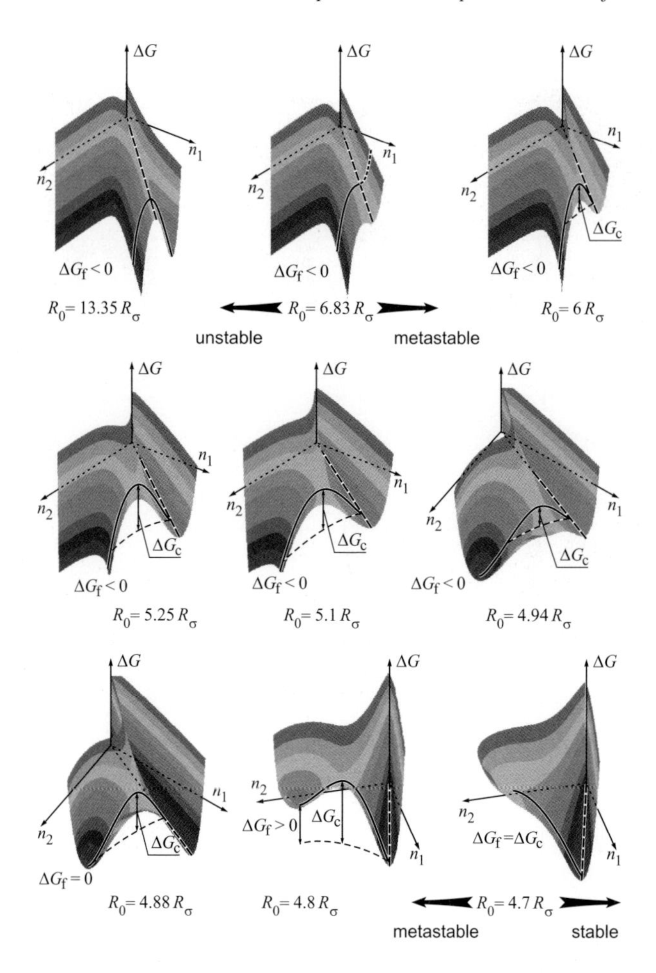

Figure 11.14: Shape of the Gibbs free energy surface for $x = 0.3$ and for different values of the domain size, R_0.

the Gibbs free energy, $\Delta G_f < 0$, as compared with the homogeneous initial state. As a consequence, once a stable state of the cluster in the ambient phase has been formed, the reverse transition is, as a rule, highly improbable. For $R_0/R_\sigma = 14.55$, the initial state of the system is also a metastable state; however, now homogeneous and heterogeneous states are characterized by the same values of the Gibbs free energy, i.e., $\Delta G_f = 0$. For this reason, the heterogeneous state can be transferred by appropriate processes back to the homogeneous initial state. Thus the inequality

$$\Delta G_f \leq 0 \tag{11.66}$$

can be considered as the condition of metastability for the homogeneous initial state. Similar processes occur with even higher probability in the range $12.66 < R_0/R_\sigma < 14.55$. Here

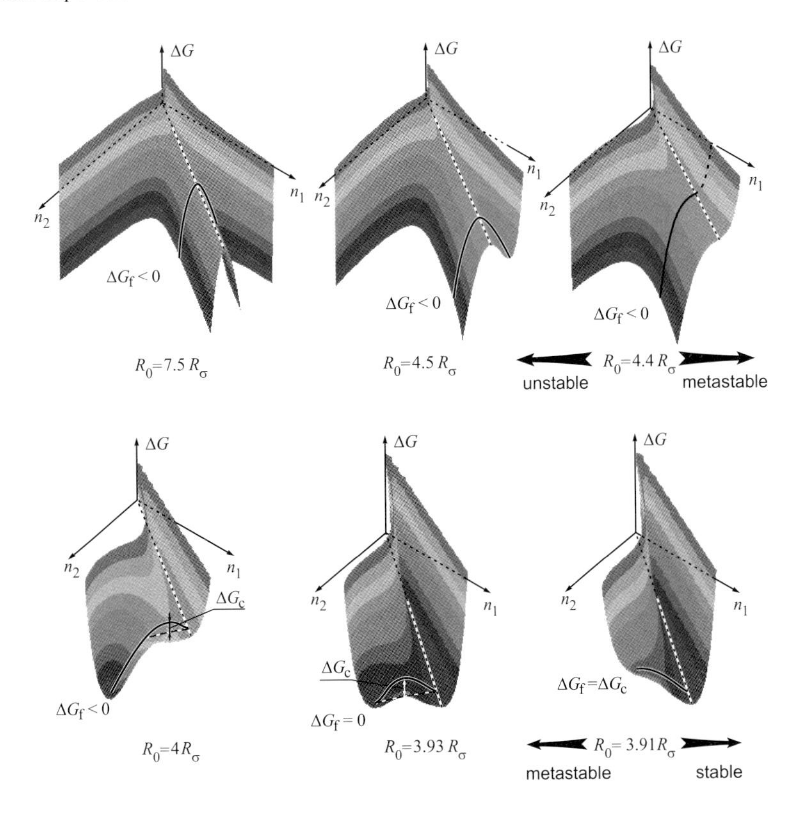

Figure 11.15: Shape of the Gibbs free energy surface for $x = 0.4$ and for different values of the domain size, R_0.

again the initial state of the system is metastable, but the final state has larger values of Gibbs' free energy as compared with the homogeneous initial state, i.e.,

$$\Delta G_c > \Delta G_f > 0 \tag{11.67}$$

holds. So, Eq. (11.67) is the condition of metastability for the heterogeneous state. And, finally, even if phase transformations may occur in a sufficiently large system, this is excluded for domain sizes $R_0/R_\sigma < 12.66$. For such system sizes, the system is to be considered here as stable.

The results discussed here for a particular value of the initial supersaturation in the range of metastable (for infinite systems) initial states – i.e., increase of the critical cluster size, R_c, and the work of critical cluster formation, ΔG_c, the increase of ΔG_f, and decrease of R_f – are general consequences of depletion effects in nucleation. They have been derived analytically in the framework of the classical Gibbs approach both for condensation in gases and phase formation in solid solutions earlier [75, 222, 229, 230, 233, 234, 262, 331]. It can be shown that the respective conclusions remain valid when the generalized Gibbs approach is employed for the description of the thermodynamics of cluster formation [261]. From a more general point of view, such dependences can be considered as consequences of the principle of le Chatelier-Braun [133, 331].

However, as to our knowledge, so far the effect of finite size on the kinetics has not been studied for the case that the process starts from unstable initial states. This task will be performed in the subsequent analysis. Shapes of the Gibbs free energy surface $\Delta G(n_{1\alpha}, n_{2\alpha})$ for $x = x_{\rm sp} \approx 0.226$ and for different values of the domain size, R_0, are presented in Figure 11.13. The shapes of the free energy are qualitatively very similar to the respective results shown in Figure 11.12.

The respective dependences, $\Delta G(n_{1\alpha}, n_{2\alpha})$, for unstable initial states with $x = 0.3$, are shown in Figure 11.14. In addition, the dependences of ΔG along the evolution path are shown in Figure 11.16 (here the path, s, is the distance in $(n_{1\alpha}/n_\sigma, n_{2\alpha}/n_\sigma)$-space, that is $ds = (dn_{1\alpha}^2 + dn_{2\alpha}^2)^{1/2}/n_\sigma$, and $s > 0$ for $x_\alpha > x$ and $s < 0$ for $x_\alpha < x$). For $R_0/R_\sigma > 6.83$, spinodal decomposition is a possible mode of evolution to the new phase. For low system sizes, here at $R_0/R_\sigma < 6.83$, a nucleation barrier arises and the system transforms to a metastable one. With the further reduction of the domain size the behavior of system is the same as for $x \leq x_{\rm sp}$. For $x = 0.4$, the surface $\Delta G(n_{1\alpha}, n_{2\alpha})$ is shown in Figure 11.15. The shapes of the thermodynamic potential surfaces are similar to the case $x = 0.3$, only the characteristic values of the system size R_0, at which the transition from spinodal decomposition to nucleation occurs, are smaller.

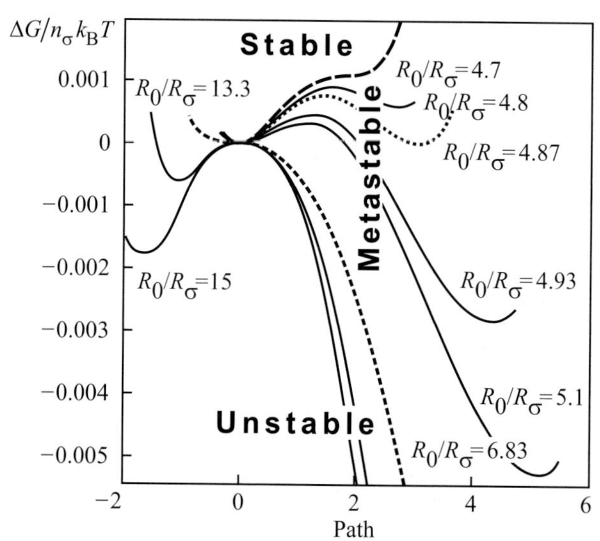

Figure 11.16: Gibbs free energy along the preferred trajectory of evolution to the new phase for initial states of the ambient phase with $x = 0.3$ and for different values of the domain size, R_0.

Equation $\Delta G_c(R_0, x) = \Delta G_f(R_0, x)$ defines the minimal domain size, $R_{0,b}(x)$, which allows nucleation in the initially homogeneous system, i.e., defines the binodal depending on the domain size. Let us define as the next step the spinodal curve for the domain of finite size.

Equation (11.63) may be rewritten as

$$\left.\frac{\partial^2 \Delta G}{\partial x_\alpha^2}\right|_{x_\alpha = x} = 2K\left(\frac{R}{R_{c,\mathrm{inf}}}\right)^2 \left\{1 - \frac{R}{R_{c,\mathrm{inf}}}\left[1 - \left(\frac{R}{R_0}\right)^3\right]\right\}\left[1 - \left(\frac{R}{R_0}\right)^3\right]^{-2}, \quad (11.68)$$

and then Eq. (11.62), which determines critical size in spinodal region, takes the form

$$R_c^4 - R_c R_0^3 + R_0^3 R_{c,\mathrm{inf}} = 0. \quad (11.69)$$

The equation has only one root for

$$R_0 = R_{0,\mathrm{sp}}(x) = \frac{4^{4/3}}{3} R_{c,\mathrm{inf}}(x) = R_\sigma \frac{2}{3} 4^{4/3}\left(4\frac{T_c}{T} - \frac{1}{x} - \frac{1}{1-x}\right)^{-1}, \quad (11.70)$$

and two real roots for $R_0 > R_{0,\mathrm{sp}}(x)$, and at $R_0 < R_{0,sp}(x)$ Eq. (26) does not have any roots. Consequently, the function $R_{0,\mathrm{sp}}(x)$ determines the minimal domain size $R_{0,\mathrm{sp}}(x)$, which allows spinodal decomposition in the system, i.e., it defines the spinodal depending on the domain size.

Dependences of minimal domain sizes $R_{0,b}/R_\sigma$ and $R_{0,\mathrm{sp}}/R_\sigma$ on the initial solute concentration, x, are presented in Figure 11.17. Metastable region is located between curves $R_{0,b}\,(x)$ and $R_{0,\mathrm{sp}}\,(x)$, unstable (spinodal) region is located to the right from $R_{0,\mathrm{sp}}\,(x)$.

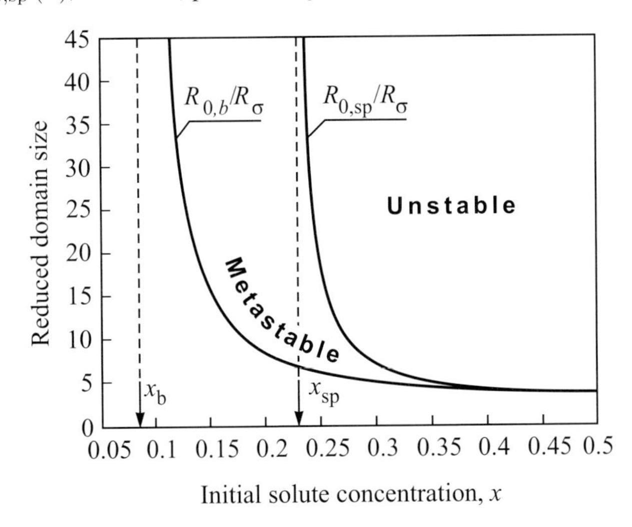

Figure 11.17: Dependence of the reduced minimal domain size, $R_{0,b}/R_\sigma$ and $R_{0,sp}/R_\sigma$, on the initial solute concentration, x.

The critical radius for the minimal domain size is determined by

$$R_{c,\mathrm{sp}}(x) \equiv R_c\left(R_{0,\mathrm{sp}}, x\right) = R_\sigma \frac{8}{3}\left(4\frac{T_c}{T} - \frac{1}{x} - \frac{1}{1-x}\right)^{-1}. \quad (11.71)$$

The dependence of the reduced critical radius R_c on the initial solute concentration x for different values of the domain size R_0 is illustrated in Figure 11.18. In the metastable region, $R_c(x)$ is determined by Eqs. (11.55), in the unstable one, by the solution of Eq. (11.71). Critical radii corresponding to the minimal domain size for the metastable region $R_{c,b}(x) \equiv R_c(R_{0,b}, x)$, and for the unstable one, $R_{c,\mathrm{sp}}(x)$, are shown in Figure 11.18 by dashed-dotted and dotted curves, respectively. Note that in the unstable region two critical radii exist: the smaller is determined by the balance between volume reduction and the increase of the thermodynamic potential due to surface formation (as for an infinite domain), the larger is determined by the effect of changes of the state parameters (depletion effect). Indeed, we see that the larger value of R_c is comparable with the domain size, R_0.

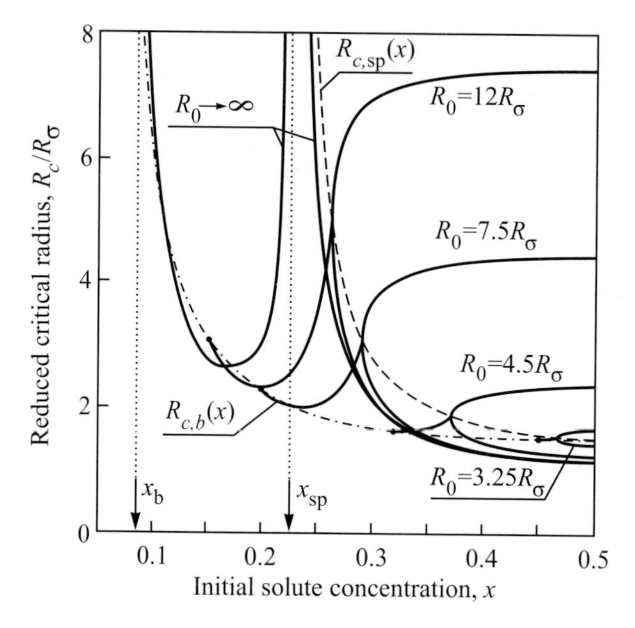

Figure 11.18: Dependence of the reduced critical radius, R_c/R_σ, on the initial solute concentration, x, for different values of the domain size.

Dependence of the minimum value of the work of critical cluster formation, $\Delta G_c/n_\sigma k_B T$, on the initial solute concentration, x, for different values of the domain size, R_0, is shown in Figure 11.19. In the region $x < x_{\mathrm{sp}}$, with domain size reduction ΔG_c increases insignificantly, while for $x > x_{\mathrm{sp}}$ $\Delta G_c = 0$ for $R_0 \to \infty$, and nonzero value of ΔG_c arises only for finite values of R_0. Dependence of the composition of the critical cluster, $x_{\alpha,c}$, on the initial solute concentration x, for different values of the domain size, R_0, is shown in Figure 11.20. We see that with growth of solute concentration $x_{\alpha,c}$ decreases down to value $x_{\alpha,c} = x$, which corresponds to the unstable region.

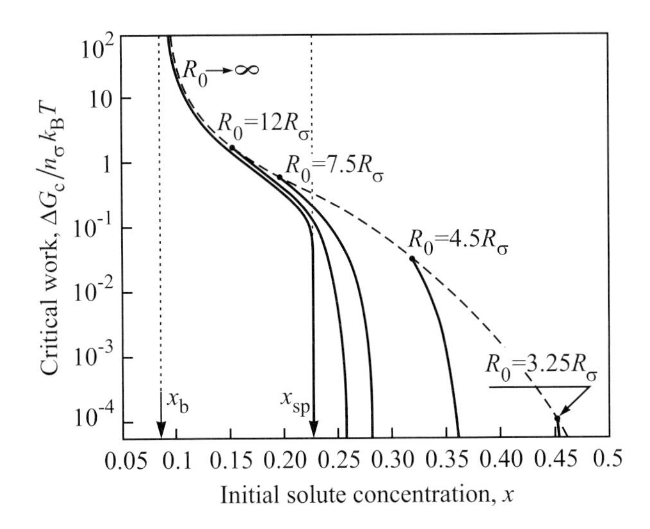

Figure 11.19: Dependence of the minimum value of the work, $\Delta G_c/(n_\sigma k_B T)$, of critical cluster formation on the initial solute concentration, x, for different values of the domain size.

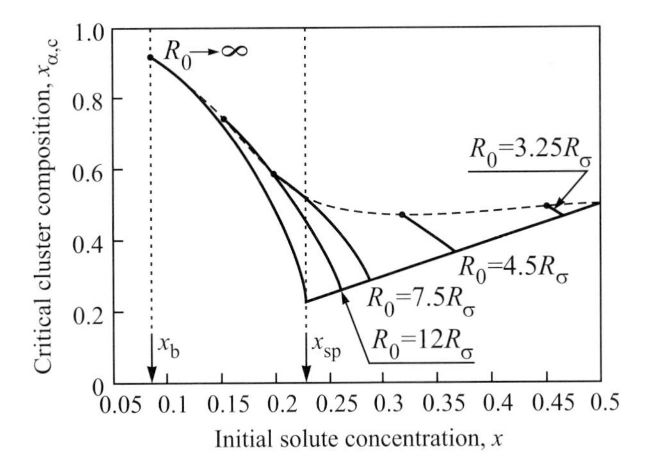

Figure 11.20: Dependence of the composition of the critical cluster, $x_{\alpha,c}$, on the initial solute concentration x, for different values of the domain size.

11.5.2 Kinetics

Having performed the respective thermodynamic analysis, we will now consider the time evolution of the clusters in segregation processes in systems of finite size. We assume that the composition in a certain region of the ambient phase is slightly shifted by a value $\delta n_{i\alpha}$ as compared with the composition of the matrix, i.e.,

$$n_{1\alpha} = n_{1\alpha,0} + \delta n_{1\alpha}, \qquad n_{2\alpha} = n_{2\alpha,0} + \delta n_{2\alpha}. \tag{11.72}$$

It is further assumed that the growth is kinetically limited (i.e., we set $\kappa = 2/3$ in Eq. (11.75)) A substitution of Eqs. (11.72) into Eqs. (11.59) yields

$$\frac{d\delta n_{1\alpha}}{dt} = -D_1(1 - x_\beta)\Theta(n_{1\alpha}, n_{2\alpha})\frac{d\Delta G}{ds}\cos\varphi\,, \tag{11.73}$$

$$\frac{d\delta n_{2\alpha}}{dt} = -D_2 x_\beta \Theta(n_{1\alpha}, n_{2\alpha})\frac{d\Delta G}{ds}\sin\varphi. \tag{11.74}$$

Here the notation

$$\Theta(n_{1\alpha}, n_{2\alpha}) = \Theta_0 n_\alpha^\kappa \tag{11.75}$$

is used, $\mathrm{d}\Delta G/\mathrm{d}s$ is the absolute value of the gradient of the function $\Delta G\,(n_{1\alpha}, n_{2\alpha})$ at values of x_α near to $x_\alpha = x$, and φ is the angle between the direction of the gradient and the axis $n_{1\alpha}$. It is defined by the equation

$$\tan\varphi = \frac{n_{2\alpha,0}}{n_{1\alpha,0} + n_{2\alpha,0}} = \frac{1-x}{x}. \tag{11.76}$$

Dividing Eq. (11.74) by Eq. (11.73) and taking into account Eq. (11.76), we obtain

$$\delta n_{1\alpha} = -\frac{D_1}{D_2}\delta n_{2\alpha}. \tag{11.77}$$

Using the variable x_α instead of $\delta n_{2\alpha}$,

$$x_\alpha = \frac{n_{2\alpha}}{n_{1\alpha} + n_{2\alpha}} = \frac{n_{2\alpha,0} + \delta n_{2\alpha}}{n_{1\alpha,0} + n_{2\alpha,0} + \left(1 - \dfrac{D_1}{D_2}\right)\delta n_{2\alpha}}\,, \tag{11.78}$$

we get the equation

$$\frac{dx_\alpha}{dt} = \Theta_0 x\,(1-x)\left(\frac{R_{c,\inf}}{R}\right)^4 \left[x D_1 + (1-x)\,D_2\right]\left(-\left.\frac{\partial^2 \Delta G}{\partial x_\alpha^2}\right|_{x=x_\alpha}\right)(x_\alpha - x)\,, \tag{11.79}$$

where R is the initial cluster radius and $\left(\partial^2\Delta G/\partial x_\alpha^2\right)\big|_{x=x_\alpha}$ is determined by Eq. (11.68). This linear equation has a solution of the form $(x_\alpha - x) \sim e^{\gamma(R)t}$, where the growth increment (or amplification factor) $\gamma(R)$ is determined via

$$\gamma\,(R) = M\left(\frac{R_{c,\inf}}{R}\right)^4\left(-\left.\frac{\partial^2\Delta G}{\partial x_\alpha^2}\right|_{x=x_\alpha}\right) \tag{11.80}$$

and

$$M = \Theta_0 x (1-x) [x D_1 + (1-x) D_2] \tag{11.81}$$

holds.

For finite systems with a domain size lower than some upper value R_{0m}

$$R_0 \le R_{0m} = 2R_{0,\text{sp}} = R_\sigma \frac{4^{7/3}}{3} \left(4\frac{T_c}{T} - \frac{1}{x} - \frac{1}{1-x} \right)^{-1}, \tag{11.82}$$

the function $\gamma = \gamma(R)$ has a maximum. The value of the maximum increases with increasing domain size. Moreover, at $R_0 \ge R_{0m}$ a second maximum of equal height arises. After this second maximum appeared, the height of the maxima does not vary any more with the further increase of the size of the domain (Figure 11.21). The growth increment reaches the maximum value for a domain size equal to

$$R_{0,\max}(R,x) = \begin{cases} \left(\frac{R}{R_\sigma}\right)^{4/3} \left(\frac{R}{R_\sigma} - \frac{3}{K}\right)^{-1/3} & \text{for} \quad R > R_{\gamma,\max} \\ R_{0m} & \text{for} \quad R \le R_{\gamma,\max}, \end{cases} \tag{11.83}$$

where $R_{\gamma,\max} = (8R_\sigma/3K)$.

The dependence of the growth increment, $\gamma(R)$, on cluster radius for various fixed domain sizes, R_0 (full curves), and for $R_0 = R_{0,\max}(R,x)$ (dashed curve) is shown in Figure 11.21 for the case $D_1 = D_2$ and $x = 0.45$, i.e., for macroscopically unstable initial states.

Equation (11.80) looks similar to the expression for the growth increment in the classical Cahn–Hilliard theory of spinodal decomposition [44, 45]. Indeed, let us introduce a *wave vector* via $k \equiv R_\sigma/R$, then Eq. (11.80) gets the form

$$\gamma(k) = Mk^4 \left(- \left. \frac{\partial^2 \Delta G}{\partial x_\alpha^2} \right|_{x=x_\alpha} \right), \tag{11.84}$$

$$\left. \frac{\partial^2 \Delta G}{\partial x_\alpha^2} \right|_{x_\alpha = x} = \frac{2K}{k^2} \left\{ 1 - \frac{1}{k} \left[1 - \left(\frac{k_0}{k} \right)^3 \right] \right\} \left[1 - \left(\frac{k_0}{k} \right)^3 \right]^{-2}. \tag{11.85}$$

Employing these relations, in Figure 11.22 the Cahn plots $\gamma(k)/k^2$ vs k^2 are shown for various fixed domain sizes, R_0 (full curves), and for $R_0 = R_{0,\max}(R,x)$ (dashed curve), where the notation $k_c = R_\sigma/R_c$ is used. In Figure 11.23, the result for $R_0 = R_{0,\max}(R,x)$ is compared with the experimental data for spinodal decomposition in the glass SiO_2–$12.5Na_2O$ [5]. The Cahn plot, obtained in this way, is different in its shape as compared with the linear classical dependence [44, 45]; it is in good agreement with the experimental data shown for comparison. Thus, the linear analysis of Eqs. (11.73) and (11.74) allows us to determine the growth increment for spinodal decomposition (Eq. (11.84)) depending on supersaturation, cluster, and domain sizes in a way giving a better agreement with the experimental data as the classical theory.

The numerical solution of Eqs. (11.73) and (11.74) allows one not only to analyze the initial states of spinodal decomposition but to trace the whole process of evolution of the

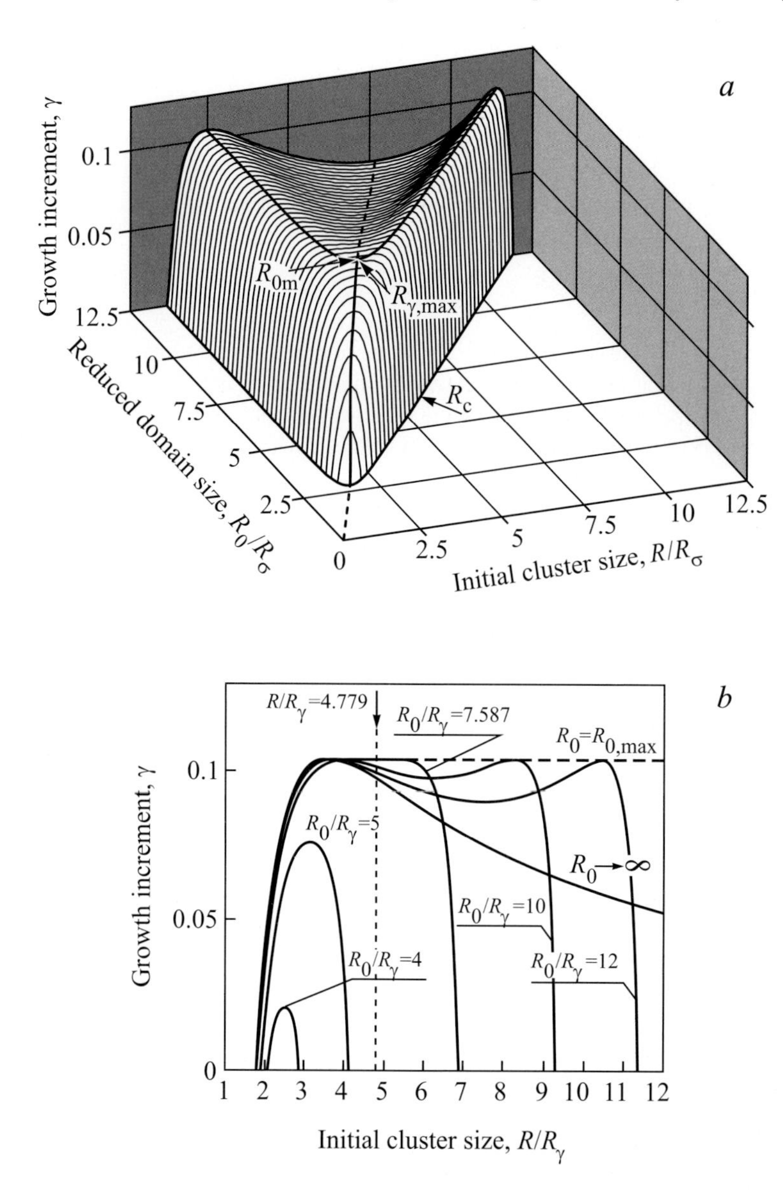

Figure 11.21: Dependence of the growth increment on the cluster radius, R/R_σ, for various domain sizes (full curves) and for $R_0 = R_{0,\max}(R, x)$ (dashed curve) for $x = 0.45$.

cluster. In doing so, we assume kinetic limited growth ($\kappa = 2/3$ in Eq. (11.75)) and again set the temperature equal to $T = 0.7T_c$. Domain size and the initial cluster radius are chosen to correspond to the maximal growth increment, i.e., $R_0 = R_{0m}$ and $R = R_{\gamma,\max}$ (see Eqs. (11.82) and (11.83)), the initial cluster composition is given by $x_\alpha|_{t=0} = x\,(1 + \delta)$.

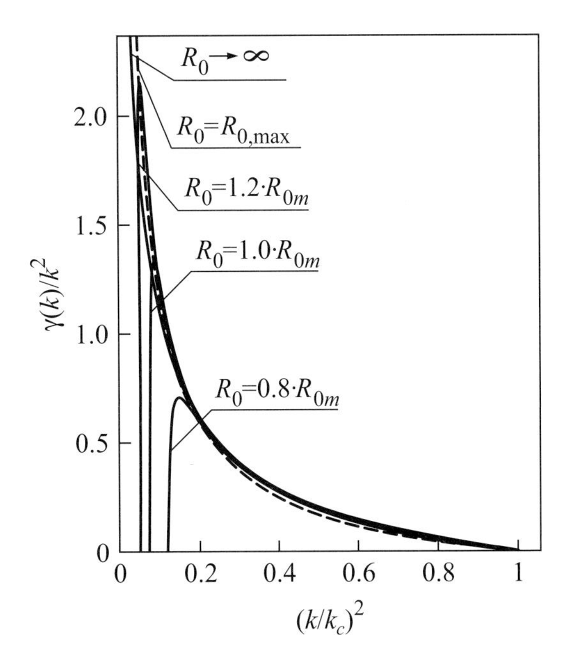

Figure 11.22: Dependence of the ratio γ/k^2 on k^2 for various domain sizes (full curves) and for $R_0 = R_{0,\max}(R, x)$ (dashed curve) for $x = 0.45$.

The results of calculations of the cluster evolution for a regular solution with a molar fraction of the segregating component in the ambient phase equal to $x = 0.45$ and $\delta = 0.01$ are presented in Figures 11.24–11.27. The shape of the Gibbs free energy surface and the trajectory of cluster evolution in the $(n_{1\alpha}/n_\sigma, n_{2\alpha}/n_\sigma)$ space (n_σ is determined by Eq. (11.53)) are shown in Figure 11.24 for different values of the partial diffusion coefficient D_1 and D_2 ((a) $D_1/D_2 = 100$, (b) $D_1/D_2 = 1$, (c) $D_1/D_2 = 0.025$, and (d) $D_1/D_2 = 0.001$; as earlier, we assume $D_1 D_2 = \text{const}$). The process starts in the point S and develops either increasing (path $S \to F$, curves (a), (b), (c), and (d)) or decreasing ($S \to F'$, curves (a'), (b'), (c'), and (d')) the concentration of the second component. For $x \neq 0.5$, the minima of Gibbs free energy, the system may approach following different paths of evolution, have different depths (for $x < 0.5$ $\Delta G_F < \Delta G_{F'}$). Preferred is the path $S \to F$, therefore we further consider only this version.

The dependences of compositions of cluster and ambient phase both on time and on cluster radius are shown in Figures 11.25 and 11.26, respectively (cf. also Figure 11.10). For the case of a quickly moving first component ($D_1/D_2 = 100$, curve (a)), the evolution along the path $S \to T$ proceeds via emission of particles of the first component from the cluster. As a result, the cluster shrinks in size (see also Figure 11.25(a)). After a time, $\tau_{\alpha f}$, the composition of the cluster almost reaches its final value, $x_\alpha \approx x_{\alpha f} \approx 0.853$ (the point T in Figures 11.24 and 11.26, also note that this state corresponds to the minimum of Gibbs free energy). During the initial time interval, $\tau < \tau_{\alpha f}$, the compositions of cluster and ambient phase approximately

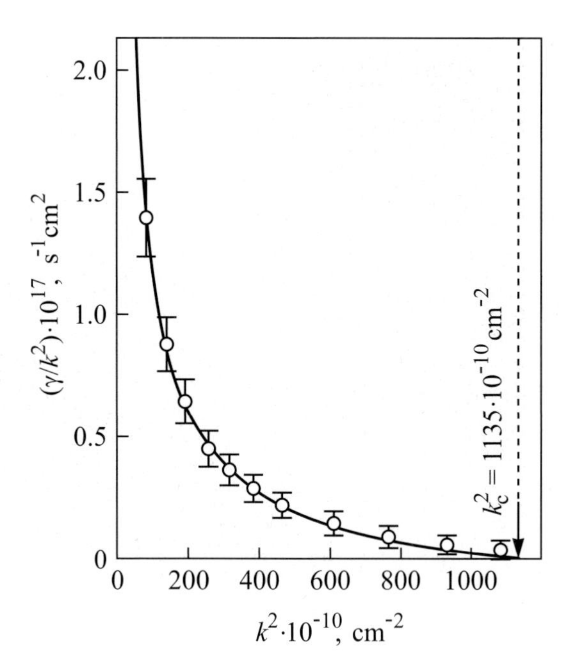

Figure 11.23: Dependence of the ratio γ/k^2 on k^2: the solid curve shows the result of calculation for $R_0 = R_{0,\max}(R, x)$ for $x = 0.45$, circles refer to experimental data for the glass SiO_2–12.5 Na_2O at 530 ° C obtained by small-angle X-ray scattering [5].

change with equal rate. This rate can be determined by the analytical expressions (11.80) with good accuracy (dashed curves in Figure 11.25). Once this stage of evolution is completed, the cluster begins to grow with approximately constant composition while the composition of the ambient phase continues to change. Since the condition of constancy of cluster composition requires attachment of atoms of both kinds in a certain well-defined proportion, the rate of evolution along the path $T \rightarrow F$ is limited by the rate of attachment of atoms of the slow second component (see Figures 11.25, 11.25(a), and 11.26). In the time $\tau_{\beta f}$, the composition of the ambient phase reaches its final value, $x_\beta \approx x_{\beta f} \approx 0.194$.

For the case of nearly equal partial diffusion coefficients, $D_2 = D_1$, the evolution proceeds similarly with the difference that the cluster size only changes insignificantly at the initial stage of evolution, $\tau < \tau_{\alpha f}$ (see Figure 11.25(b)) and the time interval $\tau_{\beta f}$ is considerably shorter as compared with the previous case. Such two-stage behavior is preserved in a wide interval of components mobility, actually only at $D_1/D_2 \approx 0.025$ cluster size and concentration begin to change monotonically down to end (see Figure 11.26, curves (c) and $(c\prime)$).

At $D_1 \ll D_2$, the situation is opposite to some extent. Along the path $S \rightarrow T'$, the cluster grows quickly due to the incorporation of atoms of the second fast component. Then, after a time τ_f, the composition of the ambient phase has almost reached its final value $x_\beta \approx x_{\beta f}$, and a slow reduction of the cluster size due to emission of atoms of the first component is found along the path $T' \rightarrow F$.

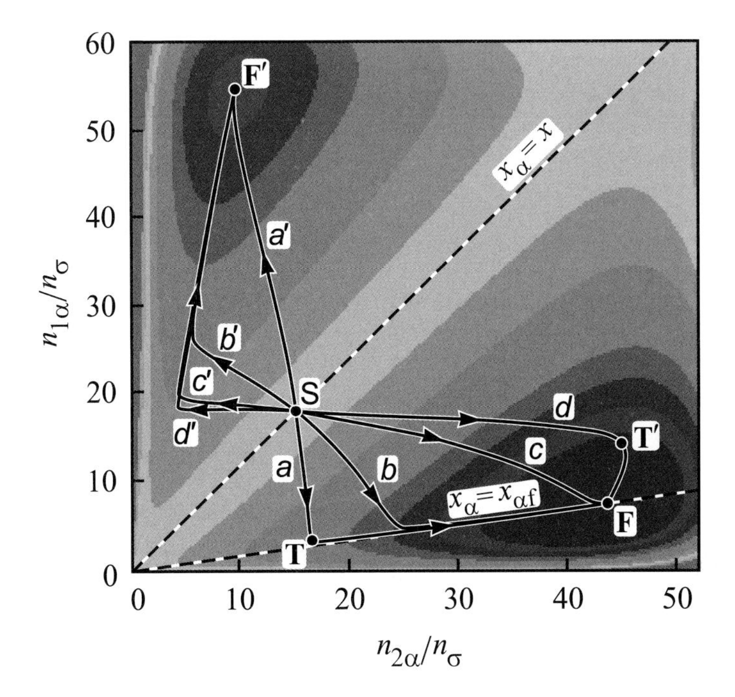

Figure 11.24: Shape of the Gibbs free energy surface and trajectory of cluster evolution in the $(n_{1\alpha}/n_\sigma, n_{2\alpha}/n_\sigma)$ space for a regular solution with a molar fraction of the segregating component in the ambient phase equal to $x = 0.45$ for different values of D_1/D_2: (a) $D_1/D_2 = 100$, (b) $D_1/D_2 = 1$, (c) $D_1/D_2 = 0.025$, and (d) $D_1/D_2 = 0.001$.

Figure 11.27 shows the dependence on the (D_1/D_2)-ratio of the characteristic times of change of cluster composition, $\tau_{\alpha f}$, of ambient phase change, $\tau_{\beta f}$, and time τ_γ. The latter parameter can be computed via the analytical expression, Eq. (11.80), resulting in

$$\tau_\gamma = \frac{1}{\gamma\left(R_{\gamma,\max}\right)} \ln\left[\frac{x_{\alpha f} - x}{x\delta}\right]. \tag{11.86}$$

The minimum time, $\min\left(\tau_{\alpha f}, \tau_{\beta f}\right)$, of change of the composition of the cluster or the ambient phase differs slightly only from τ_γ, while the full time of decomposition is determined by the maximum time $\max\left(\tau_{\alpha f}, \tau_{\beta f}\right)$, which is twice as large as τ_γ for $D_1/D_2 \approx 0.01$, and for $D_1 \gg D_2$ $\tau_{\beta f}$ larger than $\tau_{\alpha f}$ and τ_γ by more than an order of magnitude.

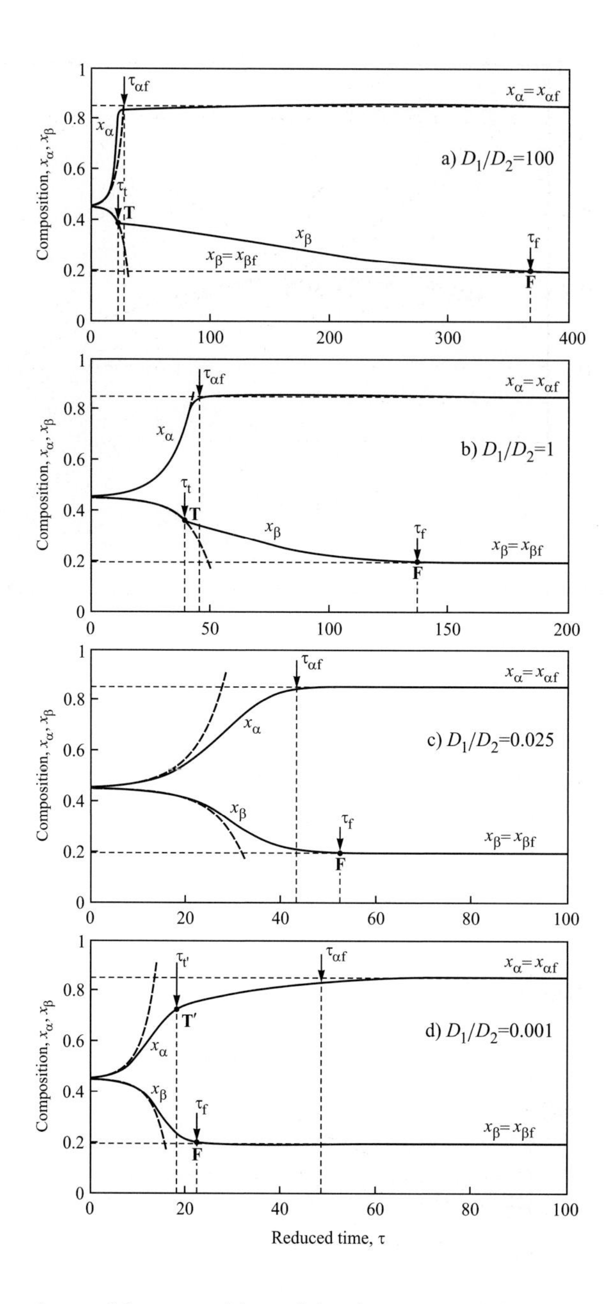

Figure 11.25: Dependence of the compositions of the cluster, x_α and the ambient phase, x_β, on time for different values of D_1/D_2: (a) $D_1/D_2 = 100$, (b) $D_1/D_2 = 1$, (c) $D_1/D_2 = 0.025$, and (d) $D_1/D_2 = 0.001$.

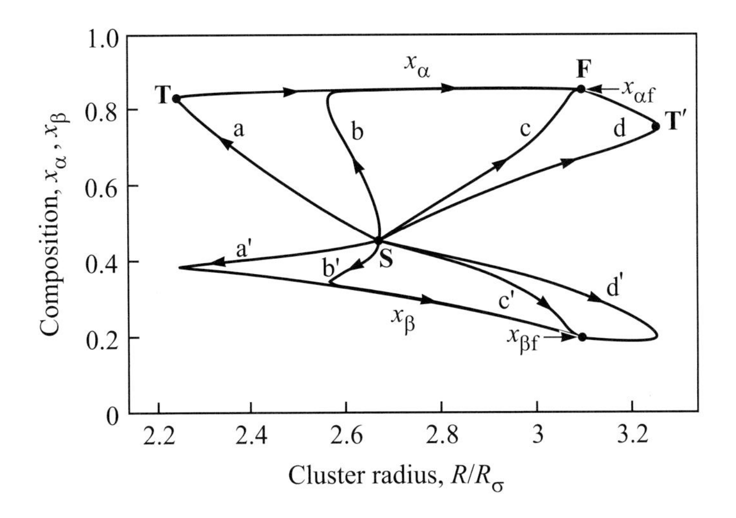

Figure 11.26: Dependence of the compositions of the cluster, x_α and the ambient phase, x_β, on the reduced cluster radius for different values of D_1/D_2: (a) $D_1/D_2 = 100$, (b) $D_1/D_2 = 1$, (c) $D_1/D_2 = 0.025$, and (d) $D_1/D_2 = 0.001$.

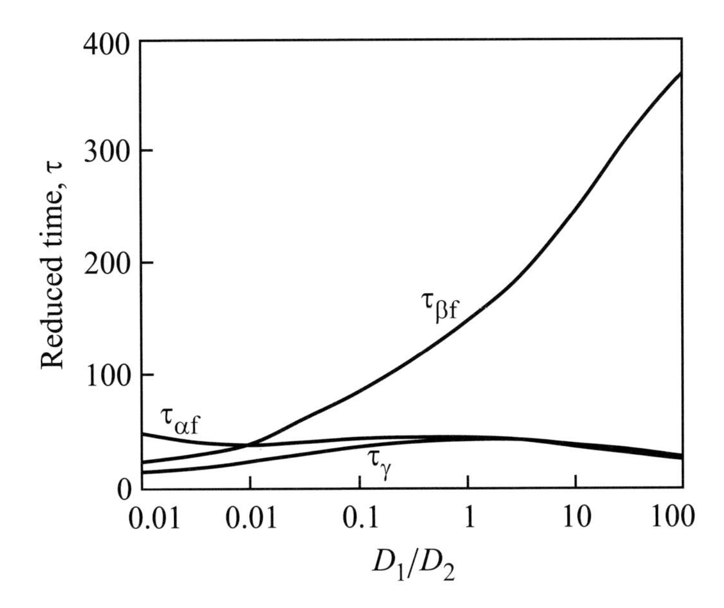

Figure 11.27: Dependence of the characteristic times $\tau_{\beta f}$, $\tau_{\alpha f}$, and τ_γ on the (D_1/D_2)-ratio.

11.5.3 Transition from Independent Cluster Growth to Coarsening

So far, we have considered phase separation in finite domains of size, R_0, considering the evolution of one cluster. However, the results of the analysis can be employed more generally allowing one to derive important conclusions about the initial stages of phase separation processes for systems of arbitrarily large sizes. The model considered above actually represents the case of an infinite domain with an ensemble of identical clusters. The analysis of such systems has already been shown to be very fruitful in previous investigations of the kinetics of phase separation in solutions when the classical Gibbs approach was employed for the thermodynamic description of the clusters and cluster ensembles [229, 230, 233, 234, 331]. The respective analysis are attempted to be generalized in future.

For a more detailed analysis of the kinetics of phase formation, the existence and evolution of the cluster-size distributions has to be taken into consideration. Independent growth of clusters of nearly the same sizes is only possible at the initial stage of the process, and such distributions are unstable. Once the depletion effects begin to dominate, the δ-shaped or Gaussian-type distribution functions are inevitably widened and the system passes into the coarsening, or competitive growth stage (Chapter 4 and [155]). In the simplest way, this can be done by solving the equations of motion of the clusters numerically. Such approach has been performed in terms of the classical description of cluster formation and evolution by a variety of authors. Here we would like to show that these analyses can be generalized by the above-mentioned approach accounting both for variations of cluster sizes and compositions. In order to illustrate these features, here we restrict the analysis to the evolution of cluster ensembles consisting only of few clusters.

Completing the analysis, we demonstrate that the transition to the competitive growth stage can be described adequately in terms of the approach employed here independently on whether the system starts the transformation from a metastable or unstable initial state. For this purposes, we consider the evolution of a system of three clusters in one domain. The clusters do not interact directly but only via consuming particles from the ambient phase. The conservation law, Eq. (11.43), then gets the form

$$n_0 = \sum_i \left(n_\alpha^{(i)} + n_\alpha^{(i)} \right) = \text{const}, \tag{11.87}$$

$$\sum_i \left(n_{1\alpha}^{(i)} + n_{2\alpha}^{(i)} \right) = \sum_i \left[n_{1\alpha}^{(i)}(0) + n_{2\alpha}^{(i)}(0) \right] = \text{const},$$

$$\sum_i \left(n_{1\beta}^{(i)} + n_{2\beta}^{(i)} \right) = \sum_i \left[n_{1\beta}^{(i)}(0) + n_{2\beta}^{(i)}(0) \right] = \text{const},$$

$$n_\alpha^{(i)} = n_{1\alpha}^{(i)} + n_{2\alpha}^{(i)}, \qquad n_\beta^{(i)} = n_{1\beta}^{(i)} + n_{2\beta}^{(i)},$$

where the indices $i = 1, 2, 3$ specify different clusters evolving in the system. The concentration of the second component in the ambient solution is then given by

$$x_\beta = \sum_i n_{2\beta}^{(i)} \left(\sum_i n_\beta^{(i)} \right)^{-1}, \tag{11.88}$$

and in the ith cluster by

$$x_\alpha^{(i)} = \frac{n_{2\alpha}^{(i)}}{n_{1\alpha}^{(i)} + n_{2\alpha}^{(i)}}. \tag{11.89}$$

The evolution of the system is determined by the set of equations

$$\frac{dn_{1\alpha}^{(i)}}{dt} = -D_1(1 - x_\beta)\Theta \frac{d}{dn_{1\alpha}^{(i)}} \Delta G\left(\left\{n_{1\alpha}^{(i)}\right\}, \left\{n_{2\alpha}^{(i)}\right\}\right), \tag{11.90}$$

$$\frac{dn_{2\alpha}^{(i)}}{dt} = -D_2 x_\beta \Theta \frac{d}{dn_{2\alpha}^{(i)}} \Delta G\left(\left\{n_{1\alpha}^{(i)}\right\}, \left\{n_{2\alpha}^{(i)}\right\}\right), \tag{11.91}$$

where

$$\frac{1}{k_B T}\Delta G\left(\left\{n_{1\alpha}^{(i)}\right\}, \left\{n_{2\alpha}^{(i)}\right\}\right) = \frac{3}{2} n_\sigma^{1/3} \sum_i \left(n_\alpha^{(i)}\right)^{2/3} \left(x_\alpha^{(i)} - x_\beta\right)^2 + \sum_i n_\alpha^{(i)} f(x_\beta, x_\alpha^{(i)}) - n_0 f(x_\beta, x). \tag{11.92}$$

As before, the functions $f(x_\beta, x_\alpha^{(i)})$ and $f(x_\beta, x)$ are determined by Eq. (11.52), and n_σ by Eq. (11.53). The domain size and initial cluster radii are assumed to correspond to the maximum growth increment, i.e., $R_0 = 3^{1/3} R_{0m}$ and $R = R_{\gamma,\max}$ (see Eqs. (11.82) and (11.83)), the initial cluster compositions are chosen as $x_{\alpha,i} = x\,(1 + \delta_i)$, where $\delta_1 = 0.004$, $\delta_2 = 1.12\delta_1$, $\delta_1 = 1.2\delta_1$ (thus, the first cluster has the lowest deviation from the initial composition, the second a larger and the third the highest one). The results of the computations are shown in Figures 11.28 and 11.29. In Figure 11.28, a cross-section of the Gibbs free energy surface and the trajectory of evolution of the first cluster is given in the $\left(n_{1\alpha}^{(1)}/n_\sigma, n_{2\alpha}^{(1)}/n_\sigma\right)$ space. In Figure 11.29, the dependence of the compositions, $x_\alpha^{(i)}$, the radii, $R^{(i)}$, of the clusters ($i = 1, 2, 3$), and the composition of ambient phase, x_β, are shown in dependence on time. For three clusters the phase space is six dimensional, therefore we only plot its two-dimensional sections for the first cluster (which is dissolved as the first one) for different moments of time (as specified in the figure).

At the first stage of the process, for $\tau < \tau_\alpha^{(i)}$ ($\tau_\alpha^{(1)} \approx 135$, $\tau_\alpha^{(2)} \approx 128$, and $\tau_\alpha^{(3)} \approx 122$), all three clusters evolve in an almost equal manner: the concentration of the second component grows, the sizes of the clusters decrease (see Figure 11.28(b)). In the initial state, the Gibbs free energy has a shape characteristic for the instability region (see Figures 11.28(a) and 11.24), but already at $\tau = \tau_b \approx 120$ a saddle point evolves being a characteristic feature of metastable states (Figure 11.28(b)). At $\tau = \tau_\alpha^{(1)}$, the concentration of the second component in the first cluster approaches the maximum value (see Figure 11.28(c), it corresponds to the path $S \to T$ in Figure 11.24). After that, the cluster begins to grow, and at $\tau = \tau_c \approx 160$ it reaches the maximum size. The Gibbs free energy then reaches a local minimum (see Figure 11.28(d)). In the case of a single cluster, the process would have finished at such state,

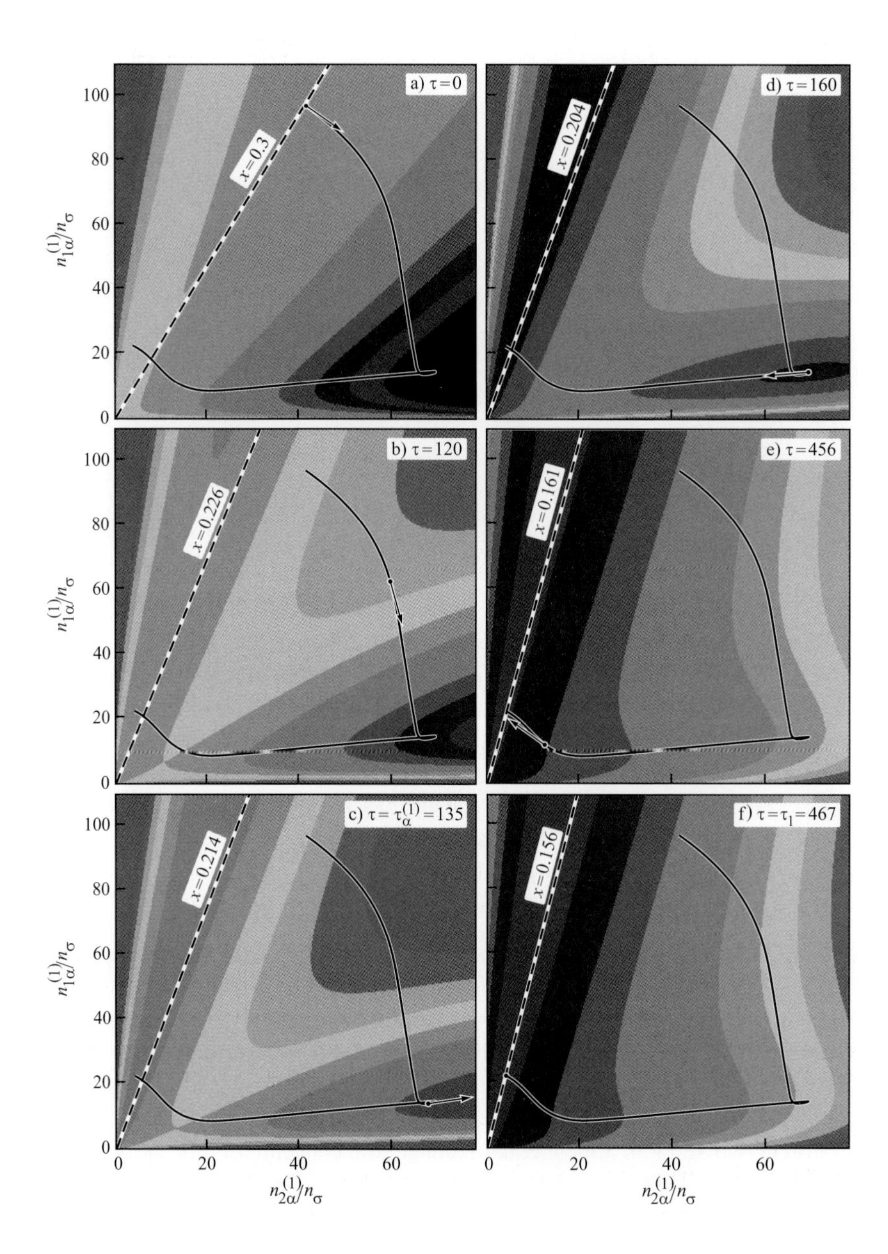

Figure 11.28: Cross-section of the Gibbs free energy surface and trajectory of evolution of the first cluster in the $\left(n_{1\alpha}^{(1)}/n_\sigma, n_{2\alpha}^{(1)}/n_\sigma\right)$-space for a regular solution with a molar fraction of the segregating component in the ambient phase equal to $x = 0.3$ for $D_1/D_2 = 1$.

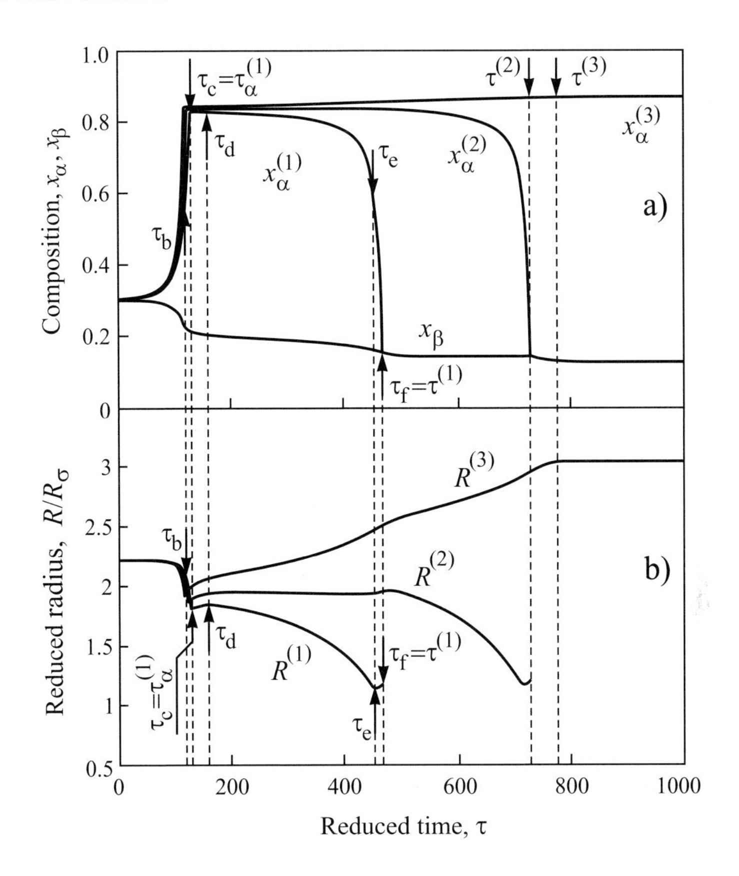

Figure 11.29: Dependence of the compositions, $x_\alpha^{(i)}$, radii, $R^{(i)}$, of the clusters ($i = 1, 2, 3$), and composition of ambient phase, x_β, on time.

however, the second and the third clusters continue to consume atoms of the second component, lowering their concentration in the ambient phase. As a result of such depletion effects, the first cluster shrinks, the concentration of the second component decreases. This process corresponds to the beginning of dissolution of the first cluster. The process is completed in a time $\tau_1 \approx 467$, when the composition of the first cluster approaches the composition of the the ambient phase, $x_\alpha^{(1)} = x_\beta$. At this moment, the radius of the cluster remains finite (see Figure 11.28(b)). The evolution of the second cluster proceeds similarly, and in time $\tau_2 \approx 729$ it is dissolved. At $\tau_3 \approx 937$ the process is finished and only one cluster remains in the domain.

11.6 Results and Discussion

In the present chapter, basic features of nucleation–growth and spinodal decomposition processes in solutions are analyzed within the framework of a thermodynamic cluster model based on the generalized Gibbs approach. This approach allows one to determine the thermodynamic potentials of clusters and ensembles of clusters in the otherwise homogeneous ambient phase for thermodynamically well-defined (cf. [258, 259]) nonequilibrium states of the considered heterogeneous systems. Hereby the cluster, representing the density and/or composition fluctuations may change with time both in size and intensive state parameters. The thermodynamic analysis is further employed as the basis for the description of the kinetics of the decomposition processes.

The thermodynamic analysis of cluster formation is performed in dependence on supersaturation for metastable and unstable initial states and domains of infinite and finite sizes. For domains of infinite sizes, in particular, the parameters of the critical clusters – size, intensive state parameters, work of critical cluster formation – are determined for metastable initial states of the solutions. It is shown that – in the framework of the generalized Gibbs approach – the notation of a critical cluster can also be extended to unstable initial states. Here the composition of the critical clusters is equal to the composition of the ambient phase and the work of critical cluster formation is equal to zero. The size of the critical clusters for unstable initial states behaves like the size of the regions with the highest amplification of density/composition differences in the classical Cahn–Hilliard approach to the description of spinodal decomposition. As shown, moreover, there is no qualitative difference between nucleation and spinodal decomposition with respect to the basic mechanism of cluster evolution. Nucleation processes, starting from thermodynamically metastable initial states, proceed qualitatively widely similar as compared with processes of phase formation governed by spinodal decomposition. As it turns out further, the classical model of nucleation is not correct in application to phase formation in solutions (cf. also [3, 254]).

As an additional step, the effect of finite domain sizes on cluster formation is analyzed. It is shown, as a general consequence, that the degree of stability of the system to phase formation increases with decreasing system size due to depletion effects. In particular, the parameters of the critical clusters depend on system size. In addition, systems of finite size may be metastable or even stable even if the infinite samples are unstable. In this case the evolution of the system starts via spinodal decomposition. Then, due to the growth of the clusters, the supersaturation decreases, and the system becomes metastable. Anyway, cluster growth continues. Finally, the supersaturation decreases to such extent that the dissolution of the clusters with smaller sizes becomes the prerequisite for the growth of the larger one, and the stage of coarsening starts [155]. Thus, the approach allows us to describe the evolution of the system from spinodal decomposition up to the coarsening stage; accounting both for changes of the sizes and the intensive state parameters of the clusters in the course of this process. An analysis of experimental results on phase separation in solutions at high supersaturations is performed in terms of the generalized cluster model showing that the generalized cluster model allows us a more correct interpretation of the dynamics of phase separation as compared with this classical theory.

In the generalized Gibbs approach, similarly to the classical theory, only the knowledge of macroscopic properties of the ambient and the newly evolving phases is required for the analysis of phase formation processes. For this reason, the approach seems to be preferable in the analysis of experimental results adding to the classical method a sound account of the change of cluster properties with size (cf., e.g., [253–255, 257–260]). The results of the analysis, as performed above, were obtained by employing the model of regular solutions. They can be quantitatively modified by a more detailed consideration of the thermodynamic properties of the real system, by incorporating additional thermodynamic factors like special properties of domain boundaries or elastic stresses, which may be of importance in a number of cases, or by accounting for peculiarities of the process of diffusion not elaborated here. Nevertheless, we believe that the scenario outlined will be valid generally for processes of segregation in solid or liquid solutions. The application of the methods and results obtained to the interpretation of experimental data in phase formation in solutions deserves special consideration. Another question to be investigated is whether the results are directly applicable for other types of phase formation processes like, for example, condensation and boiling or have to be eventually modified. These topics will be addressed in forthcoming analyses.

References

[1] Abyzov, A. S., Slezov, V. V., Tanatarov, L. V., Sov. Phys. Solid State **34(12)**, 1952 (1992).

[2] Abyzov, A. S., Slezov, V. V., Tanatarov, L. V., Radiat. Eff. Defects Solids **129**, 257 (1994).

[3] Abyzov, A. S., Schmelzer, J. W. P., J. Chem. Phys. **127**, 114504 (2007).

[4] Agranovich, V. M., Mikhlin, E. Ya., At. Energy **12**, 385 (1962) (in Russian).

[5] Andreev, N. S., Boiko, G. G., Bokov, N. A., J. Non-Cryst. Solids **5**, 41 (1970).

[6] Ardell, A. J., Nicholson, R. B., J. Phys. Chem. Sol. **27**, 1793 (1966).

[7] Ardell, A. J., Nicholson, R. B., Eshelby, J. D. Acta Met. **14**, 1295 (1966).

[8] Ardell, A. J., Acta Met. **15**, 1772 (1967).

[9] Ardell, A. J., Acta Met. **16**, 511 (1968).

[10] Ardell, A. J., J. Inst. Met. **33**, 111 (1969).

[11] Ardell, A. J., Met. Trans. **1**, 525 (1970).

[12] Ardell, A. J., Acta Met. **20**, 61 (1972).

[13] Avrami, M., J. Chem. Phys. **7**, 1103 (1939); **8**, 212 (1940); **9**, 177 (1941).

[14] Baidakov, V. G., *Explosive Boiling of Superheated Cryogenic Liquids* (Wiley-VCH, Berlin, Weinheim, 2007).

[15] Bakai, A. S., Kiryukhin, N. M., Voprosy Atomnoy Nauki i Tekhniki, Ser.: Radiation Damage Physics and Radiation Technology, 1983, issue 5(28) p. 33 (in Russian).

[16] Bakai, A. S., Pis'ma Zh. Tekh. Fis. **9**, 1477 (1983).

[17] Balibar, S., Misusaki, T., Sasaki, Y., J. Low Temp. Phys. **120**, 293 (2000).

[18] Balliger, N. K., Honeycombe, R. W., Met. Sci. **4**, 121 (1980).

[19] Barnes, R. S., Mazey, D. J., Phil. Mag. **60**, 1247 (1960).

[20] Barrett, J. C., Clement, C. F., J. Aerosol Sci. **22**, 327 (1991).

[21] Bartels, J., Schweitzer, F., Schmelzer, J., J. Non-Cryst. Solids **125**, 129 (1990).

[22] Bartels, J., Lembke, U., Pascova, R., Schmelzer, J., Gutzow, I., J. Non-Cryst. Solids **136**, 181 (1991).

[23] Bartels, J., *On the kinetics of nucleation and growth in condensed systems*, Ph.D. Thesis, University of Rostock, Rostock, 1991.

[24] Becker, R., Döring, W., Annalen der Physik **24**, 719 (1935).

[25] Becker, R., Annalen der Physik **32**, 128 (1938).

[26] Behrens, M., Stauffer, D., Annalen der Physik **2**, 105 (1993).

[27] Bennemann, K. H. Ketterson, J. B., *The Physics of Liquid and Solid Helium* (Wiley, New York, 1976).

[28] Binder, K., Stauffer, D., Adv. Phys. **25**, 343 (1976).

[29] Binder, K., Phys. Rev. **B 15**, 4425 (1977).

[30] Binder, K. In: *Stochastic Nonlinear Systems in Physics, Chemistry and Biology*. L. Arnold and R. Lefever (Eds.) (Berlin, Springer, 1981, p. 62 ff).

[31] Binder, K., J. Physique (France) **41**, C4-51 (1980).

[32] Binder, K., Kalos, M. H., J. Stat. Phys. **22**, 363 (1980).

[33] Binder, K., Rep. Progr. Phys. **50**, 783 (1987).

[34] Binder, K., Spinodal decomposition. In: *Materials Science and Technology* (VCH Verlagsgesellschaft, Weinheim, 1992).

[35] Binder, K., Spinodal decomposition. In: J. W. Cahn, P. Haase, and R. J. Kramer (Eds.), *Materials Science and Technology*, vol. 5, Phase Transformations in Materials, Chapter 7, p. 405 ff (VCH, Weinheim, 1990).

[36] Blander, M., Adv. Colloid Interface Sci. **10**, 1 (1979).

[37] Bokshtein, B. S., Bokshtein, S. Z., Zhukhovitskii, A. A., *Termodinamika i Kinetika Diffuzii v Tvyordykh Telakh (Thermodynamics and Kinetics of Diffusion in Solids)* ((in Russian), Moscow, Metallurgiya, 1974).

[38] Boyd, J. D., Nickolson, R. B., Acta Met. **19**, 1379 (1971).

[39] Brener, E. A., Marchenko, V. I., Müller-Krumbhaar, H., Spatschek, R., Phys. Rev. Lett. **84**, 4914 (2000).

[40] Brailsford, A. D., Wynblatt, P., Acta Met. **27**, 489 (1979).

[41] Brailsford, A. D., Nucl. Mater. **91**, 221 (1980).

[42] Brautman, L. J., Krock, R. H. (Eds.), *Modern Composite Materials* (Addison-Wesley, Reading, MA, 1967).

[43] Burmistrov, S., Chagovets, V., Dubovskii, L., Rudavskii, E., Satoh, T., Sheshin, G., Physica **B 284–288**, 321 (2000).

[44] Cahn, J. W., Hilliard, J. E., J. Chem. Phys. **28**, 258 (1958); **31**, 688 (1959).

[45] Cahn, J. W., Acta Met. **8**, 554 (1960).

[46] Cauvin, R., Martin, G., Phys. Rev. **B 23**, 3322 (1981).

[47] Chagovets, V., Rudavskii, E., Sheshin, G., Usherov-Marshak, I., J. Low Temp. Phys. **113**, 1005 (1998).

[48] Chassagne, J., phys. stat. solidi (a) **40**, 629 (1977).

[49] Chepelevetskij, M. L., Zh. Fiz. Khim. (USSR) **13**, 561 (1939).

[50] Christian, J. W. In: R. W. Cahn (Ed.): *Physical Metallurgy* (North-Holland, Amsterdam, 1965, pp. 227–346).

[51] Christian, J. W., *The Theory of Transformations in Metals and Alloys* (2nd Edition, Oxford University Press, Oxford, 1975).

[52] Chuistov, K. V., *Modulirovannye Struktury v Stareyushchikh Splavakh (Modulated Structures in Aged Alloys)* ((in Russian), Kiev, Naukova Dumka, 1975).

[53] Colton, J. S., Suh, N. P., Polym. Eng. Sci. **27**, 485 (1987).

[54] Cook, H. E., Acta Met. **18**, 297 (1970).

[55] Cortes, R. G., Sepulveda, A. O., J. Mater. Sci. **22**, 3880 (1987).

[56] Cowan, B., Fardis, M., Phys. Rev. **44**, 4304 (1991).

[57] Csernai, L. P., Kapusta, J. L., Phys. Rep. **131**, 223 (1986).

[58] Daneliya, E. P., Rosenberg, V. M., *Vnutrenne-Okislennye Splavy* (Internally Oxidized Alloys (in Russian), Moscow, Metallurgiya, 1975).

[59] Davies, C. K. L., Nash, P., Stevens, R. N., Acta Met. **28**, 179 (1980).

[60] Demo, P., Kozisek, Z., Phys. Rev. **B 48**, 3620 (1993); Philosophical Magazine **B 70**, 49 (1994).

[61] Desbois, J., Boisgard, R., Ngo, C., Nemeth, J., Z. Phys. **A 328**, 101 (1987).

[62] Döring, W., Z. Physikalische Chemie **36**, 371 (1937) 371; **38**, 292 (1938).

[63] Edwards, D. O., Balibar, S., Phys. Rev. **B 39**, 4083 (1989).

[64] Ehrlich, S. N., Simmons, R. D., J. Low Temp. Phys. **68**, 125 (1987).

[65] Enomoto, Y., Tokuyama, M., Kawasaki, K., Acta Met. **34**, 2119 (1986).

[66] Eselson, B. N., Mikheev, V. A., Grigor'ev, V. N., Fiz. Nizk. Temp. **2**, 1229 (1976) [Sov. J. Low Temp. Phys. **2**, 599 (1976)].

[67] Farkas, L., Z. Physikalische Chemie **125**, 236 (1927).

[68] Fischmeister, A., Grimwall, G. In: *Proc. of Conference on Sintering and Related Phenomena*, J. Kuczynski, Ed. (New York, Plenum Press, p. 119, 1973).

[69] Fisher, M. E., Physics **3**, 255 (1967).

[70] Founter, P. K., Alcab, C. B., Met. Trans. **3**, 2633 (1972).

[71] Frenkel, Ya. I., *Kinetic Theory of Liquids* (Oxford University Press, Oxford, 1946).

[72] Freudental, A. M., *Inelastic Behaviour of Engineering Materials and Structures* (Wiley, New York, 1956).

[73] Frost, H. J., Russell, K. C., Acta Met. **30**, 953 (1982).

[74] Furukawa, H., Binder, K., Phys. Rev **A 26**, 556 (1982).

[75] Furukawa, H., Phys. Rev. A.: Gen. Phys. **28**, 1729 (1983).

[76] Ganding, W., Warlimont, H., Z. Metallkunde **60**, 488 (1983).

[77] Ganshin, A. N., Grigor'ev, V. N., Maidanov, V. A., Omelaenko, N. F., Penzev, A. A., Rudavskii, E., Rybalko, A. S., Fiz. Nizk. Temp. **25**, 356 (1999) [Low Temp. Phys. **25**, 259 (1999)].

[78] Ganshin, A. N., Grigor'ev, V. N., Maidanov, V. A., Omelaenko, N. F., Penzev, A. A., Rudavskii, E. Ya., Rybalko, A. S., Tokar, Yu. A., Fiz. Nizk. Temp. **25**, 796 (1999) [Low Temp. Phys. **25**, 592, Erratum 928 (1999)].

[79] Ganshin, A., Grigor'ev, V., Maidanov, V., Rudavskii, E., Rybalko, A., Slezov, V., Syrnikov, Ye., J. Low Temp. Phys. **126**, 151 (2002).

[80] Ganshin, A., Grigor'ev, V., Maidanov, V., Omelaenko, N., Penzev, A., Rudavskii, E., Rybalko, A., J. Low Temp. Phys. **116**, 349 (1999).

[81] Gan'shin, A. N., Grigor'ev, V. N., Maidanov, V. A., Penzev, A. A., . Rudavskii, E. Ya., Rybalko, A. S., Syrnikov, Ye. V., Fiz. Nizk. Temp. **29**, 487 (2003) [Low Temp. Phys. **29**, 362 (2003)].

[82] Ganshin, A. N. et al., Low Temp. Phys. **25**, 592 (1999).

[83] Geguzin, Ya. E., *Physik des Sinterns* (Deutscher Verlag für Grundstoffindustrie, Leipzig, 1973).

[84] Gelb, L. D., Gubbins, K. E., Radhakrishnan, R., Sliwinska-Bartkowiak, M., Rep. Prog. Phys. **62**, 1573 (1999).

[85] Gibbs, J. W.: *On the Equilibrium of Heterogeneous Substances*, Collected Works, vol. 1, *Thermodynamics* (Longmans and Green, New York, London, Toronto, 1928).

[86] Grätz, H., Simmich, O., Mater. Sci. Forum **13/14**, 287 (1987).

[87] Greenwood, G. W. J., Inst. Met. **33**, 103 (1969).

[88] Grigor'ev, V. N., Maidanov, V. A., Penzev, A. A., Rudavskii, E. Ya., Rybalko, A. S., Syrnikov, Ye. V., Slezov, V. V., Low Temp. Phys. **30**, 128 (2004).

[89] Grigor'ev, V. N., Maidanov, V. A., Penzev, A. A., Rudavskii, E. Ya., Rybalko, A. S., Syrnikov, Ye. V., Fiz. Nizk. Temp. **29**, 1165 (2003) [Low Temp. Phys. **29**, 883 (2003)].

[90] Grinin, A. P., Kuni, F. M., Kurasov, V. B., Kolloidnyi Zhurnal **52**, 430, 437, 444 (1990).

[91] Grinin, A. P.: *Kinetic Theory of Formation of a Condensed Phase in a Supersaturated Vapor*, D.Sc. Thesis, St. Petersburg State University, St. Petersburg, 1993 (in Russian).

[92] Gross, D. H. E., Phys. Rep. **279**, 119 (1997).

[93] Gunton, J. D., San Miguel, M., Sahni, P. S. The dynamics of first-order phase transitions. In: *Phase Transitions and Critical Phenomena*, vol. 8, C. Domb and J. L. Lebowitz (Eds.) (Academic Press, London, New York, 1983).

[94] Gutzow, I., Schmelzer, J. W. P., *The Vitreous State: Thermodynamics, Structure, Rheology, and Crystallization* (Springer, Berlin, 1995).

[95] Gutzow, I., Schmelzer, J., Dobreva, A., J. Non-Cryst. Solids **219**, 1 (1997).

[96] Haase, R., *Thermodynamik der Irreversiblen Prozesse* (D. Steinkopf, Darmstadt, 1963).

[97] Haase, R., *Thermodynamik der Mischphasen* (Springer, Berlin, 1956).

[98] Hale, B., Kulmala, M., (Eds.): *Nucleation and Atmospheric Aerosols 2000*, Proceedings of the Fifteenth International Conference on Nucleation and Atmospheric Aerosols, Rolla, Missouri, USA, 6–11 August, 2000; AIP Conference Proceedings, vol. 534 (Melville, New York, 2000).

[99] Halley, P. R., Adams, E. D., J. Low Temp. Phys. **110**, 121 (1998).

[100] Han, J. H., Han, C. D., J. Polym. Sci. **B 28**, 711 (1990); **B 28**, 743 (1990).

[101] Heckel, K. W. Trans. AIME **233**, 1994 (1965).

[102] Heimendane, M., Thomas, J., Met. Trans. **230**, 1520 (1964).

[103] Huse, D. A., Phys. Rev. **B 34**, 7845 (1986).

[104] Jacob, K., Möller, J., J. Cryst. Growth **197**, 973 (1999).

[105] Jain, S. C., Hughes, A. E., Mater. Sci. **13**, 1611 (1978).

[106] Jam, S. C., Arora, N. D., J. Phys. Chem. Sol. **35**, 187 (1974).

[107] Jayanth, C. S., Nash, P., J. Mater. Sci. **24**, 3041 (1989).

[108] Johnson, W. C., Acta Met. **32**, 465 (1984).

[109] Johnson, W. C., Voorhees, P. W., Zupon, D. E., Metall. Trans. **20 A**, 1175 (1989).

[110] Johnson, W. C., Howe, J. M., Laughlin, D. E., Soffa, W. A. (Eds.): *Solid to Solid Phase Transformations*, Proceedings of the International Conference on Solid to Solid Phase Transformations in Inorganic Materials PTM94; Nemacolin Woodlands, Farmington, July 1994.

[111] Jones, H., Met. Sci. Eng. **5**, 1 (1969).

[112] Kämpfer, B., Lukasc, B., Paal, G., *Cosmic Phase Transitions* (Teubner-Texte zur Physik, Bd. 29, Teubner, Leipzig, 1994).

[113] Kaischew, R., Stranski, I. N., Z. Phys. Chem. **A 170**, 295 (1934).

[114] Kampmann, R., Ebel, Th., Haese, M., Wagner, R., phys. stat. solidi (b) **172**, 295 (1992).

[115] Kashchiev, D., Verdoes, D., van Rosmalen, G. M., J. Cryst. Growth **110**, 373 (1991).

[116] Kashchiev, D., Firoozabadi, A., J. Chem. Phys. **98**, 4690 (1993).

[117] Katz, J. L., Wiedersich, H., J. Colloid Interface Sci. **61**, 351 (1977).

[118] Katz, J. L., Spaepen, F., Phil. Mag. **B 37**, 137 (1978).

[119] Katz, J. L., Donohue, M. D., Adv. Chem. Phys. **40**, 137 (1979).

[120] Katz, J. L., Pure Appl. Chem. **64**, 1661 (1992).

[121] Katz, J. L., Fisk, J. A., Rudek, M. M.: *Nucleation of Single-Component Supersaturated Vapors*, In: [135], pp. 1–10.

[122] Kawasaki, K., Enomoto, Y., Physica **A 150**, 463 (1988).

[123] Kelton, K. F., Greer, A. L., Thompson, C. V., J. Chem. Phys. **79**, 6261 (1983).

[124] Khachaturyan, A. G., *Theory of Phase Transitions and Structure of Solid Solutions* (Nauka, Moscow, 1979 (in Russian)).

[125] Kiang, C. S., Phys. Rev. Letts. **24**, 47 (1970).

[126] Kingsley, S. C. J., Maidanov, V., Saunders, J., Cowan, B., J. Low Temp. Phys. **113**, 1017 (1998).

[127] Kingsley, S. C. J., Kosarev, I., Roobol, L., Maidanov, V., Saunders, J., Cowan, B., J. Low Temp. Phys. **110**, 400 (1998).

[128] Koiwa, M., Otsuka, K., Miyazaki, T. (Eds.): *Solid to Solid Phase Transformations*, Proceedings of the International Conference on Solid to Solid Phase Transformations in Inorganic Materials PTM99; Kyoto, Japan, May 24–28, 1999 (The Japan Institute of Metals, 1999).

[129] Kolmogorov, A. N., Bull. Acad. Sci. USSR (Sci. Math. Nat.) **3**, 355 (1937).

[130] Kozisek, Z., Demo, P., J. Cryst. Growth **132**, 491 (1993).

[131] Kramers, H. A., Physica **7**, 284 (1940).

[132] Kremer, K. J., J. Aerosol Sci. **9**, 243 (1977).

[133] Kubo, R., *Thermodynamics* (North-Holland, Amsterdam, 1968).

[134] Kukushkin, S. A., Osipov, A. V., Uspekhi Fiz. Nauk **168**, 1083 (1998).

[135] Kulmala, M., Wagner, P. E., (Eds.): *Nucleation and Atmospheric Aerosols 1996*, Proceedings of the Fourteenth International Conference on Nucleation and Atmospheric Aerosols, Helsinki, 26–30 August, 1996 (Pergamon Press, London, 1996).

[136] Kuni, F. M., Grinin, A. P., Kolloidnyi Journal **46**, 460 (1984) (in Russian, English translation available).

[137] Kuni, F. M., Grinin, A. P., Teoreticheskaya i Matematicheskaya Fizika **80**, 418 (1989) (in Russian, English translation available).

[138] Kuni, F. M., Ognenko, V. M., Ganyuk, L. N., Grechko, L. G., Kolloidn. Zh. **55**, 22 (1993) (in Russian).

[139] Kuni, F. M., Grinin, A. P., Kurasov, V. B.: *Lectures on Kinetic Theory of Condensation Under Dynamical Conditions* (St. Petersburg State University Press, St. Petersburg, 1996).

[140] Kuni, F. M., Grinin, A. P., Kurasov, V. B.: *Kinetics of Condensation at Gradual Creation of the Metastable State*, In: [246], pp. 160–193.

[141] Kuni, F. M., Shchekin, A. K., Grinin, A. P.: *Kinetics of Condensation on Macroscopic Solid Nuclei at Low Dynamic Vapor Supersaturation*. In: [246], pp. 208–236.

[142] Kuni, F. M., Zhuvikina, I. A., Colloid J. **64**, 166 (2002).

[143] Kuni, F. M., Zhuvikina, I. A., Grinin, A. P., Colloid J. **65**, 201 (2003).

[144] Landau, L. D., Lifshitz, E. M., *Statistical Physics* (Academy of Sciences Publishing House, Berlin, 1987).

[145] Langer, J. S., Ann. Phys. (NY) **41**, 108 (1967); **54**, 258 (1969).

[146] Langer, J. S., Bar-on, M., Miller, H. D., Phys. Rev. **A 11**, 1417 (1975).

[147] Langer, J. S.: An introduction to the kinetics of first-order phase transition. In: G. Godreche (Ed.), *Solids Far from Equilibrium*, Chapter 3, p. 297 ff. (Cambridge University Press, Cambridge, 1992).

[148] Leubner, I. H., J. Imaging Sci. **29**, 219 (1985); **31**, 145 (1987).

[149] Leubner, I. H., J. Cryst. Growth **84**, 496 (1987).

[150] Leubner, I. H., J. Phys. Chem. **91**, 6069 (1987).

[151] Levich, V. G., *Physico-Chemical Hydrodynamics* (Academy of Sciences Publishers, Moscow, 1952, in Russian).

[152] Li, Jin-Song, Maksimov, I. L., Wilemski, G., Phys. Rev. **E 61**, R 4719 (2000) and references cited therein.

[153] Lifshitz, I. M., Slezov, V. V., ZhETF (in Russian) **35**, 479 (1958); Zh. Exp. Teor.Fiz. [Sov.Phys. JETP] **35**, 475 (1958) and references cited therein.

[154] Lifshitz, I. M., Slezov, V. V., Fiz. Tverd. Tela (St. Petersburg) **1**, 1401 (1959) (in Russian).

[155] Lifshitz, I. M., Slezov, V. V., J. Phys. Chem. Solids **19**, 35 (1961).

[156] Lifshitz, I. M., Polessky, V. I., Khokhlov, V. A., Zh. Eksp. Teor. Fiz. **74**, 268 (1978) [Sov. Phys. JETP **47**, 137 (1978)].

[157] Lopez, H., Shewmon, P. G., Acta Met. **31**, 1945 (1983).

[158] Lothe, J., Pound, G. M., J. Chem. Phys. **36**, 2080 (1962).

[159] Lubetkin, S. D.: Bubble nucleation and growth. In: *Controlled Particle, Droplet, and Bubble Formation*, D. J. Wedlock (Ed.) (Butterworth-Heinemann, Oxford, 1994).

[160] Lubetkin, S. D.: *The Fundamentals of Bubble Evolution*, Chemical Society Reviews 243 (1995).

[161] Ludwig, F.-P., Bartels, J., Schmelzer, J., J. Mater. Sci. **29**, 4852 (1994).

[162] Ludwig, F.-P., Schmelzer, J., Milchev, A., J. Phase Transitions **48**, 237 (1994).

[163] Ludwig, F.-P., Schmelzer, J., Z. Phys. Chemie **192**, 155 (1995).

[164] Ludwig, F.-P., Schmelzer, J., J. Colloid Interface Sci. **181**, 503 (1996).

[165] Mackawa, M., Okumura, Y., Okuda, Y., Phys. Rev. **B 65**, 144525 (2002).

[166] Maidanov, V., Ganshin, A., Grigor'ev, V., Penzev, A., Rudavskii, E., Rybalko, A., Syrnikov, Ye., Pis'ma Zh. Eksp. Teor. Fiz. **73**, 329 (2001) [JETP Lett. **73**, 289 (2001)].

[167] Maidanov, V., Ganshin, A., Grigor'ev, V., Penzev, A., Rudavskii, E., Rybalko, A., Syrnikov, Ye., J. Low Temp. Phys. **126**, 401 (2002).

[168] Maksimov, I. L., Nishioka, K., Phys. Lett. **A 264**, 51 (1999).

[169] Mansur, L. K., Yoo, M. H., J. Nucl. Mater. **74**, 228 (1978).

[170] Marqusee, J. A. and Ross, J., J. Chem. Phys. **80**, 536 (1984).

[171] Marder, M., Phys. Rev. **A 36**, 898 (1987).

[172] Martin, J. W., Doerty, R. D., *Stability of Microstructure in Metallic Systems* (Cambridge University Press, Cambridge, 1976).

[173] Mazenko, G. F., Valls, O. T., Zhang, F. C., Phys. Rev. **B 31**, 4453 (1985).

[174] Mazenko, G. F., Valls, O. T., Zannetti, M., Phys. Rev. **B 38**, 520 (1988).

[175] Mchedlov-Petrosyan, P. O., Slezov, V. V., Dokl. Akad. Nauk Ukr. SSR **A 5**, 67 (1987).

[176] Melikhov, F. M., Trofimov, F. M., Kuni, F. M., Kolloidn. Zh. **56**, 201 (1994) (in Russian).

[177] Milchev, A., Gerroff, I., Schmelzer, J., Z. Physik **B 94**, 101 (1994).

[178] Mirold, P., Binder, K., Acta Metallurgica **25**, 1435 (1977).

[179] Miyazaki, T., Nakamura, K., Mori, H., J. Mater. Sci. **14**, 1827 (1979).

[180] Morosova, E. V., Fiz. i Khimiya Stekla **17**, 717, 726 (1991).

[181] Morosova, E. V., Fiz. i Khimiya Stekla **17**, 875 (1991).

[182] Möhlenkamp, T. In: GSI Scientific Report, 1994, p 46; U. Grundinger (Ed.), Darmstadt, 1994 and references cited therein.

[183] Möller, J., Jacob, K., Schmelzer, J., J. Phys. Chem. Solids **59**, 1097 (1998).

[184] Möller, J.: *On the theoretical description of phase transformation processes in elastic and viscoelastic media*, Ph.D. Thesis, University of Rostock, 1996 (in German).

[185] Mullin, W. J., Phys. Rev. Lett. **20**, 254 (1968).

[186] Müller, W. F. J. et al.: *Boiling Nuclei*, GSI-Nachrichten 03-95, U. Grundinger, V. Metag (Eds.), Darmstadt, 1995.

[187] Nabarro, F. R., Proc. Royal Soc. (London) **A 175**, 519 (1940).

[188] Nechiporenko, E. P., Slezov, V. V., Sagalovich, V. V., Fizika Tverd. Tela (in Russian) **14**, 1469 (1972).

[189] Neimark, A. V., Vishnyakov, A., J. Phys. Chem. **B 109**, 5962 (2005); **B 110**, 9403 (2006).

[190] Neuman, C. H., J. Chem. Phys. **60**, 4508 (1974).

[191] Nielsen, A. E., *Kinetics of Precipitation* (Pergamon Press, Oxford, 1964).

[192] Nelson, R. A., Hudson, J. A., Mazy, D. J., Nucl. Mater. **44**, 318 (1972).

[193] O'Dowd, C., Wagner, P. (Eds.): *Nucleation and Atmospheric Aerosols 2007*, Proceedings of the Seventeenth International Conference on Nucleation and Atmospheric Aerosols, Galway, Ireland, August 13–17, 2007 (Springer, Berlin, 2007).

[194] Ohishi, O., Yamamoto, H., Suzuki, M., J. Low Temp. Phys. **112**, 199 (1998).

[195] Olemski, A. I., Koplyk, I. V., Uspekhi Fiz. Nauk **165**, 1105 (1995).

[196] Olson, T., Hamill, P., J. Chem. Phys. **104**, 210 (1996).

[197] Ostapchuk, P. N., Slezov, V. V., Tur, A. V., Yanovsky, V. V., Fiz. Metallov Metalloved. **67**, 462 (1989).

[198] Oxtoby, D. W. et al., J. Chem. Phys. **89**, 7521 (1988); **94**, 4472 (1991); **103**, 3686 (1995).

[199] Pan, J., Das Gupta, S., Phys. Lett. **B 344**, 30 (1995); Phys. Rev. **C 51**, 1384 (1995).

[200] Pascova, R., Gutzow, I., Glastechnische Berichte **56**, 324 (1983).

[201] Penrose, O., Lebowitz, J. L., J. Stat. Phys. **19**, 243 (1978).

[202] Penzev, A., Ganshin, A., Grigor'ev, V., Maidanov, V., Rudavskii, E., Rybalko, A., Slezov, V., Syrnikov, Ye., J. Low Temp. Phys. **126**, 151 (2002).

[203] Poole, M., Saunders, J., Cowan, B., Phys. Rev. Lett. **100**, 075301 (2007).

[204] Portnoi, K. I., Babich, B. N., *Dispersno-Uprochnyonnye Splavy (Age-Hardening of Alloys)* ((in Russian), Moscow, Metallurgiya, 1974).

[205] Reguera, D., Reiss, H., J. Chem. Phys. **119**, 1533 (2003).

[206] Reiss, H., J. Chem. Phys. **18**, 840 (1950).

[207] Reiss, H., Katz, J. L., Cohen, E. R., J. Chem. Phys. **48**, 5553 (1968).

[208] Reiss, H. et al., J. Chem. Phys. **95**, 9209 (1991); **97**, 5766 (1992); **99**, 5374 (1993).

[209] Reiss, H., Kegel, W. K., J. Phys. Chem. **100**, 10428 (1996).

[210] Richardson, R. C., Hunt, E., Meyer, H., Phys. Rev. **138**, A1326 (1965).

[211] Robertson, B., Phys. Rev. **151**, 273 (1966).

[212] Roskosz, M., Toplis, M. J., Besson, P., Richet, P., J. Non-Cryst. Solids **351**, 1266 (2005).

[213] Roskosz, M., Toplis, M. J., Richet, P., J. Non-Cryst. Solids **352**, 180 (2006).

[214] Rowlinson, J. S.: Translation of J. D. van der Waals' The thermodynamic theory of capillarity under the hypothesis of a continuous variation of density, J. Stat. Phys. **20**, 197 (1979); German version: van der Waals, J. D., Z. Phys. Chemie **13**, 657 (1893).

[215] Ruckenstein, E., Nowakowski, B., J. Colloid Interface Sci. **137**, 583 (1990).

[216] Ruckenstein, E., Nowakowski, B., J. Chem. Phys. **94**, 1397 (1991).

[217] Rusanov, A. I., *Phasengleichgewichte und Grenzflächenerscheinungen* (Akademie-Verlag, Berlin, 1978).

[218] Russell, K. C., Progr. Mater. Science **28**, 229 (1984).

[219] Sanada, M., Maksimov, I. L., Nishioka, K., J. Cryst. Growth **199**, 67 (1999).

[220] Schmelzer, J.: *Thermodynamics of Finite Systems and the Kinetics of First-Order Phase Transitions*, D.Sc. Thesis (Habilarbeit), University of Rostock, Rostock, 1985.

[221] Schmelzer, J., Z. Physik. Chem. (Leipzig) **266**, 1057 (1985).

[222] Schmelzer, J., Schweitzer, F., Z. Phys. Chemie (Leipzig) **266**, 943 (1985); **270**, 5 (1989); **271**, 565 (1990).

[223] Schmelzer, J. W. P., Gutzow, I.: On the kinetic description of Ostwald ripening in elastic and viscoelastic media. In: *Selforganization by Non-linear Irreversible Processes*, Springer Series in Synergetics **33**, 144, W. Ebeling, H. Ulbricht (Eds.) (Springer, Berlin, 1986).

[224] Schmelzer, J., Gutzow, I., Wissenschaftliche Zeitschrift der Wilhelm-Pieck-Universität Rostock, Mathematisch-Naturwissenschaftliche Reihe **35**, 5 (1986).

[225] Schmelzer, J., Schweitzer, F., Wissenschaftliche Zeitschrift der Universität Rostock, Naturwissenschaftliche Reihe **36**, 83 (1987).
[226] Schmelzer, J., Ulbricht, H., J. Colloid Interface Sci. **117**, 325 (1987).
[227] Schmelzer, J., Schweitzer, F., J. Non-Equil. Thermodyn. **12**, 255 (1987).
[228] Schmelzer, J., Z. Phys. Chemie (Leipzig) **269**, 633 (1988).
[229] Schmelzer, J., Gutzow, I., Z. Phys. Chemie (Leipzig) **269**, 4852 (1988).
[230] Schmelzer, J., Ulbricht, H., J. Colloid Interface Sci. **117**, 325 (1987); **128**, 104 (1989).
[231] Schmelzer, J., Wiss. Z. Universität Rostock, Mathematisch-Naturwiss. Reihe **39**, 103 (1990).
[232] Schmelzer, J., Gutzow, I., Pascova, R., J. Cryst. Growth **104**, 505 (1990).
[233] Schmelzer, J., Pascova, R., Gutzow, I., phys. status solidi (a) **117**, 363 (1990).
[234] Schmelzer, J., physica status solidi (b) **161**, 173 (1990).
[235] Schmelzer, J., Milchev, A., Phys. Lett. **A 158**, 307 (1991).
[236] Schmelzer, J. W. P., Möller, J., J. Phase Transitions **38**, 261 (1992).
[237] Schmelzer, J., Möller, J., Slezov, V. V., J. Phys. Chem. Solids **56**, 1013 (1995).
[238] Schmelzer, J., Slezov, V. V., Milchev, A., J. Phase Transitions **54**, 193 (1995).
[239] Schmelzer, J. W. P., Gutzow, I., Schmelzer, J. (Jr.), J. Colloid Interface Sci. **178**, 657 (1996).
[240] Schmelzer, J. W. P., Röpke, G., Ludwig, F.-P., Phys. Rev. **C 55**, 1917 (1997).
[241] Schmelzer, J. W. P. (under participation of V. V. Slezov, I. S. Gutzow, S. Todorova, J. Schmelzer Jr.): *Bubble Formation and Growth in Viscous Liquids*, Project Report for BASF Ludwigshafen, Rostock – Ludwigshafen, 1998, p. 203.
[242] Schmelzer, J. W. P., Labudde, D., Röpke, G., Physica **A 254**, 389 (1998).
[243] Schmelzer, J. W. P., Slezov, V. V.: *Nucleation and Growth of a New Phase with a Given Stoichiometric Composition*. In: Proceedings XVIII. International Congress on Glass, The American Ceramic Society, Glass and Optical Materials Division, San Francisco, California, July 5–10, 1998.
[244] Schmelzer, J. W. P., Schmelzer, J. (Jr.), J. Colloid Interface Sci. **215**, 345 (1999).
[245] Schmelzer, J., Röpke, G., Mahnke, R., *Aggregation Phenomena in Complex Systems* (Wiley-VCH, Weinheim, 1999).
[246] Schmelzer, J. W. P., Röpke, G., Priezzhev, V. B. (Eds.), *Nucleation Theory and Applications* (Joint Institute for Nuclear Research Publishing Department, Dubna, Russia, 1999 (available also directly from the editor J. W. P. Schmelzer (further information can be received via Email juern-w.schmelzer@physik.uni-rostock.de))).
[247] Schmelzer, J. W. P., Röpke, G., Schmelzer, J. (Jr.), Slezov, V. V., Shapes of cluster size distributions evolving in nucleation–growth processes. In: J. W. P. Schmelzer, G. Röpke, V. B. Priezzhev (Eds.), *Nucleation Theory and Applications* (Joint Institute for Nuclear Research Publishing Department, Dubna, Russia, 1999 (available also directly from the editor J. W. P. Schmelzer (further information can be received via Email juern-w.schmelzer@physik.uni-rostock.de))), pp. 82–129.

[248] Schmelzer, J. W. P.: Comments on curvature dependent surface tension and nucleation theory. In: J. W. P. Schmelzer, G. Röpke, V. B. Priezzhev (Eds.), *Nucleation Theory and Applications* (Joint Institute for Nuclear Research Publishing Department, Dubna, Russia, 1999 (available also directly from the editor J. W. P. Schmelzer (further information can be received via Email juern-w.schmelzer@physik.uni-rostock.de))), pp. 268–289.

[249] Schmelzer, J. W. P., Schmelzer, J. (Jr.), Gutzow, I.: *Reconciling Gibbs and van der Waals: A New Approach to Nucleation Theory*, In: [246], pp. 237–267; J. Chem. Phys. **112**, 3820 (2000).

[250] Schmelzer, J. W. P., Vasilevskaya, T. N., Andreev, N. S.: *On the Initial Stages of Spinodal Decomposition*, In: [246], pp. 425–444.

[251] Schmelzer, J. (Jr.), Lembke, U., Kranold, R., J. Chem. Phys. **113**, 1268 (2000).

[252] Schmelzer, J. W. P.: *Dynamical Approaches to Cluster Formation and Nuclear Multifragmentation.* Proceedings of 4th Catania Relativistic Ion Studies Conference CRIS 2002, Catania, July 10–14, 2002, AIP Conference Proceedings **644**, pp. 178–187 (Melville, New York, 2002).

[253] Schmelzer, J. W. P., Gokhman, A. R., Fokin, V. M., J. Colloid Interface Sci. **272**, 109 (2004).

[254] Schmelzer, J. W. P., Abyzov, A. S., Möller, J., J. Chem. Phys. **121**, 6900 (2004).

[255] Schmelzer, J. W. P., Physics and Chemistry of Glasses **45**, 116 (2004).

[256] Schmelzer, J. W. P. (Ed.), *Nucleation Theory and Applications*, (Wiley-VCH, Berlin, Weinheim, 2005).

[257] Schmelzer, J. W. P.: *Summary and Outlook.* In: [256], pp. 447–452.

[258] Schmelzer, J. W. P., Boltachev, G. Sh., Baidakov, V. G.: *Is Gibbs' Thermodynamic Theory of Heterogeneous Systems Really Perfect?* In: [256], pp. 418–446.

[259] Schmelzer, J. W. P., Boltachev, G. Sh., Baidakov, V. G., J. Chem. Phys. **124**, 194502 (2006).

[260] Schmelzer, J. W. P., J. Non-Cryst. Solids **354**, 269 (2008).

[261] Schmelzer, J. W. P., Abyzov, A. S.: *Generalized Gibbs' Thermodynamic Analysis of Cluster Formation in Confined Systems: General Results*, in preparation.

[262] Schweitzer, F., Schmelzer, J., Rostocker Physikalische Manuskripte **10**, 32 (1987); **12**, 61 (1988).

[263] Shah, M., Galkin, O., Vekilov, P. G., J. Chem. Phys. **121**, 7505 (2004).

[264] Shewmon, P. G., *Diffusion in Solids* (McGraw-Hill, New York (1963), Metallurgiya, Moscow (1966)).

[265] Shi, G., Seinfeld, J. H., Okuyama, K., Phys. Rev. **A 41**, 2101 (1990).

[266] Shirinyan, A. S., Gusak, A. M., Phil. Mag. **A 84**, 579 (2004).

[267] Shneidman, V. A., Sov. Phys.-Tech. Phys. **57**, 131 (1987); **58**, 2202 (1988).

[268] Shneidman, V. A., Weinberg, M. C., J. Non-Cryst. Solids **160**, 89 (1993).

[269] Shrenk, R., Friz, O., Fujii, Y., Syskakis, E., and Pobell, F., J. Low Temp. Phys. **84**, 133 (1991).

[270] Shvarts, V. A., Mikhin, N. P., Rudavskii, E. Ya., Usenko, A. M., Tokar, Yu. A., Mikheev, V. A., Fiz. Nizk. Temp. **21**, 717 (1995) [Low Temp. Phys. **21**, 556 (1995)].

[271] Skripov, V. P., *Metastable Liquids* (Nauka, Moscow, 1972 (in Russian); Wiley, New York, 1974).
[272] Skripov, V. P., Skripov, P. V., Uspekhi Fiz. Nauk **128**, 193 (1979).
[273] Skripov, V. P., Faizullin, M. Z., *Crystal-Liquid-Gas Phase Transitions and Thermodynamic Similarity* (Wiley-VCH, Berlin, Weinheim, 2006).
[274] Slezov, V. V., Shikin, V. B., Fizika Tverd. Tela (in Russian) **6**, 7 (1964).
[275] Slezov, V. V., Euro Nuclear **2**, 75 (1965).
[276] Slezov, V. V., Saralidze, Z., Sov. Phys. Solid State **7**, 921 (1965).
[277] Slezov, V. V., Shikin, V. B., Fizika Tverd. Tela (in Russian) **7**, 127 (1965).
[278] Slezov, V. V., Fiz. Tverd. Tela **9**, 15 (1967).
[279] Slezov, V. V., Fizika Tverd. Tela (in Russian) **17**, 2557 (1975).
[280] Slezov, V. V., Sagalovich, V. V., Fizika Tverd. Tela (in Russian) **17**, 1497 (1975).
[281] Slezov, V. V., Sagalovich, V. V., Fizika Tverd. Tela (in Russian) **17**, 2751 (1975).
[282] Slezov, V. V., Sagalovich, V. V., Izv. Akad. Nauk SSSR (in Russian), Neorg. Materialy **12**, 1719 (1976).
[283] Slezov, V. V., Sagalovich, V. V., J. Phys. Chem. Solids **38**, 943 (1977).
[284] Slezov, V. V., J. Phys. Chem. Solids **39**, 367 (1978).
[285] Slezov, V. V., Sagalovich V. V., Tanatarov, L. V., J. Phys. Chem. Solids **39**, 705 (1978).
[286] Slezov, V. V., J. Phys. Chem. Solids **44**, 13 (1983).
[287] Slezov, V. V., Sagalovich, V. V., J. Phys. Chem. Solids **44**, 23 (1983).
[288] Slezov, V. V., Sagalovich, V. V.: *Diffuzionny Raspad Tvyordykh Rastvorov: Obzor (Diffusive Decomposition of Solid Solutions: a Review)* (in Russian) TsNIIATOMINFORM (1984).
[289] Slezov, V. V., Sagalovich, V. V., Sov. Phys. Uspekhi **151**, 67 (1987) (in Russian).
[290] Slezov, V. V., Schmelzer, J., Möller, J., J. Cryst. Growth **132**, 419 (1993).
[291] Slezov, V. V., Schmelzer, J., Sov. Phys. Solid. State **36**, 353 (1994).
[292] Slezov, V. V., Schmelzer, J., Sov. Physics Solid State **36**, 363 (1994).
[293] Slezov, V. V., Schmelzer, J. W. P., J. Phys. Chem. Solids **55**, 243 (1994).
[294] Slezov, V. V.: Theory of diffusive decomposition of solid solutions. In: *Soviet Scientific Reviews, Section* **A 17**, Physics Reviews, I. M. Khalatnikov (Ed.) (Harwood Academic Publishers, London, 1995).
[295] Slezov, V. V., Schmelzer, J., Tkach, Ya. Yu., Fizika Tverdogo Tela **37**, 3212 (1995) (in Russian, English translation available).
[296] Slezov, V. V., Fizika tverdogo tela **37**, 2879 (1995) (in Russian).
[297] Slezov, V. V., Schmelzer, J., Tkach, Ya. Yu., J. Phys. Chem. Solids **57**, 8340 (1996).
[298] Slezov, V. V., Schmelzer, J. W. P., Tkatch, Ya. Yu., J. Chem. Phys. **105**, 8340 (1996).
[299] Slezov, V. V., Fizika Tverdogo Tela **38**, 433 (1996) (in Russian, English translation available).
[300] Slezov, V. V., Schmelzer, J., Fiz. Tverd. Tela (St. Petersburg) **39**, 2210 (1997); Phys. Solid State **39**, 1971 (1997).
[301] Slezov, V. V., J. Phys. Chem. Solids **58**, 455 (1997).
[302] Slezov, V. V., Phys. Rep. **288**, 389 (1997).

[303] Slezov, V. V., Schmelzer, J., Tkatch, Ya. Yu., J. Phys. Chem. Solids **59**, 869 (1997).

[304] Slezov, V. V., Schmelzer, J. W. P., Tkatch, Ya. Yu., J. Mater. Sci. **32**, 3739 (1997).

[305] Slezov, V. V., Rogozhkin, V. V., Abyzov A. S., Phys. Solid State **40**, 601 (1998).

[306] Slezov, V. V., Schmelzer, J., J. Phys. Chem. Solids **59**, 1507 (1998).

[307] Slezov, V. V., Schmelzer, J. W. P. Kinetics of nucleation–growth processes: the first stages. In: J. W. P. Schmelzer, G. Röpke, V. B. Priezzhev (Eds.), *Nucleation Theory and Applications* (Joint Institute for Nuclear Research Publishing Department, Dubna, Russia, 1999 (available also directly from the editor J. W. P. Schmelzer (further information can be received via Email juern-w.schmelzer@physik.uni-rostock.de))), pp. 6–81.

[308] Slezov, V. V., Fiz. Tverd. Tela (St. Petersburg) **42**, 733 (2000).

[309] Slezov, V. V., Schmelzer, J. W. P., Phys. Rev. **E 65**, 031506 (2002).

[310] Slezov, V. V., J. Colloid Interface Sci. **255**, 274 (2002).

[311] Slezov, V. V., Phys. Solid State **45**, 335 (2003).

[312] Slezov, V. V., Fiz. Tverd. Tela (St. Petersburg) **45**, 317 (2003) [Phys. Solid State **45**, 335 (2003)].

[313] Slezov, V. V., Abyzov, A. S., Slezova, Zh. V., Colloid J. **66**, 575 (2004).

[314] Slezov, V. V., Abyzov, A. S., Slezova, Zh. V., Colloid J. **67**, 94 (2005).

[315] Slezov, V. V., Schmelzer, J. W. P., Abyzov, A. S.: *A New Method of Determination of the Coefficients of Emission in Nucleation Theory*. In: [256], pp. 39–73.

[316] Smirnov, B. M., Usp. Fiz. Nauk **164**, 665 (1994) (in Russian).

[317] Smith, A., Maidanov, V. A., Rudavskii, E. Ya., Grigor'ev, V. N., Slezov, V. V., Poole, M., Saunders, J., Cowan, B., Phys. Rev. **B 67**, 245314 (2003).

[318] Stauffer, D., J. Aerosol Sci. **7**, 319 (1976).

[319] Stauffer, D., *Introduction into Percolation Theory* (Taylor and Francis, London, Philadelphia, 1985).

[320] Su, C. H., Voorhees, P. W., Acta Mater. **44**, 1987 (1996).

[321] Surh, M. P., Sturgeon, J. B., Wolfer, W. G., J. Nucl. Mater. **378**, 86–97 (2008).

[322] Swift, W. M., Metallurgic Trans. **53**, 275 (1973).

[323] Thompson, M. W. *Defects and Radiation Damage in Metals*, Cambridge University, London (1969).

[324] Tokuyama, M., Kawasaki, K., Physica **A 123**, 386 (1984).

[325] Tokuyama, M., Enomoto, Y., Phys. Rev. Lett. **69**, 312 (1992); Phys. Rev. **E 47**, 1156 (1993).

[326] Tolman, R. C., J. Chem. Phys. **17**, 119, 333 (1949).

[327] Toramaru, A., J. Geophys. Res. **94**, 17.523 (1989).

[328] Trinkaus, H., Phys. Rev. **B 27**, 7372 (1983).

[329] Trinkaus, H., Yoo, M. H., Phil. Magazine **A 55**, 269 (1987).

[330] Tunitskij, N. N., Zh. Fiz. Khimii **15**, 1061 (1941) (in Russian).

[331] Ulbricht, H., Schmelzer, J., Mahnke, R., Schweitzer, F., *Thermodynamics of Finite Systems and the Kinetics of First-Order Phase Transitions* (Teubner Publ., Leipzig, 1988).

[332] Urban, K., Martin, G., Acta Met. **30**, 1209 (1982).

[333] van der Waals, J. D., Kohnstamm, Ph., *Lehrbuch der Thermodynamik* (Johann-Ambrosius-Barth Verlag, Leipzig und Amsterdam, 1908).

[334] Vasilevskaya, T. N., Andreev, N. S., Glass Phys. Chem. **22**, 510 (1996).

[335] Vedula, K. M., Heckel, R. W., Met. Trans. **1**, 9 (1970).

[336] Velasco, E., Toxvaerd, S., Phys. Rev. Lett. **71**, 388 (1993).

[337] Volmer, M., *Kinetik der Phasenbildung* (Th. Steinkopff, Dresden, 1939).

[338] von Smoluchowski, M., Phys. Z. **17**, 557, 585 (1916).

[339] von Smoluchowski, M., Z. Phys. Chemie.

[340] Voorhees, P. W., Glicksman, M. E., Acta Met. **32**, 2001, 2013 (1984). **92**, 129 (1917).

[341] Voorhees, P. W., J. Stat. Phys. **38**, 231 (1985).

[342] Wagner, C., Z. Electrochem. **B 1.65**, 581 (1961).

[343] Wagner, P. E., J. Colloid Interface Sci. **44**, 181 (1973); **53**, 439 (1975).

[344] Wakeshima, H., J. Phys. Soc. Japan **9**, 400 (1954); **9**, 407 (1954).

[345] Wan, G., Sahm, P. R., Acta Met. Mater. **38**, 967, 2367 (1990).

[346] Wanderka, N., Ramachandra, C., Wahi, R. P., Wollenberger, H., J. Nucl. Mater. **189**, 9 (1992).

[347] Ward, C. A., Levart, E., J. Appl. Phys. **56**, 491 (1984).

[348] Ward, C. A., Tikuisis, P., Venter, R. D., J. Appl. Phys. **53**, 6076 (1982).

[349] Warlimont, H., Thomas, G., Met. Sci. J. **4**, 47 (1970).

[350] Wehner, M. F., Wolfer, W. G., Philos. Mag. **A 52**, 189 (1985).

[351] Wiedersich, H., Katz, J., Adv. Colloid Interface Sci. **10**, 33 (1979).

[352] Wilemski, G., J. Chem. Phys. **103**, 1119 (1995).

[353] Wilemski, G., Wyslouzil, B. E., J. Chem. Phys. **103**, 1127 (1995); **103**, 1137 (1995); **105**, 1090 (1996).

[354] Wilt, P. M., J. Colloid Interface Sci. **112**, 530 (1986).

[355] Wu, D. T.: *Nucleation Theory*. In: Solid State Physics: Advances in Research and Applications **50**, 37 (1997).

[356] Wyslouzil, B. E., Wilemski, G., J. Chem. Phys. **103**, 1137 (1995) and references cited therein.

[357] Zeldovich, Ya. B., Sov. Phys. JETP **12**, 325 (1942).

[358] Zettlemoyer, A. C. (Ed.), *Nucleation*, Marcel Decker, New York, 1969; Nucleation phenomena, Adv. Colloid Interface Sci. **7** (1977).

Index